NEW TERTIARY MATHEMATICS

C. Plumpton
Queen Mary College, London

P. S. W. MacIlwaine

Volume 2 Part 1

Further Pure Mathematics

PERGAMON PRESS

Oxford · New York · Toronto · Sydney · Paris · Frankfurt

UK	Pergamon Press Ltd., Headington Hill Hall, Oxford OX3 0BW, England
USA	Pergamon Press Inc., Maxwell House, Fairview Park, Elmsford, New York 10523, U.S.A.
CANADA	Pergamon of Canada, Suite 104, 150 Consumers Road, Willowdale, Ontario, M2J 1P9, Canada
AUSTRALIA	Pergamon Press (Aust.) Pty. Ltd., P.O. Box 544, Potts Point, N.S.W. 2011, Australia
FRANCE	Pergamon Press SARL, 24 rue des Ecoles, 75240 Paris, Cedex 05, France
FEDERAL REPUBLIC OF GERMANY	Pergamon Press GmbH, 6242 Kronberg-Taunus, Hammerweg 6, Federal Republic of Germany

First edition 1980

British Library Cataloguing in Publication Data

New tertiary mathematics. – (Pergamon international library)
Vol. 2. Part 1: Further pure mathematics
1. Mathematics – 1961 –
I. Plumpton, Charles   II. MacIlwaine, Patrick Sydney Wilson
510   QA39.2   79–41454

ISBN 0–08–025033–5 hardcover
ISBN 0–08–021644–7 flexicover (net)
ISBN 0–08–025032–7 flexicover (non-net)
ISBN 0–08–021646–3 (4 volume set) (hard only)

Printed in Great Britain by A. Wheaton & Co. Ltd., Exeter

PERGAMON INTERNATIONAL LIBRARY
of Science, Technology, Engineering and Social Studies

*The 1000-volume original paperback library in aid of education,
industrial training and the enjoyment of leisure*

Publisher: Robert Maxwell, M.C.

NEW TERTIARY MATHEMATICS

Volume 2 Part 1
Further Pure Mathematics

THE PERGAMON TEXTBOOK
INSPECTION COPY SERVICE

An inspection copy of any book published in the Pergamon International Library will gladly be
sent to academic staff without obligation for their consideration for course adoption or
recommendation. Copies may be retained for a period of 60 days from receipt and returned if
not suitable. When a particular title is adopted or recommended for adoption for class use and
the recommendation results in a sale of 12 or more copies, the inspection copy may be retained
with our compliments. The Publishers will be pleased to receive suggestions for revised
editions and new titles to be published in this important International Library.

Some other Pergamon titles of interest

C. PLUMPTON & W. A. TOMKYS
Sixth Form Pure Mathematics
Volumes 1 & 2

Theoretical Mechanics for Sixth Forms
2nd SI Edition Volumes 1 & 2

D. T. B. MARJORAM
Exercises in Modern Mathematics

Further Exercises in Modern Mathematics

Modern Mathematics in Secondary Schools

D. G. H. B. LLOYD
Modern Syllabus Algebra

E. Œ. WOLSTENHOLME
Elementary Vectors
3rd SI Edition

Preface

IN this volume, the "Pure Mathematics" of the "core", Volume 1 Part I, is extended and developed to cover almost all the topics included in Advanced Level Pure Mathematics and Further Mathematics syllabuses of the seven Examining Boards. The aim of the book is to provide an adequate course and source book for mathematical pupils in Sixth Forms of schools, Sixth Form Colleges, Colleges of Further Education and Technical Colleges, or engaged in private study. "Modern" and "traditional" topics, notation and methods of treatment have been combined to produce the "integrated" course now in use or being developed by the Examining Boards.

Many of the chapters, with the exception of those on calculus, are independent of one another so that a student following a modular course can select topics in order of preference. However, in any treatment of this kind there must be some cross-referencing—for example, elementary properties of determinants are used in Chapter 14 which deals with vector products, whereas a fuller theory of determinants and matrices is considered in Chapter 15.

The mathematical requirements of students proceeding to university, to read mathematics (pure and/or applied), science and engineering on the one hand or computer science and social sciences, including economics and statistics, on the other hand, differ appreciably. Accordingly both analytical and algebraic topics, to the limit which may reasonably be expected of pre-university students, are considered in detail.

In the belief that worked examples are of great value, many examples of varying difficulty are worked in the text and exercises added after each major topic is covered. The miscellaneous exercises at the end of each chapter are intended to help revision of earlier work.

The sections, exercises, equations and figures are numbered to correspond to the chapters, e.g. §15:4 is the fourth section of Chapter 15; Ex. 19.5 is the exercise at the end of §19:5; equation (13.7) is the seventh (numbered) equation of Chapter 13. Only those equations to which subsequent reference is made are numbered. Certain sections are starred, not so much to imply excessive difficulty, as to indicate that they can be omitted without affecting the logical sequence of the particular chapter.

We are grateful to the following G.C.E. examination boards for permission to reproduce questions from their papers:

The University of London (L.);
The Northern Universities Joint Matriculation Board (N.);
The Oxford and Cambridge Schools Examination Board; the Schools Mathematics Project, Mathematics in Education and Industry (O.C.) (O.C.S.M.P.) (O.C.M.E.I.);
The Cambridge Local Examinations Syndicate (C.);

vi   **PREFACE**

The Oxford Delegacy (O.);
The Associated Examination Board (A.E.B.).

The questions are theirs, the answers are ours!

C. PLUMPTON
P. S. W. MACILWAINE

Contents

xii **CONTENTS**

Glossary of Symbols and Abbreviations

General

$\approx$	is approximately equal to
$=$	is equal to
$\neq$	is not equal to; and in general / through a symbol implies the negation of that symbol
$\equiv$	is identical to
$>, <$	is greater than, is less than
$\geqslant, \leqslant$	is greater than or equal to, is less than or equal to
$\gg, \ll$	is very much greater than, is very much less than
$/$	divided by, e.g. $3/4 = \frac{3}{4}$ (The stroke is called the solidus)
$:$	such that, e.g. $x : x > 2$
$\forall$	for all (relevant) values of
$\exists$	there is, there are, there exist(s)
$\sum$	the sum of, e.g. $\sum\limits_{r=1}^{3} a^r = a + a^2 + a^3$
$\prod$	the product of, e.g. $\prod\limits_{r=1}^{4} \alpha_r = \alpha_1 \alpha_2 \alpha_3 \alpha_4$
∞	"infinity"
$p, q, \ldots$	statements, propositions (which are either true or false)
$p \Rightarrow q,\ q \Leftarrow p$	if p is true, then q is true; (the truth of) p implies (the truth of) q; (the truth of) q is a necessary condtion for (the truth of) p; (the truth of) p is a sufficient condition for (the truth of) q
$p \Leftrightarrow q,\ q \Leftrightarrow p$	p implies q and q implies p; p and q are either both true or both false; p and q are equivalent; p is a necessary and sufficient condition for q; p is true if and only if q is true
$\sim p$	the negative of p, the proposition p is false
$p \therefore q$	p is true and therefore q is true

Sets

$A, B, C., \ldots$	a set is usually denoted by a capital letter
$a, b, c, \ldots$	an element of a set is usually denoted by a small (lower case) letter
$A = \{a, b, c, \ldots\}$	the set A consists of the elements $a, b, c, \ldots$

$n(A)$	the number of elements in the set A
$\in$	belongs to, is an element of (a set), e.g. $3 \in \{1, 2, 3\}$
$A \subset B, B \supset A$	A is a subset of B, A is included in B
$\mathscr{E}$	the universal set (of all elements under consideration)
$\{\ \}, \phi$	the empty (null) set
A'	the complement of A, i.e. the set of all elements not in A $A' = \{x : x \in \mathscr{E}, x \notin A\}$
$A \cup B$	the union set of A and B, i.e. the set of all elements in A or in B (or in both)
$A \cap B$	the intersection set of A and B, i.e. the set of all elements in A and in B
$\mathbb{N}$	the set of all natural numbers (positive integers) $\{1, 2, 3, \ldots\}$
$\mathbb{Z}$	the set of all integers $\{0, \pm 1, \pm 2, \ldots\}$
$\mathbb{Q}$	the set of rational numbers $\{p/q : p, q \in \mathbb{Z}, q \neq 0\}$
$\mathbb{R}$	the set of real numbers
$\mathbb{Q}^+, \mathbb{R}^+$	the set of positive rational, real numbers
$\mathbb{C}$	the set of complex numbers

Functions of a real variable

$f : x \rightarrow f(x)$	the function f maps x onto $f(x)$
gf	the composite function "f followed by g"
f^{-1}	the inverse of the function f
f'	the derivative of the function f
$\dfrac{df}{dx}, \dfrac{d^2 f}{dx^2}, \ldots$	the first, second, ... derivatives of $f(x)$ with respect to x
$\dot{x}, \ddot{x}, \ldots$	the first, second derivatives ... of x with respect to t if $x = x(t)$
$\dfrac{\partial y}{\partial u}, \dfrac{\partial^2 y}{\partial u \partial v}, \dfrac{\partial^2 y}{\partial v^2}, \ldots$	the first, second, ... partial derivatives of $y(u, v, \ldots)$ with respect to $u, v, \ldots$
δx	a small increment in x
$\delta y, \delta u, \ldots$	small increments in $y, u, \ldots$ corresponding to δx

$$\int_a^b f(x)\,dx = \left[F(x) \right]_a^b = F(b) - F(a)$$

	the area of the region enclosed by the curve $y = f(x)$, the x-axis and the ordinates $x = a$, $x = b$ is equal to $F(b) - F(a)$, where $F'(x) \equiv f(x)$
$\sqrt{x}$	the positive square root of $x (x \in \mathbb{R}^+)$
$\pm \sqrt{x}$	the positive or negative square root of x $(x \in \mathbb{R})$
$\|x\|$	the modulus of x; $\forall x > 0, \|x\| = x,$ $\forall x < 0, \|x\| = -x$
$[x]$	the greatest integer not greater than x, e.g. $[3] = 3, [-4 \cdot 2] = -5$

$n!$	factorial n; $n \in \mathbb{N}$, $n! = n(n-1)(n-2)\ldots 3.2.1$
$\log_a x$	the logarithm of x to base a
$\lg x$	the logarithm of x to base 10
$\ln x$	the logarithm of x to base e (natural or Napierian logarithm)
$\displaystyle\lim_{x \to a+} f(x)$	the limit of $f(x)$ as $x \to a$ from above
$\displaystyle\lim_{x \to a-} f(x)$	the limit of $f(x)$ as $x \to a$ from below
$O[f(x)]$	the order of magnitude of $f(x)$, e.g. $\dfrac{5}{x^2} = O\left(\dfrac{1}{x^2}\right)$

Functions of a complex variable

z	the complex number $x + iy (x, y \in \mathbb{R})$
z^*	the conjugate of z; $z = x + iy \Rightarrow z^* = x - iy$
$|z|$	the modulus of z; $r = |z| = \sqrt{(x^2 + y^2)}$
$\arg z$	the argument of z; $\arg z = \theta$; $\cos\theta = x/r$, $\sin\theta = y/r$ (the principal value of $\arg z$ is θ: $-\pi < \theta \leqslant \pi$ or $0 \leqslant \theta < 2\pi$)
$\operatorname{Re} z$	the real part of z; $\operatorname{Re}(x + iy) = x$.
$\operatorname{Im} z$	the imaginary part of z; $\operatorname{Im}(x + iy) = y$.
$f: z \to f(z) = w$	the function f maps the point representing z in the complex z-plane onto the point representing w in the complex w-plane

Vectors, matrices, algebraic structure

$\mathbf{a}, \mathbf{b}\ldots$	vectors
$\overrightarrow{PQ}$	a directed line segment representing a vector
$\mathbf{i}, \mathbf{j}, \mathbf{k}$	(unit) base vectors in mutually perpendicular directions
$\hat{\mathbf{n}}$	a unit vector in the direction of $\mathbf{n}$
a_1, a_2, a_3	components of a vector, e.g. $a = a_1\mathbf{i} + a_2\mathbf{j} + a_3\mathbf{k}$
$\mathbf{a} + \mathbf{b}$	the vector sum of two vectors
$\mathbf{a}.\mathbf{b}$	the scalar product of two vectors
$\mathbf{a} \times \mathbf{b}$	the vector product of two vectors
$\mathbf{M}$	a matrix, or the transformation represented by a matrix
$|\mathbf{M}|$, det $\mathbf{M}$	the determinant of the matrix $\mathbf{M}$
$\mathbf{M}^{\mathsf{T}}$	the transpose of the matrix $\mathbf{M}$
$\mathbf{M}^{-1}$	the inverse of the matrix $\mathbf{M}$
a_{rs}	the element in the rth row and the sth column of a matrix
$\mathbf{I}$	a unit matrix, or an identity transformation, such that $a_{rs} = 0 (r \neq s)$, $a_{rr} = 1$
$a\, \mathcal{O}\, b$	the combination of elements a, b under the operation $\mathcal{O}$
$a\, \mathcal{R}\, b$	a has the relation $\mathcal{R}$ to b
$G = (S, *)$	the group G formed from the operation $*$ defined on the set S

The Greek Alphabet

A	α	alpha	N	ν	nu
B	β	beta	Ξ	ξ	xi
Γ	γ	gamma	O	o	omicron
Δ	δ	delta	Π	π	pi
E	ε	epsilon	P	ρ	rho
Z	ζ	zeta	Σ	σ	sigma
H	η	eta	T	τ	tau
Θ	θ	theta	Y	υ	upsilon
I	ι	iota	Φ	ϕ	phi
K	κ	kappa	X	χ	chi
Λ	λ	lambda	Ψ	ψ	psi
M	μ	mu	Ω	ω	omega

13 Further Complex Numbers: De Moivre's Theorem

13:1 DE MOIVRE'S THEOREM

Equation (9.12) of Volume 1, viz.

$$r_1(\cos\theta_1 + i\sin\theta_1) \times r_2(\cos\theta_2 + i\sin\theta_2) = r_1 r_2[\cos(\theta_1 + \theta_2) + i\sin(\theta_1 + \theta_2)]$$

can be written

$$(r_1, \theta_1) \times (r_2, \theta_2) = (r_1 r_2, \theta_1 + \theta_2).$$

Similar results can be obtained for successive products, e.g.

$$(r_1 r_2, \theta_1 + \theta_2) \times (r_3, \theta_3) = (r_1 r_2 r_3, \theta_1 + \theta_2 + \theta_3).$$

In particular, when $r_1 = r_2 = r_3 = r$ and $\theta_1 = \theta_2 = \theta_3 = \theta$,

$$(r, \theta)^2 \times (r, \theta) = (r^3, 3\theta)$$

and, when $r = 1$,

$$(1, \theta)^2 \times (1, \theta) = (1, \theta)^3 = (1, 3\theta),$$

i.e.

$$(\cos\theta + i\sin\theta)^3 = \cos 3\theta + i\sin 3\theta.$$

(i) We can extend the above result $\forall n \in \mathbb{N}$ by induction thus:

$$(\cos\theta + i\sin\theta)^n = \cos n\theta + i\sin n\theta$$
$$\Rightarrow (\cos\theta + i\sin\theta)^{n+1} = (\cos n\theta + i\sin n\theta)(\cos\theta + i\sin\theta)$$
$$= (\cos n\theta\cos\theta - \sin n\theta\sin\theta) + i(\sin n\theta\cos\theta + \cos n\theta\sin\theta)$$
$$= \cos(n+1)\theta + i\sin(n+1)\theta.$$

But the result is clearly (identically) true for $n = 1$. Hence by induction

$$(\cos\theta + i\sin\theta)^n = \cos n\theta + i\sin n\theta \quad \forall n \in \mathbb{N}.$$

This is a particular case of *de Moivre's theorem*.

(ii) De Moivre's theorem for $n \in \mathbb{Z}$.

If $n = -m$, where $m \in \mathbb{N}$, then

$$(\cos\theta + i\sin\theta)^n = \frac{1}{(\cos\theta + i\sin\theta)^m}$$

$$= \frac{1}{\cos m\theta + i\sin m\theta}$$

$$= \frac{\cos m\theta - i\sin m\theta}{\cos^2 m\theta + \sin^2 m\theta}$$

$$= \cos(-m\theta) + i\sin(-m\theta)$$

$$= \cos n\theta + i\sin n\theta$$

since $-m = n$. This proves the theorem when n is a negative integer.

Clearly the result holds when $n = 0$, so we have proved de Moivre's theorem $\forall n \in \mathbb{Z}$.

(iii) De Moivre's theorem for $n \in \mathbb{Q}$.

Let $n = p/q$ where $p \in \mathbb{Z}$ and $q \in \mathbb{N}$. In this case $(\cos\theta + i\sin\theta)^n$ is many-valued.

We shall prove first that *one* of the values of $(\cos\theta + i\sin\theta)^n$ is $(\cos n\theta + i\sin n\theta)$. From (i), we have

$$\left(\cos\frac{p\theta}{q} + i\sin\frac{p\theta}{q}\right)^q = \cos p\theta + i\sin p\theta,$$

and from (i) or (ii)

$$\cos p\theta + i\sin p\theta = (\cos\theta + i\sin\theta)^p.$$

Hence $\left(\cos\dfrac{p\theta}{q} + i\sin\dfrac{p\theta}{q}\right)$ is a value of $(\cos\theta + i\sin\theta)^{p/q}$.

Thus we have proved de Moivre's theorem for $n \in \mathbb{Q}$.

Notation $z = a^{p/q}$ means that z satisfies the equation $z^q = a^p$ and therefore has many values; $\sqrt[q]{a^p}$ means the principal value of $a^{p/q}$.

The values of $(\cos\theta + i\sin\theta)^{p/q}$ *where* $p \in \mathbb{Z}$ *and* $q \in \mathbb{N}$.

If a value of $(\cos\theta + i\sin\theta)^{p/q}$ is $k(\cos\phi + i\sin\phi)$, where $k \in \mathbb{R}$, then

$$k^q(\cos\phi + i\sin\phi)^q = (\cos\theta + i\sin\theta)^p$$

$$\Rightarrow k^q(\cos q\phi + i\sin q\phi) = \cos p\theta + i\sin p\theta.$$

But if two complex numbers are equal, their moduli are also equal

$$\Rightarrow k^q = 1$$

$$\Rightarrow k = 1 \ \forall q \in \mathbb{N}$$

$$\Rightarrow \cos q\phi + i\sin q\phi = \cos p\theta + i\sin p\theta.$$

Then, equating real and imaginary parts,

$$\cos q\phi = \cos p\theta, \qquad \sin q\phi = \sin p\theta$$

$$\Leftrightarrow q\phi = p\theta + 2m\pi$$

$$\Leftrightarrow \phi = \frac{p}{q}\theta + \frac{2m}{q}\pi, \text{ where } m \in \mathbb{Z}.$$

If now we give to m the successive values $0, 1, 2, \ldots, (q-1)$, we have

$$(\cos\theta + i\sin\theta)^{p/q} = \left(1, \frac{p\theta}{q}\right), \left(1, \frac{p\theta + 2\pi}{q}\right), \ldots$$

$$= \left(1, \frac{p\theta + 2k\pi}{q}\right) \quad \text{for } k = 0, 1, 2, \ldots, (q-1).$$

These values of $(\cos\theta + i\sin\theta)^{p/q}$ are all distinct because the difference between the arguments of any two of them is less than 2π. There are no further values of $(\cos\theta + i\sin\theta)^{p/q}$, for if $m = aq + b$ where a and b are positive integers and $b < q$, then $(\cos\theta + i\sin\theta)^{p/q}$ takes the value

$$\cos\left\{\frac{p\theta + 2(aq + b)\pi}{q}\right\} + i\sin\left\{\frac{p\theta + 2(aq + b)\pi}{q}\right\},$$

i.e. $$\cos\left\{\frac{p\theta + 2b\pi}{q} + 2a\pi\right\} + i\sin\left\{\frac{p\theta + 2b\pi}{q} + 2a\pi\right\},$$

and this value is equal to the value of

$$\cos\left(\frac{p\theta + 2b\pi}{q}\right) + i\sin\left(\frac{p\theta + 2b\pi}{q}\right)$$

which is one of those listed above. We have shown, therefore, that $(\cos\theta + i\sin\theta)^{p/q}$ when p and q are integers has just q values and these values are given above.

It is worth repeating that de Moivre's theorem, as stated and proved above, refers only to a rational exponent. The theorem can be shown to be true for an irrational exponent, but in this book we shall not extent the proof beyond the case of a rational exponent.

Example 1. Solve the equation $z^5 + 1 = 0$.

$$z^5 = -1$$
$$\Rightarrow z = (-1)^{1/5} = (\cos\pi + i\sin\pi)^{1/5}$$

$$\Rightarrow z = \cos\frac{\pi}{5} + i\sin\frac{\pi}{5} = (1, \pi/5),$$

$$\text{or } \cos\frac{3\pi}{5} + i\sin\frac{3\pi}{5} = (1, 3\pi/5),$$

$$\text{or } \cos\frac{5\pi}{5} + i\sin\frac{5\pi}{5} = (1, 5\pi/5) = -1 + i0,$$

$$\text{or } \cos\frac{7\pi}{5} + i\sin\frac{7\pi}{5} = (1, 7\pi/5),$$

$$\text{or } \cos\frac{9\pi}{5} + i\sin\frac{9\pi}{5} = (1, 9\pi/5).$$

Example 2. Solve the equation

$$z^6 + 2z^3 + 2 = 0.$$

This equation can be written

$$(z^3 + 1)^2 + 1 = 0$$

$$\Rightarrow z^3 = -1 - i \text{ or } -1 + i$$

$$\Rightarrow z^3 = \sqrt{2}[\cos(-\tfrac{3}{4}\pi) + i\sin(-\tfrac{3}{4}\pi)]$$

$$\text{or } z^3 = \sqrt{2}(\cos\tfrac{3}{4}\pi + i\sin\tfrac{3}{4}\pi)$$

$$\Rightarrow z \approx 1\cdot12[\cos(-\tfrac{3}{4}\pi) + i\sin(-\tfrac{3}{4}\pi)]^{1/3}$$

$$= 1\cdot12\left[\cos\frac{(8k-3)\pi}{12} + i\sin\frac{(8k-3)\pi}{12}\right],$$

$$\text{or } z \approx 1\cdot12(\cos\tfrac{3}{4}\pi + i\sin\tfrac{3}{4}\pi)^{1/3}$$

$$= 1\cdot12\left[\cos\frac{(8k+3)\pi}{12} + i\sin\frac{(8k+3)\pi}{12}\right], \quad k = 0, 1, 2,$$

$$\Rightarrow z \approx (1\cdot12, -\pi/4) \text{ or } (1\cdot12, \pi/4)$$

$$\text{or } (1\cdot12, 5\pi/12) \text{ or } (1\cdot12, 11\pi/12)$$

$$\text{or } (1\cdot12, 13\pi/12) \text{ or } (1\cdot12, 19\pi/12).$$

Example 3. Expand $(\cos\theta + i\sin\theta)^4$ by the binomial theorem. Hence deduce expressions for (a) $\cos 4\theta$ in terms of $\cos\theta$ and (b) $\sin 4\theta$ in terms of $\sin\theta$ and $\cos\theta$. Deduce an expression for $\tan 4\theta$ in terms of $\tan\theta$.

By the binomial theorem (which we are entitled to use, since the product of complex numbers is defined according to the laws of real algebra),

$$(\cos\theta + i\sin\theta)^4 = \cos^4\theta + 4i\cos^3\theta\sin\theta - 6\cos^2\theta\sin^2\theta$$
$$- 4i\cos\theta\sin^3\theta + \sin^4\theta.$$

But by de Moivre's theorem

$$(\cos\theta + i\sin\theta)^4 = \cos 4\theta + i\sin 4\theta.$$

Equating real and imaginary parts, we have

$$\cos 4\theta = \cos^4\theta - 6\cos^2\theta\sin^2\theta + \sin^4\theta,$$
$$= \cos^4\theta - 6\cos^2\theta(1 - \cos^2\theta) + (1 - \cos^2\theta)^2$$
$$= 8\cos^4\theta - 8\cos^2\theta + 1,$$

and
$$\sin 4\theta = 4\cos^3\theta\sin\theta - 4\cos\theta\sin^3\theta$$

$$\Rightarrow \tan 4\theta = \frac{4\cos^3\theta\sin\theta - 4\cos\theta\sin^3\theta}{\cos^4\theta - 6\cos^2\theta\sin^2\theta + \sin^4\theta}$$

$$= \frac{4\tan\theta - 4\tan^3\theta}{1 - 6\tan^2\theta + \tan^4\theta}.$$

Example 4. Prove that $\cos^4\theta = \tfrac{1}{8}(\cos 4\theta + 4\cos 2\theta + 3),$

$\sin^4\theta = \tfrac{1}{8}(\cos 4\theta - 4\cos 2\theta + 3).$

If $z = \cos\theta + i\sin\theta$, then $\dfrac{1}{z} = \cos\theta - i\sin\theta$.

$$\left(z^n + \frac{1}{z^n}\right) = 2\cos n\theta,$$

$$\left(z^n - \frac{1}{z^n}\right) = 2i\sin n\theta$$

$$\Rightarrow 16\cos^4\theta = (2\cos\theta)^4 = \left(z + \frac{1}{z}\right)^4$$

$$= z^4 + 4z^2 + 6 + 4/z^2 + 1/z^4$$

$$= \left(z^4 + \frac{1}{z^4}\right) + 4\left(z^2 + \frac{1}{z^2}\right) + 6$$

$$= 2\cos 4\theta + 8\cos 2\theta + 6$$

$$\Leftrightarrow \cos^4\theta = \tfrac{1}{8}(\cos 4\theta + 4\cos 2\theta + 3).$$

Also
$$16\sin^4\theta = (2i\sin\theta)^4 = \left(z - \frac{1}{z}\right)^4$$

$$= z^4 - 4z^2 + 6 - 4/z^2 + 1/z^4$$

$$= \left(z^4 + \frac{1}{z^4}\right) - 4\left(z^2 + \frac{1}{z^2}\right) + 6$$

$$= 2\cos 4\theta - 8\cos 2\theta + 6$$

$$\Leftrightarrow \sin^4\theta = \tfrac{1}{8}(\cos 4\theta - 4\cos 2\theta + 3).$$

Example 5. Prove that $\cos\dfrac{\pi}{7} + \cos\dfrac{3\pi}{7} + \cos\dfrac{5\pi}{7} = \dfrac{1}{2}$.

The roots of the equation $z^7 + 1 = 0$ are -1, $\cos\dfrac{\pi}{7} + i\sin\dfrac{\pi}{7}$,

$$\cos\frac{3\pi}{7} + i\sin\frac{3\pi}{7}, \ \cos\frac{5\pi}{7} + i\sin\frac{5\pi}{7}, \ \cos\frac{9\pi}{7} + i\sin\frac{9\pi}{7}, \ \cos\frac{11\pi}{7} + i\sin\frac{11\pi}{7}, \ \cos\frac{13\pi}{7} + i\sin\frac{13\pi}{7}.$$

Since the coefficient of z^6 in the equation is zero, the sum of the roots of the equation is zero. The sum of the real parts of the roots is therefore zero

$$\Rightarrow \left(\cos\frac{\pi}{7} + \cos\frac{13\pi}{7}\right) + \left(\cos\frac{3\pi}{7} + \cos\frac{11\pi}{7}\right)$$

$$+ \left(\cos\frac{5\pi}{7} + \cos\frac{9\pi}{7}\right) - 1 = 0.$$

But $\cos\dfrac{\pi}{7} = \cos\dfrac{13\pi}{7}$, etc.

$$\Rightarrow 2\cos\frac{\pi}{7} + 2\cos\frac{3\pi}{7} + 2\cos\frac{5\pi}{7} = 1$$

$$\Leftrightarrow \cos\frac{\pi}{7} + \cos\frac{3\pi}{7} + \cos\frac{5\pi}{7} = \frac{1}{2}.$$

Exercise 13.1

1. Simplify $(\cos\theta - i\sin\theta)^4/(\cos\theta + i\sin\theta)^5$.

2. Simplify $(\cos\tfrac{1}{4}\pi + i\sin\tfrac{1}{4}\pi)^3\,(\cos\tfrac{3}{4}\pi - i\sin\tfrac{3}{4}\pi)^2$.

3. Simplify $\dfrac{(\cos\tfrac{1}{3}\pi + i\sin\tfrac{1}{3}\pi)^2\,(\cos\tfrac{2}{3}\pi + i\sin\tfrac{2}{3}\pi)^3}{(\cos\tfrac{1}{4}\pi + i\sin\tfrac{1}{4}\pi)^2\,(\cos\tfrac{3}{4}\pi + i\sin\tfrac{3}{4}\pi)^4}$.

4. Find the square roots of each of the following in the form $r(\cos\theta + i\sin\theta)$:

      (i) i, (ii) $3 + 4i$, (iii) $5 - 12i$, (iv) $-7 + 24i$, (v) $(1 + i)/(1 - i)$.

5. Find the cube roots of each of the following in the form $r(\cos\theta + i\sin\theta)$:

      (i) 1, (ii) i, (iii) $1 + i$, (iv) $3 + 4i$.

6. Draw a diagram to show the positions of the points representing the five fifth-roots of unity in the coordinate plane.

7. Solve the equation $z^4 - 2z^2 + 5 = 0$.

8. Solve the equation $z^6 + 1 = 0$.

9. Express $\cos 5\theta$ in terms of powers of $\cos\theta$ only and obtain an expression for $\tan 5\theta$ in terms of $\tan\theta$.

10. Obtain expressions for $\cos n\theta$ and for $\sin n\theta$ in terms of $\sin \theta$ and $\cos \theta$, when $n \in \mathbb{N}$.

11. Express $\sin^6 \theta$ and $\cos^6 \theta$ each in terms of cosines of multiples of θ.

12. Use the method suggested by question 11 to obtain the following integrals

$$\text{(i) } \int \sin^6 \theta \, d\theta, \text{ (ii) } \int \cos^8 \theta \, d\theta, \text{ (iii) } \int \sin^4 \theta \cos^2 \theta \, d\theta.$$

13. Express $\dfrac{(\cos \alpha + i \sin \alpha)(\cos \beta + i \sin \beta)}{(\cos \gamma + i \sin \gamma)(\cos \delta + i \sin \delta)}$ in the form $(\cos \theta + i \sin \theta)$.

14. Solve the equation $(z + 1)^6 = (z - 1)^6$.

15. Find all the fourth roots of $28 + 96i$.

16. If $\alpha = \cos \frac{2}{5}\pi + i \sin \frac{2}{5}\pi$, prove that

$$1 + \alpha + \alpha^2 + \alpha^3 + \alpha^4 = 0.$$

17. Prove that the roots of the equation $z^n = (z - 1)^n$, where $n \in \mathbb{N}$, are $\frac{1}{2}\{1 + i \cot (r\pi/n)\}$ where $r = 1, 2, 3, \ldots, (n - 1)$.

18. Show that $(-1 + i \sqrt{3})^n + (-1 - i \sqrt{3})^n \in \mathbb{R}$, where $n \in \mathbb{N}$, and find its value in terms of trigonometrical functions.

19. Show that $\cos n\theta$ and $\sin n\theta / \sin \theta$, where $n \in \mathbb{N}$, are expressible as polynomials in $\cos \theta$ and obtain these polynomials in the case $n = 6$.   (L.)

20. Show that $z^2 - 2z \cos 2\pi/7 + 1$ is a factor of $z^7 - 1$, and write down the two other real quadratic factors of $z^7 - 1$.   (L.)

21. Express in the form $a + ib$,

$$\text{(i) } (1 + \cos \theta + i \sin \theta)^n, \qquad \text{(ii) } \left(\frac{\sin \theta + i \cos \theta}{\sin \theta - i \cos \theta}\right)^n.$$

22. Use de Moivre's theorem to show that

$$\tan 5\theta = \frac{5t - 10t^3 + t^5}{1 - 10t^2 + 5t^4},$$

where $t = \tan \theta$.

13:2 THE EXPONENTIAL FORM OF A COMPLEX NUMBER

We have shown in § 13:1 that, for $n \in \mathbb{Q}$,

$$(\cos \theta + i \sin \theta)^n = \cos n\theta + i \sin n\theta,$$

i.e. if

$$f(\theta) = \cos \theta + i \sin \theta,$$

then

$$\{f(\theta)\}^n = f(n\theta).$$

This is a property of a function which obeys the index laws, as for example if $f(x) = a^x$; then

$$\{f(x)\}^n = a^{nx} = f(nx).$$

It is reasonable, therefore, to expect $\cos \theta + i \sin \theta$ to be associated with such a function as a^θ.

In chapter 11 we obtained the results

$$\cos \theta = 1 - \frac{\theta^2}{2!} + \frac{\theta^4}{4!} - \frac{\theta^6}{6!} + \cdots,$$

$$\sin \theta = \theta - \frac{\theta^3}{3!} + \frac{\theta^5}{5!} - \frac{\theta^7}{7!} + \cdots,$$

whence we have

$$\cos \theta + i \sin \theta = 1 + i\theta - \frac{\theta^2}{2!} - \frac{i\theta^3}{3!} + \frac{\theta^4}{4!} + \frac{i\theta^5}{5!} - \frac{\theta^6}{6!} - \frac{i\theta^7}{7!} + \cdots$$

$$= 1 + i\theta + \frac{(i\theta)^2}{2!} + \frac{(i\theta)^3}{3!} + \frac{(i\theta)^4}{4!} + \frac{(i\theta)^5}{5!} + \frac{(i\theta)^6}{6!} + \frac{(i\theta)^7}{7!} + \cdots$$

assuming that rearrangement of the sum of the two infinite series is permissible.
This is of the same form as the expansion

$$e^x = 1 + x + \frac{x^2}{2!} + \frac{x^3}{3!} + \cdots$$

where x is a *real variable*, but since the expansion was obtained for a real variable only it is not possible to *deduce* that $\cos \theta + i \sin \theta = e^{i\theta}$; indeed $e^{i\theta}$ has not yet been defined.

If we write $\exp z$ for $1 + z + \frac{z^2}{2!} + \frac{z^3}{3!} + \cdots$, where $z \in \mathbb{C}$,

then $\exp(i\theta) = \cos \theta + i \sin \theta$

and $\{\exp(i\theta)\}^n = \exp(ni\theta).$

Also, from equation (9.12),

$$\exp(i\theta) \times \exp(i\phi) = \exp[i(\theta + \phi)]$$

and, from equation (9.13),

$$\exp(i\theta) \div \exp(i\phi) = \exp[i(\theta - \phi)].$$

We therefore *define* $e^{i\theta}$ as $\exp(i\theta)$ so that

$$e^{i\theta} = 1 + i\theta + \frac{(i\theta)^2}{2!} + \frac{(i\theta)^3}{3!} + \cdots$$

and $e^{i\theta}$, thus defined, obeys the laws which e^x, where $x \in \mathbb{R}$, obeys.

Then $e^{i\theta} = \cos \theta + i \sin \theta.$ (13.1)

It follows that $e^{i(\theta + 2n\pi)} = e^{i\theta} e^{i2n\pi}$

$$= e^{i\theta}(\cos 2n\pi + i \sin 2n\pi)$$

$$= e^{i\theta} \ (n = 0, \pm 1, \pm 2 \ldots).$$

Also, since

$$(\cos \theta + i \sin \theta)^{-1} = (\cos \theta - i \sin \theta) = \cos(-\theta) + i \sin(-\theta),$$

$$e^{-i\theta} = 1/e^{i\theta}.$$

We can summarize these postulates, and the deductions we have made from them here, in the conclusion that complex numbers, expressed by definition as powers of e, obey the same laws of computation as do real numbers similarly expressed.

Example 1. $3 + 4i = 5(\cos\theta + i\sin\theta)$ where $\sin\theta = 4/5$, $\cos\theta = 3/5$ and therefore $\theta \approx 0\cdot927$

$$\Rightarrow 3 + 4i \approx 5e^{(0\cdot927)i}.$$

(θ *must* now be expressed in radians, since this is the assumption involved in the use of the series for sine and cosine.)

Example 2. Evaluate the integrals $I_1 = \int_0^{\pi/2} e^{2\theta}\cos\theta\,d\theta$ and $I_2 = \int_0^{\pi/2} e^{2\theta}\sin\theta\,d\theta$.

We make the assumption here that the rules of integration are applicable to complex quantities.

$$I_1 + iI_2 = \int_0^{\pi/2} e^{2\theta}e^{i\theta}\,d\theta = \int_0^{\pi/2} e^{(2+i)\theta}\,d\theta$$

$$= \left[\frac{e^{(2+i)\theta}}{2+i}\right]_0^{\pi/2} = \frac{1}{5}\left[(2-i)e^{(2+i)\theta}\right]_0^{\pi/2}$$

$$= \frac{1}{5}\left[e^{2\theta}(2-i)(\cos\theta + i\sin\theta)\right]_0^{\pi/2}$$

$$= \frac{1}{5}\left[e^{2\theta}\{(2\cos\theta + \sin\theta) - i(\cos\theta - 2\sin\theta)\}\right]_0^{\pi/2}.$$

Equating real and imaginary parts, we find

$$I_1 = \frac{1}{5}\left[e^{2\theta}(2\cos\theta + \sin\theta)\right]_0^{\pi/2} = (e^\pi - 2)/5,$$

$$I_2 = -\frac{1}{5}\left[e^{2\theta}(\cos\theta - 2\sin\theta)\right]_0^{\pi/2} = (2e^\pi + 1)/5.$$

Example 3. *Rotation of axes in the coordinate plane.*

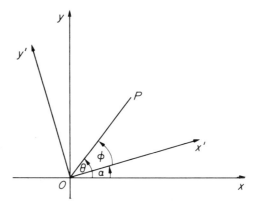

FIG. 13.1.

Suppose the rectangular coordinates of a point P are (x, y) when referred to one set of axes Oxy and (x', y') when referred to another set $Ox'y'$ inclined at an angle α, see Fig. 13.1, to the first set. Let $z = x + iy$ and $z' = x' + iy'$; then, since $|z| = OP = |z'|$ and $\arg z = \theta = \phi + \alpha = \arg z' + \alpha$,

$$z = z'e^{i\alpha}$$

or

$$z' = ze^{-i\alpha}$$

$$\Rightarrow x' = x \cos \alpha + y \sin \alpha, \quad y' = -x \sin \alpha + y \cos \alpha,$$
$$x = x' \cos \alpha - y' \sin \alpha, \quad y = x' \sin \alpha + y' \cos \alpha.$$

Example 4. Sum the series $\cos \theta + \cos 2\theta + \ldots + \cos n\theta$. Hence find the sum of the series $\sin \theta + 2 \sin 2\theta + \ldots + n \sin n\theta$.

If

$$C = \cos \theta + \cos 2\theta + \ldots + \cos n\theta$$

and

$$S = \sin \theta + \sin 2\theta + \ldots + \sin n\theta,$$

then

$$C + iS = e^{i\theta} + e^{2i\theta} + \ldots + e^{in\theta} = e^{i\theta} \frac{(1 - e^{in\theta})}{(1 - e^{i\theta})}$$

$$\Rightarrow C + iS = \frac{e^{i\theta} \cdot e^{in\theta/2}(e^{-in\theta/2} - e^{in\theta/2})}{e^{i\theta/2}(e^{-i\theta/2} - e^{i\theta/2})} = \frac{e^{(n+1)i\theta/2}(-2i)\sin \frac{1}{2}n\theta}{(-2i)\sin \frac{1}{2}\theta}$$

$$\Rightarrow C = \frac{\sin \frac{1}{2}n\theta \cos \frac{1}{2}(n+1)\theta}{\sin \frac{1}{2}\theta} \quad \text{and} \quad S = \frac{\sin \frac{1}{2}n\theta \sin \frac{1}{2}(n+1)\theta}{\sin \frac{1}{2}\theta}.$$

This example shows that it is often helpful to do the manipulations with complex numbers before separating out the real and imaginary parts. The second sum required is $-dC/d\theta$.

Exercise 13.2

1. Express each of the following in the form $re^{i\theta}$ (giving the principal value of θ):

(i) $1+i$, (ii) $1-i$, (iii) i, (iv) $\dfrac{1-i}{1+i}$, (v) $1-i\sqrt{3}$.

2. Express each of the following in the form $a + ib$:

(i) $e^{-i\pi/4}$, (ii) $e^{i\pi}$, (iii) $e^{-i\pi}$, (iv) $e^{1+i\pi/3}$, (v) e^{x+iy}.

3. Express $1 + e^{i\theta}$ in the form $re^{i\phi}$.

4. Express $e^{i\theta} + e^{-i\theta}$ in the form $re^{i\phi}$.

5. (i) Find, in the form $re^{i\theta}$, all the roots of the equation

$$(z^4 + 1)(z^2 + 1) = 0.$$

(ii) If $z = \cos \theta + i \sin \theta$ and $w = 2z - 3 - 3/z$, express $|w|^2$ in terms of $\cos \theta$ and hence show that the greatest value of $|w|^2$ is $275/8$. (L.)

6. (i) Find all the solutions of the equation $z^5 = 1$, where $z \in \mathbb{C}$. Hence show that

$$\frac{z^5 - 1}{z - 1} = \left(z^2 - 2z \cos \frac{2\pi}{5} + 1\right)\left(z^2 - 2z \cos \frac{4\pi}{5} + 1\right).$$

(ii) If $z = \cos \pi/n + i \sin \pi/n$, where $n \in \mathbb{N}$, $n \geqslant 2$, show that $(1 + z)^n$ is purely imaginary. (L.)

7. (i) Solve the equation $z^3 + i = 0$, giving each of the roots in the form $a + bi$, where a and b are real.

(ii) Show that $(1 - e^{ix})(1 - e^{-ix}) = 2(1 - \cos x)$.

Show also that

$$1 + \cos x + \cos 2x + \ldots + \cos nx$$

is the real part of

$$1 + e^{ix} + e^{2ix} + \ldots + e^{nix}$$

and deduce that, if $\cos x \neq 1$, the sum of the first series is

$$\frac{1}{2}\left\{1 + \frac{\cos nx - \cos(n+1)x}{1 - \cos x}\right\}.$$

(L.)

13:3 TRANSFORMATIONS OF THE COMPLEX PLANE

If $w = u + iv$, $z = x + iy$, then an equation

$$w = f(z),$$

(13.2)

where $f(z)$ is a function in a domain D of the z-plane, i.e. the plane of the coordinate axes Oxy, relates a point in the domain D to a point in some domain D' of the w-plane, in which the coordinate axes are $O'uv$. The domain D is said to be *mapped* on D' by the transformation (13.2) and a curve C in D maps into a curve C' in D'. The mapping of D on D' is said to be *one-one* if, to each point of D there corresponds one and only one point of D'. [However, mappings are not necessarily one–one.]

It can be shown that, provided $f(z)$ satisfies certain conditions, the angle between two curves in the z-plane intersecting at P is equal to the angle at P', the image of P, between the corresponding curves in the w-plane. Because angles are conserved in this way, such transformations are called *conformal transformations*.

In real analysis we use two axes, one Ox to represent the variable x, and the other Oy to represent the value of a function $y = f(x)$; the properties of the function are then represented by its graph. In complex analysis we use two complex planes, one to represent the variable z, and the other to represent the value w of a function $w = f(z)$; the properties of the (conformal) transformation then represent the properties of the function. Many of the functions in complex variable theory bear the same names as the simple functions of real analysis, e.g. sine, logarithm, exponential, and the properties of the functions of a complex variable are extensions of the properties of the "same" function of a real variable, e.g. the two branches of $z^{1/2}$ are extensions of the two values $\pm \sqrt{x}$. In illustration of this we consider now some special transformations and state or obtain the relation between corresponding curves C, C' in the z- and w-planes respectively.

(1) The transformation

$$w = z + b$$

merely corresponds to a translation of the origin, for, if $b = \alpha + i\beta$, where α and β are real, then

$$u = x + \alpha, \quad v = y + \beta$$

implying that the origin O' in the w-plane is the point $(-\alpha, -\beta)$ in the z-plane.

(2) By comparing modulus and argument it is easy to see that the transformation

$$w = az$$

corresponds to a constant magnification and a constant rotation, the magnification factor being $|a|$ and the anticlockwise rotation angle being arg a.

(3) The transformation

$$w = az + b,$$

the combination of (1), (2) above, corresponds to a change of origin followed by a constant rotation and a constant magnification.

(4) The transformation

$$w = k^2/z,$$

where k is real. Writing $w = R\,e^{i\phi}$, $z = r\,e^{i\theta}$ we find $R = k^2/r$, $\phi = -\theta$. Hence we obtain C' from C by noting that a region inside the circle $|z| = k$ maps into a region outside $|w| = k$, and that $\phi = -\theta$ involves reflection in the real axis. For example, the shaded areas in Fig. 13.2(i) transform into the correspondingly shaded areas in Fig. 13.2(ii). This is a case in which the region *outside* C is transformed into the region *inside* C'.

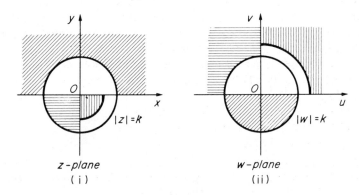

z-plane   w-plane
(i)   (ii)

Fig. 13.2.

(5) The *bilinear*, or *Möbius transformation*

$$w = \frac{az + b}{cz + d}, \quad ad \neq bc,$$

consists of a combination of the transformations (1)–(4) considered above. *The bilinear transformation transforms finite circles into circles.* Further, it is the most general transformation for which one value of z gives one and only one value of w and conversely, i.e. *the transformation is one-one from the z-plane to the w-plane.* The techniques of the bilinear transformation are illustrated in the following example.

Example. If $w = \dfrac{1 + iz}{i + z}$, where $w = u + iv$ and $z = x + iy$, show that the circle $|w| = k$, where $k < 1$, transforms into a circle in the z-plane, whose centre is on the imaginary axis.

Show that the semi-circular region, in the upper half of the w-plane, bounded by the circle $|w| = \frac{1}{2}$ and the real axis $v = 0$, transforms into a region bounded by two circles in the z-plane, and indicate this region in a sketch.

When $|w| = k$,

$$|1 + iz| = k|i + z|$$
$$\Leftrightarrow |z - i| = k|z + i|. \tag{1}$$

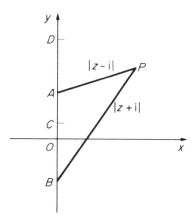

FIG. 13.3.

If P lies on this locus and A, B are the points i, $-$i, respectively, Fig. 13.3, equation (1) implies that $PA = kPB$. Therefore, if $k < 1$, the locus of P is the circle (of Apollonius) on CD as diameter where C and D are the points $(1-k)i/(1+k)$, $(1+k)i/(1-k)$ dividing AB internally and externally in the ratio $k:1$.

Alternatively (1) can be written

$$(z-i)(z^*+i) = k^2(z+i)(z^*-i)$$
$$\Leftrightarrow x^2 + y^2 - 2y + 1 = k^2(x^2 + y^2 + 2y + 1)$$
$$\Leftrightarrow (1-k^2)(x^2+y^2) - 2(1+k^2)y + 1 - k^2 = 0,$$

which is the circle obtained above.

When $k = \frac{1}{2}$, F', D', H', the points $\frac{1}{2}$, $\frac{1}{2}$i, $-\frac{1}{2}$ respectively of the w-plane, transform into F, D, H, the points $(4+3i)/5$, $3i$, $(-4+3i)/5$, respectively, of the z-plane, Fig. 13.4.

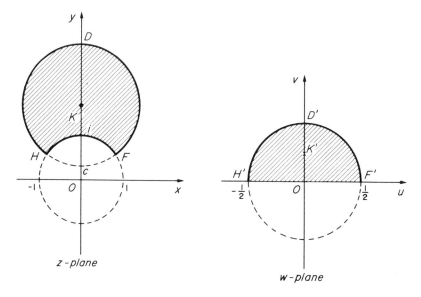

FIG. 13.4.

The locus in the z-plane corresponding to the real axis in the w-plane is

$$\text{Im}\left(\frac{1+iz}{i+z}\right) = \frac{1}{2}\left(\frac{1+iz}{i+z} - \frac{1-iz^*}{-i+z^*}\right) = 0$$

$$\Rightarrow (1+iz)(z^*-i) - (1-iz^*)(z+i) = 0$$

$$\Leftrightarrow zz^* = 1$$

$$\Leftrightarrow x^2 + y^2 = 1.$$

Further, since the origin of the w-plane corresponds to $z = +i$ of the z-plane the line $H'\,F'$ transforms into the minor arc HF of the unit circle. Further, the point K', $\frac{1}{4}i$, of the w-plane transforms into the point K, $5i/3$, of the z-plane. Accordingly, the interior of the semi-circle $|w| = 1$, Im $w > 0$, transforms into the interior of the contour FDH as shown in Fig. 13.4.

---

(6) The transformation

$$w = z + a^2/z,$$

where a is real, transforms the circle $|z| = a$ into the straight line joining the points $\pm 2a$. Other circles concentric with the origin are all transformed into confocal ellipses as illustrated in the following example.

Example. The z-plane is mapped on to the w-plane by the transformation $w = z + 1/z$.
 If $z = re^{i\theta}$, $w = u + iv$, prove that the circles $r = $ constant ($\neq 1$) in the z-plane are transformed into ellipses in the w-plane with foci at the points ± 2, and that the circle $r = 1$ is transformed into that portion of the real axis for which $-2 \leqslant u \leqslant 2$. Show that the point $w = \frac{1}{2}(3+i)$ corresponds to two points in the z-plane and find them.

Writing $z = re^{i\theta}$, $0 \leqslant \theta < 2\pi$, we find

$$w = z + 1/z = (r+1/r)\cos\theta + i(r-1/r)\sin\theta$$

$$\Leftrightarrow u = (r+1/r)\cos\theta, \quad v = (r-1/r)\sin\theta. \tag{1}$$

Equations (1) are the parametric equations of an ellipse of semi-axes $(r+1/r)$, $(r-1/r)$ which is described once as θ increases from 0 to 2π. With the usual notation for an ellipse

$$a = (r+1/r), \ b = (r-1/r)$$

and the relation $b^2 = a^2(1-e^2)$ gives $a^2e^2 = 4$ so that the foci $(\pm ae, 0)$ are the points $(\pm 2, 0)$. In fact the family of circles $|z| = r$, where $r > 0$, transforms into a family of confocal ellipses.
 The circle $r = 1$, $0 \leqslant \theta \leqslant 2\pi$, transforms into

$$u = 2\cos\theta, \quad v = 0, \quad 0 \leqslant \theta \leqslant 2\pi,$$

i.e. that portion of the real axis for which $-2 \leqslant u \leqslant 2$. Corresponding points are shown in Fig. 13.5.

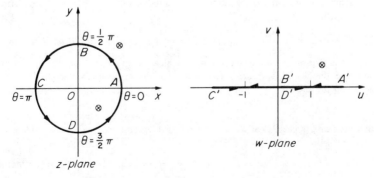

FIG. 13.5.

The interior of the unit circle in the z-plane transforms into the whole of the w-plane (cut between the points ± 2). The exterior of the unit circle in the z-plane transforms into this same region in the w-plane. When $w = \frac{1}{2}(3 + i)$,

$$z + 1/z = \tfrac{1}{2}(3 + i)$$
$$\Leftrightarrow 2z^2 - (3 + i)z + 2 = [z - (1 + i)][2z - (1 - i)] = 0.$$

Hence the point $\frac{1}{2}(3 + i)$ in the w-plane corresponds to the two points $(1 + i)$, $-\frac{1}{2}(1 - i)$ in the z-plane. (See Fig. 13.5.)

---

(7) The transformation

$$w = z^{\pi/\alpha}, \quad \alpha \geqslant \tfrac{1}{2},$$

transforms the wedge of the z-plane bounded by the lines $\theta = 0$, $\theta = \alpha$, and the arc of the circle $|z| = r$ into the area within the upper half of the semicircle $|w| = r^{\pi/\alpha}$.

In particular, when $r \to \infty$ the infinite wedge in the z-plane transforms into the upper half of the w-plane.

Exercise 13.3

1. Show that the linear transformation $w = -2iz + 5i$ is equivalent to a rotation about the origin followed by a magnification and then by a translation.

Find the point in the complex plane which is mapped on itself by this transformation. Show that the transformation is equivalent to a rotation about this point followed by a magnification.   (L.)

2. Prove that, if z is a complex number and z^* is its conjugate, then $zz^* = |z|^2$.

Given that $f(t) = (a + bt)/(1 + t)$ where a, b are complex numbers and t is a real or complex variable prove that, if $|w| = r$ and $r \neq 1$, then

$$|f(w) - f(-r^2)|^2 = |a - b|^2 r^2/(r^2 - 1)^2.$$

Hence show that, if z varies in the Argand diagram so that

$$\left| \frac{z - a}{z - b} \right| = k,$$

where k is a constant and $k \neq 1$, then the locus of z is a circle. Find the centre and radius of the circle in terms of k, a, b and $|a - b|$.

3. The function w given by $w = z^2$, where $w = u + iv$ and $z = x + iy$, maps the complex z-plane to the complex w-plane. Find the image in the w-plane of the following sets of points in the z-plane:

(i) $\{x + iy : x = 1\}$,   (ii) $\{x + iy : y = 1\}$,   (iii) $\{x + iy : y > 0\}$.

Show how examples (i) and (ii) can be used to illustrate the property of conformal transformations that "angles are preserved".

4. The transformation $w = \dfrac{z + i}{iz + 2}$, $z \neq 2i$, maps the complex number $z = x + iy$ onto the complex number $w = u + iv$. Find the two points in the complex plane which are invariant under the transformation.

Show that if z lies on the imaginary axis, then w also lies on the imaginary axis, and that if z lies on the real axis, then w lies on the circle

$$2u^2 + 2v^2 + v - 1 = 0.$$

Sketch this circle and indicate clearly onto which part the *positive* real axis of the z-plane is mapped.   (N.)

5. The transformation $w = \dfrac{1}{z}$ maps the z-plane to the w-plane; $z = x + iy = re^{i\theta}$, where $r \neq 0$, and $w = u + iv$. Express each of x and y in terms of u and v, and each of u and v in terms of r and θ.

(i) Find the image in the w-plane of the line $y = 2x + 1$. Show the line and its image in sketches of the two planes.

(ii) The point P in the z-plane describes the circle $r = 2$ in the anticlockwise direction. Describe the movement of the image P'.

(iii) The point Q in the z-plane describes the curve $r = 2\cos\theta$ in the anticlockwise direction for $0 \leqslant \theta < \frac{1}{2}\pi$. Describe the movement of the image Q', and show the curve and its image in sketches. (N.)

6. Sketch the region S in the Argand diagram which lies inside the boundary

$$
\begin{aligned}
z &= k, & 0 \leqslant k \leqslant 1, \\
z &= 1 + li, & 0 \leqslant l \leqslant 1, \\
z &= m + i, & 0 \leqslant m \leqslant 1, \\
z &= ni, & 0 \leqslant n \leqslant 1.
\end{aligned}
$$

The point z is mapped onto the point w by the transformation $w = iz^2$. Sketch the boundary of the region into which S is mapped and show the region into which S is mapped. Find the point or points of the z-plane which map onto $w = -\frac{1}{2}$. (N.)

7. A transformation T of the complex z-plane into itself is given by the function $f : z \rightarrow z^2$. The point P represents z and Q represents $f(z)$. The locus S of P is the circle on OA as diameter where O is the origin and A represents $\frac{2}{3} + 0i$. State the polar equation of S with O as pole and the real axis as initial line.

Show that the locus $T(S)$ of Q has the polar equation $9r = 2(1 + \cos\theta)$. Sketch S and $T(S)$ and find their three points of intersection.

Show that if a line through O cuts $T(S)$ again at Q and Q' then QQ' is of fixed length. (N.)

8. Show that if $z = x + iy$ and $w = u + iv$, the transformation

$$
w = \frac{z - 3i}{z - 4}
$$

transforms the circle through the points $z_1 = 0$, $z_2 = 4$, $z_3 = 3i$ into the line $u = 0$ and that the inside of this circle corresponds to the half plane $u < 0$.

Show also that the circle $|w| = 1$ corresponds to a diameter of the circle in the z-plane.

9. Show that the locus of points in the Argand diagram representing complex numbers

$$
z = \tfrac{1}{2}(\sqrt{3} + i) + k(-1 + i\sqrt{3}),
$$

where $k \in \mathbb{R}$, is a straight line which is a tangent to the circle $|z| = 1$. Describe the images of the line and the circle under the transformation

$$
w = (1 + i\sqrt{3})z + 3i,
$$

and give equations for these images. (L.)

Miscellaneous Exercise 13

1. Without using tables or a calculator, simplify

$$
\frac{\left(\cos\dfrac{\pi}{9} + i\sin\dfrac{\pi}{9}\right)^4}{\left(\cos\dfrac{\pi}{9} - i\sin\dfrac{\pi}{9}\right)^5}.
$$

(L.)

2. If $z = 2e^{-i\pi/3}$, express z, z^2 and $1/z$ in the form $x + yi$, where x and y are real. State the modulus and argument of each of these complex numbers, giving the argument in each case as an angle θ such that $-\pi < \theta \leqslant \pi$. Represent the three numbers in an Argand diagram. (L.)

3. Express $\sqrt{3} - i$ in the form $r(\cos\theta + i\sin\theta)$. Hence, or otherwise, show that $(\sqrt{3} - i)^9$ can be expressed as ci, where c is real, and give the value of c. (L.)

4. (i) If $z = \cos\theta + i\sin\theta$, show that

$$
z^n + z^{-n} = 2\cos n\theta \text{ and } z^n - z^{-n} = 2i\sin n\theta.
$$

Hence deduce that $\cos^6 \theta + \sin^6 \theta = \frac{1}{8}(3 \cos 4\theta + 5)$.

(ii) Given that $1, \omega, \omega^2$ are the cube roots of unity, prove that

$$(a + b + c)(a + \omega b + \omega^2 c)(a + \omega^2 b + \omega c) = a^3 + b^3 + c^3 - 3abc. \qquad \text{(L.)}$$

5. Solve completely the equation

$$z^6 + 2z^3 + 4 = 0,$$

giving each root in the form $r(\cos \theta + i \sin \theta)$, where $r > 0$ and $0 \leqslant \theta < 2\pi$. (N.)

6. (i) Find the modulus of the complex number

$$\frac{z - 3i}{1 + 3iz}$$

when $|z| = 1$.

(ii) Find the region of the Argand diagram in which z must lie when

$$\left| \frac{z - 3 + 2i}{1 - 3z - 2iz} \right| > 1. \qquad \text{(N.)}$$

7. If

$$z_1 = \frac{1 + 3i}{2 + i}, \quad z_2 = \frac{3 - i}{2 - i},$$

find the modulus and argument of each of the three complex numbers z_1, z_2 and z_1/z_2, giving the arguments in degrees and minutes between $-180°$ and $+180°$.

(ii) If $z = \cos \theta + i \sin \theta$, prove that

$$\frac{z - 1}{z + 1} = i \tan \tfrac{1}{2}\theta.$$

By taking P, a point on the unit circle, to represent z in the Argand diagram, give a geometric interpretation to your result.

8. Show that

$$\cos 7\theta = 64 \cos^7 \theta - 112 \cos^5 \theta + 56 \cos^3 \theta - 7 \cos \theta.$$

Prove that

(i) $\cos \dfrac{\pi}{7} \cos \dfrac{3\pi}{7} \cos \dfrac{5\pi}{7} = -\dfrac{1}{8}$,

(ii) $\cos^4 \dfrac{\pi}{7} + \cos^4 \dfrac{3\pi}{7} + \cos^4 \dfrac{5\pi}{7} = \dfrac{13}{16}$. (N.)

9. (i) If $|z| < 1$, show that $|z - 3 - 4i| > 4$.

(ii) If the point $z (\equiv x + iy)$ describes the circle $(x - 1)^2 + y^2 = 1$, find the equation of the path described by the point $1/z$.

(iii) Show that the roots of the equation

$$z^6 - 4z^3 + 8 = 0$$

lie on a circle of radius $\sqrt{2}$. Make a diagram showing the circle and the six roots. (N.)

10. (i) Find the modulus and argument of each root of the equation

$$z^4 - iz^2 \sqrt{3} - 1 = 0.$$

(ii) Solve the simultaneous equations

$$|z - 1| = 2\sqrt{2}, \quad |z - 1 - i| = |z|. \qquad \text{(N.)}$$

11. (i) Solve the simultaneous equations

$$|z + 1| + |z - 1| = 4, \quad \arg(iz) = \pi.$$

(ii) Find the complex number z such that $\arg(z-1) = \pi/4$ and $\left|z^2\right| = 5$.

(iii) Sketch the loci in the Argand diagram given by the equations

$$\text{(a)} \ \left|z-1-i\right| = \left|z\right|, \quad \text{(b)} \ \left|z+1\right| = 2,$$

and find the values of z that satisfy both equations.   (L.)

12. (i) Show that $z = 1+i$ is a root of the equation

$$z^4 + 3z^2 - 6z + 10 = 0.$$

Find the other roots of this equation.

(ii) Sketch the curve in the Argand diagram defined by

$$\left|z-1\right| = 1, \qquad \operatorname{Im} z \geqslant 0.$$

Find the value of z at the point P in which this curve is cut by the line $\left|z-1\right| = \left|z-2\right|$. Find also the values of $\arg z$ and $\arg(z-2)$ at P.   (L.)

13. The point P represents the complex number z in the Argand diagram. Find the locus of the point representing the number $2z/(z-1)$ when P moves round the circle $\left|z\right| = 1$.

Describe the locus defined by each of the following equations, and illustrate each locus in an Argand diagram.

(a) $\left|z+1\right|^2 + \left|z-1\right|^2 = 4,$
(b) $\left|z+i\right| + \left|z-i\right| = 3,$
(c) $\arg(z-1) = \arg(z+1).$ (L.)

14. Prove that if $\sin x$ is not zero

$$\cos x + \cos 3x + \ldots + \cos(2n-1)x = \tfrac{1}{2}\sin 2nx \operatorname{cosec} x.$$

Evaluate the integrals

$$\text{(a)} \ \int_{\pi/4}^{\pi/2} \sin 8x \operatorname{cosec} x \, dx, \quad \text{(b)} \ \int_{0}^{\pi/4} \sin 8x \sec x \, dx.$$ (L.)

15. (i) If the point representing the complex number z in an Argand diagram describes the circle $\left|z-1\right| = 1$, show that the point representing the complex number $1/z$ describes a straight line.

(ii) Show that the equation

$$e^z = z+a,$$

where a is real, has two real roots if $a > 1$ and no real root if $a < 1$. Show also that, for any real a, this equation has no root of the form iv, where v is real and non-zero.   (L.)

16. Find the sum of the first n terms of the series $1 + z + z^2 + z^3 + \ldots$. Show that, if $z = \tfrac{1}{2}(\cos\theta + i\sin\theta)$, the sum to infinity is

$$\frac{2(2-\cos\theta) + 2i\sin\theta}{5-4\cos\theta}.$$ (N.)

17. (i) Complex numbers z_1 and z_2 are given by the formulae

$$z_1 = R_1 + i\omega L, \quad z_2 = R_2 - \frac{i}{\omega C},$$

and z is given by the formula

$$\frac{1}{z} = \frac{1}{z_1} + \frac{1}{z_2}.$$

Find the value of ω for which z is a real number.

(ii) Use De Moivre's theorem to prove that, if

$$2\cos\theta = x + \frac{1}{x},$$

then

$$2 \cos n\theta = x^n + \frac{1}{x^n}.$$

Hence, or otherwise, solve the equation

$$5x^4 - 11x^3 + 16x^2 - 11x + 5 = 0. \tag{O.C.}$$

18. (i) If the roots of the equation $z^3 - 1 = 0$ are z_1, z_2 and z_3 prove that

(a) $z_1 z_2 z_3 = 1$,
(b) $|z_1 - z_2| = |z_2 - z_3|$,

and find a possible value for

$$\arg(z_2 - z_3) - \arg(z_1 - z_2).$$

(ii) If $z = \cos\theta + i\sin\theta$ show that $z^4 + z^{-4} = 2\cos 4\theta$. By expanding $(z + z^{-1})^4$ deduce that

$$\cos 4\theta = 8\cos^4\theta - 8\cos^2\theta + 1. \tag{L.}$$

19. Prove that the roots of the equation

$$z^{n-1} + z^{n-2} + z^{n-3} + \ldots + 1 = 0$$

are the values of

$$\cos\frac{2r\pi}{n} + i\sin\frac{2r\pi}{n},$$

where $r = 1, 2, 3, \ldots, n-1$.

If $\alpha = \cos\frac{2\pi}{13} + i\sin\frac{2\pi}{13}$, prove that

$$\alpha + \alpha^3 + \alpha^4 + \alpha^9 + \alpha^{10} + \alpha^{12} \quad \text{and} \quad \alpha^2 + \alpha^5 + \alpha^6 + \alpha^7 + \alpha^8 + \alpha^{11}$$

are the roots of the quadratic

$$z^2 + z - 3 = 0.$$

20. The complex numbers z, w are represented in the complex plane by the points $P(x, y)$ and $P'(u, v)$. State in geometrical terms the transformation of the plane $P \to P'$ given by $w = z(\cos\theta + i\sin\theta)$. Verify your statement by deriving from $w = z(\cos\theta + i\sin\theta)$ the matrix form

$$\begin{pmatrix} u \\ v \end{pmatrix} = \mathbf{M} \begin{pmatrix} x \\ y \end{pmatrix}.$$

What geometrical transformation is given by

$$w - c = (z - c)(\cos\theta + i\sin\theta),$$

where c is a complex constant?

If the transformation $z \to w$ corresponds to a turn of $120°$ about the point $(1, 0)$, and the transformation $w \to w'$ corresponds to a turn of $60°$ about the point $(-3, 0)$, show that $w' = -z + 2\sqrt{3}i$ and hence express in geometrical terms the single transformation equivalent to the two transformations in the order given.

21. A point P is represented by the complex number z which satisfies the equation $|z - i| = k|z + i|$. Give a geometrical interpretation of this equation. Show that the equation can be expressed in the form

$$zz^*(1 - k^2) + i(z - z^*)(1 + k^2) + (1 - k^2) = 0,$$

where z^* stands for the complex conjugate of z. Deduce that, if $k \neq 1$, the locus of P is a circle.

Let $w = f(z)$ be the transformation of the complex plane into itself whose equation is

$$\frac{w - i}{w + i} = i\left(\frac{z - i}{z + i}\right)$$

and let z_0 be a complex number with non-zero imaginary part which is neither i nor $-$ i. Show that the points z_n defined by

$$z_1 = f(z_0), \quad z_2 = f(z_1), \dots, z_n = f(z_{n-1}), \dots$$

all lie on a circle. What happens in the case where z_0 has zero imaginary part? Prove also that under this transformation f, the half-plane $\operatorname{Im}(z) > 0$ maps into the half-plane $\operatorname{Im}(w) > 0$.   (O.C.S.M.P.)

14 Further Vectors: Coordinate Geometry in Three Dimensions

14:1 THE VECTOR PRODUCT

The vector product of two vectors $\mathbf{a}$ and $\mathbf{b}$ is written $\mathbf{a} \times \mathbf{b}$ and is defined as a vector $\mathbf{v}$ of modulus $|\mathbf{a}| \, |\mathbf{b}| \sin \theta$, where θ is the angle between the positive directions of $\mathbf{a}$ and $\mathbf{b}$. [Sometimes the notation $\mathbf{a} \wedge \mathbf{b}$ is used for the vector product.] The vector $\mathbf{v}$ is perpendicular to the plane of $\mathbf{a}$ and $\mathbf{b}$, in a direction which is determined by the following rule:

If θ is the angle of rotation from the positive direction of $\mathbf{a}$ to the positive direction of $\mathbf{b}$, the direction of $\mathbf{v}$ is the direction in which the point of a right-handed corkscrew would move when the corkscrew was rotated through the angle θ, Fig. 14.1. Thus

$$\mathbf{a} \times \mathbf{b} = \hat{\mathbf{v}} ab \sin \theta, \tag{14.1}$$

where $\hat{\mathbf{v}}$ is the unit vector in the direction shown in Fig. 14.1.

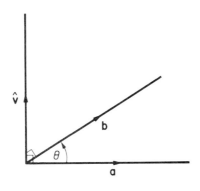

Fig. 14.1.

We add here further properties of the vector product:

(1) $$\mathbf{a} \times \mathbf{b} = -\mathbf{b} \times \mathbf{a}. \tag{14.2}$$

This follows directly from the definition because the direction of the rotation from $\mathbf{b}$ to $\mathbf{a}$ is opposite to that from $\mathbf{a}$ to $\mathbf{b}$, and shows that the vector product is *non-commutative*.

(2) The distributive law

$$\mathbf{a} \times (\mathbf{b} + \mathbf{c}) = \mathbf{a} \times \mathbf{b} + \mathbf{a} \times \mathbf{c}. \tag{14.3}$$

and its generalized form

$$\left(\sum_{i=1}^{m} \mathbf{p}_i \right) \times \left(\sum_{j=1}^{n} \mathbf{q}_j \right) = \sum_{i=1}^{m} \sum_{j=1}^{n} \mathbf{p}_i \times \mathbf{q}_j \tag{14.3a}$$

can be shown to hold for vector products.

(3) If $\mathbf{a} \times \mathbf{b} = \mathbf{0}$, then $|\mathbf{a}| \, |\mathbf{b}| \sin \theta = 0$. Hence, if neither of the vectors is a null vector, $\sin \theta = 0$, i.e. $\theta = 0$ or π and the vectors are parallel.

(4) It follows directly from the definition that $\mathbf{a} \times \mathbf{a} = \mathbf{0}$.

(5) With the usual notation for unit vectors along the coordinate axes

$$\mathbf{i} \times \mathbf{i} = \mathbf{0}, \quad \mathbf{j} \times \mathbf{j} = \mathbf{0}, \quad \mathbf{k} \times \mathbf{k} = \mathbf{0},$$
$$\mathbf{i} \times \mathbf{j} = \mathbf{k} = -\mathbf{j} \times \mathbf{i}, \quad \mathbf{j} \times \mathbf{k} = \mathbf{i} = -\mathbf{k} \times \mathbf{j}, \tag{14.4}$$
$$\mathbf{k} \times \mathbf{i} = \mathbf{j} = -\mathbf{i} \times \mathbf{k}.$$

(6) When we express $\mathbf{a} \times \mathbf{b}$ in terms of components, by use of equations (14.3a), (14.4) we see that

$$\mathbf{a} \times \mathbf{b} = (a_1\mathbf{i} + a_2\mathbf{j} + a_3\mathbf{k}) \times (b_1\mathbf{i} + b_2\mathbf{j} + b_3\mathbf{k})$$
$$= \mathbf{i}(a_2b_3 - a_3b_2) + \mathbf{j}(a_3b_1 - a_1b_3) + \mathbf{k}(a_1b_2 - a_2b_1). \tag{14.5}$$

This is frequently written as a determinant (see §15:1)

$$\mathbf{a} \times \mathbf{b} = \begin{vmatrix} \mathbf{i} & \mathbf{j} & \mathbf{k} \\ a_1 & a_2 & a_3 \\ b_1 & b_2 & b_3 \end{vmatrix} \tag{14.5a}$$

of which (14.5) is the expansion. Reversing the positions of the rows of this determinant changes the sign and shows the *non-commutative* property of the vector product $\mathbf{a} \times \mathbf{b}$.

Example 1. If $\mathbf{a} = 2\mathbf{i} - 3\mathbf{j} + 4\mathbf{k}$, $\mathbf{b} = 3\mathbf{i} + 4\mathbf{j} - 7\mathbf{k}$, then

$$\mathbf{a} \times \mathbf{b} = \begin{vmatrix} \mathbf{i} & \mathbf{j} & \mathbf{k} \\ 2 & -3 & 4 \\ 3 & 4 & -7 \end{vmatrix}$$
$$= 5\mathbf{i} + 26\mathbf{j} + 17\mathbf{k}.$$

Example 2. $(\mathbf{a} \times \mathbf{b})^2 = |\mathbf{a} \times \mathbf{b}|^2 = a^2b^2 \sin^2 \theta = a^2b^2 - a^2b^2 \cos^2 \theta$
$$= (a_1^2 + a_2^2 + a_3^2)(b_1^2 + b_2^2 + b_3^2) - (a_1b_1 + a_2b_2 + a_3b_3)^2$$
$$= a^2b^2 - (\mathbf{a}.\mathbf{b})^2.$$

But

$$\mathbf{a} \times \mathbf{b} = (a_2b_3 - a_3b_2)\mathbf{i} + (a_3b_1 - a_1b_3)\mathbf{j} + (a_1b_2 - a_2b_1)\mathbf{k}$$
$$\Rightarrow (\mathbf{a} \times \mathbf{b})^2 = (a_2b_3 - a_3b_2)^2 + (a_3b_1 - a_1b_3)^2 + (a_1b_2 - a_2b_1)^2.$$

Hence

$$(a_1^2 + a_2^2 + a_3^2)(b_1^2 + b_2^2 + b_3^2) - (a_1b_1 + a_2b_2 + a_3b_3)^2$$
$$= (a_2b_3 - a_3b_2)^2 + (a_3b_1 - a_1b_3)^2 + (a_1b_2 - a_2b_1)^2.$$

Since **a** and **b** are arbitary vectors their components may have any values whatever. The last relation is known as *Lagrange's identity*.

Example 3. Find the unit vector perpendicular to both $2\mathbf{i}-\mathbf{j}+\mathbf{k}$ and $3\mathbf{i}+4\mathbf{j}-\mathbf{k}$.

The vector product of the two vectors has the required direction. Let

$$\mathbf{a} = (2\mathbf{i}-\mathbf{j}+\mathbf{k}) \times (3\mathbf{i}+4\mathbf{j}-\mathbf{k}) = -3\mathbf{i}+5\mathbf{j}+11\mathbf{k}.$$

Hence the required unit vector is

$$\hat{\mathbf{a}} = \frac{\mathbf{a}}{|\mathbf{a}|} = \frac{-3\mathbf{i}+5\mathbf{j}+11\mathbf{k}}{(9+25+121)^{1/2}} = -\frac{3\mathbf{i}}{\sqrt{155}} + \frac{5\mathbf{j}}{\sqrt{155}} + \frac{11\mathbf{k}}{\sqrt{155}}.$$

PLANE AREAS IN THREE DIMENSIONS

Using vector products leads to the concept of the *vector area* of a triangle. The magnitude of the vector product $\overrightarrow{OA} \times \overrightarrow{OB}$ is $OA.OB \sin\theta = 2\Delta$, where Δ is the area of the triangle OAB. Writing the vector product $\overrightarrow{OA} \times \overrightarrow{OB}$ gives a sense of rotation $O \to A \to B$ around the triangle (see Fig. 14.2). This sense of rotation is linked by the r.h. corkscrew rule with the positive direction of $\overrightarrow{OA} \times \overrightarrow{OB}$. Therefore, we ascribe a vector area **S** to the triangle OAB given by

$$\mathbf{S} = \tfrac{1}{2}(\overrightarrow{OA} \times \overrightarrow{OB}). \tag{14.6}$$

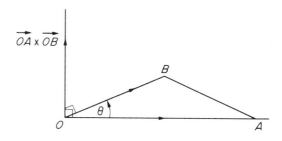

FIG. 14.2.

In using a directed line segment as a vector we prescribe the positive sense along the line and obtain the components by orthogonal projection of the *line* segment on the *coordinate lines* (or axes) of the frame of reference. In representing a directed area by a vector the positive sense of rotation around the perimeter is prescribed and we obtain the components by orthogonal projection of the *area* segment on the *coordinate planes*. Our investigation has shown that the vector product gives the vector area **S** for the special shape of a triangle, in terms of the line segments which constitute two of its sides, by formula (14.6). Similarly if $ABCD$ is a parallelogram its vector area is $\overrightarrow{AB} \times \overrightarrow{AD}$.

Example 1. If A, B are the points with position vectors $2i - 3j + 4k$, $3i + 4j - 7k$ respectively, then, by example 1, p. 423,

$$\overrightarrow{OA} \times \overrightarrow{OB} = 5i + 26j + 17k.$$

The area of the triangle OAB is therefore

$$\tfrac{1}{2}|\overrightarrow{OA} \times \overrightarrow{OB}| = \tfrac{1}{2}\sqrt{(5^2 + 26^2 + 17^2)} = \tfrac{1}{2}\sqrt{(990)}.$$

Example 2. A triangle is projected orthogonally onto a plane which is perpendicular to the direction $\hat{n}$; find the area of the projection.

Since no frame of reference is specified, we choose a frame such that the coordinate plane Oyz is parallel to the given plane. Then $\hat{n} \equiv k$. The area of the projected triangle is $S_3 = S.k = S.\hat{n}$. The last form does not involve the frame of reference and therefore is true in general. Hence the projected area is $S.\hat{n}$.

Example 3. If $\overrightarrow{OA} = 2j + k$ and $\overrightarrow{OB} = 2i + 3j$, where i, j, k are mutually orthogonal unit vectors, calculate

(a) $\cos AOB$,
(b) the area of the triangle OAB,
(c) the area of the projection of the triangle AOB on to a plane normal to the unit vector k.   (L.)

(a) The unit vectors in the directions of $\overrightarrow{OA}$ and $\overrightarrow{OB}$ are $(2j + k)/\sqrt{5}$ and $(2i + 3j)/\sqrt{13}$ respectively

$$\Rightarrow \cos AOB = (2j + k).(2i + 3j)/\sqrt{65} = 6/\sqrt{65}.$$

(b) The vector area of the triangle OAB is

$$S = \tfrac{1}{2}(2j + k) \times (2i + 3j) = \tfrac{1}{2}(-3i + 2j - 4k)$$
$$\Rightarrow |S| = S = \tfrac{1}{2}(9 + 4 + 16)^{1/2} = \tfrac{1}{2}\sqrt{29}.$$

(c) The area of projection is $S.k = -2$. (The negative sign means that the sense of rotation OAB on the projected figure is the negative sense, $Oy \to Ox$, in the plane $z = 0$.)

---

*Example 4. A tetrahedron has its vertices at $A(0, 0, 1)$, $B(3, 0, 1)$, $C(2, 3, 1)$, and $D(1, 1, 2)$. Find the angles between the faces ABC and BCD, between the edges AB, AC, and between the edge BC and the face ADC.   (L.)

The vectors representing the edges are

$$\overrightarrow{AB} = 3i, \quad \overrightarrow{AC} = 2i + 3j, \quad \overrightarrow{BC} = -i + 3j, \quad \overrightarrow{BD} = -2i + j + k.$$

Therefore the vector areas of the faces ABC and BCD are, respectively,

$$S_1 = \tfrac{1}{2}3i \times (2i + 3j) = \tfrac{9}{2}k,$$
$$S_2 = \tfrac{1}{2}(-i + 3j) \times (-2i + j + k) = \tfrac{1}{2}(3i + j + 5k).$$

The unit vectors normal to these faces are, respectively,

$$\hat{n}_1 = k \quad \text{and} \quad \hat{n}_2 = (3i + j + 5k)/\sqrt{35}.$$

The angle α between the faces is given by

$$\cos \alpha = \hat{n}_1.\hat{n}_2 = 5/\sqrt{35} = \sqrt{(5/7)}.$$

Unit vectors along the edges AB, AC are, respectively,

$$i, \ (2i + 3j)/\sqrt{13}.$$

Therefore the angle β between the edges is given by

$$\cos \beta = i.(2i + 3j)/\sqrt{13} = 2/\sqrt{13}.$$

The vector area of the face ADC is

$$S_3 = \tfrac{1}{2}\overrightarrow{AD} \times \overrightarrow{AC} = \tfrac{1}{2}(\mathbf{i}+\mathbf{j}+\mathbf{k}) \times (2\mathbf{i}+3\mathbf{j}) = \tfrac{1}{2}(-3\mathbf{i}+2\mathbf{j}+\mathbf{k}),$$

and the unit normal to this face is $\hat{\mathbf{n}}_3 = (-3\mathbf{i}+2\mathbf{j}+\mathbf{k})/\sqrt{14}$. The unit vector along BC is $(-\mathbf{i}+3\mathbf{j})/\sqrt{10}$, and therefore the angle between BC and the normal to ADC is γ where

$$\cos\gamma = (-\mathbf{i}+3\mathbf{j}).(-3\mathbf{i}+2\mathbf{j}+\mathbf{k})/\sqrt{140} = 9/\sqrt{140}.$$

Hence the angle between BC and the plane ADC is $\pi/2 - \gamma$, i.e. $\sin^{-1}(9/\sqrt{140})$.

Exercise 14.1

1. Find a unit vector which is perpendicular to both of the vectors $\mathbf{j}+4\mathbf{k}$ and $3\mathbf{i}+2\mathbf{j}+4\mathbf{k}$. Find also the angle between this unit vector and the vector $(\mathbf{i}-\mathbf{j}-\mathbf{k})$.

2. Given that $\mathbf{r} = \mathbf{a}+\mathbf{b}$, $\mathbf{s} = \mathbf{a}-\mathbf{b}$ and $|\mathbf{a}| = |\mathbf{b}| = c$, show that $\mathbf{r}.\mathbf{s} = 0$ and that

$$|\mathbf{r} \times \mathbf{s}| = 2(c^4 - (\mathbf{a}.\mathbf{b})^2)^{1/2}.$$

3. The vectors $\mathbf{u}$ and $\mathbf{v}$ are given by

$$\mathbf{u} = 2\mathbf{i}-\mathbf{j}+2\mathbf{k}, \ \mathbf{v} = a\mathbf{i}+b\mathbf{k}$$

and

$$\mathbf{u} \times \mathbf{v} = \mathbf{i}+c\mathbf{k}.$$

Find a, b and c. Find also the cosine of the angle between $\mathbf{u}$ and $\mathbf{v}$.

4. The non-zero vectors $\mathbf{a}$, $\mathbf{b}$ are such that $\mathbf{a} \times \mathbf{b} = \mathbf{0}$. State what is implied by this equation about the directions of $\mathbf{a}$ and $\mathbf{b}$.

The three non-zero vectors, $\mathbf{p}$, $\mathbf{q}$ and $\mathbf{r}$ satisfy the equation $\mathbf{p} \times \mathbf{q} = \mathbf{r} \times \mathbf{p}$. Deduce that $\mathbf{q}+\mathbf{r} = k\mathbf{p}$, where k is a scalar. If also $\mathbf{p} \times \mathbf{q} = \mathbf{q} \times \mathbf{r} \neq \mathbf{0}$, prove that $\mathbf{p}+\mathbf{q}+\mathbf{r} = \mathbf{0}$.

5. The point A has position vector $\mathbf{a}$ referred to the origin O and $\overrightarrow{OU} = \mathbf{u}$ is a unit vector. The point B is the reflection of A in the line OU. Show that the position vector $\mathbf{b}$ of B is given by

$$\mathbf{b} = 2(\mathbf{a}.\mathbf{u})\mathbf{u} - \mathbf{a}.$$

Show also that

(i) $\mathbf{b} \times \mathbf{a} = 2(\mathbf{a}.\mathbf{u})(\mathbf{b} \times \mathbf{u})$,
(ii) $\mathbf{b} \times \mathbf{u} = \mathbf{u} \times \mathbf{a}$.

6. No three of the points A, B, C, D are collinear and

$$\overrightarrow{AB} \times \overrightarrow{AC} = \overrightarrow{AC} \times \overrightarrow{AD}.$$

Prove that
(a) A, B, C, D are coplanar,
(b) AC bisects BD.

7. The points A, B, C have coordinates $(1, 2, 3)$, $(2, 4, 1)$, $(3, 3, 1)$, respectively, referred to a system of mutually perpendicular axes. Find $\overrightarrow{AB}.\overrightarrow{AC}$ and the elements of $\overrightarrow{AB} \times \overrightarrow{AC}$. Hence find $B\hat{A}C$ and the area of the triangle ABC.   (L.)

14:2 LINES, PLANES AND SPHERES

(1) The equation of a straight line

Vector equations of lines in the plane Oxy were obtained in §8:5. Here the derivation is repeated for convenience in three dimensions.

To reach a point on a straight line a moving point, starting from the origin, can go from O to A, a point on the line, and then proceed along the line. Suppose $\mathbf{a}$ is the postion of a point A on the line, $\hat{\mathbf{e}}$ is a unit vector along the direction of the line, and s is the distance AP (Fig. 14.3). Then the position vector of P is

$$\mathbf{r} = \mathbf{a} + s\hat{\mathbf{e}}. \tag{14.7}$$

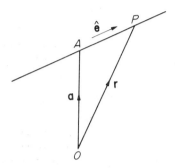

FIG. 14.3.

This is the equation of a straight line in vectors. If we express this in terms of components, those of $\overrightarrow{OP}$ are (x, y, z) where

$$\frac{x - a_1}{l_1} = \frac{y - a_2}{l_2} = \frac{z - a_3}{l_3} = s. \tag{14.7a}$$

Here (l_1, l_2, l_3), the components of $\hat{\mathbf{e}}$, are the *direction cosines* (see §8:4) of the line. If $\mathbf{b}$, *not* a unit vector, is parallel to the line then the displacement $\overrightarrow{AP}$ is a multiple of $\mathbf{b}$, and we can still write the equation of the line in the form

$$\mathbf{r} = \mathbf{a} + \mathbf{b}t, \tag{14.7b}$$

where t is a parameter which varies as P moves along the line. This gives rise to

$$\frac{x - a_1}{b_1} = \frac{y - a_2}{b_2} = \frac{z - a_3}{b_3} (= t). \tag{14.7c}$$

In equations (14.7b) and (14.7c) the parameter t is *not* equal to the distance AP unless $|\mathbf{b}| = 1$. Each of these equations can be taken as the standard form of the equation of a line but of course in neither case is unique, since A can be any point on the line.

Example 1. Find vector and cartesian equations of the line through the points $A(1, -2, 1)$, $B(0, -2, 3)$. Find also the coordinates of the point in which this line cuts the plane Oyz.

The vector $\overrightarrow{AB} = (0 - 1, -2 + 2, 3 - 1) = (-1, 0, 2) = -\mathbf{i} + 0\mathbf{j} + 2\mathbf{k}$ is parallel to the line and the point A lies on the line. But

$$\overrightarrow{OA} = (1, -2, 1) = \mathbf{i} - 2\mathbf{j} + \mathbf{k}.$$

Hence equation (14.7b) becomes

$$\mathbf{r} = \overrightarrow{OA} + t\overrightarrow{AB} = \mathbf{i} - 2\mathbf{j} + \mathbf{k} + t(-\mathbf{i} + 0\mathbf{j} + 2\mathbf{k}).$$

The corresponding cartesian equations [cf. equation (14.7c)] are

$$\frac{x-1}{-1} = \frac{y+2}{0} = \frac{z-1}{2} \quad (=t).$$

Here the zero in the denominator of the second ratio is to be interpreted as indicating that the numerator of this ratio vanishes also, i.e. for all points on the line, $y + 2 = 0$.

The parametric equations of the line can be written

$$x = 1 - t, \quad y = -2, \quad z = 1 + 2t.$$

This line meets the plane Oyz where $x = 0$, i.e. where $t = 1$. Hence the coordinates of the required point are $(0, -2, 3)$.

---

*Example 2. Find equations for the line which passes through the origin and intersects each of the lines

$$\frac{x-1}{-1} = \frac{y+2}{2} = \frac{z-3}{1},$$

$$\frac{x+1}{3} = \frac{y-1}{2} = \frac{z+1}{-1}.$$

Find also the coordinates of its point of intersection with the first of the given lines.

This problem can be done without the use of vectors but here we use vector methods to illustrate the simplicity and straightforwardness of the vectorial approach.

Vector equations for the two lines are

$$\mathbf{r}_1 = \mathbf{i} - 2\mathbf{j} + 3\mathbf{k} + p(-\mathbf{i} + 2\mathbf{j} + \mathbf{k})$$

and

$$\mathbf{r}_2 = -\mathbf{i} + \mathbf{j} - \mathbf{k} + q(3\mathbf{i} + 2\mathbf{j} - \mathbf{k}).$$

If OP_1P_2 is a straight line, then there is a number λ such that $\mathbf{r}_1 = \lambda\mathbf{r}_2$, where $\mathbf{r}_1$ and $\mathbf{r}_2$ are the position vectors of P_1 and P_2 and the parameters p and q have appropriate values. We write the vector equation $\mathbf{r}_1 - \lambda\mathbf{r}_2 = \mathbf{0}$ in terms of components and obtain

$$1 - p - \lambda(-1 + 3q) = 0, \qquad -2 + 2p - \lambda(1 + 2q) = 0,$$
$$3 + p - \lambda(-1 - q) = 0$$
$$\Leftrightarrow (1 + \lambda) - p - 3\lambda q = 0, \quad -(2 + \lambda) + 2p - 2\lambda q = 0,$$
$$(3 + \lambda) + p + \lambda q = 0. \qquad (1)$$

We regard these as three equations in the unknowns $p, \lambda q$ which are satisfied simultaneously. For consistency

$$\begin{vmatrix} 1+\lambda & -1 & -3 \\ -(2+\lambda) & 2 & -2 \\ 3+\lambda & 1 & 1 \end{vmatrix} = \begin{vmatrix} 1+\lambda & -1 & -3 \\ \lambda & 0 & -8 \\ 4+2\lambda & 0 & -2 \end{vmatrix} = \begin{vmatrix} \lambda & -8 \\ 4+2\lambda & -2 \end{vmatrix}$$

$$= 14\lambda + 32 = 0$$

$$\Leftrightarrow \lambda = \frac{16}{7}.$$

With this value of λ the equations (1) become

$$9 + 7p - 48q = 0, \quad 2 + 14p + 32q = 0, \quad 5 + 7p - 16q = 0,$$

which are all satisfied by $p = -\frac{3}{7}$ and $q = \frac{1}{8}$. These values of p and q give for the position vectors of P_1 and P_2

$$\mathbf{r}_1 = \tfrac{1}{7}(10\mathbf{i} - 20\mathbf{j} + 18\mathbf{k})$$

and

$$\mathbf{r}_2 = \tfrac{1}{8}(-5\mathbf{i} + 10\mathbf{j} - 9\mathbf{k}),$$

i.e. the coordinates of P_1, P_2 are $(\frac{10}{7}, -\frac{20}{7}, \frac{18}{7})$, $(-\frac{5}{8}, \frac{5}{4}, -\frac{9}{8})$ respectively. Both these points lie on the line whose equation is

$$\mathbf{r} = s(5\mathbf{i} - 10\mathbf{j} + 9\mathbf{k}),$$

which is a vector equation of the line through the origin intersecting both the given lines. Its cartesian equations can be written

$$\frac{x}{5} = \frac{y}{-10} = \frac{z}{9}.$$

---

(2) The equation of a plane

Three or more free vectors are said to be *coplanar* when they are all parallel to one plane. For example, the vectors $\mathbf{b} = \overrightarrow{AB}$, $\mathbf{c} = \overrightarrow{AC}$, $\mathbf{d} = \overrightarrow{AD}$ are coplanar if A, B, C, D all lie in one plane.

Provided that $\overrightarrow{AB}$ and $\overrightarrow{AC}$ are not parallel to each other and D lies in the plane ABC, then (Fig. 14.4)

$$\mathbf{d} = \overrightarrow{AD} = \overrightarrow{AK} + \overrightarrow{KD},$$

where $\overrightarrow{KD}$ is parallel to $\overrightarrow{AC}$. Hence,

$$\mathbf{d} = s\mathbf{b} + t\mathbf{c},$$

where the scalars $s = AK/AB$, $t = KD/AC$. This relation must be satisfied if $\mathbf{b}, \mathbf{c}, \mathbf{d}$ are coplanar. Conversely, if there are scalars s, t such that $\mathbf{d}$ can be expressed in terms of $\mathbf{b}, \mathbf{c}$ in this manner, then $\mathbf{b}, \mathbf{c}, \mathbf{d}$ are coplanar.

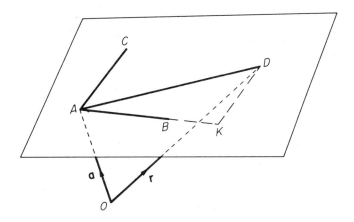

FIG. 14.4.

The position of D in Fig. 14.4 can be varied by giving s, t different values so that an arbitrary point of the plane has position vector of the form

$$\mathbf{r} = \mathbf{a} + s\mathbf{b} + t\mathbf{c}, \tag{14.8}$$

where O is the origin of position vectors and $\overrightarrow{OD} = \mathbf{r}$. This is the equation of a plane in vector notation. If the plane passes through the origin, then $\mathbf{a} = \mathbf{0}$. Note that the vector equation of a plane, a two-dimensional locus, involves two parameters (s and t), whereas the vector equation of a line, a one-dimensional locus, involves one parameter only.

Example. Find vector and cartesian equations of the plane through the three points A, B, C with position vectors $\mathbf{a}, \mathbf{b}, \mathbf{c}$.

The vectors $\mathbf{b} - \mathbf{a}$ and $\mathbf{c} - \mathbf{a}$ both lie in the plane. Therefore equation (14.8) becomes, in this case,

$$\mathbf{r} = \mathbf{a} + s(\mathbf{b} - \mathbf{a}) + t(\mathbf{c} - \mathbf{a}). \tag{1}$$

Writing $\mathbf{r} = x\mathbf{i} + y\mathbf{j} + z\mathbf{k}$, $\mathbf{a} = a_1\mathbf{i} + a_2\mathbf{j} + a_3\mathbf{k}$, etc., the components of equation (1) can be written

$$(x - a_1) + s(a_1 - b_1) + t(a_1 - c_1) = 0,$$
$$(y - a_2) + s(a_2 - b_2) + t(a_2 - c_2) = 0,$$
$$(z - a_3) + s(a_3 - b_3) + t(a_3 - c_3) = 0.$$

These equations are consistent, when considered as equations for s, t, if [see p. 466.]

$$\begin{vmatrix} x - a_1 & a_1 - b_1 & a_1 - c_1 \\ y - a_2 & a_2 - b_2 & a_2 - c_2 \\ z - a_3 & a_3 - b_3 & a_3 - c_3 \end{vmatrix} = 0; \tag{2}$$

which is an equation of the plane ABC. This equation can be written in the form

$$\begin{vmatrix} x & a_1 & b_1 & c_1 \\ y & a_2 & b_2 & c_2 \\ z & a_3 & b_3 & c_3 \\ 1 & 1 & 1 & 1 \end{vmatrix} = 0; \tag{3}$$

this is the symmetrical form for the equation of a plane and, after expansion of the determinant, is equivalent to

$$A_1 x + B_1 y + C_1 z + D_1 = 0, \tag{4}$$

where A_1, B_1, C_1, D_1 are constants. Thus, the equation of a plane is linear in the coordinates (x, y, z) of a variable point P on the plane. Conversely one linear equation in (x, y, z) represents a plane.

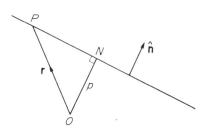

FIG. 14.5.

The use of a vector product associates a plane area with a vector perpendicular to the plane. Suppose that $\hat{\mathbf{n}}$ is a unit vector normal to a plane (the unit normal) and $\mathbf{r}$ is the position vector of an arbitrary point P on the plane; N is the foot of the perpendicular from O to the plane. Figure 14.5, which is a section of a three-dimensional figure, indicates that

$$\mathbf{r} \cdot \hat{\mathbf{n}} = p, \tag{14.9}$$

where $p = ON$ is the perpendicular distance of O from the plane. If the sense $O \to N$ is the same as that of $\hat{\mathbf{n}}$, p is positive (or O is on the *negative* side of the plane); if the sense $O \to N$ is opposite to that of $\hat{\mathbf{n}}$, then p is negative. Equation (14.9) is the *normal* form for the equation of a plane and, when expressed in cartesian elements, becomes

$$lx + my + nz = p, \tag{14.10}$$

where $\hat{\mathbf{n}} = i\mathbf{l} + j\mathbf{m} + k\mathbf{n}$, i.e. (l, m, n) are the direction cosines of the normal to the plane. If the direction ratios of the normal to the plane are $a{:}b{:}c$ so that $\mathbf{n} = a\mathbf{i} + b\mathbf{j} + c\mathbf{k}$, then equation (14.9) and (14.10) can be written

$$\mathbf{r}.\mathbf{n} = q \tag{14.11}$$
$$\Rightarrow ax + by + cz = q \tag{14.12}$$

(cf. the result of the example of p. 430).

The connection between the forms (14.11) or (14.10) and the form (14.8) for the equation of a plane is obtained by the use of a vector product. In equation (14.8) the vectors $\mathbf{b}$ and $\mathbf{c}$ lie in the plane so that the vector product

$$\mathbf{n} = \mathbf{b} \times \mathbf{c} \text{ or } \hat{\mathbf{n}} = (\mathbf{b} \times \mathbf{c})/(bc \sin \theta) \tag{14.13}$$

is normal to the plane because, by definition the vector product is perpendicular to both $\mathbf{b}$ and $\mathbf{c}$. Therefore,

$$\mathbf{r}.\mathbf{n} = \mathbf{a}.\mathbf{n} + s(\mathbf{b}.\mathbf{n}) + t(\mathbf{c}.\mathbf{n}) = \mathbf{a}.\mathbf{n}$$
$$\Rightarrow \mathbf{r}.\mathbf{n} = q$$

since $\mathbf{b}.\mathbf{n} = 0 = \mathbf{c}.\mathbf{n}$ and we have written $\mathbf{a}.\mathbf{n} = q$. Division by $|\mathbf{n}| = bc \sin \theta$ gives the form (14.9).

Example 1. Find the perpendicular distance from a point A, position vector $\mathbf{r}_0$, to the plane $\mathbf{r}.\hat{\mathbf{n}} = p$. Interpret the sign of the result and express the distance in terms of cartesian coordinates.

The equation of the plane through A parallel to the given plane is $(\mathbf{r} - \mathbf{r}_0).\hat{\mathbf{n}} = 0$. The equations of the planes are therefore $\mathbf{r}.\hat{\mathbf{n}} = \mathbf{r}_0.\hat{\mathbf{n}}$ and $\mathbf{r}.\hat{\mathbf{n}} = p$. Therefore (see Fig. 14.6) $OM = \mathbf{r}_0.\hat{\mathbf{n}}$ and $ON = p$. Hence the required perpendicular distance is

$$MN = ON - OM = p - \mathbf{r}_0.\hat{\mathbf{n}}. \tag{1}$$

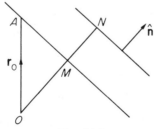

FIG. 14.6.

MN is positive if it is drawn in the positive sense of $\hat{\mathbf{n}}$.

The cartesian form of equation (1) is

$$MN = p - (lx_0 + my_0 + nz_0),$$

where (l, m, n) are the direction cosines of the normal $\hat{\mathbf{n}}$ and (x_0, y_0, z_0) are the coordinates of A. Using the notation of equations (14.11) and (14.12), the perpendicular distance of the point $\mathbf{r}_0$ from the plane $\mathbf{r}.\mathbf{n} = q$ is easily shown to be

$$(q - \mathbf{r}_0.\mathbf{n})/|\mathbf{n}| = \{q - (ax_0 + by_0 + cz_0)\}/\sqrt{(a^2 + b^2 + c^2)}.$$

Example 2. Find the acute angle between the planes Γ_1, Γ_2,

$$3x + 4y + 2z - 7 = 0, \quad 2x - 3y + z - 9 = 0. \tag{1}$$

Find also an equation of the plane Π which passes through the line of intersection of the above planes and also through the point $(1, 2, 2)$.

The angle between the planes Γ_1, Γ_2 is equal to the angle between their normals. But the normals to Γ_1, Γ_2 have direction cosines $(3/\sqrt{29}, 4/\sqrt{29}, 2/\sqrt{29})$, $(2/\sqrt{14}, -3/\sqrt{14}, 1/\sqrt{14})$. Hence the angle θ between the planes is given by

$$\cos\theta = (6 - 12 + 2)/(\sqrt{14}.\sqrt{29}) = -4/\sqrt{406}.$$

This gives the obtuse angle between Γ_1, Γ_2; the acute angle is $\cos^{-1}(4/\sqrt{406})$.
To find an equation of the plane Π we consider the equation

$$(3x + 4y + 2z - 7) + \lambda(2x - 3y + z - 9) = 0, \tag{1}$$

where λ is a parameter. This equation, being linear in x, y, z, is the equation of a plane. Further it is satisfied by all points for which $3x + 4y + 2z - 7 = 0$ *and* $2x - 3y + z - 9 = 0$. Hence it passes through the line of intersection of Γ_1 and Γ_2. An equation of the form (1) is said to represent a *pencil (or sheaf) of planes*, for, by varying λ, different planes through a fixed line are obtained. We can choose λ to satisfy one condition. Substituting $x = 1$, $y = 2$, $z = 2$ in equation (1) gives the value of λ corresponding to the member Π of this pencil of planes which passes through $(1, 2, 2)$. We find $\lambda = 8/11$ and the required plane Π reduces to

$$49x + 20y + 30z = 149.$$

Example 3. Show that for all numerical values of λ the point $(3 + \lambda, 5 + 2\lambda, 2 + 3\lambda)$ is on the line through $A(3, 5, 2)$ perpendicular to the plane, Π,

$$x + 2y + 3z - 5 = 0.$$

Perpendiculars AP and BQ are drawn through the points $A(3, 5, 2)$ and $B(-7, -1, 0)$ to the given plane. Find the coordinates of P and Q. If M is the middle point of PQ find the equations of the line Γ in the plane, passing through M and perpendicular to PQ.

The line through A perpendicular to Π has direction ratios $1:2:3$. Hence its equations can be written

$$\frac{x-3}{1} = \frac{y-5}{2} = \frac{z-2}{3} = t$$

and the required result follows.
This line meets Π where

$$(3 + t) + 2(5 + 2t) + 3(2 + 3t) - 5 = 0 \Rightarrow t = -1,$$

so that P is the point $(2, 3, -1)$. Similarly Q is the point $(-6, 1, 3)$.
The line PQ has direction ratios $4:1:-2$ and the mid-point of PQ is $(-2, 2, 1)$. Hence the equations of the line Γ can be written

$$\frac{x+2}{l} = \frac{y-2}{m} = \frac{z-1}{n},$$

where $4l + m - 2n = 0$, $l + 2m + 3n = 0$. These last two equations express the facts that Γ is respectively perpendicular to PQ and to the normal to the plane Π. Hence $l:m:n = 1:-2:1$ and equations for Γ are

$$\frac{x+2}{1} = \frac{y-2}{-2} = \frac{z-1}{1}.$$

Example 4. Find in standard form an equation of the line of intersection of the two planes

$$x + 6y + z + 15 = 0, \quad 3x + 2y - z + 1 = 0.$$

We solve this problem by two methods.

1. In vector notation the planes are given by

$$\mathbf{r}.(\mathbf{i}+6\mathbf{j}+\mathbf{k}) = -15, \quad \mathbf{r}.(3\mathbf{i}+2\mathbf{j}-\mathbf{k}) = -1.$$

The direction of the line of intersection of these planes is perpendicular to the two vectors which are normal to the planes, i.e. it is parallel to the vector product of these two (normal) vectors. Hence the line has the direction of the vector

$$(\mathbf{i}+6\mathbf{j}+\mathbf{k}) \times (3\mathbf{i}+2\mathbf{j}-\mathbf{k}) = -8\mathbf{i}+4\mathbf{j}-16\mathbf{k}.$$

The unit vector in this direction is

$$(-2\mathbf{i}+\mathbf{j}-4\mathbf{k})/\sqrt{21}.$$

Either by inspection, or by finding the intersection with, say, $z = 0$, we must now find one point on the line. A trial shows that $2\mathbf{i}-3\mathbf{j}+\mathbf{k}$ (or $x = 2$, $y = -3$, $z = 1$) satisfies the equations of both planes. Hence, an equation of the line is

$$\mathbf{r} = (2\mathbf{i}-3\mathbf{j}+\mathbf{k})+\lambda(-2\mathbf{i}+\mathbf{j}-4\mathbf{k})$$

where λ is a variable scalar.

If we put $z = 0$ in the equations of the planes and solve for x, y we obtain

$$x = 3/2, \quad y = -11/4,$$

so that the equation of the line becomes

$$\mathbf{r} = (6\mathbf{i}-11\mathbf{j})/4+\mu(-2\mathbf{i}+\mathbf{j}-4\mathbf{k}).$$

(The variable μ is related to λ by $\mu = \lambda - 1/4$.)

2. The second method regards the equations of the planes as simultaneous equations in x, y which we solve, finding x, y in terms of z. Thus we write the given equations in the form

$$\frac{x}{-6(z-1)-2(z+15)} = \frac{y}{3(z+15)+(z-1)} = \frac{1}{2-18}$$

$$\Rightarrow x = \tfrac{1}{2}z+\tfrac{3}{2}, \quad y = -\tfrac{1}{4}z-\tfrac{11}{4}.$$

This solution, in which z can be given any value, gives corresponding values of x, y, z for points on the line of intersection. We can obtain a vector form by writing

$$\mathbf{r} = x\mathbf{i}+y\mathbf{j}+z\mathbf{k} = z\left(\frac{\mathbf{i}}{2}-\frac{\mathbf{j}}{4}+\mathbf{k}\right)+\frac{3\mathbf{i}}{2}-\frac{11\mathbf{j}}{4}.$$

Since z is arbitrary we write $z = -4\mu$ on the r.h. side and obtain

$$\mathbf{r} = \frac{3\mathbf{i}}{2}-\frac{11\mathbf{j}}{4}-\mu(2\mathbf{i}-\mathbf{j}+4\mathbf{k}),$$

as before.

The process of solving for x, y is equivalent to the operation of finding the vector product in the former method.

---

(3) The equation of a sphere

The surface of a sphere is the locus of points which are at a fixed distance from the centre. If the centre is at (f, g, h) and the radius is a, the coordinates of an arbitrary point on the sphere must satisfy the equation

$$(x-f)^2 + (y-g)^2 + (z-h)^2 = a^2 \qquad (14.14)$$

$$\Leftrightarrow x^2 + y^2 + z^2 - 2fx - 2gy - 2hz + c = 0, \qquad (14.15)$$

where

$$a^2 = f^2 + g^2 + h^2 - c. \qquad (14.16)$$

The characteristics of equation (14.15) which show that it represents a sphere are

(i) the coefficients of x^2, y^2 and z^2 are equal;

(ii) there are no terms involving the products yz, zx, xy.

If these conditions are satisfied, then equation (14.15) may be put into the form of equation (14.14) showing that it represents a sphere. *Provided that the coefficients of* x^2, y^2 *and* z^2 *are unity*, f, g, h of equation (14.15) are the coordinates of the centre and the independent term c gives the radius through equation (14.16). In particular the sphere of radius a with centre at the origin has equation

$$x^2 + y^2 + z^2 = a^2.$$

All the coordinate planes are planes of symmetry: the origin is a centre of symmetry and all the coordinate axes are axes of symmetry.

In vector form, the equation of a sphere, whose centre has position vector $\mathbf{r}_0$ and whose radius is R, can be written

$$(\mathbf{r} - \mathbf{r}_0)^2 = R^2. \tag{14.17}$$

If the position vectors of the ends A, B of a diameter of a sphere are $\mathbf{a}$, $\mathbf{b}$ respectively and $\mathbf{r}$ is the position vector of an arbitrary point P on the sphere, then AP and BP are perpendicular

$$\Rightarrow \overrightarrow{AP} . \overrightarrow{BP} = 0$$

$$\Rightarrow (\mathbf{r} - \mathbf{a}) . (\mathbf{r} - \mathbf{b}) = 0, \tag{14.18}$$

which is a vector equation of the sphere.

As shown by the following examples, most problems concerning spheres are best solved by expressing analytically the simple geometrical properties of the sphere.

Two important geometrical properties of planes and spheres are:

(i) The curve of intersection of a plane and a sphere is a circle.

(ii) The curve of intersection of two spheres is a circle.

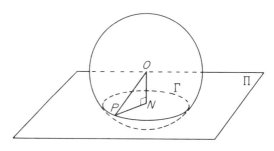

FIG. 14.7.

To prove theorem (i), consider the curve of intersection, Γ say, of the plane Π, with the sphere of centre O, and radius a, Fig. 14.7. Let N be the foot of the perpendicular from O to Π and let P be any point on Γ. Then $OP = a$ (by the fundamental property of the sphere) and, by Pythagoras' theorem,

$$PN^2 = OP^2 - ON^2 = a^2 - ON^2 = \text{constant}.$$

Therefore P is a fixed distance from a fixed point in the plane Π, i.e. P lies on a circle centre N and radius $\sqrt{(a^2 - ON^2)}$.

(ii) Consider now the intersection of two spheres centres O_1, O_2 and radii a_1, a_2 respectively, Fig. 14.8, and let Q be an arbitrary point on the curve, C, of intersection of the spheres. Let M be the foot of the perpendicular from Q to $O_1 O_2$. Then the triangle $O_1 Q O_2$ has sides of fixed length ($O_1 Q = a_1$, $O_2 Q = a_2$, $O_1 O_2 = $ distance between the centres of the spheres). Therefore M is a fixed point on $O_1 O_2$ and so, as Q varies, it (Q) always lies in a plane through M perpendicular to $O_1 O_2$. Therefore the curve C lies in a plane and by theorem (i) is a circle with centre M.

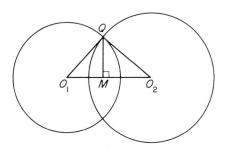

FIG. 14.8.

Example 1. Find the radius of the circle of intersection of the sphere

$$x^2 + y^2 + z^2 + 5y + z - 15 = 0$$

and the plane

$$4x + 3y + 5z - 15 = 0.$$

Find also the coordinates of the centre of this circle. Show that the sphere of least radius passing through this circle has equation

$$x^2 + y^2 + z^2 - 4x + 2y - 4z = 0.$$

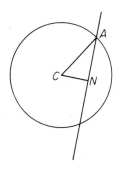

FIG. 14.9.

The equation of the sphere can be writeen

$$x^2 + (y + \tfrac{5}{2})^2 + (z + \tfrac{1}{2})^2 = \tfrac{43}{2}.$$

Hence the centre of the sphere is $(0, -5/2, -1/2)$ and the radius is $\sqrt{(43/2)}$. The perpendicular distance of the centre C from the plane is (Fig. 14.9)

$$CN = \frac{15 + \frac{15}{2} + \frac{5}{2}}{(16 + 9 + 25)^{1/2}} = \frac{5}{\sqrt{2}},$$

see Example 1, p. 431. The radius of the circle is NA where $NA^2 = CA^2 - CN^2 = \frac{43}{2} - \frac{25}{2} = 9 \Rightarrow NA = 3$.

The line CN passes through C and is in the direction $4\mathbf{i} + 5\mathbf{j} + 5\mathbf{k}$ (the normal to the plane). The equations of the line CN are

$$\frac{x}{4} = \frac{y + \frac{5}{2}}{3} = \frac{z + \frac{1}{2}}{5}(= p).$$

Substituting $x = 4p$, $y = 3p - 5/2$, $z = 5p - \frac{1}{2}$ into the equation of the plane gives $p = \frac{1}{2}$. Therefore, the centre N of the circle is $(2, -1, 2)$.

The sphere of least radius through this circle has its centre at N and radius 3, i.e. the given circle is a great circle of this sphere, whose equation is therefore

$$(x - 2)^2 + (y + 1)^2 + (z - 2)^2 = 9$$
$$\Leftrightarrow x^2 + y^2 + z^2 - 4x + 2y - 4z = 0,$$

which is the required result.

Example 2. Show that an equation of the sphere which has its centre at the point $(2, 3, -1)$ and touches the line

$$\frac{x + 1}{-5} = \frac{y - 8}{3} = \frac{z - 4}{4} \quad \text{is} \quad x^2 + y^2 + z^2 - 4x - 6y + 2z + 5 = 0.$$

The line joining the centre, C, of the sphere and the point of contact is perpendicular to the given line. An arbitrary point P on the line has coordinates $(-1 - 5p, 8 + 3p, 4 + 4p)$. The direction ratios of PC are $3 + 5p$: $-5 - 3p$: $-5 - 4p$. This direction is perpendicular to the line if

$$0 = -5(3 + 5p) + 3(-5 - 3p) + 4(-5 - 4p) = -50 - 50p$$

$$\Rightarrow p = -1.$$

Hence the foot N of the perpendicular from C to the given line is $(4, 5, 0)$ and the radius of the required sphere is $PN = (4 + 4 + 1)^{1/2} = 3$. Therefore an equation of the sphere is

$$(x - 2)^2 + (y - 3)^2 + (z + 1)^2 = 9$$
$$\Leftrightarrow x^2 + y^2 + z^2 - 4x - 6y + 2z + 5 = 0.$$

Example 3. Find equations for the two spheres, which contain the circle

$$x^2 + y^2 + z^2 - 4x + 8z + 6 = 0, \quad 2x + 2y + 4z + 1 = 0,$$

and touch the plane $x = 0$. (Note that the *circle* is given by two equations.)

The equation

$$x^2 + y^2 + z^2 - 4x + 8z + 6 + \lambda(2x + 2y + 4z + 1) = 0 \tag{1}$$

represents a surface through the intersection of the sphere $x^2 + y^2 + z^2 - 4x + 8z + 6 = 0$ and the plane $2x + 2y + 4z + 1 = 0$, for all values of λ. This surface is also a sphere since the coefficients of x^2, y^2 and z^2 are all unity and there are no terms in xy, yz or zx. If this sphere touches $x = 0$, the distance of its centre from the plane $x = 0$ (i.e. the x-coordinate of the centre) equals its radius. The equation of the sphere is

$$x^2 + y^2 + z^2 - 2(2 - \lambda)x + 2\lambda y + 2(4 + 2\lambda)z + 6 + \lambda = 0. \tag{2}$$

Hence, for it to touch $x = 0$,

$$a^2 = (2 - \lambda)^2 = (2 - \lambda)^2 + \lambda^2 + (4 + 2\lambda)^2 - 6 - \lambda$$
$$\Rightarrow 5\lambda^2 + 15\lambda + 10 = 5(\lambda + 1)(\lambda + 2) = 0 \Rightarrow \lambda = -1, \, -2.$$

Equations of the possible spheres are

$$x^2 + y^2 + z^2 - 6x - 2y + 4z + 5 = 0, \quad x^2 + y^2 + z^2 - 8x - 4y + 4 = 0.$$

Exercise 14.2

1. Find vector and cartesian equations for the line joining the points $(-1, 1, -2)$ and $(4, 2, 5)$.

2. Find vector and cartesian equations for the plane which contains the points $(3, -4, 1)$, $(2, 2, -1)$ and $(-3, 1, 2)$.

3. Find vector and cartesian equations for the line joining the points with position vectors $2\mathbf{i} - \mathbf{j} + 2\mathbf{k}$ and $4\mathbf{i} - 3\mathbf{j} + 3\mathbf{k}$. Find also the position vector of the point where this line cuts the plane whose vector equation is

$$\mathbf{r} = 2\mathbf{i} + s\mathbf{j} + t\mathbf{k}.$$

4. Find a cartesian equation for the plane which passes through the points $(2, 2, 1)$, $(2, 3, 2)$, $(-1, 3, 0)$.

5. Show that the points $(1, 0, 1)$, $(3, 2, 0)$, $(4, 4, 1)$, $(2, 2, 2)$ are coplanar.

6. Find cartesian equations for the line through the points $(1, 2, 3)$, $(2, 1, 5)$ and show that this line lies in the plane $x + 3y + z - 10 = 0$.

7. Find an equation for the plane which goes through the origin and is perpendicular to each of the planes given by

$$2x - y + 3z - 1 = 0, \quad x + 2y + z = 0.$$

8. Find an equation for the plane which goes through the point $M(2, -1, 1)$ and is perpendicular to the two planes given by

$$2x - z + 1 = 0, \quad y = 0.$$

9. The direction of a straight line AB is determined by the unit vector $\hat{\mathbf{n}}$ and $\overrightarrow{AP}$ is the position vector relative to A of any point P. Prove that the perpendicular distance from P to AB is given by the magnitude of $\overrightarrow{AP} \times \hat{\mathbf{n}}$. Hence, or otherwise, find the perpendicular distance of the point $(3, 1, 2)$ from the straight line

$$\frac{x-2}{3} = \frac{y-3}{2} = \frac{z-1}{2}.$$

10. Find vector and cartesian equations for the plane which passes through the point with position vector $2\mathbf{i} + 3\mathbf{j} - \mathbf{k}$ and is perpendicular to the vector $-3\mathbf{i} + 4\mathbf{j} + 2\mathbf{k}$.

11. Find the cosine of the acute angle between the two planes $3x + 2y - 4z = 7$ and $6x - 3y + 2z = 4$.

12. Find the direction cosines of the line of intersection of the plane passing through the points $(1, 1, 0)$, $(1, -2, 1)$, $(-1, 0, 1)$, and the plane $x + y - z = 2$.

Find also an equation for the plane which is parallel to the given line and which contains the origin and the point $(3, 2, 1)$.

13. B and C are the points $(2, 1, 0)$ and $(1, 0, 2)$ respectively. If A is a point in the plane $x = 0$ such that the triangle ABC is equilateral, prove that there are two possible positions of A and that the line joining them passes through the origin O. If the two positions of A are A_1 and A_2, show that the acute angle between the planes $A_1 A_2 B$ and $A_1 A_2 C$ is $\cos^{-1} \frac{12}{7}$.

14. Find an equation for the plane which passes through the origin and contains the straight line

$$\frac{x-1}{2} = \frac{y-2}{1} = \frac{z-3}{-2}.$$

Find equations for the straight line meeting the axis of x at right angles, whose orthogonal projection on this plane coincides with the above straight line.

15. Prove that the sphere

$$x^2 + y^2 + z^2 - 2x - 2y - 2z + 1 = 0$$

touches the coordinate axes and find the coordinates of the points of contact. Find also the centre and the radius of the circle formed by the intersection of the sphere and the plane through these points of contact.

16. Show that the sphere

$$x^2 + y^2 + z^2 - (4 - \lambda)x - 2(1 - \lambda)y - 2(1 + \lambda)z + 2(1 - \lambda) = 0$$

passes through the intersection of

$$x^2 + y^2 + z^2 - 4x - 2y - 2z + 2 = 0 \quad \text{and} \quad x + 2y - 2z - 2 = 0.$$

A circle of radius 2 and centre $(2, 1, 1)$ is drawn on the plane $x + 2y - 2z - 2 = 0$; find the radius and centre of the sphere which passes through this circle and through the origin of coordinates.

17. Show that the equation $x^2 + y^2 + z^2 - 2x - 4y + 1 = 0$ is the equation of a sphere S, and determine the centre and radius of S.

Find an equation for the sphere S' of radius 3 which touches S externally at the point $(0, 1, \sqrt{2})$, and an equation for the common tangent plane to S and S' at this point.

18. The equation of a sphere with centre C is $x^2 + y^2 + z^2 + 2x = 0$. The equation of a plane p is $x + y + z - 2 = 0$. Find the coordinates of the foot A of the perpendicular from C to p. Deduce that the plane does not intersect the sphere.

19. Show that the equation $(\mathbf{r} - \mathbf{r}_0)^2 = R^2$ defines a sphere with centre $C(\mathbf{r}_0)$ and radius R, i.e. that this equation is satisfied by the radius-vector of a point $M(\mathbf{r})$ if and only if M lies on this sphere.

20. Find the radii-vectors of the points of intersection of a straight line and sphere given by $\mathbf{r} = \mathbf{a}t$ and $\mathbf{r}^2 = R^2$ respectively. Find the cartesian coordinates of these points if $\mathbf{a} = (l, m, n)$.

21. Find the radii-vectors of the points of intersection of a straight line and sphere given by $\mathbf{r} = \mathbf{r}_0 + \mathbf{a}t$ and $(\mathbf{r} - \mathbf{r}_0)^2 = R^2$ respectively. Find their cartesian coordinates if $\mathbf{r}_0 = (x_0, y_0, z_0)$, $\mathbf{a} = (l, m, n)$.

22. Find an equation for the plane which touches the sphere given by $(\mathbf{r} - \mathbf{r}_0)^2 = R^2$ at the point $M(\mathbf{r}_1)$.

23. Find an equation for the sphere, centre $C(\mathbf{r}_1)$, which touches the plane given by $\mathbf{r} \cdot \mathbf{n} + D = 0$. Write down an equation of this sphere in cartesian coordinates if $\mathbf{r}_1 = (x_1, y_1, z_1)$, $\mathbf{n} = (\alpha, \beta, \gamma)$.

24. Prove that the lines l_1, l_2 with equations

$$\frac{x+2}{2} = \frac{y-5}{-1} = \frac{z+3}{2},$$

$$\frac{x-1}{1} = \frac{y-1}{2} = \frac{z+2}{3},$$

respectively intersect and find the coordinates of the point of intersection P.

Find: (i) an equation for the plane Π_1 containing l_1 and l_2;

   (ii) an equation for the plane Π_2 which passes through the line l_1 and which is perpendicular to the plane Π_1;

   (iii) an equation for the plane which passes through the origin and contains the line of intersection of Π_1 and Π_2.

25. The position vectors of the points A and B are given by

$$\mathbf{a} = 5\mathbf{i} + \mathbf{j} + 2\mathbf{k}, \qquad \mathbf{b} = -\mathbf{i} + 7\mathbf{j} + 8\mathbf{k},$$

respectively. Show that the two lines

$$\mathbf{r} = \mathbf{a} + \lambda\mathbf{l}, \qquad \mathbf{r} = \mathbf{b} + \mu\mathbf{m},$$

where $\mathbf{l} = -4\mathbf{i} + \mathbf{j} - \mathbf{k}$, $\mathbf{m} = 2\mathbf{i} - 5\mathbf{j} - 7\mathbf{k}$, intersect, and find the position vector of the point of intersection K. If D has position vector $3\mathbf{i} + 7\mathbf{j} - 2\mathbf{k}$, show that the line KD is perpendicular to the line AB.

26. (i) The direction of a straight line through the origin O is determined by the unit vector $\mathbf{n}$. The position vector relative to O of a point P is $\mathbf{r}$. Prove that the perpendicular distance from P to the line is given by the magnitude of $\mathbf{r} \times \mathbf{n}$. Hence, or otherwise, find the perpendicular distance from the point $(3, 1, 2)$ to the straight line

$$\frac{x-2}{3} = \frac{y-3}{2} = \frac{z-1}{2}.$$

(ii) Find vector and cartesian equations of the plane which passes through the point with position vector $2\mathbf{i} + 3\mathbf{j} - \mathbf{k}$ and is perpendicular to the vector $-3\mathbf{i} + 4\mathbf{j} + 2\mathbf{k}$.   (L.)

27. Prove that, provided $\mathbf{b} \cdot \mathbf{n} \neq 0$, the line $\mathbf{r} = \mathbf{a} + t\mathbf{b}$ and the plane $\mathbf{r} \cdot \mathbf{n} = p$ intersect in a point whose position vector is $\mathbf{a} + (p - \mathbf{a} \cdot \mathbf{n})\mathbf{b}/(\mathbf{b} \cdot \mathbf{n})$.

28. From a point whose position vector from the origin is $\mathbf{s}$, the perpendicular is drawn to a plane. If this perpendicular is represented by the vector $\mathbf{p}$, find, in scalar product form, a vector equation of the plane.

From the point $\mathbf{s}_1 = \mathbf{i} + \mathbf{j} + \mathbf{k}$ the perpendicular drawn to a plane Π_1 is $\mathbf{i} - \mathbf{j} + \mathbf{k}$. From the point $\mathbf{s}_2 = 3\mathbf{i} - \mathbf{j} - \mathbf{k}$ the perpendicular drawn to a plane Π_2 is $\mathbf{i} + 2\mathbf{j} + \mathbf{k}$. Find an equation for a plane Π_3 containing the line of intersection of the planes Π_1 and Π_2 and passing through the mid-point of the line joining $\mathbf{s}_1$ to $\mathbf{s}_2$. Find also the length of the perpendicular from the origin on to the plane Π_3.   (L.)

29. Find the direction cosines of the line l common to the planes

$$x + y + z - 3 = 0, \quad x + y + 2z + 1 = 0.$$

Show that for all values of the parameter λ the equation

$$(x + y + z - 3) + \lambda(x + y + 2z + 1) = 0$$

represents a plane passing through the line l. Find the equation of the plane P which passes through the line l and through the origin.

Obtain equations of the straight line through the origin which lies in the plane P and is perpendicular to the line l.   (L.)

30. Find an equation for the plane which contains the points $P(1, -3, -4)$ and $Q(-1, 0, 5)$ and is perpendicular to the plane $3x + 7y - 2z + 5 = 0$.

Find parametric equations for the locus of points in the plane $3x + 7y - 2z + 5 = 0$ equidistant from P and Q.   (L.)

31. Obtain an equation for the plane which is perpendicular to the vector $\mathbf{i} + 2\mathbf{j} + 2\mathbf{k}$ and which passes through the point P with position vector $2\mathbf{i} + \mathbf{j} + \mathbf{k}$.

Obtain also an equation of the plane containing the origin, the point P and the point with position vector $\mathbf{i}$.   (L.)

32. (i) Show that the lines given by the equations

$$\frac{x-2}{1} = \frac{y}{-2} = \frac{z-1}{2} \quad \text{and} \quad \frac{x+1}{-2} = \frac{y-3}{1} = \frac{z+3}{-2}$$

meet, and find an equation for the plane which contains them.

(ii) Find the radius of the circle of intersection of the sphere

$$x^2 + y^2 + z^2 + 5y + z - 15 = 0$$

and the plane

$$4x + 3y + 5z - 15 = 0.$$ (L.)

33. A plane equally inclined to the coordinate axes cuts a sphere which passes through the origin in a circle of radius 2 and centre $(1, 2, -1)$. Find the distance between the centres of the circle and the sphere. Find, also, an equation of the tangent plane through the origin.

34. Find an equation for the sphere which passes through the points $(0, 0, 0)$, $(2a, 0, 0)$, $(0, 2b, 0)$, $(0, 0, 2c)$. Determine its centre and radius.

Find the equations of the two spheres each of which passes through the points $(0, 0, 0)$, $(2, 0, 0)$, $(0, 2, 0)$ and touches the line $x = y = z - 1$.

Show that the two spheres intersect orthogonally.

14:3 THE INTERSECTION OF THREE PLANES

When the determinant

$$\Delta = \begin{vmatrix} a_1 & b_1 & c_1 \\ a_2 & b_2 & c_2 \\ a_3 & b_3 & c_3 \end{vmatrix}$$

does not vanish, the planes

$$a_1 x + b_1 y + c_1 z + d_1 = 0,$$
$$a_2 x + b_2 y + c_2 z + d_2 = 0,$$
$$a_3 x + b_3 y + c_3 z + d_3 = 0$$

have just one point which is common to all three of them and which is given by the solution of the three equations (see p. 467).

If $\Delta = 0$, the equations

$$a_1 l + b_1 m + c_1 n = 0,$$
$$a_2 l + b_2 m + c_2 n = 0,$$
$$a_3 l + b_3 m + c_3 n = 0$$

have a set of solutions which are not all zero, of the form

$$l:m:n = k_1:k_2:k_3,$$

where k_1, k_2, k_3 are constants. There exists, therefore, a line with direction ratios $l:m:n$ which is perpendicular to the normal of each of the three planes. This line is parallel to each of the three planes and therefore, if $\Delta = 0$, the three planes are all parallel to the same line and the following possibilities arise:

(1) Two of the planes are parallel or all three of the planes are parallel.
(2) The three planes have a common line of intersection.
(3) The three planes intersect in pairs along three parallel lines and form a triangular prism.

(We exclude from consideration the trivial case in which one of the given equations is a multiple of another so that two of the planes are coincident.)

In any particular problem the above three cases can be distinguished as follows:

Case (1) arises when there exists a relation of the form

$$\frac{a_1}{a_2} = \frac{b_1}{b_2} = \frac{c_1}{c_2} \neq \frac{d_1}{d_2}$$

so that two of the planes have a common normal and do not coincide.

In the other two cases we can find the line of intersection of two of the planes. If the coordinates of an arbitrary point on this line (expressed parametrically) satisfy identically the equation of the third plane, then the three planes have a common line of intersection [Case (2)]; otherwise the planes intersect along parallel distinct lines.

Example 1. Consider

$$x + y + z - 4 = 0,$$
$$2x - 3y - z + 3 = 0,$$
$$x + 2y + 3z - 9 = 0.$$

Here

$$\Delta = \begin{vmatrix} 1 & 1 & 1 \\ 2 & -3 & -1 \\ 1 & 2 & 3 \end{vmatrix} = -7$$

and hence the planes intersect in one point. Solution of the equations gives this point as $(1, 1, 2)$.

Example 2. Consider

$$2x - y - z + 4 = 0, \tag{1}$$
$$x + y - z - 1 = 0, \tag{2}$$
$$5x - 4y - 2z + 13 = 0. \tag{3}$$

Here

$$\Delta = \begin{vmatrix} 2 & -1 & -1 \\ 1 & 1 & -1 \\ 5 & -4 & -2 \end{vmatrix} = 0.$$

Since the direction ratios of the normals to the planes are distinct no two of the planes are parallel, and the planes intersect in lines which are parallel or coincident. But the line of intersection of (1) and (2) has equations

$$\frac{x+5}{2} = \frac{y}{1} = \frac{z+6}{3} \tag{4}$$

$$\Leftrightarrow x = 2\lambda - 5, \quad y = \lambda, \quad z = 3\lambda - 6.$$

These values of (x, y, z) identically satisfy equation (3) and hence the planes have the common line of intersection (4).

Example 3. Consider

$$2x - 3y + 5z - 4 = 0,$$
$$x - y + 3z - 4 = 0,$$
$$4x - 6y + 10z - 5 = 0,$$

where

$$\Delta = \begin{vmatrix} 2 & -3 & 5 \\ 1 & -1 & 3 \\ 4 & -6 & 10 \end{vmatrix} = 0.$$

The first and third of these planes are parallel and thus the planes have no common point. (The second plane is not parallel to the other two, but meets them in parallel lines.)

Example 4. Consider

$$x + 4y - z = 0$$
$$2x + 8y - 2z + 5 = 0,$$
$$3x + 12y - 3z + 1 = 0.$$

The three planes are all parallel and have no common point.

Example 5. Consider

$$4x + y + z - 10 = 0, \tag{1}$$
$$6x + 3y + z - 4 = 0, \tag{2}$$
$$x - 5y + 2z - 3 = 0. \tag{3}$$

Here

$$\Delta = \begin{vmatrix} 4 & 1 & 1 \\ 6 & 3 & 1 \\ 1 & -5 & 2 \end{vmatrix} = 0$$

and hence the planes intersect in lines which are parallel or coincident. The line of intersection of (1) and (2) is

$$\frac{x}{1} = \frac{y+3}{-1} = \frac{z-13}{-3}$$

$$\Leftrightarrow x = \lambda, \quad y = -\lambda + 3, \quad z = -3\lambda + 13.$$

These coordinates do not satisfy equation (3). Hence the planes do *not* have a common line of intersection. Thus the three planes intersect in pairs along three parallel lines with direction ratios $1: -1: -3$.

Exercise 14.3

In each of questions 1–10 examine the nature of the intersection of the set of planes given.
  (i) If the planes have a common point, find the coordinates of that point.
  (ii) If the planes have a common line, find equations for the line in the form

$$\frac{x-a}{l} = \frac{y-b}{m} = \frac{z-c}{n}.$$

(iii) If the planes intersect in parallel lines, find direction ratios of those lines.

1. $x + y + z - 5 = 0$; $2x + 3y + 4z - 6 = 0$; $5x + 6y + 7z + 3 = 0$.

2. $2x - 3y + z - 4 = 0$; $x + y - z - 10 = 0$; $2x + 5y + 2z + 11 = 0$.

3. $3x + y - z - 5 = 0$; $x - y - z + 1 = 0$; $5x + 3y - z - 11 = 0$.

4. $2x - 3y + z - 8 = 0$; $x + 5y - z + 1 = 0$; $4x - 6y + 2z = 0$.

5. $x - 2y + 3z - 4 = 0$; $3x - 4y + 5z - 6 = 0$; $7x - 8y + 9z - 11 = 0$.

6. $x - 3y - 5z - 10 = 0$; $2x + y + z - 2 = 0$; $x + 3y - z = 0$.

7. $x + 4y - 2z = 0$; $2x + 8y - 4z + 5 = 0$; $5x + 20y - 10z + 1 = 0$.

8. $x - 4y - 4z + 9 = 0$; $5x - 8y - 2z + 21 = 0$; $2x - 2y + z + 6 = 0$.

9. $2x + y + z + 5 = 0$; $x - 2y - 3z + 5 = 0$; $2x + 5y + z + 1 = 0$.

10. $7x - 2y + 3z - 5 = 0$; $17x - 2y + 13z + 3 = 0$; $10x - y + 8z = 0$.

11. Find the number of solutions of the equations

$$x - 2y + 3z = q,$$
$$x + y - 2z = 8,$$
$$px - 3y + 4z = 10,$$

in each of the cases:

(a) $p \neq 3, q = 1$;   (b) $p = 3, q = 1$;   (c) $p = 3, q = 2$.

Show that in one of these cases the equations represent three planes with a common line of intersection, and find equations of this line.
State how the planes meet in the other two cases.   (L.)

12. Solve the equations

$$x - 2y + 3z = 0,$$
$$3x + ay + z = 4,$$
$$3x - y + 2az = b,$$

in terms of a and b.

Discuss the solutions, if any, in the special cases (i) $a = 2, b = 2\frac{1}{2}$; (ii) $a = -3\frac{1}{2}, b = -1$. Interpret these special cases geometrically.   (L.)

13. Find the solution set of the simultaneous equations

$$x + 2y + 3z = 10$$
$$2x + 3y - 4z = 8$$
$$ax + by + cz = d$$

in each of the following cases:

(a) when $a = b = c = 1, d = 6$,

(b) when $a = 0, b = 1, c = 10, d = 12$,

(c) when $a = 0, b = 1, c = 10, d = 3$.

If the solution set is infinite, express the values of x and y in terms of z.
Interpret your answer in each case geometrically.   (L.)

14. Find the values of λ for which the equations

$$x + 3y + 5z = k,$$
$$\lambda x - y - 9z = 1,$$
$$\lambda^2 x + 5y + z = 7,$$

do not possess a *unique* solution. For *one* of these values of λ, find the value of k for which a solution exists and solve the equations completely. Interpret your solution geometrically.   (L.)

15. (i) Solve the following equations:

$$3x + 4y + z = 7,$$
$$2x - 5y + 3z = -10,$$
$$x + 2y + z = 1.$$

Give a geometrical interpretation of your solution.

(ii) Show that the following simultaneous equations are not independent and give a geometrical interpretation of the nature of the solution:

$$x + 2y + 5z = 6,$$
$$3x - y + 6z = 2,$$
$$9x - 10y + 9z = -10.$$

16. By considering their geometrical interpretation, or otherwise, show that the simultaneous equations

$$x - 2y + 2z = 4,$$
$$2x - y + 2z = 5,$$
$$8x - 7y + 10z = k$$

have a non-unique solution when $k = 23$ but have no solution for other values of k.

Obtain the general solution of the three equations when $k = 23$. Hence find the solution of the simultaneous equations

$$x - 2y + 2z = \ 4,$$
$$2x - y + 2z = \ 5,$$
$$8x - 7y + 10z = 23,$$
$$4x^2 - 9y^2 + 2z^2 + 9 = \ 0.$$

14:4 TRIPLE PRODUCTS

(1) The triple scalar product

This is the scalar product of a vector $\mathbf{a}$ and the vector product $\mathbf{b} \times \mathbf{c}$, i.e. $\mathbf{a}.\mathbf{b} \times \mathbf{c}$ or $\mathbf{b} \times \mathbf{c}.\mathbf{a}$. Because the vector product is non-commutative

$$\mathbf{a}.\mathbf{b} \times \mathbf{c} = -\mathbf{a}.\mathbf{c} \times \mathbf{b} = -\mathbf{c} \times \mathbf{b}.\mathbf{a}. \tag{14.19}$$

Using formulae (14.4) and (14.5) we find

$$\mathbf{a}.\mathbf{b} \times \mathbf{c} = \mathbf{a} \times \mathbf{b}.\mathbf{c} = \begin{vmatrix} a_1 & a_2 & a_3 \\ b_1 & b_2 & b_3 \\ c_1 & c_2 & c_3 \end{vmatrix} = \Delta. \tag{14.20}$$

We can form similar triple scalar products from the same three vectors, using $\mathbf{c} \times \mathbf{a}$ with $\mathbf{b}$ and $\mathbf{a} \times \mathbf{b}$ with $\mathbf{c}$, and we find that all the possible triple scalar products formed when the cyclic order of the factors $\mathbf{a}$, $\mathbf{b}$, $\mathbf{c}$ is maintained are equal to Δ. Those in which the cyclic order is reversed are equal to $-\Delta$. (This corresponds to one of the properties of a determinant.) The complete list of products is

$$\Delta = \mathbf{a} \times \mathbf{b}.\mathbf{c} = \mathbf{b} \times \mathbf{c}.\mathbf{a} = \mathbf{c} \times \mathbf{a}.\mathbf{b}$$
$$= \mathbf{a}.\mathbf{b} \times \mathbf{c} = \mathbf{b}.\mathbf{c} \times \mathbf{a} = \mathbf{c}.\mathbf{a} \times \mathbf{b} \tag{14.21}$$

and

$$-\Delta = \mathbf{a} \times \mathbf{c}.\mathbf{b} = \mathbf{b} \times \mathbf{a}.\mathbf{c} = \mathbf{c} \times \mathbf{b}.\mathbf{a}$$
$$= \mathbf{a}.\mathbf{c} \times \mathbf{b} = \mathbf{b}.\mathbf{a} \times \mathbf{c} = \mathbf{c}.\mathbf{b} \times \mathbf{a}. \tag{14.22}$$

These formulae show that the positions of the "dot" and "cross" product signs within the triple product may be interchanged.

We now give a direct geometrical interpretation of the triple scalar product. Suppose that the three vectors $\mathbf{a}$, $\mathbf{b}$, $\mathbf{c}$ represent three concurrent edges of a parallelepiped, Fig. 14.10. Then the vector product $\mathbf{b} \times \mathbf{c}$ represents the vector area of one face and the triple product $\mathbf{a}.\mathbf{b} \times \mathbf{c}$ gives the volume of the parallelepiped. If $\mathbf{a}$ points towards the positive side of the face OBC the volume is positive. The other triple products with the same cyclic order give the volume of the same figure, the evaluation starting with a different face. Reversal of the cyclic order changes the sign of the volume because it

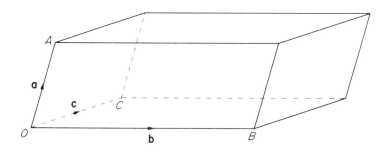

FIG. 14.10.

interchanges the positive and negative sides of the face OBC (or of whichever face is being used). This result is also stated by saying that, if $\mathbf{a}\,.\,\mathbf{b}\times\mathbf{c}$ is positive, then the three vectors form a *right-handed* triad; if $\mathbf{a}\,.\,\mathbf{b}\times\mathbf{c}$ is negative, they form a *left-handed* triad. The vectors $\mathbf{a}$, $\mathbf{b}$, $\mathbf{c}$ form a right-handed triad if the figure $OABC$ can be continuously deformed, by altering the angles at O by amounts less than $\frac{1}{2}\pi$ without altering the sense of any of the vectors, so that OA, OB, OC coincide with the positive directions of a r.h. rectangular frame of reference.

If the triangle OBC is taken as the base of a tetrahedron whose opposite vertex is A, then the volume of the tetrahedron is

$$V = \tfrac{1}{3}\mathbf{a}\,.\,[\tfrac{1}{2}(\mathbf{b}\times\mathbf{c})] = \tfrac{1}{6}(\mathbf{a}\,.\,\mathbf{b}\times\mathbf{c}). \tag{14.23}$$

The same considerations apply to the sign of this volume as to the volume of the parallelepiped. If the points A, B, C, D have coordinates (x_1, y_1, z_1), (x_2, y_2, z_2), (x_3, y_3, z_3), (x_4, y_4, z_4) respectively, then the volume of the tetrahedron $ABCD$ is

$$V = \tfrac{1}{6}\overrightarrow{AB}\,.\,\overrightarrow{AC}\times\overrightarrow{AD} = \tfrac{1}{6}\begin{vmatrix} x_2 - x_1 & y_2 - y_1 & z_2 - z_1 \\ x_3 - x_1 & y_3 - y_1 & z_3 - z_1 \\ x_4 - x_1 & y_4 - y_1 & z_4 - z_1 \end{vmatrix} \tag{14.23a}$$

$$\Leftrightarrow V = \tfrac{1}{6}\begin{vmatrix} 1 & x_1 & y_1 & z_1 \\ 1 & x_2 & y_2 & z_2 \\ 1 & x_3 & y_3 & z_3 \\ 1 & x_4 & y_4 & z_4 \end{vmatrix}, \tag{14.23b}$$

where, in each case, the numerical value of the determinant gives the scalar volume of the tetrahedron.

Example 1. The edges OP, OQ, OR of a tetrahedron $OPQR$ are the vectors $\mathbf{a}$, $\mathbf{b}$ and $\mathbf{c}$ respectively, where $\mathbf{a} = 2\mathbf{i} + 4\mathbf{j}$, $\mathbf{b} = 2\mathbf{i} - \mathbf{j} + 3\mathbf{k}$ and $\mathbf{c} = 4\mathbf{i} - 2\mathbf{j} + 5\mathbf{k}$. Evaluate $\mathbf{b}\times\mathbf{c}$ and deduce that OP is perpendicular to the plane OQR.

Write down the length of OP and the area of the triangle OQR and hence the volume of the tetrahedron. Verify your result by evaluating $\mathbf{a}\,.\,(\mathbf{b}\times\mathbf{c})$.   (L.)

$$\mathbf{b}\times\mathbf{c} = (2\mathbf{i} - \mathbf{j} + 3\mathbf{k})\times(4\mathbf{i} - 2\mathbf{j} + 5\mathbf{k}) = \mathbf{i} + 2\mathbf{j} = \tfrac{1}{2}\mathbf{a}.$$

Now $\mathbf{b}\times\mathbf{c}$ is perpendicular to the plane OQR; therefore the vector $\mathbf{a}$, and hence also OP, are perpendicular to the plane OQR.

The length $OP = |\overrightarrow{OP}| = |\mathbf{a}| = 2\sqrt{5}$. Also $|\mathbf{b}\times\mathbf{c}| = \sqrt{5}$ so that the area of the triangle OQR is $\tfrac{1}{2}\sqrt{5}$. Hence the volume of the tetrahedron is

$$\tfrac{1}{3}OP\,.\,(\text{area }OQR) = \tfrac{1}{3}(2\sqrt{5})\,(\tfrac{1}{2}\sqrt{5}) = \tfrac{5}{3}.$$

Also

$$\mathbf{a}\,.\,\mathbf{b}\times\mathbf{c} = \begin{vmatrix} 2 & 4 & 0 \\ 2 & -1 & 3 \\ 4 & -2 & 5 \end{vmatrix} = 10$$

$$\Rightarrow V = \tfrac{1}{6}(\mathbf{a}\,.\,\mathbf{b}\times\mathbf{c}) = \tfrac{5}{3}.$$

Example 2. Prove that the perpendicular distance of the point A with position vector $\mathbf{a}$ from the line whose vector equation is $\mathbf{r} = \mathbf{b} + \lambda\hat{\mathbf{e}}$, where $\hat{\mathbf{e}}$ is a unit vector, is $|(\mathbf{a} - \mathbf{b})\times\hat{\mathbf{e}}|$.

Find the shortest distance from $A\,(1, 3, -2)$ to the line joining $(2, 1, -2)$ and $(-1, 1, 2)$.   (L.)

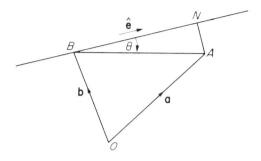

FIG. 14.11.

Suppose O is the origin of position vectors, Fig. 14.11, and B is the point with position vector $\mathbf{b}$ so that $\overrightarrow{OB} = \mathbf{b}$. Then $\overrightarrow{BA} = \mathbf{a} - \mathbf{b}$ and the required perpendicular distance AN is $BA \sin \theta$. But $|\overrightarrow{BA} \times \hat{\mathbf{e}}| = |\overrightarrow{BA}| |\hat{\mathbf{e}}| \sin \theta = BA \sin \theta$

$$\Rightarrow AN = |(\mathbf{a} - \mathbf{b}) \times \hat{\mathbf{e}}|. \tag{1}$$

The line joining the points $(2, 1, -2), (-1, 1, 2)$ has equations

$$\frac{x-2}{-3} = \frac{y-1}{0} = \frac{z+2}{4} \tag{2}$$

or, expressed in suitable vector form, $[\hat{\mathbf{e}} = (-3\mathbf{i} + 4\mathbf{k})/5]$,

$$\mathbf{r} = 2\mathbf{i} + \mathbf{j} - 2\mathbf{k} + \lambda(-3\mathbf{i} + 4\mathbf{k})/5.$$

Hence, using equation (1), the perpendicular distance

$$AN = |\{(\mathbf{i} + 3\mathbf{j} - 2\mathbf{k}) - (2\mathbf{i} + \mathbf{j} - 2\mathbf{k})\} \times (-3\mathbf{i} + 4\mathbf{k})/5|$$
$$= |(8\mathbf{i} + 4\mathbf{j} + 6\mathbf{k})/5| = \tfrac{1}{5}\sqrt{116}.$$

Example 3. Find the length of the shortest distance between the two lines

$$\mathbf{r} = \mathbf{a} + p\mathbf{l}, \mathbf{r} = \mathbf{b} + q\mathbf{m}.$$

By pure geometry it can be shown that the shortest distance is perpendicular to both lines. Hence a unit vector in this direction is

$$\hat{\mathbf{n}} = \frac{\mathbf{l} \times \mathbf{m}}{|\mathbf{l} \times \mathbf{m}|}.$$

A vector joining arbitrary points A and B, one on each of the lines, is $\mathbf{a} - \mathbf{b} + p\mathbf{l} - q\mathbf{m}$. The shortest distance, d, is the projection of this vector on to the direction $\hat{\mathbf{n}}$

$$\Rightarrow d = (\mathbf{a} - \mathbf{b} + p\mathbf{l} - q\mathbf{m}).\hat{\mathbf{n}} = \frac{(\mathbf{a} - \mathbf{b}).\mathbf{l} \times \mathbf{m}}{|\mathbf{l} \times \mathbf{m}|}.$$

Example 4. Prove that the equations of any two straight lines may be expressed in the form

$$y = kx, \quad z = c; \quad y = -kx, \quad z = -c.$$

Referred to an arbitrary set of axes the equations of two skew lines involve twelve constants. By suitable choice of axes this number can be reduced to two.

Suppose (Fig. 14.12) that AA', BB' are two skew lines and the length of the common perpendicular is $2c$. Then, if the origin O is taken as the mid-point of the perpendicular and Oz is directed along it, the lines AA',

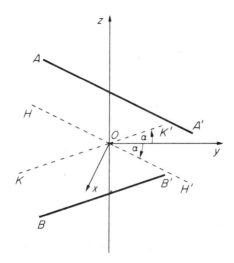

FIG. 14.12.

BB' lie in the planes $z = c$, $z = -c$ respectively. Suppose that the projections of AA', BB' on the plane $z = 0$ are HH', KK' respectively. Then we choose Ox, Oy to be the bisectors of the angles between HH' and KK' and, if $H\hat{O}K = 2\alpha$ so that $H'\hat{O}y = \alpha$, the equations of the projections HH', KK' are $y = x\cot\alpha$, $y = -x\cot\alpha$ respectively. Hence writing $k = \cot\alpha$ the equations of the lines are

$$y = kx, \quad z = c; \quad y = -kx, \quad z = -c.$$

Exercise 14.4(a)

1. A tetrahedron has its vertices at the points $O(0, 0, 0)$, $A(1, 1, 2)$, $B(-1, 2, -1)$, and $C(0, -1, 3)$. By consideration of the vector product $\overrightarrow{AC} \times \overrightarrow{AB}$, or otherwise, determine (i) the area of the face ABC and (ii) the unit vector normal to the face ABC.

Hence, or otherwise, determine the volume of the tetrahedron.

2. Find the volume of the tetrahedron whose vertices are the points $(1, 3, -1)$, $(2, 2, 3)$, $(4, 2, -2)$, $(3, 7, 4)$.

3. Show that the points $(3, 4, 0)$, $(-1, -2, -2)$, $(2, 8, 3)$, $(-5, 3, 3)$ are coplanar and find the equation of their plane.

4. Find the shortest distance between a diagonal of a rectangular parallelepiped, of edges a, b, c, and the edges which do not meet that diagonal.

5. Find the perpendicular distance from the point $(2, 2, 1)$ to the line of intersection of the two planes

$$2x - y - z = 3, \quad 3x - y - 3z = 4.$$

6. The coordinates of the four points A, B, C, D are respectively $(1, 2, 1)$, $(-1, 0, 2)$, $(2, 1, 3)$ and $(3, -1, 1)$. Find the shortest distance between the lines AB and CD.

7. Show that the shortest distance between the lines

$$\frac{x+4}{3} = \frac{y-3}{-2} = \frac{z+6}{5}$$

and $x - 2y - z = 0$, $x - 10y - 3z = -7$, is $\frac{1}{2}\sqrt{14}$ and find the coordinates of its end points.

8. Find the shortest distance between the straight lines with vector equations

$$\mathbf{r} = 3\mathbf{i} + s\mathbf{j} - \mathbf{k}, \quad \mathbf{r} = 9\mathbf{i} - 2\mathbf{j} - \mathbf{k} + t(\mathbf{i} - 2\mathbf{j} + \mathbf{k}). \tag{L.}$$

9. If a variable line intersects both the lines

$$y = kx, \quad z = c \quad \text{and} \quad y = -kx, \quad z = -c$$

and is equally inclined to them, prove that it lies on one or other of the surfaces

$$yz = kcx, \quad kzx = cy. \tag{L.}$$

*(2) *The triple vector product*

The vector product of **a** and **b** × **c**, i.e. **a** × (**b** × **c**), is a triple vector product. The position of the brackets in this expression is very important. The vector **a** × (**b** × **c**) is perpendicular both to **a** and to **b** × **c**; since it is perpendicular to **b** × **c** it lies in the plane of **b** and **c**. On the other hand, (**a** × **b**) × **c**, being perpendicular to **a** × **b**, lies in the plane of **a** and **b**. These are in general different planes. Hence **a** × (**b** × **c**) ≠ (**a** × **b**) × **c**.

The formula for expanding **a** × (**b** × **c**) can be obtained by writing out components and rearranging the expressions. The result is

$$\mathbf{a} \times (\mathbf{b} \times \mathbf{c}) = \mathbf{b}(\mathbf{a} . \mathbf{c}) - \mathbf{c}(\mathbf{a} . \mathbf{b}). \tag{14.24}$$

Formula (14.24) can be established by the use of components as follows. Choose the coordinate axes so that Ox is directed along **a**, and the plane Oxy is the plane of **a** and **b**. Then we can write

$$\mathbf{a} = a_1 \mathbf{i}, \quad \mathbf{b} = b_1 \mathbf{i} + b_2 \mathbf{j}, \quad \mathbf{c} = c_1 \mathbf{i} + c_2 \mathbf{j} + c_3 \mathbf{k}$$
$$\Rightarrow \mathbf{b} \times \mathbf{c} = b_2 c_3 \mathbf{i} - b_1 c_3 \mathbf{j} + (b_1 c_2 - b_2 c_1) \mathbf{k}$$
$$\Rightarrow \mathbf{a} \times (\mathbf{b} \times \mathbf{c}) = (a_1 b_2 c_1 - a_1 b_1 c_2) \mathbf{j} - a_1 b_1 c_3 \mathbf{k}.$$

But

$$(\mathbf{a} . \mathbf{c}) \mathbf{b} - (\mathbf{a} . \mathbf{b}) \mathbf{c} = a_1 c_1 \mathbf{b} - a_1 b_1 \mathbf{c}$$
$$= (a_1 b_2 c_1 - a_1 b_1 c_2) \mathbf{j} - a_1 b_1 c_3 \mathbf{k}.$$

However, a relation between vectors is independent of any particular frame of reference and hence, since **a** × (**b** × **c**) and (**a** . **c**)**b** − (**a** . **b**)**c** have the same components referred to the frame chosen above, equation (14.24) follows at once.

Example 1. Find the vector **x** which satisfies the equations

$$\mathbf{x} \times \mathbf{a} = \mathbf{b}, \quad \mathbf{x} . \mathbf{c} = p$$

in which p is a given scalar and $\mathbf{a} . \mathbf{c} \neq 0$.   (L.)

Taking the vector product of the first equation with **c** gives

$$(\mathbf{b} \times \mathbf{c}) = (\mathbf{x} \times \mathbf{a}) \times \mathbf{c} = \mathbf{a}(\mathbf{x} . \mathbf{c}) - \mathbf{x}(\mathbf{a} . \mathbf{c}) = p\mathbf{a} - \mathbf{x}(\mathbf{a} . \mathbf{c})$$
$$\Rightarrow \mathbf{x} = (p\mathbf{a} - \mathbf{b} \times \mathbf{c})/(\mathbf{a} . \mathbf{c}).$$

Example 2. Find the vector **x** and the scalar λ which satisfy the equations

$$\mathbf{a} \times \mathbf{x} = \mathbf{b} + \lambda \mathbf{a}, \quad \mathbf{a} . \mathbf{x} = 2,$$

where

$$\mathbf{a} = \mathbf{i} + 2\mathbf{j} - \mathbf{k}, \quad \mathbf{b} = 2\mathbf{i} - \mathbf{j} + \mathbf{k}. \tag{L.}$$

The given equations are equivalent to four scalar equations which have the three components of $\mathbf{x}$ and λ as the unknowns.

Taking the vector product of the first equation with $\mathbf{a}$ gives

$$\mathbf{b} \times \mathbf{a} + \lambda(\mathbf{a} \times \mathbf{a}) = (\mathbf{a} \times \mathbf{x}) \times \mathbf{a} = \mathbf{x}\mathbf{a}^2 - \mathbf{a}(\mathbf{x}.\mathbf{a})$$

$$\Rightarrow \mathbf{x} = \frac{\mathbf{b} \times \mathbf{a} + 2\mathbf{a}}{\mathbf{a}^2}.$$

In the special case given

$$\mathbf{x} = \frac{(-\mathbf{i} + 3\mathbf{j} + 5\mathbf{k}) + 2(\mathbf{i} + 2\mathbf{j} - \mathbf{k})}{6} = \frac{1}{6}(\mathbf{i} + 7\mathbf{j} + 3\mathbf{k}).$$

Taking the scalar product of the first equation with $\mathbf{a}$ gives

$$0 = \mathbf{a}.\mathbf{a} \times \mathbf{x} = \mathbf{a}.\mathbf{b} + \lambda \mathbf{a}^2$$

$$\Leftrightarrow \lambda = -\frac{\mathbf{a}.\mathbf{b}}{\mathbf{a}^2} = -\frac{(2 - 2 - 1)}{6} = \frac{1}{6}.$$

---

*Exercise 14.4(b)

1. $Oxyz$ is a right-handed rectangular cartesian frame of reference and $\mathbf{i}, \mathbf{j}, \mathbf{k}$ are unit vectors parallel to the axes Ox, Oy, Oz respectively. Given the vectors

$$\overrightarrow{OA} = \mathbf{a} = x\mathbf{i} - 6\mathbf{j} - 3\mathbf{k}, \quad \overrightarrow{OB} = \mathbf{b} = 4\mathbf{i} + 3\mathbf{j} - \mathbf{k}, \quad \overrightarrow{OC} = \mathbf{c} = \mathbf{i} - 3\mathbf{j} + 2\mathbf{k},$$

find the value of x for which

    (i) $\mathbf{a}$ and $\mathbf{b}$ are perpendicular vectors,

    (ii) $\mathbf{a}, \mathbf{b}$ and $\mathbf{c}$ are coplanar vectors.

When $x = 2$, find

    (iii) a unit vector perpendicular to both $\mathbf{a}$ and $\mathbf{b}$,

    (iv) the area of the triangle ABC,

    (v) $\mathbf{a} \times (\mathbf{b} \times \mathbf{c})$.

2. (i) The point A lies at the extremity of the position vector $\mathbf{a}$ measured from an origin O. Show that the position vector of any point P lying on the line through A parallel to a vector $\mathbf{b}$ may be expressed as

$$\mathbf{r} = \mathbf{a} + p\mathbf{b},$$

where p is a scalar parameter, and explain the geometrical significance of p.

    (ii) Show that the component of the vector $\mathbf{a}$ parallel to the vector $\mathbf{b}$ is $\mathbf{c}$ where

$$\mathbf{b}^2\mathbf{c} = (\mathbf{a}.\mathbf{b})\mathbf{b}.$$

    (iii) The vector $\mathbf{c}$ and the scalar λ satisfy the equations

$$\mathbf{a} \times \mathbf{c} = \lambda\mathbf{a} + \mathbf{b},$$
$$\mathbf{a}.\mathbf{c} = 1,$$

where $\mathbf{a} = \mathbf{i} + 2\mathbf{j}, \mathbf{b} = 2\mathbf{i} + \mathbf{j} - 2\mathbf{k}$. Find the value of λ, and express $\mathbf{c}$ in terms of $\mathbf{i}, \mathbf{j}$ and $\mathbf{k}$.   (L.)

3. Solve the following equations for $\mathbf{x}$.

    (i) $\mathbf{x} \times \mathbf{a} = \mathbf{b}$,   where $\mathbf{a}.\mathbf{b} = 0$.

    (ii) $\lambda\mathbf{x} + \mathbf{x} \times \mathbf{c} = \mathbf{a}$.

    (iii) $a\mathbf{x} + (\mathbf{x}.\mathbf{u})\mathbf{v} = \mathbf{w}$,

considering the cases in which $a + \mathbf{u}.\mathbf{v}$ is not zero and is zero separately.   (L.)

4. The unit vectors $\mathbf{p}, \mathbf{q}$ make the respective angles α, β with a unit vector $\mathbf{k}$, and the angle between the plane containing $\mathbf{p}, \mathbf{k}$ and that containing $\mathbf{q}, \mathbf{k}$ is θ. If

$$\mathbf{u} = \mathbf{k} \times \mathbf{p} \quad \text{and} \quad \mathbf{v} = \mathbf{k} \times \mathbf{q},$$

prove that

$$\mathbf{u}.\mathbf{v} = \sin\alpha \sin\beta \cos\theta.$$

Show also that $\mathbf{u} \times \mathbf{v}$ is parallel to $\mathbf{k}$ and of magnitude $\sin\alpha \sin\beta \sin\theta$.

5. If the vector $\mathbf{x}$ and the scalar m satisfy the equations

$$\mathbf{a} \times \mathbf{x} = m\mathbf{a} + \mathbf{b}, \quad \mathbf{a}.\mathbf{x} = 1,$$

where $\mathbf{a} = \mathbf{i} - 2\mathbf{j}$, $\mathbf{b} = 2\mathbf{i} + \mathbf{j} - 2\mathbf{k}$, find the value of m and the expression for the vector $\mathbf{x}$ in terms of $\mathbf{i}, \mathbf{j}, \mathbf{k}$.

6. Solve for $\mathbf{x}$ the equations

(i) $\mathbf{x} = \mathbf{a} \times \mathbf{x}$,

(ii) $\mathbf{x} - \mathbf{a} = \mathbf{a} \times \mathbf{x}$,

(iii) $\mathbf{a} \times \mathbf{x} = \mathbf{b} \times \mathbf{x}$,

(iv) $(\mathbf{a} \times \mathbf{x}) \times (\mathbf{b} \times \mathbf{x}) = \mathbf{0}$,

where $\mathbf{a}, \mathbf{b}$ are constant unequal non-zero vectors, but possibly being scalar multiples of each other.

Miscellaneous Exercise 14

1. A tetrahedron $OABC$ has its vertices at the points $O\,(0, 0, 0)$, $A\,(1, 2, -1)$, $B\,(-1, 1, 2)$ and $C\,(2, -1, 1)$. Write down expressions for $\overrightarrow{AB}$ and $\overrightarrow{AC}$ in terms of the mutually orthogonal unit vectors $\mathbf{i}, \mathbf{j}$ and $\mathbf{k}$ and find $\overrightarrow{AB} \times \overrightarrow{AC}$. Deduce the area of the triangle ABC. Find also the length of the perpendicular from O to the plane ABC. Hence, or otherwise, find the volume of the tetrahedron $OABC$. (L.)

2. The position vectors of the points A, B, C are $\mathbf{a}, \mathbf{b}, \mathbf{c}$, where

$$\mathbf{a} = 2\mathbf{i} - 4\mathbf{j} - 3\mathbf{k}, \quad \mathbf{b} = 6\mathbf{i} + 4\mathbf{k},$$
$$\mathbf{c} = (-2\mathbf{i} + \mathbf{j} + 8\mathbf{k}) + t\,(4\mathbf{i} + 4\mathbf{j} + 7\mathbf{k}).$$

Calculate the area of the triangle ABC and explain the geometrical significance of the fact that this area is independent of t.

The points P, Q, R are the feet of the perpendiculars drawn from A, B, C to the plane $\mathbf{r}.(2\mathbf{i} - \mathbf{j} + 2\mathbf{k}) = 0$. Find the area of the triangle PQR. (L.)

3. Relative to an origin O, the fixed points A, B, C have position vectors $\mathbf{a}, \mathbf{b}, \mathbf{c}$, where

$$\mathbf{a} = -\mathbf{i} + 4\mathbf{j} + 2\mathbf{k}, \; \mathbf{b} = 6\mathbf{j} - \mathbf{k}, \; \mathbf{c} = 3\mathbf{i} + 7\mathbf{j} - 6\mathbf{k}.$$

Find equations for
(a) the plane ABC,
(b) the line L_1 which passes through A and B,
(c) the line L_2 which passes through C and is perpendicular to the plane $2x - y - 2z = 0$.
Show that L_1 meets L_2 and find the position vector of the point of intersection. (L.)

4. The position vectors of the vertices A, B, C of a tetrahedron $OABC$ with respect to O as origin are

$$\overrightarrow{OA} = 2\mathbf{i} - \mathbf{j}, \quad \overrightarrow{OB} = \mathbf{j} + \mathbf{k}, \quad \overrightarrow{OC} = \mathbf{i} + 3\mathbf{j} - \mathbf{k}.$$

Find the angle between (a) the edges AB, AC, (b) the faces OAB, OAC.
Prove that BC is perpendicular to the plane OAB, and hence prove that the volume of $OABC$ is $3/2$. (L.)

5. ABC is the base of a pyramid $VABC$. The foot of the perpendicular from V to the base lies at M, the midpoint of AC. If the position vectors of A, B, C are $(\mathbf{i} + \mathbf{j} + 2\mathbf{k})$, $(-\mathbf{i} + 2\mathbf{j} + 2\mathbf{k})$, $(3\mathbf{i} + 3\mathbf{j} - \mathbf{k})$ respectively, and the volume of the pyramid is 9, find
(a) a cartesian equation for the plane ABC,
(b) the area of the triangle ABC,
(c) the height of the pyramid,
(d) a vector equation for VM,
(e) the coordinates of V, given that V lies on the same side of the plane ABC as the origin. (L.)

6. The point $P\,(14 + 2\lambda, 5 + 2\lambda, 2 - \lambda)$ lies on a fixed straight line for all values of λ. Find cartesian equations

for this line and find the cosine of the acute angle between this line and the line $x = z = 0$.

Show that the line $2x = -y = -z$ is perpendicular to the locus of P. Hence, or otherwise, find the equation of the plane containing the origin and all possible positions of P. (L.)

7. Find the direction cosines of the line l common to the planes

$$x + y + z - 3 = 0, \quad x + y + 2z + 1 = 0.$$

Show that for all values of the parameter λ the equation

$$(x + y + z - 3) + \lambda (x + y + 2z + 1) = 0$$

represents a plane passing through the line l. Find the equation of the plane P which passes through the line l and through the origin.

Obtain equations for the straight line through the origin which lies in the plane P and is perpendicular to the line l. (L.)

8. Three planes have equations

$$x - 6y - z = 5,$$
$$3x + 2y + z = -1,$$
$$5x + pz = q.$$

Show that
(a) the planes have a common point of intersection unless $p = 1$,
(b) when $p = 1$, $q = 2$, the planes intersect in pairs in three parallel lines,
(c) when $p = 1$, $q = 1$, the planes have a common line of intersection.
Give equations for the line of intersection in (c). (L.)

9. Find the most general form for the vector $\mathbf{u}$ when

$$\mathbf{u} \times (\mathbf{i} + \mathbf{j} + 2\mathbf{k}) = \mathbf{i} - \mathbf{j}. \tag{N.}$$

10. The vectors $\mathbf{a}$, $\mathbf{b}$ and $\mathbf{c}$ are such that $\mathbf{a} \neq 0$ and

$$\mathbf{a} \times \mathbf{b} = 2\mathbf{a} \times \mathbf{c}.$$

Show that

$$\mathbf{b} - 2\mathbf{c} = \lambda\mathbf{a},$$

where λ is a scalar.
Given that

$$|\mathbf{a}| = |\mathbf{c}| = 1, |\mathbf{b}| = 4,$$

and the angle between $\mathbf{b}$ and $\mathbf{c}$ is $\cos^{-1}(\tfrac{1}{4})$, show that

$$\lambda = +4 \quad \text{or} \quad -4.$$

For each of these cases find the cosine of the angle between $\mathbf{a}$ and $\mathbf{c}$. (N.)

11. Show that the line

$$\frac{x - 1}{4} = \frac{y - 2}{5} = \frac{z - 3}{6}$$

and the line

$$\frac{x - 4}{1} = \frac{y - 5}{2} = \frac{z - 6}{3} \quad \text{intersect.}$$

Find an equation for the plane in which they lie and the coordinates of their point of intersection. (L.)

12. Prove, by the use of vectors, that the lines joining the mid-points of opposite sides of a skew quadrilateral (i.e. whose sides do not lie in one plane) bisect each other.

If $\mathbf{a}$, $\mathbf{b}$, $\mathbf{c}$, $\mathbf{d}$ are the sides of a skew quadrilateral taken in order, show that the area of the parallelogram formed by the mid-points of the sides is equal to the magnitude of

$$\tfrac{1}{4}(\mathbf{b} \times \mathbf{c} + \mathbf{c} \times \mathbf{d} + \mathbf{d} \times \mathbf{b}).$$

13. Show that the curve given by

$$\mathbf{r} = \left(\frac{a}{\sqrt{2}} \sin t\right)(\mathbf{i} + \mathbf{j}) + (a \cos t)\,\mathbf{k},$$

where $0 \leqslant t < 2\pi$, is a circle of radius a, and find the vector equation of the plane in which it lies.
Show that the curve given by

$$\mathbf{r} = (a \sin^2 p)\mathbf{i} + (a \sin p \cos p)\mathbf{j} + (a \cos p)\mathbf{k},$$

where $0 \leqslant p < 2\pi$, lies on a sphere of radius a.
Find the position vectors of the four points common to the two curves.

14. Prove that if a point P in three-dimensional Euclidean space, with origin O, has coordinates which satisfy the equation

$$yz + zx + xy = 0,$$

then any point on the line OP has coordinates which also satisfy the equation.
Prove that if $P(x, y, z)$ is such that

$$x + y + z = 3d$$

and

$$yz + zx + xy = 0,$$

then

$$(x - d)^2 + (y - d)^2 + (z - d)^2 = 6d^2.$$

Deduce that any such point P lies on a circle centre (d, d, d). What is the radius of this circle?

(O.C.S.M.P.)

15. A point P is moving in the (x, y) plane so that at time t its position vector is $\mathbf{r}$ and its acceleration is

$$-25\mathbf{r} - 6\dot{\mathbf{r}} + 60\mathbf{u},$$

where $\mathbf{u}$ is a rotating unit vector making an angle $5t$ radians with the positive x-axis.
Initially $x = 3$ and $\dot{x} = 5$. Show that

$$x = e^{-3t}(3 \cos 4t + \sin 4t) + 2 \sin 5t.$$

Show that when t is large P travels approximately in a circle.   (N.)

16. A right circular cone has its vertex at the point $(4, -5, 3)$ and the centre of its base at the point $(0, 1, -1)$.
Write down
(a) equations for the axis of the cone,
(b) an equation for the plane containing the base of the cone.
If the line

$$\frac{x - 4}{3} = \frac{y + 5}{-8} = \frac{z - 3}{2}$$

is a generator of the cone, find the coordinates of the point where this generator meets the base, and deduce that the volume of the cone is $6\pi\sqrt{17}$.   (L.)

15 Determinants and Matrices

15:1 PROPERTIES OF DETERMINANTS

The idea of a determinant as a number associated with a matrix is introduced in elementary work.

$$\mathbf{M} = \begin{pmatrix} a & b \\ c & d \end{pmatrix} \Leftrightarrow |\mathbf{M}| = \begin{vmatrix} a & b \\ c & d \end{vmatrix} = ad - bc.$$

The definition of a determinant is such that there is a determinant "of order n" associated with every $n \times n$ matrix; but determinants have uses and properties apart from this association, and we shall consider some of these.

A determinant Δ_n of order n comprises a "square" array of elements arranged in n rows and n columns. We shall denote by a_{rs} the element in the rth row and the sth column. The value of the determinant is defined as

$$\sum_{s=1}^{n} (-1)^{s+1}_{a_{1s}} M_{1s},$$

where M_{1s}, the "minor" of a_{1s}, is the determinant (of order $n-1$) which is left when we strike out the row and column through a_{1s}. Thus a determinant of order n is defined in terms of determinants of order $n-1$, a determinant of order $n-1$ in terms of determinants of order $n-2$, and so on. We call $(-1)^{s+1} M_{1s}$ the *co-factor* of a_{1s}, written A_{1s}, so that if Δ_n is the determinant of $\mathbf{A}_n$,

$$\Delta_n = |\mathbf{A}_n| = \sum_{s=1}^{n} a_{1s} A_{1s}.$$

Sometimes det $\mathbf{A}$ is used for $|\mathbf{A}|$. The modulus signs round the matrix $\mathbf{A}_n$ are used to denote the determinant Δ_n of the matrix $\mathbf{A}_n$.

Examples. (i) $\mathbf{A}_3 \equiv \begin{pmatrix} a_{11} & a_{12} & a_{13} \\ a_{21} & a_{22} & a_{23} \\ a_{31} & a_{32} & a_{33} \end{pmatrix}$

$$|\mathbf{A}_3| = \Delta_3 = a_{11} \begin{vmatrix} a_{22} & a_{23} \\ a_{32} & a_{33} \end{vmatrix} - a_{12} \begin{vmatrix} a_{21} & a_{23} \\ a_{31} & a_{33} \end{vmatrix} + a_{13} \begin{vmatrix} a_{21} & a_{22} \\ a_{31} & a_{32} \end{vmatrix}$$

$$= a_{11}(a_{22}a_{33} - a_{23}a_{32}) - a_{12}(a_{21}a_{33} - a_{23}a_{31}) + a_{13}(a_{21}a_{32} - a_{22}a_{31})$$

[as previously defined].

$$\text{(ii) } \Delta_4 = a_{11} \begin{vmatrix} a_{22} & a_{23} & a_{24} \\ a_{32} & a_{33} & a_{34} \\ a_{42} & a_{43} & a_{44} \end{vmatrix} - a_{12} \begin{vmatrix} a_{21} & a_{23} & a_{24} \\ a_{31} & a_{33} & a_{34} \\ a_{41} & a_{43} & a_{44} \end{vmatrix}$$

$$+ a_{13} \begin{vmatrix} a_{21} & a_{22} & a_{24} \\ a_{31} & a_{32} & a_{34} \\ a_{41} & a_{42} & a_{44} \end{vmatrix} - a_{14} \begin{vmatrix} a_{21} & a_{22} & a_{23} \\ a_{31} & a_{32} & a_{33} \\ a_{41} & a_{42} & a_{43} \end{vmatrix}.$$

In listing some important properties of determinants, we shall not generally give proofs, and the student should verify that the results hold at least for Δ_3.

(1) The *transpose* of Δ_n, where rows and columns are interchanged, is equal to Δ_n;

$$\text{e.g.} \quad \begin{vmatrix} a_{11} & a_{12} & a_{13} \\ a_{21} & a_{22} & a_{23} \\ a_{31} & a_{32} & a_{33} \end{vmatrix} = \begin{vmatrix} a_{11} & a_{21} & a_{31} \\ a_{12} & a_{22} & a_{32} \\ a_{13} & a_{23} & a_{33} \end{vmatrix}.$$

Because of this property, we can replace "row" by "column" in subsequent properties without affecting their truth value.

(2) Every interchange of a pair of rows of Δ_n changes the sign of Δ_n but not its numerical value;

$$\text{e.g.} \quad \begin{vmatrix} a_{31} & a_{32} & a_{33} \\ a_{21} & a_{22} & a_{23} \\ a_{11} & a_{12} & a_{13} \end{vmatrix} = -\Delta_3 \quad \text{and} \quad \begin{vmatrix} a_{12} & a_{13} & a_{11} \\ a_{22} & a_{23} & a_{21} \\ a_{32} & a_{33} & a_{31} \end{vmatrix} = \Delta_3.$$

(row 1 and row 3 interchanged)   (col. 1 and col. 3, col. 2 and col. 3 interchanged.)

(3) If two rows of Δ_n are identical, then $\Delta_n = 0$; e.g.

$$\begin{vmatrix} a_{11} & a_{12} & a_{12} \\ a_{21} & a_{22} & a_{22} \\ a_{31} & a_{32} & a_{32} \end{vmatrix} = 0 \text{ (col. 2} \equiv \text{col. 3)}.$$

(4) If every element in a row of Δ_n has a factor k, then k is a factor of Δ_n and can be "taken out" of the row; e.g.

$$\begin{vmatrix} a_{11} & 3a_{12} & a_{13} \\ a_{21} & 3a_{22} & a_{23} \\ a_{31} & 3a_{32} & a_{33} \end{vmatrix} = 3\Delta_3.$$

(5) Two determinants *both of order n* can be "added" if all corresponding rows are the same in each determinant except for one pair, by adding corresponding elements of the unlike rows; e.g.

$$\begin{vmatrix} 4 & -3 & 9 \\ 2 & 7 & 0 \\ -1 & -5 & 4 \end{vmatrix} + \begin{vmatrix} 4 & -3 & 9 \\ 2 & 7 & 0 \\ 3 & -1 & 2 \end{vmatrix} = \begin{vmatrix} 4 & -3 & 9 \\ 2 & 7 & 0 \\ 2 & -6 & 6 \end{vmatrix}$$

(adding corresponding elements in row 3).

(6) The value of Δ_n is unchanged if any multiple of the elements of one row are added

to (or subtracted from) the corresponding elements of another row; e.g.

$$\begin{vmatrix} a_{11}-5a_{13} & a_{12} & a_{13} \\ a_{21}-5a_{23} & a_{22} & a_{23} \\ a_{31}-5a_{33} & a_{32} & a_{33} \end{vmatrix} = \Delta_3.$$

This property can often be used to ease the evaluation of a determinant, as in Example 1 below.

Example 1. $\begin{vmatrix} 133 & 38 & 95 \\ -15 & 8 & 20 \\ -3 & -4 & -11 \end{vmatrix} = \underset{\text{rule (4)}}{} \begin{vmatrix} 19 & 7 & 2 & 5 \\ -15 & 8 & 20 \\ -3 & -4 & -11 \end{vmatrix}$

$\underset{\text{row } 2-4 \times \text{row } 1}{=} \underset{\text{rule (6)}}{} 19 \begin{vmatrix} 7 & 2 & 5 \\ -43 & 0 & 0 \\ -3 & -4 & -11 \end{vmatrix} = \underset{\text{rule (2)}}{} -19 \begin{vmatrix} -43 & 0 & 0 \\ 7 & 2 & 5 \\ -3 & -4 & -11 \end{vmatrix}$

$$= -19 \times [-43(-22+20)] = -1634.$$

(Note the advantage of reducing as many elements as possible to zero.)

---

(7) By definition

$$\sum_{s=1}^{n} a_{1s} A_{1s} = \Delta_n;$$

to replace the first row by the rth row of Δ_n, keeping the other rows in the same order, involves $(r-1)$ changes of sign, by (2).

Hence

$$(-1)^{r-1} \sum_{s=1}^{n} a_{rs} A_{rs} = \Delta_n \quad \text{for } r = 1, 2, \ldots, n.$$

Now if we replace a_{1s} by a_{2s} throughout the first row of Δ_n; the new determinant is zero by (3); but the cofactors of a_{1s} are unaltered by the replacement.

Hence

$$\sum_{s=1}^{n} a_{2s} A_{1s} = 0,$$

and generalising,

$$\sum_{s=1}^{n} a_{ps} A_{qs} = \pm \Delta_n (p = q), \qquad \sum_{s=1}^{n} a_{ps} A_{qs} = 0(p \neq q).$$

For example,

$$\sum_{s=1}^{3} a_{2s} A_{1s} = a_{21}(a_{22} a_{33} - a_{23} a_{32}) - a_{22}(a_{21} a_{33} - a_{23} a_{31})$$
$$+ a_{23}(a_{21} a_{32} - a_{22} a_{31}) = 0.$$

It is often possible to apply the factor theorem to determinants.

Example 2. Factorize $\Delta = \begin{vmatrix} bc & ca & ab \\ a^2 & b^2 & c^2 \\ 1 & 1 & 1 \end{vmatrix}$.

(a) Δ is symmetrical in a, b and c because the cyclic interchange $a \to b, b \to c, c \to a$ gives the determinant

$$\begin{vmatrix} ca & ab & bc \\ b^2 & c^2 & a^2 \\ 1 & 1 & 1 \end{vmatrix}$$

and this determinant, being the result of two interchanges of columns from Δ [rule (2)], is equal to Δ. The factors as a whole of Δ are therefore symmetrical in a, b and c.

(b) The substitution $a = b$ in Δ gives the determinant

$$\begin{vmatrix} bc & cb & b^2 \\ b^2 & b^2 & c^2 \\ 1 & 1 & 1 \end{vmatrix}$$

and this determinant is equal to zero because the first and second columns are identical [rule (3)]

$$\Rightarrow (a - b) \text{ and, by symmetry, } (b - c) \text{ and } (c - a) \text{ are factors of } \Delta.$$

(c) Δ is of the fourth degree in a, b and c, and therefore Δ has a fourth linear factor which must be a symmetrical one. The only such factor is $(a + b + c)$

$$\Rightarrow \Delta \equiv k(a - b)(b - c)(c - a)(a + b + c), \text{ where } k \text{ is constant.}$$

Inspection of Δ shows that the term $+ b^3 c$ occurs in the expansion. Hence $k = -1$ and

$$\Delta = -(a - b)(b - c)(c - a)(a + b + c).$$

Example 3. Factorize $\Delta = \begin{vmatrix} x^2 & x & 1 \\ 1 & x^2 & x \\ x & 1 & x^2 \end{vmatrix}$.

Taking col. $1 + $ col. $2 + $ col. 3 and extracting the common factor $(x^2 + x + 1)$ from the first column gives

$$\Delta = (x^2 + x + 1) \begin{vmatrix} 1 & x & 1 \\ 1 & x^2 & x \\ 1 & 1 & x^2 \end{vmatrix} .$$

The factor theorem shows $(x - 1)$ to be a factor of Δ and hence, taking col. 2–col. 3 and extracting the factor $(x - 1)$,

$$\Delta = (x^2 + x + 1)(x - 1) \begin{vmatrix} 1 & 1 & 1 \\ 1 & x & x \\ 1 & -(x + 1) & x^2 \end{vmatrix} .$$

The factor theorem now shows $(x - 1)$ to be a factor and hence, taking row 1–row 2 and extracting the factor $(x - 1)$,

$$\Delta = (x^2 + x + 1)(x - 1)^2 \begin{vmatrix} 0 & -1 & -1 \\ 1 & x & x \\ 1 & -(x + 1) & x^2 \end{vmatrix}$$

$$\Leftrightarrow \Delta = (x^2 + x + 1)(x - 1)^2 (x^2 - x + x + 1 + x)$$

$$= (x^2 + x + 1)^2 (x - 1)^2.$$

Exercise 15.1

Evaluate

1. $\begin{vmatrix} 1 & 3 & 7 \\ 2 & 4 & 9 \\ 3 & 5 & 9 \end{vmatrix}$ **2.** $\begin{vmatrix} -18 & 1 & 9 \\ 5 & -6 & -3 \\ 3 & -2 & 6 \end{vmatrix}$ **3.** $\begin{vmatrix} 2 & -3 & 5 \\ 10 & -15 & 24 \\ 8 & -12 & 21 \end{vmatrix}$ **4.** $\begin{vmatrix} -4 & 12 & 5 \\ -5 & 18 & 6 \\ -6 & 27 & 7 \end{vmatrix}$

5. $\begin{vmatrix} 8 & -2 & -4 \\ 7 & 1 & -2 \\ 6 & 4 & 0 \end{vmatrix}$ **6.** $\begin{vmatrix} 21 & 5 & -6 \\ 10 & 3 & -3 \\ 25 & 7 & -8 \end{vmatrix}$ **7.** $\begin{vmatrix} 17 & 13 & 12 \\ 7 & 7 & 7 \\ 5 & 11 & 6 \end{vmatrix}$ **8.** $\begin{vmatrix} 26 & 16 & 7 \\ -1 & 7 & 4 \\ 80 & 71 & 32 \end{vmatrix}$

9. $\begin{vmatrix} 1 & x & x^2+1 \\ 1 & y & xy \\ 1 & -(x+y) & -x^2 \end{vmatrix}$ **10.** $\begin{vmatrix} a & 2b & -c \\ -3a & 4b & -4c \\ -a & 6b & -6c \end{vmatrix}.$

11. $\begin{vmatrix} -2 & 1 & -1 & 2 \\ 1 & -2 & 1 & -6 \\ 3 & 3 & 2 & 10 \\ 4 & 1 & -3 & -11 \end{vmatrix}$ **12.** $\begin{vmatrix} 3 & 1 & -1 & -2 \\ 4 & -2 & 1 & -6 \\ -1 & 3 & 2 & 10 \\ 0 & 1 & -3 & -11 \end{vmatrix}$ **13.** $\begin{vmatrix} 3 & -2 & -1 & -2 \\ 4 & 1 & 1 & -6 \\ -1 & 3 & 2 & 10 \\ 0 & 4 & -3 & -11 \end{vmatrix}$

14. $\begin{vmatrix} 3 & -2 & 1 & -2 \\ 4 & 1 & -2 & -6 \\ -1 & 3 & 3 & 10 \\ 0 & 4 & 1 & -11 \end{vmatrix}$ **15.** $\begin{vmatrix} 3 & -2 & 1 & -1 \\ 4 & 1 & -2 & 1 \\ -1 & 3 & 3 & 2 \\ 0 & 4 & 1 & -3 \end{vmatrix}$

Factorise each of the following determinants:

16. $\begin{vmatrix} x^2 & y^2 & z^2 \\ x & y & z \\ 1 & 1 & 1 \end{vmatrix}$ **17.** $\begin{vmatrix} x^3 & y^3 & 1 \\ x & y & 1 \\ 1 & 1 & 1 \end{vmatrix}$ **18.** $\begin{vmatrix} x^3 & y^3 & z^3 \\ x^2 & y^2 & z^2 \\ 1 & 1 & 1 \end{vmatrix}$ **19.** $\begin{vmatrix} ab & bc & ca \\ c & a & b \\ 1 & 1 & 1 \end{vmatrix}$

20. $\begin{vmatrix} a & b & c \\ c & a & b \\ b & c & a \end{vmatrix}$ **21.** $\begin{vmatrix} x & xy & x^2 \\ y & yz & y^2 \\ z & zx & z^2 \end{vmatrix}$ **22.** $\begin{vmatrix} a-x & 2a & 2a \\ 2(x-c) & x-2c-a & 2(x-c) \\ 2c & 2c & c-x \end{vmatrix}$

23. $\begin{vmatrix} -x & x+y & x+z \\ x+y & -y & y+z \\ y+z & x+z & -z \end{vmatrix}$ **24.** $\begin{vmatrix} c & a-b & -c \\ b-2c & -2a-b & a-c \\ c & a+b & c \end{vmatrix}.$

25. Solve the equation $\begin{vmatrix} x & 2 & 3 \\ 3 & 3+x & 5 \\ 7 & 8 & 8+x \end{vmatrix} = 0.$

26. Solve the equation

$$\begin{vmatrix} x+5 & 3 & 2 \\ 2 & x+3 & 5 \\ 1 & 2 & x+7 \end{vmatrix} = 0.$$

27. Show that the equation

$$\begin{vmatrix} x+1 & 2 & 5 \\ 3 & -x & 1 \\ 1 & -2 & x+1 \end{vmatrix} = 0$$

has only one real root, and find that root.

28. Solve the equation

$$\begin{vmatrix} x & 2 & 3 \\ 4 & 1 & x \\ 1 & 2+x & 2 \end{vmatrix} = 0.$$

29. (a) Solve for λ,

$$\begin{vmatrix} 3-\lambda & -5 & -1 \\ 2 & 2-\lambda & 0 \\ 2 & 5 & 1 \end{vmatrix} = 0.$$

(b) Factorise completely

$$\Delta = \begin{vmatrix} x & y & z \\ x^2 & y^2 & z^2 \\ x^3 & y^3 & z^3 \end{vmatrix}.$$

30. (i) Express the determinant

$$\begin{vmatrix} 0 & a & b & c \\ a & 0 & c & b \\ b & c & 0 & a \\ c & b & a & 0 \end{vmatrix}$$

as the product of four linear factors.

(ii) Find the roots (real or complex) of the equation

$$\begin{vmatrix} x^2 & 1 & 2 \\ 4 & 2-2x & x \\ 1 & 5-4x & 2x \end{vmatrix} = 0.$$

(C.)

31. Prove that

$$\begin{vmatrix} 1 & 1 & 1 & 1 & 1 \\ 1 & x & x^2 & x^3 & x^4 \\ 1 & x^2 & x^4 & x^6 & x^8 \\ 1 & x^3 & x^6 & x^9 & x^{12} \\ 1 & x^4 & x^8 & x^{12} & x^{16} \end{vmatrix} = x^{10}(x-1)^4(x^2-1)^3(x^3-1)^2(x^4-1).$$

(C.)

32. Prove that $a^2+b^2+c^2$ is a factor of the determinant

$$\begin{vmatrix} a^2 & (b+c)^2 & bc \\ b^2 & (c+a)^2 & ca \\ c^2 & (a+b)^2 & ab \end{vmatrix}.$$

Hence, or otherewise, factorise the determinant completely.   (C.)

33. Given that all the elements of the determinant

$$\begin{vmatrix} a_1 & a_2 & a_3 & a_4 \\ a_4 & a_1 & a_2 & a_3 \\ a_3 & a_4 & a_1 & a_2 \\ a_2 & a_3 & a_4 & a_1 \end{vmatrix}$$

are real, express the determinant as the product of two real linear factors and one real quadratic factor.

(C.)

34. Without exapanding the determinant show that $(ab+cd+ef)$ is a factor of

$$\begin{vmatrix} 0 & a & f & c \\ a & 0 & d & -e \\ -f & d & 0 & -b \\ c & e & -b & 0 \end{vmatrix}$$

Deduce or show otherwise that the determinant is equal to $(ab+cd+ef)^2$.   (C.)

35. (i) If $\alpha, \beta, \gamma, \delta$ are the roots of the equation

$$px^4+qx^3+rx^2+x+1 = 0,$$

prove that

$$\begin{vmatrix} 1+\alpha & 1 & 1 & 1 \\ 1 & 1+\beta & 1 & 1 \\ 1 & 1 & 1+\gamma & 1 \\ 1 & 1 & 1 & 1+\delta \end{vmatrix} = 0.$$

(ii) Prove that

$$\begin{vmatrix} 1 & a & a^2 & a^3 + bcd \\ 1 & b & b^2 & b^3 + cda \\ 1 & c & c^2 & c^3 + dab \\ 1 & d & d^2 & d^3 + abc \end{vmatrix} = 0.$$

(C.)

15:2 MATRICES—PROPERTIES AND SPECIAL TYPES

An $m \times n$ matrix $A = (a_{rs})$ is a rectangular array of elements a_{rs} arranged in m rows and n columns, thus

$$A = \begin{pmatrix} a_{11} & a_{12} & \cdots & a_{1n} \\ a_{21} & a_{22} & \cdots & a_{2n} \\ \cdots & \cdots & \cdots & \cdots \\ a_{m1} & a_{m2} & \cdots & a_{mn} \end{pmatrix}.$$

For convenience we shall list a number of definitions, though their usefulness may not be apparent until later.

(1) We *define* the operation of *scalar multiplication* by

$$kA = \begin{pmatrix} ka_{11} & ka_{12} & \cdots & ka_{1n} \\ ka_{21} & ka_{22} & \cdots & ka_{2n} \\ \cdots & \cdots & \cdots & \cdots \\ ka_{m1} & ka_{m2} & \cdots & ka_{mn} \end{pmatrix},$$

i.e. kA is obtained from A on multiplying *every* element of the matrix A by k.

Compare with property (4) of determinants, where $k|A|$ means that k is a factor of the elements of a *single* row or column.

(2) A *zero matrix* $\mathbf{0}$ is a matrix of which all the elements are 0's; e.g.

$$\begin{pmatrix} 0 & 0 \\ 0 & 0 \\ 0 & 0 \end{pmatrix}.$$

(3) We *define* the sum of two matrices $A = (a_{ij})$, $B = (b_{ij})$ to be

$$A + B = \begin{pmatrix} a_{11} + b_{11} & a_{12} + b_{12} & \cdots & a_{1n} + b_{1n} \\ a_{21} + b_{21} & a_{22} + b_{22} & \cdots & a_{2n} + b_{2n} \\ \cdots & \cdots & \cdots & \cdots \\ a_{m1} + b_{m1} & a_{m2} + b_{m2} & \cdots & a_{mn} + b_{mn} \end{pmatrix} = B + A,$$

i.e. we add corresponding elements in the two matrices. We can only do this if the two matrices A and B have the same order, i.e. if each contains m rows and n columns. We cannot, therefore, add the matrices

$$\begin{pmatrix} 1 & 2 \\ 3 & 4 \end{pmatrix}, \quad \begin{pmatrix} 1 & 2 & 3 \\ 4 & 5 & 6 \\ 7 & 8 & 9 \end{pmatrix}, \quad \begin{pmatrix} 1 & 2 \\ 3 & 4 \\ 5 & 6 \end{pmatrix},$$

since they are all of different orders.

From this definition it is clear that addition on matrices is commutative, since $\mathbf{A} + \mathbf{B} = \mathbf{B} + \mathbf{A}$, and associative, since

$$(\mathbf{A} + \mathbf{B}) + \mathbf{C} = \mathbf{A} + (\mathbf{B} + \mathbf{C}).$$

(4) We *define* the difference between two matrices, i.e. *subtraction*, by the relations

$$\mathbf{A} - \mathbf{B} = \mathbf{A} + (-1)\mathbf{B},$$

where we have added to $\mathbf{A}$ the scalar multiple (-1) of $\mathbf{B}$. We define *equal matrices* to be such that $\mathbf{A} - \mathbf{B} = \mathbf{0}$, where $\mathbf{0}$ stands for the zero matrix of the same order as $\mathbf{A}$ and $\mathbf{B}$.

(5) If $\mathbf{A} = (a_{ik})$, $i = 1, 2, \ldots, m$, $k = 1, 2, \ldots, n$,
    $\mathbf{B} = (b_{kj})$, $k = 1, 2, \ldots, n$, $j = 1, 2, \ldots, p$,
then the product $\mathbf{C} = \mathbf{AB}$ is defined by $\mathbf{C} = (c_{ij})$, where

$$c_{ij} = \sum_{k=1}^{n} a_{ik} b_{kj}.$$

To form the product $\mathbf{AB}$ we can speak of *premultiplying* $\mathbf{B}$ by $\mathbf{A}$ or of *postmultiplying* $\mathbf{A}$ by $\mathbf{B}$.

Note that

(i) $\mathbf{C}$ is a matrix with m rows and p columns;
(ii) for the product $\mathbf{AB}$ to exist, the number of columns of $\mathbf{A}$ must be the same as the number of rows of $\mathbf{B}$;
(iii) the fact that $\mathbf{AB}$ exists does not even imply the existence of $\mathbf{BA}$ and, in general, $\mathbf{AB} \neq \mathbf{BA}$. Thus multiplication on matrices is not commutative, but the associative law holds, i.e. $\mathbf{A}(\mathbf{BD}) = (\mathbf{AB})\mathbf{D}$, provided of course that the matrices involved are compatible for multiplication.

(6) An *identity* or *unit matrix* $\mathbf{I}$ is a *square* matrix (a_{rs}) such that $a_{rs} = 0$ for $r \neq s$, $a_{rs} = 1$ for $r = s$; e.g.

$$\begin{pmatrix} 1 & 0 & 0 & 0 \\ 0 & 1 & 0 & 0 \\ 0 & 0 & 1 & 0 \\ 0 & 0 & 0 & 1 \end{pmatrix}.$$

The elements a_{rr} of a square matrix form the *leading diagonal*.

(7) A *singular matrix* $\mathbf{M}$ is a square matrix such that $|\mathbf{M}| = 0$.

(8) The *inverse matrix* $\mathbf{M}^{-1}$ of a *square non-singular* matrix $\mathbf{M}$ is such that

$$\mathbf{M}^{-1}\mathbf{M} = \mathbf{MM}^{-1} = \mathbf{I}.$$

A singular matrix has no inverse.

(9) The *transpose* of an $m \times n$ matrix $\mathbf{Q}$ is an $n \times m$ matrix $\mathbf{Q}^T$ such that the rows and columns of $\mathbf{Q}$ have been interchanged; hence a_{rs} for $\mathbf{Q}$ is a_{sr} for $\mathbf{Q}^T$; e.g.

$$\mathbf{Q} = \begin{pmatrix} a_{11} & a_{12} & a_{13} \\ a_{21} & a_{22} & a_{23} \\ a_{31} & a_{32} & a_{33} \\ a_{41} & a_{42} & a_{43} \end{pmatrix} \Leftrightarrow \mathbf{Q}^T = \begin{pmatrix} a_{11} & a_{21} & a_{31} & a_{41} \\ a_{12} & a_{22} & a_{32} & a_{42} \\ a_{13} & a_{23} & a_{33} & a_{43} \end{pmatrix}.$$

Sometimes a dash (prime) is used to denote the transpose, e.g. $\mathbf{A}'$.

(10) The *adjoint* (adj $\mathbf{A}$) of an $n \times n$ matrix $\mathbf{A}$ (a_{rs}) is the transpose of the matrix formed from the cofactors of a_{rs} in $|a_{rs}|$.

(11) A matrix $\mathbf{M}$ is symmetric if $\mathbf{M} = \mathbf{M}^T$; e.g.

$$\begin{pmatrix} a_{11} & a_{12} & a_{13} \\ a_{12} & a_{22} & a_{23} \\ a_{13} & a_{23} & a_{23} \end{pmatrix} \text{ is symmetric;}$$

clearly a symmetric matrix must be square.

(12) A *diagonal matrix* $\mathbf{M} = (a_{rs})$ is such that $a_{rs} = 0$ when $r \neq s$; e.g.

$$\begin{pmatrix} 2 & 0 & 0 & 0 \\ 0 & 0 & 0 & 0 \\ 0 & 0 & -7 & 0 \\ 0 & 0 & 0 & 4 \end{pmatrix}.$$

Such a matrix may be written (diag. 2, 0, -7, 4).

If $\mathbf{D} = (\text{diag. } a_{11}, a_{22}, \dots, a_{nn})$, $\mathbf{D}^k = (\text{diag. } a_{11}{}^k, a_{22}{}^k, \dots, a_{nn}{}^k)$.

(13) A *triangular matrix* is such that all its elements *below* the leading diagonal (*upper* triangular) or *above* the leading diagonal (*lower* triangular) are 0's; e.g.

$$\begin{pmatrix} 9 & -1 & 0 \\ 0 & 3 & 11 \\ 0 & 0 & 1 \end{pmatrix} \text{ is upper triangular,} \qquad \begin{pmatrix} -5 & 0 & 0 \\ 8 & 2 & 0 \\ -1 & -4 & 1 \end{pmatrix} \text{ is lower triangular.}$$

(14) An *orthogonal matrix* is a square matrix such that the row and column vectors are mutually perpendicular, and of unit magnitude; e.g.

$$\begin{pmatrix} \tfrac{3}{5} & -\tfrac{4}{5} \\ \tfrac{4}{5} & \tfrac{3}{5} \end{pmatrix}, \quad \begin{pmatrix} \tfrac{1}{3} & \tfrac{2}{3} & \tfrac{2}{3} \\ \tfrac{2}{3} & \tfrac{1}{3} & -\tfrac{2}{3} \\ \tfrac{2}{3} & -\tfrac{2}{3} & \tfrac{1}{3} \end{pmatrix} \text{ are each orthogonal.}$$

A matrix with perpendicular vectors can be "normalised" to make it orthogonal, e.g.

$$\begin{pmatrix} 5 & -12 \\ 12 & 5 \end{pmatrix} \text{ becomes } \begin{pmatrix} \tfrac{5}{13} & -\tfrac{12}{13} \\ \tfrac{12}{13} & \tfrac{5}{13} \end{pmatrix}.$$

It will be shown later that if $\mathbf{P}$ is an orthogonal matrix, $\mathbf{P}^T = \mathbf{P}^{-1}$; and this property can be used as the definition of orthogonality.

15:3 TO FIND THE INVERSE OF A SQUARE NON-SINGULAR MATRIX

Let $\mathbf{M}$ be the $n \times n$ matrix (a_{rs}). Consider the product of $\mathbf{M}$ with its adjoint matrix $(\text{adj } \mathbf{M}) = (A_{sr})$.

$$(a_{rs})(A_{sr}) = \begin{pmatrix} \sum\limits_{r=1}^{n} a_{1r}A_{1r} & \sum\limits_{r=1}^{n} a_{1r}A_{2r} \cdots \cdots \\ \sum\limits_{r=1}^{n} a_{2r}A_{1r} & \sum\limits_{r=1}^{n} a_{2r}A_{2r} \cdots \cdots \\ \cdots \cdots \cdots \cdots \cdots \cdots \cdots \\ \cdots \cdots \cdots \cdots \cdots \cdots \cdots \end{pmatrix}.$$

But from §15:1(7),

$$\sum_{r=1}^{n} a_{sr}A_{sr} = |\mathbf{M}| \quad \forall s \in \{1, 2, \ldots, n\}$$

$$\sum_{r=1}^{n} a_{sr}A_{qr} = 0, \quad (q \neq s).$$

Hence $(a_{rs})(A_{sr}) = |\mathbf{M}|\mathbf{I}$ and similarly $(A_{sr})(a_{rs}) = |\mathbf{M}|\mathbf{I}$. It follows that, if $|\mathbf{M}| \neq 0$, an inverse of $\mathbf{M}$ is

$$\frac{1}{|\mathbf{M}|}(\text{adj } \mathbf{M}) = \mathbf{M}^{-1}.$$

We now prove that this inverse is unique. For if $\mathbf{B}$ is another matrix such that $\mathbf{BM} = \mathbf{MB} = \mathbf{I}$,

$$\mathbf{B} - \mathbf{M}^{-1} = \mathbf{I}(\mathbf{B} - \mathbf{M}^{-1}) = \mathbf{M}^{-1}\mathbf{M}(\mathbf{B} - \mathbf{M}^{-1}) = \mathbf{M}^{-1}(\mathbf{MB} - \mathbf{I}) = \mathbf{M}^{-1}(\mathbf{I} - \mathbf{I}) = 0$$
$$\Rightarrow \mathbf{B} = \mathbf{M}^{-1}.$$

Example 1. $\begin{pmatrix} a & b \\ c & d \end{pmatrix}^{-1} = \dfrac{1}{ad - bc} \begin{pmatrix} d & -b \\ -c & a \end{pmatrix}$ provided $ad - bc \neq 0$.

Example 2. $\begin{pmatrix} -8 & -2 & -1 \\ 0 & 3 & -4 \\ -7 & -1 & -2 \end{pmatrix}^{-1} = \dfrac{1}{3}\begin{pmatrix} -10 & -3 & 11 \\ 28 & 9 & -32 \\ 21 & 6 & -24 \end{pmatrix}.$

Check $\begin{pmatrix} -8 & -2 & -1 \\ 0 & 3 & -4 \\ -7 & -1 & -2 \end{pmatrix}\begin{pmatrix} -10 & -3 & 11 \\ 28 & 9 & -32 \\ 21 & 6 & -24 \end{pmatrix} = 3\mathbf{I}.$

Other methods of finding an inverse matrix will be discussed later.

We can now "divide" by a non-singular matrix $\mathbf{A}$. To do this we multiply by the inverse, $\mathbf{A}^{-1}$. It should be noted that, if $\mathbf{AB} = \mathbf{C}$, then

$$\mathbf{C}^{-1} = (\mathbf{AB})^{-1} = \mathbf{B}^{-1}\mathbf{A}^{-1}.$$

This follows from the relations

$$\mathbf{C}^{-1}(\mathbf{AB}) = \mathbf{B}^{-1}\mathbf{A}^{-1}\mathbf{AB} = \mathbf{B}^{-1}\mathbf{IB} = \mathbf{B}^{-1}\mathbf{B} = \mathbf{I},$$

and

$$(\mathbf{AB})\mathbf{C}^{-1} = \mathbf{ABB}^{-1}\mathbf{A}^{-1} = \mathbf{AIA}^{-1} = \mathbf{AA}^{-1} = \mathbf{I}.$$

By induction this result can be generalised to

$$(\mathbf{A}_1\mathbf{A}_2 \ldots \mathbf{A}_n)^{-1} = \mathbf{A}_n^{-1} \ldots \mathbf{A}_2^{-1}\mathbf{A}_1^{-1}.$$

Example 1. The inverse of the transpose of a non-singular matrix is the transpose of the inverse, since $(\text{adj } \mathbf{A})^T = \text{adj } \mathbf{A}^T$ and $\det \mathbf{A} = \det \mathbf{A}^T$.

Therefore
$$(\mathbf{A}^T)^{-1} = (\mathbf{A}^{-1})^T.$$

Example 2. The inverse of a symmetric non-singular matrix is also symmetric. This follows from the previous result. For a symmetric matrix $\mathbf{A} = \mathbf{A}^T$.

Therefore
$$(\mathbf{A}^{-1})^T = (\mathbf{A}^T)^{-1} = \mathbf{A}^{-1}.$$

Exercise 15.3

1. Calculate the inverse of the matrix

$$\begin{pmatrix} 1 & 2 & 3 \\ -3 & 1 & 0 \\ 2 & 0 & 1 \end{pmatrix}.$$ (O.C.S.M.P.)

2. Find the matrix A such that $\mathbf{AB} = \mathbf{C}$ where

$$\mathbf{B} = \begin{pmatrix} 1 & 3 & 5 \\ 2 & 4 & 6 \end{pmatrix} \quad \text{and} \quad \mathbf{C} = (3 \quad 5 \quad 7).$$ (O.C.S.M.P.)

3. A 2×2 matrix $\mathbf{M}$ has the property that

$$\mathbf{M}\begin{pmatrix} a \\ b \end{pmatrix} = \begin{pmatrix} p \\ q \end{pmatrix} \quad \text{and} \quad \mathbf{M}\begin{pmatrix} c \\ d \end{pmatrix} = \begin{pmatrix} r \\ s \end{pmatrix}.$$

Prove that, provided a, b, c, d satisfy a certain restrictions,

$$\mathbf{M} = \mathbf{V}\mathbf{U}^{-1},$$

where

$$\mathbf{U} = \begin{pmatrix} a & c \\ b & d \end{pmatrix} \quad \text{and} \quad \mathbf{V} = \begin{pmatrix} p & r \\ q & s \end{pmatrix}.$$

State the restriction on a, b, c, d.   (O.C.S.M.P.)

4. Express x and t in terms of y and z if the matrix

$$\mathbf{A} = \begin{pmatrix} x & y \\ z & t \end{pmatrix}$$

satisfies the relation

$$\mathbf{A}^2 - 2\mathbf{A} + \mathbf{I} = 0,$$

where

$$I = \begin{pmatrix} 1 & 0 \\ 0 & 1 \end{pmatrix}, \quad \mathbf{0} = \begin{pmatrix} 0 & 0 \\ 0 & 0 \end{pmatrix}.$$

Verify that, if x, y, z, t are real, then y and z, if not zero, are opposite in sign.

5. Let

$$A = \begin{pmatrix} 1 & 1 & 2 \\ 0 & 1 & 2 \\ 2 & 2 & 5 \end{pmatrix}, \quad B = \begin{pmatrix} 1 & 1 & 2 \\ 0 & 1 & 2 \\ 0 & 0 & 1 \end{pmatrix}, \quad C = \begin{pmatrix} 1 & 0 & 0 \\ 0 & 1 & 2 \\ 0 & 0 & 1 \end{pmatrix}.$$

Find elementary matrices E_1, E_2 and E_3 such that

$$E_1 A = B, \quad E_2 B = C \quad \text{and} \quad E_3 C = I,$$

where I is the unit matrix. Hence, or otherwise, find the inverse matrix of A.   (O.C.S.M.P.)

6. The matrix X is said to be a real square root of the matrix A if $X^2 = A$ and if all the elements of X are real. Show that A has

(i) four real square roots if $A = \begin{pmatrix} 4 & 0 \\ 6 & 9 \end{pmatrix}$,

(ii) two real square roots if $A = \begin{pmatrix} 4 & 0 \\ 6 & 4 \end{pmatrix}$,

(iii) an infinite number of real square roots if $A = \begin{pmatrix} 4 & 0 \\ 0 & 4 \end{pmatrix}$.   (C.)

7. The polynomial $ax^2 + bx + c$ can be represented by the vector

$$\begin{pmatrix} a \\ b \\ c \end{pmatrix}.$$

Find a 3×3 matrix M which premultiplies this vector to give the vector representing the derivative of the polynomial. Check that $M^3 = \mathbf{0}$ and comment on the significance of this result.   (O.C.S.M.P.)

8. P and Q are two non-singular 2×2 matrices and P^T and Q^T are the transposes of P and Q respectively. Show that $(PQ)^T = Q^T P^T$.

If also $P^{-1} = P^T$ and $Q^{-1} = Q^T$, show that $(PQ)^{-1} = (PQ)^T$. Find all possible values of the determinant of (PQ).   (L.)

*9. The matrix $M(\theta)$ is defined by

$$M(\theta) = \begin{pmatrix} \cosh \theta - k \sinh \theta & (1 + k) \sinh \theta \\ (1 - k) \sinh \theta & \cosh \theta + k \sinh \theta \end{pmatrix}.$$

(a) Show that

$$M(\theta) M(\phi) = M(\theta + \phi) = M(\phi) M(\theta).$$

(b) Find the value of the determinant of $M(\theta)$ and hence write down the inverse of $M(\theta)$.

(c) Show that $\{M(\theta)\}^n = M(n\theta)$ for all positive integers n.   (L.)

(See § 16:7 for the definitions of $\cosh \theta$ and $\sinh \theta$.)

10. (i) If $A = \begin{pmatrix} 2 & 3 \\ 5 & 8 \end{pmatrix}$, find numbers p and q such that $A(A - pI) = qI$, where I is the 2×2 identity matrix.

(ii) Find the inverse of the matrix

$$M = \begin{pmatrix} 2 & 1 & 1 \\ 1 & -1 & 2 \\ 3 & 2 & -1 \end{pmatrix}.$$

If $\mathbf{MX} = \mathbf{N}$, where $\mathbf{X} = \begin{pmatrix} x \\ y \\ z \end{pmatrix}$ and $\mathbf{N} = \begin{pmatrix} a \\ 2a \\ b \end{pmatrix}$, find x, y and z in terms of a and b.   (L.)

11. If $\mathbf{A}(x)$ denotes the matrix

$$\begin{pmatrix} 2-x & 2x-2 \\ 1-x & 2x-1 \end{pmatrix},$$

prove that $\mathbf{A}(x)$ is singular if and only if $x = 0$. Prove also that $\mathbf{A}(x)\mathbf{A}(y) = \mathbf{A}(xy)$ and hence show that the square of $\mathbf{A}(-1)$ is the identity matrix.

Find the inverse of $\mathbf{A}(2)$.   (L.)

12. If

$$\mathbf{a} \equiv \begin{pmatrix} a \\ b \\ c \end{pmatrix},$$

so that the transpose $\mathbf{a}'$ is given by

$$\mathbf{a}' = (a, b, c),$$

write down the 3×3 matrix $\mathbf{aa}'$ and prove that the determinant of this matrix has value zero.

Obtain the analogous result when

$$\mathbf{a} \equiv \begin{pmatrix} a & 0 \\ b & 1 \\ c & 2 \end{pmatrix}.$$ (O.C.S.M.P.)

13. The letters $\mathbf{A}, \mathbf{B}, \mathbf{C}, \mathbf{D}, \mathbf{E}$ denote 2×2 matrices with entries $a_{ij}, b_{ij}, c_{ij}, d_{ij}, e_{ij}$, respectively; for instance,

$$\mathbf{A} = \begin{pmatrix} a_{11} & a_{12} \\ a_{21} & a_{22} \end{pmatrix}.$$

The identity and zero matrices are denoted by $\mathbf{I}$ and $\mathbf{O}$. Thus the 4×4 matrix

$$\mathbf{M} = \begin{pmatrix} a_{11} & a_{12} & 1 & 0 \\ a_{21} & a_{22} & 0 & 1 \\ 0 & 0 & b_{11} & b_{12} \\ 0 & 0 & b_{21} & b_{22} \end{pmatrix}$$

can also be written as

$$\mathbf{M} = \begin{pmatrix} \mathbf{A} & \mathbf{I} \\ \mathbf{O} & \mathbf{B} \end{pmatrix}.$$

Verify, by using the usual algebraic rules for 4×4 matrices, that

$$\begin{pmatrix} \mathbf{A} & \mathbf{I} \\ \mathbf{O} & \mathbf{B} \end{pmatrix}\begin{pmatrix} \mathbf{C} & \mathbf{D} \\ \mathbf{O} & \mathbf{E} \end{pmatrix} = \begin{pmatrix} \mathbf{AC} & \mathbf{AD}+\mathbf{E} \\ \mathbf{O} & \mathbf{BE} \end{pmatrix}.$$

Using the principle of mathematical induction, prove that

$$\mathbf{M}^n = \begin{pmatrix} \mathbf{A}^n & \mathbf{P}_n \\ \mathbf{O} & \mathbf{B}^n \end{pmatrix},$$

where

$$\mathbf{P}_n = \mathbf{A}^{n-1} + \mathbf{A}^{n-2}\mathbf{B} + \ldots + \mathbf{A}^{n-r}\mathbf{B}^{r-1} + \ldots + \mathbf{B}^{n-1}.$$

Find $\mathbf{K}^n$ when

$$\mathbf{K} = \begin{pmatrix} 1 & k \\ 0 & 1 \end{pmatrix}.$$

and so evaluate $\mathbf{M}^n$ when

$$\mathbf{M} = \begin{pmatrix} 1 & 1 & 1 & 0 \\ 0 & 1 & 0 & 1 \\ 0 & 0 & 1 & 2 \\ 0 & 0 & 0 & 1 \end{pmatrix}.$$

(O.C.S.M.P.)

15:4 SOLUTION OF HOMOGENEOUS LINEAR EQUATIONS

We shall consider a set of n linear (first degree) equations in n unknowns

$$a_{11}x_1 + a_{12}x_2 + \ldots + a_{1n}x_n = b_1,$$
$$\ldots\ldots\ldots\ldots\ldots\ldots\ldots\ldots\ldots\ldots\ldots$$
$$\ldots\ldots\ldots\ldots\ldots\ldots\ldots\ldots\ldots\ldots\ldots$$
$$a_{n1}x_1 + a_{n2}x_2 + \ldots + a_{nn}x_n = b_n,$$

which can be written in matrix form

$$\begin{pmatrix} a_{11} & \ldots\ldots & a_{1n} \\ & \ldots\ldots\ldots & \\ & \ldots\ldots\ldots & \\ a_{n1} & \ldots\ldots & a_{nn} \end{pmatrix} \begin{pmatrix} x_1 \\ \cdot \\ \cdot \\ x_n \end{pmatrix} = \begin{pmatrix} b_1 \\ \cdot \\ \cdot \\ b_n \end{pmatrix}$$

or $$\mathbf{Ax} = \mathbf{b}.$$

We shall first consider the case $\mathbf{b} = \mathbf{0}$, making the equations *homogeneous* in $x_1, x_2, \ldots, x_n$.

If $|\mathbf{A}| \neq 0$, $\mathbf{A}$ has an inverse $\mathbf{A}^{-1}$

$$\Rightarrow \mathbf{A}^{-1}\mathbf{Ax} = \mathbf{A}^{-1}\mathbf{0} = \mathbf{0} \Rightarrow \mathbf{x} = \mathbf{0}.$$

Thus if $\mathbf{A}$ is non-singular, the equations have only the *trivial* solution

$$x_1 = x_2 = \ldots = x_n = 0.$$

Example. For the equations

$$5x - y + 7z = 0, \tag{1}$$
$$3x + 8y - 2z = 0, \tag{2}$$
$$3x + 8y + z = 0, \tag{3}$$

$$\begin{vmatrix} 5 & -1 & 7 \\ 3 & 8 & -2 \\ 3 & 8 & 1 \end{vmatrix} = \begin{vmatrix} 5 & -1 & 7 \\ 3 & 8 & -2 \\ 0 & 0 & 3 \end{vmatrix} = 129 \neq 0.$$

Clearly from (2) and (3) $z = 0 \Rightarrow x = y = 0$.

Now suppose that $\mathbf{A}$ is singular so that $|\mathbf{A}| = 0$, so that, by §15:1(7),

$$|\mathbf{A}| = \sum_{r=1}^{n} a_{1r}A_{1r} = 0$$

and also
$$\sum_{r=1}^{n} a_{sr}A_{1r} = 0 \quad \text{for } s = 2, 3, \ldots, n.$$

Hence $x_1 = A_{11}, x_2 = A_{12}, \ldots, x_n = A_{1n}$ is a non-trivial solution of the equations provided that the cofactors are not all zero. Similarly any set of cofactors of $\mathbf{A}$, $\{A_{s1}, A_{s2}, \ldots, A_{sn}\}$, will be a solution, and clearly $\{kA_{s1}, \ldots, kA_{sn}\}$ will be a solution $\forall k \in \mathbb{R}$.

If all the determinants of order $(n - r)$ in $|\mathbf{A}|$ are zero, the equations reduce to $(n - r - 1)$ independent equations.

Example 1. Show that the equations
$$
\begin{aligned}
3x + y + 2z &= 0, \\
x - 2y - 4z &= 0, \\
7x + 3y + 6z &= 0,
\end{aligned}
$$

are consistent.

The determinant $\Delta = \begin{vmatrix} 3 & 1 & 2 \\ 1 & -2 & -4 \\ 7 & 3 & 6 \end{vmatrix} = 0$ since the last column is twice the second column. The cofactors

of the last column are, $17, -2, -7$ respectively. If the three equations are respectively multiplied by $17, -2, -7$ and added the coefficients of x, y and z vanish; this shows that the equations are consistent. In fact

$$\text{adj } \mathbf{A} = \begin{pmatrix} 0 & 0 & 0 \\ -34 & 4 & 14 \\ 17 & -2 & -7 \end{pmatrix}.$$

Each column of adj $\mathbf{A}$ is a multiple of $\begin{pmatrix} 0 \\ 2 \\ -1 \end{pmatrix}$

and so the solution of the equations is given by $x = 0$, $y = 2k$, $z = -k$, where k is arbitrary.

Example 2. If
$$
\begin{aligned}
x - 2y + z &= 0, \\
-3x + 6y - 3z &= 0, \quad \equiv \mathbf{Ax} = \mathbf{0}, \\
7x - 14y + 7z &= 0,
\end{aligned}
$$

all the determinants of order 2 in $|\mathbf{A}| = \begin{vmatrix} 1 & -2 & 1 \\ -3 & 6 & -3 \\ 7 & -14 & 7 \end{vmatrix}$ are zero.

The equations reduce to one independent equation, with solution set $\{\lambda, \mu, 2\mu - \lambda\} \ \forall \ \lambda, \mu$.

15:5 SOLUTION OF NON-HOMOGENEOUS LINEAR EQUATIONS

We now consider a set of n non-homogeneous linear equations in n unknowns
$$\mathbf{Ax} = \mathbf{B} \quad (\mathbf{B} \neq \mathbf{0}).$$
$$\text{If } |\mathbf{A}| \neq 0, \ \exists \mathbf{A}^{-1} : \mathbf{A}^{-1}\mathbf{Ax} = \mathbf{A}^{-1}\mathbf{B}$$
$$\Rightarrow \mathbf{x} = \mathbf{A}^{-1}\mathbf{B} = \frac{1}{|\mathbf{A}|}(\text{adj } \mathbf{A})\mathbf{B}$$

$$\Rightarrow x_1 = \frac{A_{11}b_1 + A_{21}b_2 + \ldots + A_{n1}b_n}{|\mathbf{A}|}.$$

It can be seen that

$$A_{11}b_1 + \ldots + A_{n1}b_n = \begin{vmatrix} a_{12} & a_{13} & \ldots & a_{1n} & -b_1 \\ a_{22} & a_{23} & \ldots & a_{2n} & -b_2 \\ \ldots & \ldots & \ldots & \ldots & \ldots \\ \ldots & \ldots & \ldots & \ldots & \ldots \\ a_{n2} & a_{n3} & \ldots & a_{nn} & -b_n \end{vmatrix}.$$

Similarly,

$$x_2 = -\frac{1}{|\mathbf{A}|} \begin{vmatrix} a_{11} & a_{13} & \ldots & a_{1n} & -b_1 \\ a_{21} & a_{23} & \ldots & a_{2n} & -b_2 \\ \ldots & \ldots & \ldots & \ldots & \ldots \\ \ldots & \ldots & \ldots & \ldots & \ldots \\ a_{n1} & a_{n2} & \ldots & a_{nn} & -b_n \end{vmatrix}.$$

Hence, provided that $|\mathbf{A}| \neq 0$, there is a unique solution

$$\frac{x_1}{\begin{vmatrix} \ \end{vmatrix}} = \frac{-x_2}{\begin{vmatrix} \ \end{vmatrix}} = \ldots = \frac{(-1)^n}{|\mathbf{A}|}.$$

One may say that this form of solution is "easier said than done", in the sense that determinants are easily written down but cumbersome to evaluate, particularly for $n > 3$; it is therefore advisable to have other methods of solution available.

15:6 SOLUTION OF NON-HOMOGENEOUS LINEAR EQUATIONS BY THE ECHELON METHOD

We are accustomed to solve such equations by using elementary operations, such as multiplication of an equation by a chosen number, addition or subtraction of equations, to achieve successive elimination of unknowns.

Geometrical interpretations of the solution of three linear equations in three unknowns are considered in §14:3.

Example 1.

$$3x_1 - x_2 + 7x_3 = 2, \tag{1}$$
$$2x_1 + 2x_2 + x_3 = 1, \tag{2}$$
$$7x_1 - 2x_2 - 3x_3 = 4. \tag{3}$$

$$3x_1 - x_2 + 7x_3 = 2,$$

$3 \times (2) - 2 \times (1)$

$$8x_2 - 11x_3 = -1, \tag{4}$$

$2 \times (3) - 7 \times (2)$

$$-18x_2 - 13x_3 = 1. \tag{5}$$

$$3x_1 - x_2 + 7x_3 = 2,$$
$$8x_2 - 11x_3 = -1,$$

$9 \times (4) + 4 \times (5)$ $\qquad\qquad\qquad -151x_3 = -5.$ $\qquad\qquad\qquad$ (6)

Hence we can find x_3 from (6), and then by substitution x_2 and then x_1. Such elementary operations are conveniently carried out in matrix form, and are then suitable for mechanical calculation. Starting with the matrix formed from the coefficients of $x_1, x_2, \ldots$ in the original equations, we can perform an elementary operation by premultiplication with a matrix obtained by carrying out the same elementary operation on the unit matrix; e.g. to add $3 \times (2)$ to $5 \times (3)$ we premultiply the matrix of coefficients by

$$\begin{pmatrix} 1 & 0 & 0 \\ 0 & 1 & 0 \\ 0 & 3 & 5 \end{pmatrix}.$$

Successive matrices obtained in this way are called *equivalent*; the equations can be solved when a triangular matrix equivalent to the original matrix of coefficients has been obtained.

The method is called the *echelon method* or *Gaussian elimination*.

Example 2. The equations $x_1 + 3x_2 + 6x_3 = 17,$ $\qquad\qquad\qquad\qquad$ (1)

$\qquad\qquad\qquad\quad 2x_1 + 8x_2 + 16x_3 = 42,$ $\qquad\qquad\qquad\qquad$ (2)

$\qquad\qquad\qquad\quad 5x_1 + 21x_2 + 45x_3 = 91$ $\qquad\qquad\qquad\qquad$ (3)

$$\equiv \begin{pmatrix} 1 & 3 & 6 \\ 2 & 8 & 16 \\ 5 & 21 & 45 \end{pmatrix} \begin{pmatrix} x_1 \\ x_2 \\ x_3 \end{pmatrix} = \begin{pmatrix} 17 \\ 42 \\ 91 \end{pmatrix}$$

$\begin{matrix} (2) - 2 \times (1) \Leftrightarrow \\ (3) - 5 \times (1) \end{matrix}$ $\begin{pmatrix} 1 & 0 & 0 \\ -2 & 1 & 0 \\ -5 & 0 & 1 \end{pmatrix} \begin{pmatrix} 1 & 3 & 6 \\ 2 & 8 & 16 \\ 5 & 21 & 45 \end{pmatrix} \begin{pmatrix} x_1 \\ x_2 \\ x_3 \end{pmatrix} = \begin{pmatrix} 1 & 0 & 0 \\ -2 & 1 & 0 \\ -5 & 0 & 1 \end{pmatrix} \begin{pmatrix} 17 \\ 42 \\ 91 \end{pmatrix}$

$$\Leftrightarrow \begin{pmatrix} 1 & 3 & 6 \\ 0 & 2 & 4 \\ 0 & 6 & 15 \end{pmatrix} \begin{pmatrix} x_1 \\ x_2 \\ x_3 \end{pmatrix} = \begin{pmatrix} 17 \\ 8 \\ 6 \end{pmatrix}$$

$\qquad\qquad\qquad\qquad\qquad\qquad\qquad\qquad\qquad\qquad\qquad\qquad\qquad\qquad$ (1)
$\qquad\qquad\qquad\qquad\qquad\qquad\qquad\qquad\qquad\qquad\qquad\qquad\qquad\qquad$ (4)
$\qquad\qquad\qquad\qquad\qquad\qquad\qquad\qquad\qquad\qquad\qquad\qquad\qquad\qquad$ (5)

$\begin{matrix} \Leftrightarrow \\ (5) - 3 \times (4) \end{matrix}$ $\begin{pmatrix} 1 & 0 & 0 \\ 0 & 1 & 0 \\ 0 & -3 & 1 \end{pmatrix} \begin{pmatrix} 1 & 3 & 6 \\ 0 & 2 & 4 \\ 0 & 6 & 15 \end{pmatrix} \begin{pmatrix} x_1 \\ x_2 \\ x_3 \end{pmatrix} = \begin{pmatrix} 1 & 0 & 0 \\ 0 & 1 & 0 \\ 0 & -3 & 1 \end{pmatrix} \begin{pmatrix} 17 \\ 8 \\ 6 \end{pmatrix}$

$$\Leftrightarrow \begin{pmatrix} 1 & 3 & 6 \\ 0 & 2 & 4 \\ 0 & 0 & 3 \end{pmatrix} \begin{pmatrix} x_1 \\ x_2 \\ x_3 \end{pmatrix} = \begin{pmatrix} 17 \\ 8 \\ -18 \end{pmatrix}$$

$$\Rightarrow 3x_3 = -18 \Leftrightarrow x_3 = -6 \Rightarrow 2x_2 - 24 = 8 \Leftrightarrow x_2 = 16$$

$$\Rightarrow x_1 + 48 - 36 = 17 \Leftrightarrow x_1 = 5.$$

By continuing the process of premultiplication until the triangular matrix becomes a diagonal one, we can find the inverse of the original matrix, thus:

$\begin{matrix} 2 \times (1) - 3 \times (2) \\ 3 \times (2) - 4 \times (3) \end{matrix}$ $\begin{pmatrix} 2 & -3 & 0 \\ 0 & 3 & -4 \\ 0 & 0 & 1 \end{pmatrix} \begin{pmatrix} 1 & 3 & 6 \\ 0 & 2 & 4 \\ 0 & 0 & 3 \end{pmatrix} = \begin{pmatrix} 2 & 0 & 0 \\ 0 & 6 & 0 \\ 0 & 0 & 3 \end{pmatrix},$

$$\begin{matrix} \frac{1}{2} \times (1) \\ \frac{1}{6} \times (2) \\ \frac{1}{3} \times (3) \end{matrix} \quad \begin{pmatrix} \frac{1}{2} & 0 & 0 \\ 0 & \frac{1}{6} & 0 \\ 0 & 0 & \frac{1}{3} \end{pmatrix} \begin{pmatrix} 2 & 0 & 0 \\ 0 & 6 & 0 \\ 0 & 0 & 3 \end{pmatrix} = \begin{pmatrix} 1 & 0 & 0 \\ 0 & 1 & 0 \\ 0 & 0 & 1 \end{pmatrix}.$$

Hence the product of all the premultiplying matrices must equal the inverse of the original matrix

$$\Rightarrow \begin{pmatrix} 1 & 3 & 6 \\ 2 & 8 & 16 \\ 5 & 21 & 45 \end{pmatrix}^{-1} = \begin{pmatrix} \frac{1}{2} & 0 & 0 \\ 0 & \frac{1}{6} & 0 \\ 0 & 0 & \frac{1}{3} \end{pmatrix} \begin{pmatrix} 2 & -3 & 0 \\ 0 & 3 & -4 \\ 0 & 0 & 1 \end{pmatrix} \begin{pmatrix} 1 & 0 & 0 \\ 0 & 1 & 0 \\ 0 & -3 & 1 \end{pmatrix} \begin{pmatrix} 1 & 0 & 0 \\ -2 & 1 & 0 \\ -5 & 0 & 1 \end{pmatrix}$$

$$= \begin{pmatrix} \frac{1}{2} & 0 & 0 \\ 0 & \frac{1}{6} & 0 \\ 0 & 0 & \frac{1}{3} \end{pmatrix} \begin{pmatrix} 8 & -3 & 0 \\ -10 & 15 & -4 \\ 1 & -3 & 1 \end{pmatrix} = \begin{pmatrix} 4 & -\frac{3}{2} & 0 \\ -\frac{5}{3} & \frac{5}{2} & -\frac{2}{3} \\ \frac{1}{3} & -1 & \frac{1}{3} \end{pmatrix}.$$

If the set of equations does not have a unique solution, this will be shown by the appearance of a row of zeros in one of the equivalent matrices, as illustrated by Example 3 following.

Example 3. The equations

$$-2x_1 + 5x_2 + 6x_3 = 1, \tag{1}$$
$$x_1 + 4x_2 + 7x_3 = 3, \tag{2}$$
$$8x_1 - 7x_2 - 4x_3 = 3 \tag{3}$$

$$\begin{matrix} \Leftrightarrow \\ 2 \times (2) + (1) \\ (3) - 8 \times (2) \end{matrix} \begin{pmatrix} 1 & 0 & 0 \\ 1 & 2 & 0 \\ 0 & -8 & 1 \end{pmatrix} \begin{pmatrix} -2 & 5 & 6 \\ 1 & 4 & 7 \\ 8 & -7 & -4 \end{pmatrix} \begin{pmatrix} x_1 \\ x_2 \\ x_3 \end{pmatrix} = \begin{pmatrix} 1 & 0 & 0 \\ 1 & 2 & 0 \\ 0 & -8 & 1 \end{pmatrix} \begin{pmatrix} 1 \\ 3 \\ 3 \end{pmatrix}$$

$$\Leftrightarrow \begin{pmatrix} -2 & 5 & 6 \\ 0 & 13 & 20 \\ 0 & -39 & -60 \end{pmatrix} \begin{pmatrix} x_1 \\ x_2 \\ x_3 \end{pmatrix} = \begin{pmatrix} 1 \\ 7 \\ -21 \end{pmatrix} \tag{4}$$
$$\tag{5}$$

$$\begin{matrix} \Leftrightarrow \\ (5) + 3 \times (4) \end{matrix} \begin{pmatrix} -2 & 5 & 6 \\ 0 & 13 & 20 \\ 0 & 0 & 0 \end{pmatrix} \begin{pmatrix} x_1 \\ x_2 \\ x_3 \end{pmatrix} = \begin{pmatrix} 1 \\ 7 \\ 0 \end{pmatrix}.$$

On examining the equations, we can see that they are linearly dependent, since $2 \times (2) - 3 \times (1) \equiv (3)$. Putting $x_3 = k$, we have

$$2x_1 - 5x_2 = -1 + 6k$$
$$x_1 + 4x_2 = 3 - 7k$$

$$\Rightarrow x_1 = \tfrac{1}{13}(1 - k), \ x_2 = \tfrac{1}{13}(7 - 20k).$$

This gives an infinite solution set $\forall k \in \mathbb{R}$.

Exercise 15.6

1. Find the inverse of the matrix

$$\begin{pmatrix} 2 & -1 & 3 \\ 5 & 4 & -3 \\ 3 & -2 & -1 \end{pmatrix}.$$

Hence, or otherwise, solve the equations

$$
\begin{aligned}
2x - y + 3z &= -25, \\
5x + 4y - 3z &= -1, \\
3x - 2y - z &= -17.
\end{aligned}
$$

2. By using elementary row operations to reduce the matrix

$$
A = \begin{pmatrix} 1 & -2 & a \\ 1 & -3 & 3 \\ 1 & -4 & 5 \end{pmatrix}
$$

to upper triangular form, show that the equation

$$
A \begin{pmatrix} x \\ y \\ z \end{pmatrix} = \begin{pmatrix} p \\ q \\ 1 \end{pmatrix}
$$

has a unique solution unless $a = 1$, but has no solution when $a = 1$ unless $p - 2q + 1 = 0$.
Find the solution sets when

(i) $a = 2$, $p = q = 1$;

(ii) $a = 1$, $p = -1$ and $q = 0$;

and interpret your solutions geometrically.   (O.C.S.M.P.)

3. By suitable operations on the rows, transform the matrix

$$
\begin{pmatrix} 3 & 4 & -1 & 1 \\ 2 & -5 & -4 & 6 \\ 1 & -2 & -3 & -1 \end{pmatrix}
$$

to reduced echelon form.

Hence, or otherwise, solve the equations

$$
\begin{aligned}
3x + 4y - z &= 1, \\
2x - 5y - 4z &= 6, \\
x - 2y - 3z &= -1.
\end{aligned}
$$

4. Solve the equations:

$$
\begin{aligned}
x + y + 3z &= a, \\
-2x + 5y + 3z &= b, \\
-x + 3y + 2z &= c
\end{aligned}
$$

for x, y and z. Use your solution to write down the inverse of

$$
\begin{pmatrix} 1 & 1 & 3 \\ -2 & 5 & 3 \\ -1 & 3 & 2 \end{pmatrix}
$$

and check that it is the inverse matrix.

Find k such that the three equations

$$
\begin{aligned}
x + y + 3z &= 1, \\
-2x + 5y + 3z &= b, \\
3x + 7y + kz &= 4,
\end{aligned}
$$

are satisfied by the same values of x, y, z as the original three equations, with $a = 1$, $c = 0$ and $b \neq -\frac{1}{4}$.
        (O.C.S.M.P.)

5. Find two matrices **A** and **B** such that $\mathbf{AB} = \mathbf{M}$ where

$$\mathbf{M} = \begin{pmatrix} 1 & 2 & -3 \\ 3 & 2 & 5 \\ 4 & 5 & 3 \end{pmatrix}, \quad \mathbf{A} = \begin{pmatrix} 1 & 0 & 0 \\ \otimes & 1 & 0 \\ \otimes & \otimes & 1 \end{pmatrix}, \quad \mathbf{B} = \begin{pmatrix} \otimes & \otimes & \otimes \\ 0 & \otimes & \otimes \\ 0 & 0 & \otimes \end{pmatrix}$$

and $\otimes$ represent numbers to be found.

If **p, q** are column vectors, write the equations

$$\begin{aligned} x + 2y - 3z &= 2, \\ 3x + 2y + 5z &= 0, \\ 4x + 5y + 3z &= -1 \end{aligned}$$

in the form $\mathbf{ABp} = \mathbf{q}$.

Show that

$$\mathbf{A}^{-1} = \begin{pmatrix} 1 & 0 & 0 \\ -3 & 1 & 0 \\ -\frac{7}{4} & -\frac{3}{4} & 1 \end{pmatrix}$$

and, by writing the equation in the form $\mathbf{Bp} = \mathbf{A}^{-1}\mathbf{q}$, find the elements of **p**.

6. Discuss the various cases which arise in the solution of the simultaneous equations

$$\begin{aligned} ax + y + z &= p, \\ x + ay + z &= q, \\ x + y + az &= r, \end{aligned}$$

where a, p, q and r are constants $\in \mathbb{R}$.

Obtain expressions for x, y and z in all cases where this is possible.

7. What condition must be satisfied by the constants a, b, c, l, m and n in order for the system of equations

$$\begin{aligned} -ny + mz &= a, \\ nx \qquad -lz &= b, \\ -mx + ly \qquad &= c \end{aligned}$$

to have solutions?

If this condition is satisfied, show that any solution of the equations will then also satisfy the additional equation $ax + by + cz = 0$, and write down the most general solution in this case.

Do any of these solutions also satisfy the equation

$$lx + my + nz = 0? \qquad \text{(O.C.S.M.P.)}$$

8. Consider the following equations for x_1, x_2, x_3:

$$\begin{aligned} x_1 - x_2 + 2x_3 &= y_1, \\ 2x_1 \qquad + 2x_3 &= y_2, \\ x_1 - 3x_2 + 4x_3 &= y_3. \end{aligned}$$

Show that, if solutions for x_1, x_2, x_3 exist, then (y_1, y_2, y_3) must satisfy $3y_1 = y_2 + y_3$.

If (y_1, y_2, y_3) is a triple for which solutions exist, find an expression for the set of solutions in terms of y_1 and y_2. \qquad (O.C.S.M.P.)

9. Find the value of k for which the equations

$$\begin{aligned} w + 2x - y - 3z &= 0, \\ w - x + 2y + 3z &= 0, \\ w \qquad + y + kz &= 0 \end{aligned}$$

have just two linearly independent solutions. Determine the solution set of the equations for this value of k.

\qquad (O.C.S.M.P.)

10. Find the inverse of the matrix

$$\mathbf{M} = \begin{pmatrix} 1 & -1 & 1 \\ 3 & -9 & 5 \\ 1 & -3 & 3 \end{pmatrix}.$$

Rewrite the set of equations

$$\begin{aligned} x - y + z &= 3, \\ 3x - 9y + 5z &= 6, \\ x - 3y + 3z &= 13 \end{aligned}$$

in a form using the matrix **M**, and hence, or otherwise, solve the equations.

15:7 LINEAR TRANSFORMATIONS

A transformation $(x, y) \to T(x, y)$ is said to be *linear* if

$$T(ax + by) = aT(x) + bT(y),$$

a, b being constants.

In elementary work we consider linear transformations mapping (x, y) onto (x', y') with equations of the form

$$\begin{aligned} x' &= ax + by, \\ y' &= cx + dy. \end{aligned}$$

It can be shown that such transformations map straight lines onto straight lines, and parallel straight lines onto parallel straight lines. By abstracting the coefficients of x and y in the equations, we can say that any 2×2 matrix $\begin{pmatrix} a & b \\ c & d \end{pmatrix}$ represents such a transformation.

Since

$$\begin{pmatrix} a & b \\ c & d \end{pmatrix}\begin{pmatrix} 0 \\ 0 \end{pmatrix} = \begin{pmatrix} 0 \\ 0 \end{pmatrix}, \quad \begin{pmatrix} a & b \\ c & d \end{pmatrix}\begin{pmatrix} 1 \\ 0 \end{pmatrix} = \begin{pmatrix} a \\ c \end{pmatrix}, \quad \begin{pmatrix} a & b \\ c & d \end{pmatrix}\begin{pmatrix} 0 \\ 1 \end{pmatrix} = \begin{pmatrix} b \\ d \end{pmatrix},$$

it is often convenient to apply a transformation to the "unit square" $OABC$ (see Fig. 15.1), either to identify a transformation from its matrix, or to find the matrix of a

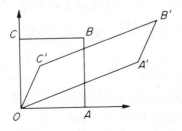

FIG. 15.1.

transformation. The square $OABC$ maps onto the parallelogram $OA'B'C'$, since parallel lines → parallel lines; to verify this,

$$B\begin{pmatrix}1\\1\end{pmatrix} \to \begin{pmatrix}a & b\\c & d\end{pmatrix}\begin{pmatrix}1\\1\end{pmatrix} = \begin{pmatrix}a+b\\c+d\end{pmatrix} \Rightarrow B' \equiv (a+b,\ c+d).$$

Example. Find the matrices representing
(a) rotation about O through $90°$ clockwise,
(b) reflection in the line $y = x$.

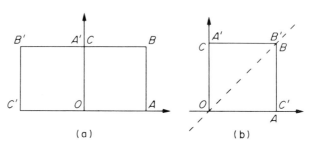

FIG. 15.2.

(a) $A\,(1, 0) \to A'\,(0, 1),\quad C\,(0, 1) \to C'\,(-1, 0) \Rightarrow$ matrix is $\begin{pmatrix}0 & -1\\1 & 0\end{pmatrix}$.

(b) $A\,(1, 0) \to A'\,(0, 1),\quad C\,(0, 1) \to C'\,(1, 0) \Rightarrow$ matrix is $\begin{pmatrix}0 & 1\\1 & 0\end{pmatrix}$

(see Fig. 15.2).

The scale factor for area under

$$\begin{pmatrix}a & b\\c & d\end{pmatrix}$$

is given by the numerical value of

$$\begin{vmatrix}a & b\\c & d\end{vmatrix} = |ad - bc|; \text{ for example, } \begin{pmatrix}3 & 0\\0 & 3\end{pmatrix}$$

represents an enlargement centre O, scale factor 3, so that the scale factor for area is 9.

If $\qquad\qquad\qquad \begin{vmatrix}a & b\\c & d\end{vmatrix} = 0,\text{ then } \dfrac{a}{c} = \dfrac{b}{d},$

so that $A'\,(a, c)$ and $C'\,(b, d)$ lie on the line $y = \dfrac{c}{a}x$ through O, and every point of the plane maps onto a point of this line. It is not possible to invert such a transformation. Note that in this case the scale factor for area is 0.

Linear transformations in three dimensions, where the origin is invariant, will be represented by 3×3 matrices and we can make use of the "unit cube" (Fig. 15.3) to

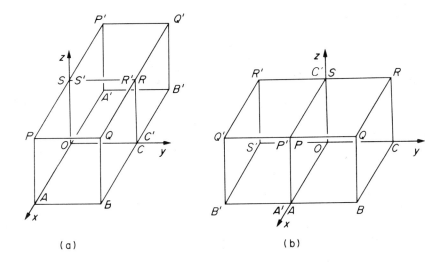

(a)   (b)

FIG. 15.3.

identify or interpret simple transformations, for example, under reflection in the yz plane ($x = 0$),

$$A(1, 0, 0) \rightarrow A'(-1, 0, 0), \ C(0, 1, 0) \rightarrow C(0, 1, 0), \ S(0, 0, 1) \rightarrow S(0, 0, 1).$$

[see Fig. 15.3(a)], and the matrix of the transformation is

$$\begin{pmatrix} -1 & 0 & 0 \\ 0 & 1 & 0 \\ 0 & 0 & 1 \end{pmatrix}.$$

Under a $+90°$ turn about Ox, $A \rightarrow A'$, $C \rightarrow C'(0, 0, 1)$, $S \rightarrow S'(0, -1, 0)$, [see Fig. 15.3(b)], and the matrix of the transformation is

$$\begin{pmatrix} 1 & 0 & 0 \\ 0 & 0 & -1 \\ 0 & 1 & 0 \end{pmatrix}.$$

Successive transformations are represented by the product of their respective matrices in corresponding order;

$$\text{e.g. where } \mathbf{Q} = \begin{pmatrix} 0 & -1 \\ 1 & 0 \end{pmatrix}$$

represents an anticlockwise quarter-turn about O, and

$$\mathbf{M} = \begin{pmatrix} 0 & 1 \\ 1 & 0 \end{pmatrix}$$

represents reflection in $y = x$,

$$MQ = \begin{pmatrix} 0 & 1 \\ 1 & 0 \end{pmatrix}\begin{pmatrix} 0 & -1 \\ 1 & 0 \end{pmatrix} = \begin{pmatrix} 1 & 0 \\ 0 & -1 \end{pmatrix}$$

represents Q followed by M and is equivalent to reflection in Ox [Fig. 15.4(a)], whereas

$$QM = \begin{pmatrix} 0 & -1 \\ 1 & 0 \end{pmatrix}\begin{pmatrix} 0 & 1 \\ 1 & 0 \end{pmatrix} = \begin{pmatrix} -1 & 0 \\ 0 & 1 \end{pmatrix}$$

represents M followed by Q and is equivalent to reflection in Oy [Fig. 15.4(b)].

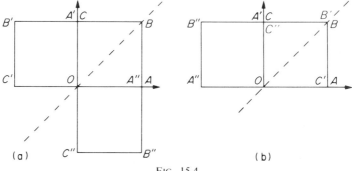

FIG. 15.4.

Products of transformations are not in general commutative; but the product of a transformation and its inverse, leaving the original figure unaltered, will be commutative: $\mathbf{T}^{-1}\mathbf{T} = \mathbf{T}\mathbf{T}^{-1} = \mathbf{I}$.

Example. Consider the result of two successive transformations with matrix $\begin{pmatrix} 2 & -1 \\ -4 & 2 \end{pmatrix}$.

The matrix is singular, and maps all points of the plane onto points of the line $y + 2x = 0$.

$$\begin{pmatrix} 2 & 1 \\ -4 & -2 \end{pmatrix}\begin{pmatrix} 2 & 1 \\ -4 & -2 \end{pmatrix} = \begin{pmatrix} 0 & 0 \\ 0 & 0 \end{pmatrix};$$

hence points on $y + 2x = 0$ are mapped onto 0.

This illustrates the fact that $\mathbf{A}^2 = \mathbf{0} \not\Rightarrow \mathbf{A} = \mathbf{0}$, or more generally $\mathbf{A}.\mathbf{B} = \mathbf{0} \not\Rightarrow \mathbf{A} = \mathbf{0}$ or $\mathbf{B} = \mathbf{0}$, an important distinction between multiplication of numbers and "multiplication" of matrices.

15:8 ORTHOGONAL MATRICES

An orthogonal matrix is a square matrix such that its row and column vectors are mutually perpendicular, and of unit length; examples are,

$$\begin{pmatrix} 1 & 0 & 0 \\ 0 & -1 & 0 \\ 0 & 0 & -1 \end{pmatrix}, \begin{pmatrix} -\dfrac{1}{2} & \dfrac{\sqrt{3}}{2} \\ \dfrac{\sqrt{3}}{2} & \dfrac{1}{2} \end{pmatrix}.$$

It can be proved from this definition that if $\mathbf{P}$ is an orthogonal matrix, then $\mathbf{P}^T = \mathbf{P}^{-1}$ as follows:

$$\text{For if } \mathbf{P} \equiv \begin{pmatrix} a_{11} \, a_{12} \, \ldots \, a_{1n} \\ \ldots\ldots\ldots\ldots \\ \ldots\ldots\ldots\ldots \\ a_{n1} \, a_{n2} \, \ldots \, a_{nn} \end{pmatrix}, \ \mathbf{P}^T\mathbf{P} = \begin{pmatrix} a_n^2 + \ldots + a_{n1}^2 & 0 \ldots \\ 0 & \ldots\ldots\ldots\ldots\ldots\ldots \\ & \ldots\ldots\ldots\ldots\ldots\ldots \\ & \ldots\ldots\ldots\ldots\ldots\ldots \end{pmatrix}$$

$$= \mathbf{P}\mathbf{P}^T.$$

$$\text{But } a_{11}^2 + \ldots + a_{n1}^2 = 1 \Rightarrow \mathbf{P}^T\mathbf{P} = \mathbf{P}\mathbf{P}^T = \mathbf{I}.$$

Since we have proved $\mathbf{P}^{-1}$ to be unique, $\mathbf{P}^T = \mathbf{P}^{-1}$.

Clearly a 2×2 orthogonal matrix represents a transformation which maps the unit square $OABC$ onto a unit square $OA'B'C'$, since if

$$\begin{pmatrix} a & b \\ c & d \end{pmatrix}$$

is orthogonal $OA' \perp OC'$ and $OA' = OC' = 1$. Such a transformation must be an isometry, i.e. one which keeps lengths and angles invariant—and the only isometries for which O is invariant are rotations and reflections, or combinations of these.

Now since $a^2 + b^2 = c^2 + d^2 = a^2 + c^2 = b^2 + d^2 = 1$ we may take $a = d = \pm \cos \theta$, $b = c = \pm \sin \theta$, choosing signs so that $ac + bd = 0$. Essentially the only distinct orthogonal 2×2 matrices are

$$\begin{pmatrix} \cos \theta & -\sin \theta \\ \sin \theta & \cos \theta \end{pmatrix} \quad \text{and} \quad \begin{pmatrix} \cos \theta & \sin \theta \\ \sin \theta & -\cos \theta \end{pmatrix},$$

since, for instance,

$$\begin{pmatrix} \cos \theta & \sin \theta \\ -\sin \theta & \cos \theta \end{pmatrix} = \begin{pmatrix} \cos(-\theta) & -\sin(-\theta) \\ \sin(-\theta) & \cos(-\theta) \end{pmatrix}.$$

The matrix

$$\begin{pmatrix} \cos \theta & -\sin \theta \\ \sin \theta & \cos \theta \end{pmatrix}$$

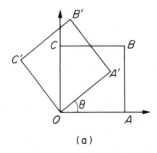

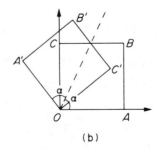

(a)   (b)

FIG. 15.5.

represents rotation about O through θ (anticlockwise), as illustrated by Fig. 15.5(a).
Writing $2\alpha = \theta$, the matrix

$$\begin{pmatrix} \cos 2\alpha & \sin 2\alpha \\ \sin 2\alpha & -\cos 2\alpha \end{pmatrix}$$

represents reflection in the line $y = x \tan \alpha$, as illustrated by Fig. 15.5(b). We have used
the fact that rotation about O through θ and then through ϕ is clearly equivalent to
rotation through $(\theta + \phi)$ to obtain the addition formulae for $\cos (\theta + \phi)$, etc.

$$\begin{pmatrix} \cos (\theta + \phi) & -\sin (\theta + \phi) \\ \sin (\theta + \phi) & \cos (\theta + \phi) \end{pmatrix} = \begin{pmatrix} \cos \phi & -\sin \phi \\ \sin \phi & \cos \phi \end{pmatrix} \begin{pmatrix} \cos \theta & -\sin \theta \\ \sin \theta & \cos \theta \end{pmatrix}$$

$$\Rightarrow \cos (\theta + \phi) = \cos \theta \cos \phi - \sin \theta \sin \phi, \text{ etc.}$$

A rotation followed by a reflection will be represented by a matrix product such as

$$\begin{pmatrix} \cos \theta & \sin \theta \\ \sin \theta & -\cos \theta \end{pmatrix} \begin{pmatrix} \cos \phi & -\sin \phi \\ \sin \phi & \cos \phi \end{pmatrix} = \begin{pmatrix} \cos (\theta - \phi) & \sin (\theta - \phi) \\ \sin (\theta - \phi) & -\cos (\theta - \phi) \end{pmatrix},$$

[Fig. 15.6(a)], representing a rotation followed by a reflection; similarly for a reflection
followed by a rotation, [Fig. 15.6(b)].

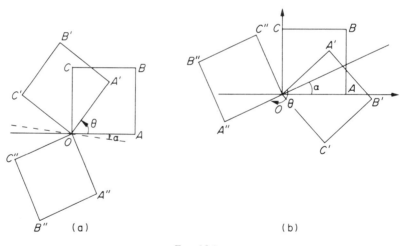

(a)   (b)

FIG. 15.6.

Two successive reflections are represented by

$$\begin{pmatrix} \cos \theta & \sin \theta \\ \sin \theta & -\cos \theta \end{pmatrix} \begin{pmatrix} \cos \phi & \sin \phi \\ \sin \phi & -\cos \phi \end{pmatrix} = \begin{pmatrix} \cos (\theta - \phi) & -\sin (\theta - \phi) \\ \sin (\theta - \phi) & \cos (\theta - \phi) \end{pmatrix}$$

which is a rotation (see Fig. 15.7).
   Orthogonal matrices are useful, for instance, in reducing the equations of conics to
standard forms.

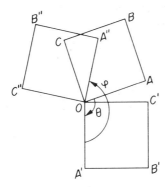

FIG. 15.7.

Exercise 15.8

1. $\mathbf{T} = \begin{pmatrix} 0 & 1 \\ -1 & 0 \end{pmatrix}$ operates on the vector $\begin{pmatrix} x \\ y \end{pmatrix}$. Name the transformation effected by each of

(i) $\mathbf{T}\begin{pmatrix} x \\ y \end{pmatrix}$ (ii) $\mathbf{T}^2\begin{pmatrix} x \\ y \end{pmatrix}$ (iii) $\mathbf{T}^4\begin{pmatrix} x \\ y \end{pmatrix}$. (L.)

2. The unit square, whose vertices are the points $(0, 0)$, $(1, 0)$, $(0, 1)$ and $(1, 1)$, is denoted by U, and

$$\mathbf{M}_1 = \begin{pmatrix} 1 & 0 \\ 0 & 2 \end{pmatrix}, \quad \mathbf{M}_2 = \begin{pmatrix} 2 & 2 \\ 1 & 4 \end{pmatrix}, \quad \mathbf{M}_3 = \begin{pmatrix} -1 & 1 \\ 0 & 0 \end{pmatrix}.$$

When the column vectors $\begin{pmatrix} 0 \\ 0 \end{pmatrix}$, etc., of the vertices of U are premultiplied by $\mathbf{M}_1$, $\mathbf{M}_2$ and $\mathbf{M}_3$, the resulting images of the square are called U_1, U_2 and U_3 respectively.

(i) Show U_1, U_2 and U_3 on separate sketches, making clear the coordinates of each point.

(ii) Calculate the matrix which transforms U_1 into U_2. (O.C.S.M.P.)

3. Show that an enlargement of the plane, with centre (a, b) and scale factor k, is given by the matrix equation

$$\begin{pmatrix} x' \\ y' \\ 1 \end{pmatrix} = \begin{pmatrix} k & 0 & a(1-k) \\ 0 & k & b(1-k) \\ 0 & 0 & 1 \end{pmatrix} \begin{pmatrix} x \\ y \\ 1 \end{pmatrix}.$$

Hence, or otherwise, show that the product of two enlargements, with given centres and scale factors, is an enlargement whose scale factor is the product of the given scale factors and whose centre lies on the line joining the given centres.

4. Prove that every 2×2 matrix $\mathbf{A}$ with real elements, satisfying the equation $\mathbf{A}\mathbf{A}^T = \mathbf{I}$, where $\mathbf{I}$ is the 2×2 unit matrix and $\mathbf{A}^T$ is the transpose of $\mathbf{A}$, may be written in one of the two forms:

$$\mathbf{M}(\theta) = \begin{pmatrix} \cos\theta & \sin\theta \\ -\sin\theta & \cos\theta \end{pmatrix}, \quad \mathbf{N}(\theta) = \begin{pmatrix} \cos\theta & \sin\theta \\ \sin\theta & -\cos\theta \end{pmatrix}.$$

Describe geometrically, giving your reasons, the transformations of vectors in two dimensions given by the matrices $\mathbf{M}(0)$, $\mathbf{M}(\tfrac{1}{2}\pi)$, $\mathbf{N}(0)$, $\mathbf{N}(\tfrac{1}{2}\pi)$.

5. An isometry of the plane takes A to X, B to Y and C to Z, where A, B, C, X, Y are the points $(0, 3)$, $(3, -1)$, $(2, 2)$, $(4, 0)$ and $(-1, 0)$ respectively. Show that the distance YZ is $\sqrt{10}$ units.

Calculate the angle between XY and YZ. Deduce that the equation of the line YZ is $y = \pm\tfrac{1}{3}(x+1)$. Account for the ambiguity in sign. (O.C.S.M.P.)

6. Explain why only one of the transformations defined below is linear:

$$T_1(x, y, z) = (x + 2y - 2z, \; 3x + 3, \; 2x - 2y + 5z),$$
$$T_2(x, y, z) = (x + 2y - 2z, \; 3x + 3z, \; 2x - 2y + 5z),$$
$$T_3(x, y, z) = (x + 2y - 2z, \; 3x + 3z^2, \; 2x - 2y + 5z).$$

Write down the matrix of the linear transformation and show that under this transformation
(a) all points are mapped onto a particular plane, whose equation should be given;
(b) all points on the line

$$\frac{x-1}{-1} = \frac{2y+2}{3} = \frac{z}{1}$$

are·mapped onto the same point, whose coordinates should be given;
(c) all points which are mapped onto the origin lie on a certain straight line, whose equations should be
given.   (L.)

7. A transformation T of the plane reflects each point in the line $y = x + 1$. Write down equations for T.
Hence, or otherwise, show that the transformation U of reflection in the line $y = -x - 1$ has the matrix
equation

$$\begin{pmatrix} x' \\ y' \end{pmatrix} = \begin{pmatrix} 0 & -1 \\ -1 & 0 \end{pmatrix} \begin{pmatrix} x \\ y \end{pmatrix} - \begin{pmatrix} 1 \\ 1 \end{pmatrix}.$$

Find the equations of the combined transformations
(a) UT (i.e. T followed by U),
(b) VUT, where V is a translation parallel to the x-axis through a distance of 2 units (in the positive sense).
Show that VUT is a rotation about the origin and give the angle of this rotation.   (L.)

8. Obtain the matrix $\mathbf{T}_1$ such that

$$\mathbf{T}_1 \begin{pmatrix} x \\ y \end{pmatrix} = \begin{pmatrix} x_1 \\ y_1 \end{pmatrix},$$

where $\begin{pmatrix} x_1 \\ y_1 \end{pmatrix}$ is the vector obtained by rotating $\begin{pmatrix} x \\ y \end{pmatrix}$ about the origin O through a positive angle α.
Obtain the matrix $\mathbf{T}_2$ such that

$$\mathbf{T}_2 \begin{pmatrix} x \\ y \end{pmatrix} = \begin{pmatrix} x_2 \\ y_2 \end{pmatrix},$$

where (x_2, y_2) is the reflection of the point (x, y) in the x-axis. Hence, or otherwise, find the matrix of the
transformation which represents a reflection in the line $4y = 3x$.   (L.)

9. Ox, Oy are rectangular axes in a plane. State the matrix representations of the following transformations
of the plane into itself:
(a) reflection in a line through O making an angle θ with Ox (θ being measured in the anti-clockwise
direction),
(b) anti-clockwise rotation about O through an angle θ.
Linear transformations T_1 and T_2 are reflections of the plane in the lines $y = x$ and $y = x \tan(\pi/3)$
respectively. Write down the matrix representations of T_1 and T_2, find the matrix representation of the
combined transformation T_2T_1 and interpret the combined transformation geometrically.
A transformation T_3 is a magnification from the origin by a factor 2; T_4 is a translation with vector

$$\begin{pmatrix} -\sqrt{3} \\ -1 \end{pmatrix}.$$

Find the matrix of the transformation $T_3T_2T_1$ and the image of the point $(0, 1)$ under the transformation
$T_4T_3T_2T_1$.   (L.)

10. Find the matrix which corresponds to the reflection M_1 in the line $y = -x \tan \frac{1}{2}\alpha$ and the matrix
which corresponds to the reflection M_2 in the line $y = x \tan \frac{1}{2}\alpha$.
By multiplying these matrices show that the combination of these reflections, in the order first M_1 and then
M_2, is the rotation about the origin counter clockwise through an angle 2α.   (O.)

11. The point P has position vector $\begin{pmatrix} x \\ y \end{pmatrix}$ referred to the origin O. When OP is rotated anticlockwise

through an angle θ about O, the position vector of P becomes $\begin{pmatrix} x_1 \\ y_1 \end{pmatrix}$.

If $\mathbf{T}_1 \begin{pmatrix} x \\ y \end{pmatrix} = \begin{pmatrix} x_1 \\ y_1 \end{pmatrix}$, show that $\mathbf{T}_1 = \begin{pmatrix} \cos\theta & -\sin\theta \\ \sin\theta & \cos\theta \end{pmatrix}$.

If $\begin{pmatrix} x_2 \\ y_2 \end{pmatrix}$ is the reflection of $\begin{pmatrix} x \\ y \end{pmatrix}$ in the x-axis and $\mathbf{T}_2 \begin{pmatrix} x \\ y \end{pmatrix} = \begin{pmatrix} x_2 \\ y_2 \end{pmatrix}$, find $\mathbf{T}_2$.

If $\mathbf{T}_\theta = \mathbf{T}_1 \mathbf{T}_2 \mathbf{T}_1^{-1}$ and $\mathbf{T}_\theta \begin{pmatrix} x \\ y \end{pmatrix} = \begin{pmatrix} x_3 \\ y_3 \end{pmatrix}$, find $\mathbf{T}_\theta$ and show that $\begin{pmatrix} x_3 \\ y_3 \end{pmatrix}$ is the reflection of $\begin{pmatrix} x \\ y \end{pmatrix}$ in the line

$y = x \tan\theta$.

If $\mathbf{T}_3 = \begin{pmatrix} \cos\phi & -\sin\phi \\ \sin\phi & \cos\phi \end{pmatrix}$ and $\mathbf{T}_\phi = \mathbf{T}_3 \mathbf{T}_2 \mathbf{T}_3^{-1}$, show that $\mathbf{T}_\phi \mathbf{T}_\theta$ represents a rotation about O and find the
angle of rotation.

12. If $P_1(x_1, y_1)$ is the reflection of $P(x, y)$ in the line $y = x \tan\alpha$ and

$$\mathbf{M}_1 \begin{pmatrix} x \\ y \end{pmatrix} = \begin{pmatrix} x_1 \\ y_1 \end{pmatrix},$$

show that

$$\mathbf{M}_1 = \begin{pmatrix} \cos 2\alpha & \sin 2\alpha \\ \sin 2\alpha & -\cos 2\alpha \end{pmatrix}.$$

If $P_2(x_2, y_2)$ is the reflection of $P(x, y)$ in the line $y = x \tan\beta$,

$$\mathbf{M}_2 \begin{pmatrix} x \\ y \end{pmatrix} = \begin{pmatrix} x_2 \\ y_2 \end{pmatrix} \text{ and } \mathbf{M}_2 \mathbf{M}_1 \begin{pmatrix} x \\ y \end{pmatrix} = \begin{pmatrix} x_3 \\ y_3 \end{pmatrix},$$

show that $P_3(x_3, y_3)$ can be obtained by rotating OP about the origin O and state the angle through which OP
would be rotated.

13. A transformation T of the plane is the product $T_3 T_2 T_1$ of the three transformations T_1, T_2, T_3, where
(a) T_1 is reflection in the line $y = x$,
(b) T_2 is an anticlockwise rotation about the origin through $60°$,
(c) T_3 is magnification, from the origin, by a factor 2.
Express each of T_1, T_2, T_3 in the matrix form

$$\begin{pmatrix} x' \\ y' \end{pmatrix} = \begin{pmatrix} a & b \\ c & d \end{pmatrix} \begin{pmatrix} x \\ y \end{pmatrix},$$

where a, b, c, d are constants, and deduce that the transformation T is given by

$$\begin{pmatrix} x' \\ y' \end{pmatrix} = \begin{pmatrix} -\sqrt{3} & 1 \\ 1 & \sqrt{3} \end{pmatrix} \begin{pmatrix} x \\ y \end{pmatrix}.$$

Find the gradients of the two lines through the origin each of which is transformed into itself by T.

15:9 EIGENVECTORS AND EIGENVALUES

The matrix

$$\begin{pmatrix} 3 & 0 \\ 0 & 2 \end{pmatrix}$$

stretches the unit square to three times its length in the x-direction and twice its length
in the y-direction. The actual scale factors 3 and 2 are not important in this context; we

are interested in the fact that the directions OA, OA' and OC, OC' are unaltered, whereas in general the transformation alters the direction of a vector as well as its magnitude; e.g. in Fig. 15.8,

$$\overrightarrow{OB} \equiv \begin{pmatrix} 1 \\ 1 \end{pmatrix} \rightarrow \begin{pmatrix} 3 & 0 \\ 0 & 2 \end{pmatrix}\begin{pmatrix} 1 \\ 1 \end{pmatrix} = \begin{pmatrix} 3 \\ 2 \end{pmatrix} \equiv \overrightarrow{OB'}.$$

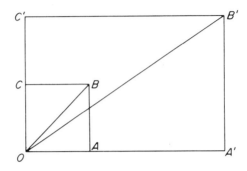

FIG. 15.8.

A vector whose *direction* is unaltered by a certain transformation is called an *eigenvector* (sometimes a *latent vector*) of that transformation, or of the matrix representing that transformation; "eigen" is the German word for equal, but remember that it is equality of *direction* to which the word refers, not of magnitude. Since the direction of a vector $\mathbf{x}$ depends only on the ratios $x_1:x_2: \ldots :x_n$, the direction is not affected when the vector is multiplied by a scalar. Thus, to find an eigenvector for a transformation with matrix

$$\begin{pmatrix} a & b \\ c & d \end{pmatrix} \text{ we require a vector } \begin{pmatrix} p \\ q \end{pmatrix}$$

such that

$$\begin{pmatrix} a & b \\ c & d \end{pmatrix}\begin{pmatrix} p \\ q \end{pmatrix} = \lambda \begin{pmatrix} p \\ q \end{pmatrix}, \text{ where } \lambda \text{ is a scalar,}$$

$$\Leftrightarrow (a - \lambda)p + bq = 0,$$
$$cp + (d - \lambda)q = 0.$$

For a non-trivial solution of these linear homogeneous equations in p and q we require (see p. 466)

$$\begin{vmatrix} a - \lambda & b \\ c & d - \lambda \end{vmatrix} = 0$$

$$\Leftrightarrow \lambda^2 - (a + d)\lambda + ad - bc = 0.$$

This equation for λ. is called the *characteristic equation* of the matrix

$$\begin{pmatrix} a & b \\ c & d \end{pmatrix}.$$

To generalise, if A is an $n \times n$ matrix, and I is the $n \times n$ identity matrix, the characteristic equation of A is $|A - \lambda I| = 0$. The roots of the characteristic equation are called the *eigenvalues* (or *latent roots*) of the matrix. Note that the sum of the eigenvalues is equal to the sum of the elements in the leading diagonal of the matrix—called the *trace* of the matrix.

Example 1. Find eigenvectors for the transformation with matrix $\begin{pmatrix} 1 & -2 \\ 1 & 4 \end{pmatrix}$.

$$\text{Let } \begin{pmatrix} 1 & -2 \\ 1 & 4 \end{pmatrix} \begin{pmatrix} p \\ q \end{pmatrix} = \lambda \begin{pmatrix} p \\ q \end{pmatrix}$$

$$\Rightarrow \lambda^2 - 5\lambda + 6 = 0 \Leftrightarrow \lambda = 2, 3.$$

For $\lambda = 2$, we have $p + 2q = 0$; hence $\begin{pmatrix} 2 \\ -1 \end{pmatrix}$, or, of course, $k \begin{pmatrix} 2 \\ -1 \end{pmatrix}$, where k is any non-zero scalar, is an eigenvector.

For $\lambda = 3$, we have $p + q = 0$, hence $k \begin{pmatrix} 1 \\ -1 \end{pmatrix}$ is an eigenvector.

$$\text{Check: } \begin{pmatrix} 1 & -2 \\ 1 & 4 \end{pmatrix} \begin{pmatrix} 2 \\ -1 \end{pmatrix} = \begin{pmatrix} 4 \\ -2 \end{pmatrix} = 2 \begin{pmatrix} 2 \\ -1 \end{pmatrix};$$

$$\begin{pmatrix} 1 & -2 \\ 1 & 4 \end{pmatrix} \begin{pmatrix} 1 \\ -1 \end{pmatrix} = \begin{pmatrix} 3 \\ -3 \end{pmatrix} = 3 \begin{pmatrix} 1 \\ -1 \end{pmatrix}.$$

Example 2. (i) Consider the matrix $\begin{pmatrix} 0 & 1 \\ -1 & 0 \end{pmatrix}$. The characteristic equation is $\lambda^2 + 1 = 0$, giving no (real) eigenvalues. The matrix represents a negative quarter-turn about O. Clearly no rotation about O can have real eigenvalues except a half-turn $\begin{pmatrix} -1 & 0 \\ 0 & -1 \end{pmatrix}$ for which $\lambda^2 + 2\lambda + 1 = 0$, giving the one real eigenvalue $\lambda = -1$.

(ii) Consider the matrix $\begin{pmatrix} 2 & 0 \\ 0 & 2 \end{pmatrix}$. The characteristic equation is

$$\lambda^2 - 4\lambda + 4 = 0 = (\lambda - 2)^2. \text{ Then } \lambda = 2 \text{ gives the eigenvector } \begin{pmatrix} 0 \\ 0 \end{pmatrix}$$

which is meaningless. The matrix represents an enlargement centre O, which of course leaves the direction of any vector unaltered.

Exercise 15.9

1. Obtain the characteristic equation, the eigenvalues and the corresponding eigenvectors of the matrix

$$A = \begin{pmatrix} 4 & 1 \\ 2 & 3 \end{pmatrix}.$$

2. Find the eigenvalues and eigenvectors of the matrix

$$A = \begin{pmatrix} 25 & 40 \\ -12 & -19 \end{pmatrix}.$$

Express the vector $\begin{pmatrix} 3 \\ -2 \end{pmatrix}$ as a linear combination of the eigenvectors and deduce that

$$A^n \begin{pmatrix} 3 \\ -2 \end{pmatrix} = 5^n \begin{pmatrix} -2 \\ 1 \end{pmatrix} + \begin{pmatrix} 5 \\ -3 \end{pmatrix}$$

for all integers n.

3. Verify that the matrix

$$\begin{bmatrix} \dfrac{1}{2} & -\dfrac{1}{2} & \dfrac{1}{\sqrt{2}} \\[2mm] \dfrac{1}{2} & -\dfrac{1}{2} & -\dfrac{1}{\sqrt{2}} \\[2mm] \dfrac{1}{\sqrt{2}} & \dfrac{1}{\sqrt{2}} & 0 \end{bmatrix}$$

is orthogonal, and determine its eigenvalues.

4. Show that 2 is an eigenvalue of the matrix

$$\begin{pmatrix} 3 & 4 & 4 \\ -1 & -2 & -4 \\ 1 & 1 & 3 \end{pmatrix}$$

and find all the other eigenvalues. Determine the eigenvectors corresponding to each eigenvalue.

5. Find the three values of λ for which a non-zero vector $\mathbf{x}$ can be found such that

$$\begin{pmatrix} 2 & -2 & 3 \\ 1 & 1 & 1 \\ 1 & 3 & -1 \end{pmatrix} \mathbf{x} = \lambda \mathbf{x},$$

and find the vector of unit modulus corresponding to each value of λ.   (L.)

6. Find the eigenvalues of the matrix

$$\mathbf{A} = \begin{pmatrix} 2 & -2 & 3 \\ 1 & 1 & 1 \\ 1 & 3 & -1 \end{pmatrix}.$$

Describe the effect on points on the line $x = y = z$ of the linear transformation defined by

$$\begin{pmatrix} x_2 \\ y_2 \\ z_2 \end{pmatrix} = \mathbf{A} \begin{pmatrix} x_1 \\ y_1 \\ z_1 \end{pmatrix},$$

and find the matrix of the inverse transformation.

Explain the nature of the transformation with matrix $(\mathbf{A} - 3\mathbf{I})$, where $\mathbf{I}$ is the unit matrix of order 3.   (L.)

7. Show that the characteristic equation of the matrix

$$\mathbf{A} = \begin{pmatrix} 1 & -4 & 1 \\ -4 & 1 & 1 \\ 4 & 4 & 4 \end{pmatrix}$$

has two equal roots.

In a linear transformation the point with position vector

$$\begin{pmatrix} x \\ y \\ z \end{pmatrix} \text{ becomes the point with position vector } \mathbf{A} \begin{pmatrix} x \\ y \\ z \end{pmatrix}.$$

Show that the straight line $x = y = -z$ is transformed into itself. Find the equation of the plane which is transformed into itself, but which does not contain this line. (L.)

8. The linear transformation of the plane consisting of reflection in the line $y = 2x$ has matrix **M**. Write down eigenvectors of **M** corresponding to the eigenvalues 1 and -1. Use your answers to calculate

$$M\begin{pmatrix} 1 \\ 0 \end{pmatrix} \text{ and } M\begin{pmatrix} 0 \\ 1 \end{pmatrix}.$$

Hence determine **M**. (O.C.S.M.P.)

9. Find the latent roots and latent vectors of the matrix **A**, where

$$A = \begin{pmatrix} 9 & -2 \\ 5 & 2 \end{pmatrix}.$$

Hence by the substitution $x = ae^{kt}$, where

$$x = \begin{pmatrix} x_1 \\ x_2 \end{pmatrix} \text{ and } a = \begin{pmatrix} a_1 \\ a_2 \end{pmatrix},$$

find solutions in matrix form of the equations

$$\ddot{x}_1 = 9x_1 - 2x_2,$$
$$\ddot{x}_2 = 5x_1 + 2x_2.$$

15:10 REDUCTION OF A MATRIX TO DIAGONAL FORM

If the $n \times n$ matrix **A** has n *distinct eigenvalues*, $\lambda_1, \lambda_2, \ldots, \lambda_n$, we can use the corresponding eigenvectors, $e_1, e_2, \ldots, e_n$, to form the columns of a matrix **B**, and hence obtain from **A** a diagonal $n \times n$ matrix **D** (see p. 461). For by definition

$$AB = A(e_1 e_2 \ldots e_n) = (\lambda_1 e_1 \ \lambda_2 e_2 \ldots \lambda_n e_n).$$

But $B^{-1}B = I$, and **AB** only differs from **B** in having scalar multiples $\lambda_1, \lambda_2, \ldots, \lambda_n$ of its columns.

Hence $B^{-1}AB = (\text{diag } \lambda_1, \lambda_2, \ldots, \lambda_n) = D.$

Example. Let $A = \begin{pmatrix} 7 & 9 \\ -2 & -4 \end{pmatrix}$. The characteristic equation of **A** is

$$\lambda^2 - 3\lambda - 10 = 0 \Leftrightarrow \lambda_1 = -2, \lambda_2 = 5.$$

The corresponding eigenvectors are $\begin{pmatrix} 1 \\ -1 \end{pmatrix}, \begin{pmatrix} 9 \\ -2 \end{pmatrix}$ respectively.

$$AB = \begin{pmatrix} 7 & 9 \\ -2 & -4 \end{pmatrix}\begin{pmatrix} 1 & 9 \\ -1 & -2 \end{pmatrix} = \begin{pmatrix} -2 \times 1 & 9 \times 5 \\ -2 \times -1 & -2 \times 5 \end{pmatrix},$$

$$B^{-1}AB = \frac{1}{7}\begin{pmatrix} -2 & -9 \\ 1 & 1 \end{pmatrix}\begin{pmatrix} -2 \times 1 & 9 \times 5 \\ -2 \times -1 & -2 \times 5 \end{pmatrix} = \frac{1}{7}\begin{pmatrix} 7 \times -2 & 0 \\ 0 & 7 \times 5 \end{pmatrix} = \begin{pmatrix} -2 & 0 \\ 0 & 5 \end{pmatrix} = D.$$

Notice that it is not generally possible to diagonalise a matrix if its characteristic equation has any repeated roots.

This "diagonalising" process has various useful applications; for instance, it may be used to calculate powers of a matrix, remembering that if $\mathbf{D} = \text{diag}(d_1, d_2, \ldots, d_n)$, $\mathbf{D}^k = (\text{diag } d_1{}^k, d_2{}^k, \ldots, d_n{}^k)$.

For $\mathbf{B}^{-1}\mathbf{A}\mathbf{B} = \mathbf{D} \Rightarrow (\mathbf{B}^{-1}\mathbf{A}\mathbf{B})(\mathbf{B}^{-1}\mathbf{A}\mathbf{B}) = \mathbf{D}^2$.

But matrices under "multiplication" obey the associative law

$$\Rightarrow \mathbf{B}^{-1}\mathbf{A}\mathbf{I}\mathbf{A}\mathbf{B} = \mathbf{D}^2 = \mathbf{B}^{-1}\mathbf{A}^2\mathbf{B}.$$

Similarly, $\mathbf{B}^{-1}\mathbf{A}^3\mathbf{B} = \mathbf{D}^3$ and, by induction,

$$\mathbf{B}^{-1}\mathbf{A}^k\mathbf{B} = \mathbf{D}^k \text{ for } k \in \mathbb{N}$$

$$\Leftrightarrow \mathbf{A}^k = \mathbf{B}\mathbf{D}^k\mathbf{B}^{-1}.$$

Example. If $\mathbf{A} = \begin{pmatrix} 7 & 9 \\ -2 & -4 \end{pmatrix}$, calculate $\mathbf{A}^4$.

$$\mathbf{A}^4 = \begin{pmatrix} 1 & 9 \\ -1 & -2 \end{pmatrix}\begin{pmatrix} (-2)^4 & 0 \\ 0 & 5^4 \end{pmatrix}\frac{1}{7}\begin{pmatrix} -2 & -9 \\ 1 & 1 \end{pmatrix}$$

$$= \frac{1}{7}\begin{pmatrix} 1 & 9 \\ -1 & -2 \end{pmatrix}\begin{pmatrix} -32 & -144 \\ 625 & 625 \end{pmatrix} = \begin{pmatrix} 799 & 783 \\ -174 & -158 \end{pmatrix}.$$

It can be shown that when $\mathbf{A}$ is symmetric, the diagonalising matrix $\mathbf{B}$, such that $\mathbf{B}^{-1}\mathbf{A}\mathbf{B} = \mathbf{D}$, will be such that its column vectors are mutually perpendicular, and hence will be orthogonal provided that the eigenvectors of $\mathbf{A}$ are normalised. Consider, for example, the 2×2 symmetric matrix $\mathbf{A} = \begin{pmatrix} a & h \\ h & b \end{pmatrix}$, of which the characteristic equation is

$$\lambda^2 - (a+b)\lambda + ab - h^2 = 0$$

with roots λ_1, λ_2, so that $\lambda_1 + \lambda_2 = a+b$, $\lambda_1\lambda_2 = ab - h^2$.

$$\mathbf{B} = \begin{pmatrix} \lambda_1 - a & h \\ h & \lambda_2 - b \end{pmatrix},$$

where we may assume that the eigenvectors are normalised. The scalar product of the column (and row) vectors is $h(\lambda_1 + \lambda_2 - a - b) = 0$, and hence $\mathbf{B}$ is orthogonal.

An example of the occurrence of a symmetric matrix is the equation of a conic $ax^2 + 2hxy + by^2 + c = 0$. The expression $ax^2 + 2hxy + by^2$ is a *quadratic form*, homogeneous of the second degree in x and y, which can be written

$$(x\ y)\begin{pmatrix} a & h \\ h & b \end{pmatrix}\begin{pmatrix} x \\ y \end{pmatrix}.$$

If we diagonalise the matrix $\begin{pmatrix} a & h \\ h & b \end{pmatrix}$ as above, we can regard $\mathbf{B}$ as a transformation matrix such that

$$(x, y) \to (x', y') \text{ where } (x\ y)\mathbf{B}^{-1} = (x'\ y') \text{ and } \mathbf{B}\begin{pmatrix} x \\ y \end{pmatrix} = \begin{pmatrix} x' \\ y' \end{pmatrix}.$$

Since the eigenvectors of $\mathbf{A}$ are perpendicular, they represent possible directions for coordinate axes; also since **B** is orthogonal, we know that it represents either a rotation about O or a reflection (see p. 477). We shall find that in this case it is the rotation required for the new coordinate axes to be in the direction of the eigenvectors.

Example. Determine the nature of the conic $5x^2 + 4xy + 8y^2 = 9$.

The characteristic equation of $\begin{pmatrix} 5 & 2 \\ 2 & 8 \end{pmatrix}$ is $\begin{vmatrix} 5-\lambda & 2 \\ 2 & 8-\lambda \end{vmatrix} = 0$

$$\Leftrightarrow \lambda^2 - 13\lambda + 36 = 0 \Leftrightarrow \lambda = 4, 9.$$

The corresponding eigenvectors are

$$\begin{pmatrix} 2 \\ -1 \end{pmatrix}, \ \begin{pmatrix} 1 \\ 2 \end{pmatrix}$$

which when normalised become

$$\begin{pmatrix} 2/\sqrt{5} \\ -1/\sqrt{5} \end{pmatrix}, \begin{pmatrix} 1/\sqrt{5} \\ 2/\sqrt{5} \end{pmatrix}. \text{ The matrix } \begin{pmatrix} 2/\sqrt{5} & 1/\sqrt{5} \\ -1/\sqrt{5} & 2/\sqrt{5} \end{pmatrix}$$

represents a clockwise rotation through the acute angle $\tan^{-1} 2$ (see p. 477).

The equation $5x^2 + 4xy + 8y^2 = 9$ becomes

$$(x\ y) \begin{pmatrix} 2/\sqrt{5} & -1/\sqrt{5} \\ 1/\sqrt{5} & 2/\sqrt{5} \end{pmatrix} \begin{pmatrix} 5 & 2 \\ 2 & 8 \end{pmatrix} \begin{pmatrix} 2/\sqrt{5} & 1/\sqrt{5} \\ -1/\sqrt{5} & 2/\sqrt{5} \end{pmatrix} \begin{pmatrix} x \\ y \end{pmatrix} = 9$$

$$\Leftrightarrow (x\ y) \begin{pmatrix} 4 & 0 \\ 0 & 9 \end{pmatrix} \begin{pmatrix} x \\ y \end{pmatrix} = 9 \Leftrightarrow 4x^2 + 9y^2 = 9 \Leftrightarrow \frac{x^2}{(9/4)} + y^2 = 1.$$

This is the equation of an ellipse with semi-axes $\frac{3}{2}$, 1.

Exercise 15.10

1. Find the eigenvalues of the matrix **M**, where

$$\mathbf{M} = \begin{pmatrix} 3 & 4 \\ 4 & 3 \end{pmatrix}.$$

Find the corresponding eigenvectors and verify that they are orthogonal. Form an orthogonal matrix **P** such that $\mathbf{P}^{-1}\mathbf{MP}$ is a diagonal matrix, the entries being the eigenvalues.   (O.C.S.M.P.)

2. Find a matrix **P** such that $\mathbf{B} = \mathbf{PAP}^{-1}$, where **B** is a diagonal matrix and

$$\mathbf{A} = \begin{pmatrix} 1 & \sqrt{5} \\ \sqrt{5} & -3 \end{pmatrix}.$$

3. In the matrix equation $\mathbf{Ax} = \lambda\mathbf{x}$, **A** represents the

$$2 \times 2 \text{ matrix } \begin{pmatrix} 3 & 4 \\ -1 & -2 \end{pmatrix},$$

$$\mathbf{x} \text{ the column vector } \begin{pmatrix} x_1 \\ x_2 \end{pmatrix}$$

and λ is a scalar. Prove that the equation has a non-zero solution **x** if, and only if, $\lambda = 2$ or -1. If

$$\mathbf{y} = \begin{pmatrix} y_1 \\ y_2 \end{pmatrix}, \quad \mathbf{z} = \begin{pmatrix} z_1 \\ z_2 \end{pmatrix}$$

are solutions corresponding to $\lambda = 2,\ -1$ respectively, and if the matrix $\mathbf{X}$ is defined by

$$\mathbf{X} = \begin{pmatrix} y_1 & z_1 \\ y_2 & z_2 \end{pmatrix},$$

find $\mathbf{X}^{-1}$ and prove that

$$\mathbf{X}^{-1}\mathbf{A}\mathbf{X} = \begin{pmatrix} 2 & 0 \\ 0 & -1 \end{pmatrix}.$$

15:11 THE CAYLEY–HAMILTON THEOREM

This theorem states that any $n \times n$ matrix $\mathbf{A}$ satisfies its characteristic equation $|\mathbf{A} - \lambda\mathbf{I}| = 0$. It may well appear "obvious" that $|\mathbf{A} - \mathbf{A}\mathbf{I}| = 0$; Cayley himself is said never to have felt obliged to produce a proof of the result, which he regarded as self-evident. We shall only indicate some ways in which the theorem can be applied, using 2×2 matrices for simplicity, though the methods are general.

Since powers of a matrix are often required, it is desirable to have an alternative method available when a matrix cannot be diagonalised (see p. 485). The Cayley–Hamilton theorem provides such an alternative.

Example 1. To find $\mathbf{A}^n$, $n \in \mathbb{N}$, where $\mathbf{A} = \begin{pmatrix} -5 & -6 \\ 1 & 2 \end{pmatrix}$.

The characteristic equation of $\mathbf{A}$ is $\lambda^2 + 3\lambda - 4 = 0$; hence

$$\mathbf{A}^2 + 3\mathbf{A} - 4\mathbf{I} = \mathbf{0}$$

$$\Rightarrow \mathbf{A}^2 = 4\mathbf{I} - 3\mathbf{A} = \begin{pmatrix} 19 & 18 \\ -3 & -2 \end{pmatrix}$$

$$\Rightarrow \mathbf{A}^3 = \mathbf{A}(4\mathbf{I} - 3\mathbf{A}) = 4\mathbf{A} - 3(4\mathbf{I} - 3\mathbf{A}) = 13\mathbf{A} - 12\mathbf{I} = \begin{pmatrix} -77 & -78 \\ 13 & 14 \end{pmatrix}$$

$$\Rightarrow \mathbf{A}^4 = \mathbf{A}(13\mathbf{A} - 12\mathbf{I}) = 13(4\mathbf{I} - 3\mathbf{A}) - 12\mathbf{A} = 52\mathbf{I} - 51\mathbf{A}.$$

Clearly for $n \in \mathbb{N}$, $\mathbf{A}^n$ can be expressed in the form $p\mathbf{I} + q\mathbf{A}$, where p and q are scalars.

Example 2. To find $\mathbf{A}^{-1}$, and $\mathbf{A}^{-n}$ for $n \in \mathbb{N}$, where $\mathbf{A} = \begin{pmatrix} -2 & 0 \\ 7 & -1 \end{pmatrix}$.

The characteristic equation of $\mathbf{A}$ is $\lambda^2 + 3\lambda + 2 = 0$; hence

$$\mathbf{A}^2 + 3\mathbf{A} + 2\mathbf{I} = \mathbf{0}.$$

Since $\mathbf{A}$ is non-singular, $\mathbf{A}^{-1}(\mathbf{A}^2 + 3\mathbf{A} + 2\mathbf{I}) = \mathbf{0}$

$$\Rightarrow \mathbf{A} + 3\mathbf{I} + 2\mathbf{A}^{-1} = \mathbf{0} \Leftrightarrow \mathbf{A}^{-1} = -\tfrac{1}{2}(\mathbf{A} + 3\mathbf{I}) = -\tfrac{1}{2}\begin{pmatrix} 1 & 0 \\ 7 & 2 \end{pmatrix}$$

$$\Rightarrow \mathbf{A}^{-2} = \mathbf{A}^{-1}\mathbf{A}^{-1} = -\tfrac{1}{2}(\mathbf{I} + 3\mathbf{A}^{-1})$$

$$= -\tfrac{1}{2}\mathbf{I} + \tfrac{3}{4}(\mathbf{A} + 3\mathbf{I}) = \tfrac{1}{4}(3\mathbf{A} + 7\mathbf{I}).$$

Hence $\mathbf{A}^{-n}$ for $n \in \mathbb{N}$ can be expressed in the form $r\mathbf{A} + s\mathbf{I}$.

Exercise 15.11

1. Given that

$$A = \begin{pmatrix} 0 & 2 & 3 \\ 2 & 0 & 0 \\ 1 & -1 & 0 \end{pmatrix}$$

evaluate A^{-1} and A^2, and show that $A^2 + 6A^{-1} = 7I$, where I is the unit matrix of order 3. Deduce that $A^3 - 7A + 6I = 0$ and use this equation to find A^3.

Verify your results by using the Cayley–Hamilton theorem.

2. Show that if the matrix A and the vector x are given by

$$A = \begin{pmatrix} 1 & -1 & 4 \\ 2 & 4 & -2 \\ -1 & -1 & 6 \end{pmatrix}, \quad x = \begin{pmatrix} 1 \\ 0 \\ 1 \end{pmatrix},$$

then the vector Ax is a multiple of x.

Obtain the characteristic equation of the matrix A, and hence find eigenvalues. Use your results to calculate A^3 and A^{-1}.

3. A linear transformation T mapping three-dimensional space into itself is defined by

$$T(x, y, z) = (4x - 2y - 2z, \ x + y - z, \ 2x - 2y).$$

(a) Write down the matrix M of T and show that it satisfies the equation

$$M^2 - 3M + 2I = 0,$$

where 0 and I are the zero and unit matrices of order 3 respectively. Hence show that M^{-1} exists and is equal to $\frac{1}{2}(3I - M)$. Find the elements of M^{-1} and state the image of (x, y, z) under the inverse transformation T^{-1}.

(b) Show that there is one line through the origin every point of which is mapped to itself by T, and find equations of this line.   (L.)

Miscellaneous Exercise 15

1. Show that the matrix

$$A = \begin{pmatrix} 1 & 2 \\ 4 & 3 \end{pmatrix}$$

satisfies the equation

$$A^2 - 4A - 5I = 0,$$

where I and 0 denote the 2×2 identity and zero matrices respectively.

Prove by induction that, for each positive integer n, there are real numbers b_n and c_n such that

$$A^n = b_n A + c_n I.$$

Hence or otherwise find the matrix B, where

$$B = A^4 - 3A^3 - 7A^2 - 10A - 6I.$$

2. The matrix

$$A = \begin{pmatrix} p & q \\ r & s \end{pmatrix}$$

has the property that $AA^T = I$. Prove that
(i) if $p = 0$, then $s = 0$ and $q^2 = r^2 = 1$,
(ii) if $p \neq 0$, then $s = \pm p$ and $r = \mp q$.

3. Find three values of k, and for each one a corresponding non-zero vector

$$\begin{pmatrix} x \\ y \\ z \end{pmatrix},$$

which satisfy

$$\begin{pmatrix} 1 & 0 & 0 \\ 2 & 2 & 0 \\ -2 & 2 & 3 \end{pmatrix} \begin{pmatrix} x \\ y \\ z \end{pmatrix} = k \begin{pmatrix} x \\ y \\ z \end{pmatrix}.$$

With the values of k in increasing order, denote the corresponding vectors you have found by $\mathbf{a}$, $\mathbf{b}$, $\mathbf{c}$ and denote the 3×3 matrix by $\mathbf{M}$. Express

$$\begin{pmatrix} 1 \\ 1 \\ 1 \end{pmatrix}$$

in terms of $\mathbf{a}$, $\mathbf{b}$, $\mathbf{c}$ and *hence* show that

$$\mathbf{M}^3 \begin{pmatrix} 1 \\ 1 \\ 1 \end{pmatrix} = \begin{pmatrix} 1 \\ 22 \\ 63 \end{pmatrix}.$$

Show also that values of λ, μ, ν can be found such that

$$(\mathbf{M}^3 + \lambda \mathbf{M}^2 + \mu \mathbf{M} + \nu \mathbf{I}) \mathbf{p} = \mathbf{0}$$

for all vectors $\mathbf{p}$.   (O.C.S.M.P.)

4. (i) Solve the equations

$$\begin{aligned} px + y - z &= 1, \\ -x + py + z &= 0, \\ -x + y - pz &= 1, \end{aligned}$$

in the two cases (a) $p \neq 0$, (b) $p = 0$.

(ii) If

$$\mathbf{M} = \begin{pmatrix} -1 & 0 \\ 0 & 1 \end{pmatrix}, \quad \mathbf{I} = \begin{pmatrix} 1 & 0 \\ 0 & 1 \end{pmatrix},$$

prove that $\mathbf{M}^{2n} = \mathbf{I}$ for every positive integer n.

Write down, with elements expressed in terms of n, a 2×2 matrix $\mathbf{N}$, not equal to $\mathbf{I}$, such that $\mathbf{N}^{2n+1} = \mathbf{I}$.   (L.)

5. If

$$\mathbf{M} = \begin{pmatrix} 1 & -1 & k \\ 4 & 7 & 3 \\ -1 & 12 & -2 \end{pmatrix},$$

evaluate, in terms of k, the determinant of the matrix $\mathbf{M}$.

If $\mathbf{x} = \begin{pmatrix} x \\ y \\ z \end{pmatrix}$, solve the equations

(a) $\mathbf{Mx} = \begin{pmatrix} 1 \\ 11 \\ 21 \end{pmatrix}$ when $k = 2$,

(b) $\mathbf{Mx} = \begin{pmatrix} 0 \\ 0 \\ 0 \end{pmatrix}$ when $k = 1$.

Interpret the result (b) geometrically. (L.)

6. Find triangular matrices $\mathbf{L}$ and $\mathbf{U}$ of the form

$$\mathbf{L} = \begin{pmatrix} 1 & 0 & 0 \\ p & 1 & 0 \\ q & r & 1 \end{pmatrix}, \quad \mathbf{U} = \begin{pmatrix} a & b & c \\ 0 & d & e \\ 0 & 0 & f \end{pmatrix}$$

such that the product $\mathbf{LU}$ equals $\mathbf{A}$, where $\mathbf{A}$ is the matrix

$$\begin{pmatrix} 3 & 1 & 2 \\ 6 & 1 & 5 \\ 9 & 1 & 5 \end{pmatrix}.$$

Find the inverse of the matrix $\mathbf{L}$, and use it to solve the equation

$$\mathbf{A} \begin{pmatrix} x \\ y \\ z \end{pmatrix} = \begin{pmatrix} 1 \\ 1 \\ 4 \end{pmatrix}.$$ (L.)

7. If

$$\mathbf{M} = \begin{pmatrix} 1 & -1 & k \\ 4 & 7 & 3 \\ 3 & 19 & 1 \end{pmatrix},$$

find det $\mathbf{M}$ in terms of k.
If $k = 2$ solve

$$\mathbf{Mx} = \mathbf{0} \quad \text{and} \quad \mathbf{Mx} = \begin{pmatrix} 3 \\ 8 \\ 16 \end{pmatrix}.$$

8. (i) Solve the equation

$$\begin{pmatrix} 1 & -3 & a \\ 2 & a & -1 \\ 0 & 2 & -3 \end{pmatrix} \begin{pmatrix} x \\ y \\ z \end{pmatrix} = \begin{pmatrix} 10 \\ 9 \\ -1 \end{pmatrix}$$

by premultiplying on each side by a suitable matrix, if the constant a is not equal to 16.
Show that if $a = 16$ there is an infinite set of solutions, and express x and y in terms of z. (L.)

9. (i) Find the matrix $\mathbf{A}$ such that $\mathbf{AB} = \mathbf{A} + \mathbf{B}$, where $\mathbf{B}$ is the matrix

$$\begin{pmatrix} 3 & 0 & 0 \\ 0 & 5 & 0 \\ 1 & 0 & 2 \end{pmatrix}.$$

(ii) Evaluate the determinant of the coefficients in the simultaneous equations

$$x - (4k - 3)y + 2z = 0,$$
$$kx - (2k - 1)y + (k + 1)z = 0,$$
$$(2k + 2)x + 3ky + (k + 2)z = 0.$$

Find all the solutions of the equations when $k = 1$. Show that, if $k \neq 1$, the only real solution is $x = y = z = 0$. (L.)

10. (i) Find all the matrices **A** such that **AB** = **BA** and **A**^{2} = **I**, where **I** is the unit matrix of order 2 and

$$\mathbf{B} = \begin{pmatrix} 1 & -1 \\ 0 & 2 \end{pmatrix}.$$

(ii) Find the product **AB** of the matrices

$$\mathbf{A} = \begin{pmatrix} 2 & 6 & 4 \\ 1 & 3 & -2 \\ -1 & 5 & 2 \end{pmatrix}, \quad \mathbf{B} = \begin{pmatrix} 2 & 1 & -3 \\ 0 & 1 & 1 \\ 1 & -2 & 0 \end{pmatrix}.$$

Use the result to solve the system of equations

$$\begin{aligned} 2x + y - 3z &= 10, \\ y + z &= -4, \\ x - 2y &= 11. \end{aligned}$$

Check your solution.   (L.)

11. Matrices **A**, **D** are defined by

$$\mathbf{A} = \begin{pmatrix} 1 & -3 & -3 \\ -8 & 6 & -3 \\ 8 & -2 & 7 \end{pmatrix}, \mathbf{D} = \begin{pmatrix} 1 & 0 & 0 \\ 0 & 4 & 0 \\ 0 & 0 & 9 \end{pmatrix}.$$

For each of the eigenvalues 1, 4, 9 of **A** find a corresponding eigenvector. Write down a matrix **P** such that **P**$^{-1}$**AP** = **D**, and calculate **P**$^{-1}$.

Write down a matrix **C** such that **C**^{2} = **D**. Hence, or otherwise, find a matrix **B** such that **B**^{2} = **A**.

(O.C.S.M.P.)

12. Find vectors **u** and **v** such that **Au** = λ**u** and **Av** = μ**v**, where **A** is the matrix

$$\begin{pmatrix} 1 & 2 \\ 1 & 1 \end{pmatrix}$$

and λ and μ are unequal constants.

Express the vector $\begin{pmatrix} 1 \\ 0 \end{pmatrix}$ as the sum of multiples of **u** and **v**.

A sequence of vectors is defined by the iteration $\mathbf{x}_{n+1} = \mathbf{Ax}_n$, with $\mathbf{x}_0 = \begin{pmatrix} 1 \\ 0 \end{pmatrix}$. Express $\mathbf{x}_n$ in terms of **u** and **v**, and show that the ratio of the two elements of $\mathbf{x}_n$ tends to a limit as n tends to infinity.   (L.)

13. Find numbers a, b and c so that the product **BA** of the matrices

$$\mathbf{B} = \begin{pmatrix} 1 & 0 & 0 \\ a & 1 & 0 \\ b & c & 1 \end{pmatrix}, \quad \mathbf{A} = \begin{pmatrix} 1 & 2 & 3 \\ -2 & 1 & 4 \\ 2 & 1 & 1 \end{pmatrix}$$

should have only zeros below the leading diagonal.

Hence, or otherwise, solve the equation

$$\mathbf{Ax} = \rho,$$

where **x** is the column vector $\begin{pmatrix} x_1 \\ x_2 \\ x_3 \end{pmatrix}$ and ρ is the column vector $\begin{pmatrix} 1 \\ -1 \\ 1 \end{pmatrix}$.   (L.)

14. Matrices **R**, **S**, **T** are given by

$$\mathbf{R} = \begin{pmatrix} 0 & 1 \\ 1 & 0 \end{pmatrix}, \quad \mathbf{S} = \begin{pmatrix} 0 & -i \\ i & 0 \end{pmatrix}, \quad \mathbf{T} = \begin{pmatrix} 1 & 0 \\ 0 & -1 \end{pmatrix},$$

where $i^2 = -1$. Show that:

(i) $R^2 = S^2 = T^2 = I$;

(ii) for each of the matrices, $X^* = X$ where * denotes the operation of transposing and taking the complex conjugate;

(iii) $TR = -RT = iS$, and find RS, SR, ST, TS;

(iv) every matrix X for which $X^* = X$ can be expressed in the form

$$X = x_0 I + x_1 R + x_2 S + x_3 T,$$

where x_0, x_1, x_2, x_3 are real numbers, and the determinant of X is then $x_0^2 - x_1^2 - x_2^2 - x_3^2$.

15. Find the solution set of the equations

$$\begin{pmatrix} 1 & 3 & a \\ 2 & -1 & -5 \\ 1 & 1 & 2 \end{pmatrix} \begin{pmatrix} x \\ y \\ z \end{pmatrix} = \begin{pmatrix} 4 \\ b \\ 1 \end{pmatrix}$$

in the following cases:

(i) $a = 9$, $b = -1$, (ii) $a = 8$, $b = -1$,

(iii) $a = 8$, $b = -2\frac{1}{2}$.

For what value of a does the equation

(i) $\begin{pmatrix} a & -5 & 2 \\ 3 & -1 & 1 \\ 1 & 2 & 1 \end{pmatrix} \begin{pmatrix} x \\ y \\ z \end{pmatrix} = \begin{pmatrix} 4 \\ 7 \\ 8 \end{pmatrix}$ not have a unique solution;

(ii) $\begin{pmatrix} a & -10 & 4 \\ 3 & -1 & 1 \\ 1 & 2 & 1 \end{pmatrix} \begin{pmatrix} x \\ y \\ z \end{pmatrix} = \begin{pmatrix} 4 \\ 7 \\ 9 \end{pmatrix}$ not have a unique solution?　　　　　(O.C.S.M.P.)

16. If $A = \begin{pmatrix} 11 & -25 \\ 4 & -9 \end{pmatrix}$ and n is a positive integer, prove by induction that

$$A^n = \begin{pmatrix} 1 + 10n & -25n \\ 4n & 1 - 10n \end{pmatrix}.$$　　　　(L.)

17. Find the matrix M^2 where M is given by

$$M = \begin{pmatrix} \cos\theta & k\sin\theta \\ (1/k)\sin\theta & -\cos\theta \end{pmatrix},$$

and k is non-zero.

Given that $MX = \lambda X$, where λ is a constant, and X is a non-null column matrix with elements x_1, x_2, show, by premultiplying this equation by M, or otherwise, that $\lambda = \pm 1$.

Show that the values of X corresponding to these two values of λ are

$$\begin{pmatrix} \mu_1 k \cos\frac{1}{2}\theta \\ \mu_1 \sin\frac{1}{2}\theta \end{pmatrix} \quad \text{and} \quad \begin{pmatrix} -\mu_2 k \sin\frac{1}{2}\theta \\ \mu_2 \cos\frac{1}{2}\theta \end{pmatrix}$$

respectively, where μ_1 and μ_2 are arbitrary constants, and find the values k if the vectors are to be orthogonal.

(N.)

18. The transformation with matrix T maps the point $P(x, y)$ of a plane onto the point $P'(x', y')$, where

$$\begin{pmatrix} x' \\ y' \end{pmatrix} = T \begin{pmatrix} x \\ y \end{pmatrix}.$$

By considering the images of the points $A(1, 0)$ and $B(0, 1)$, or otherwise, determine the geometrical transformation represented by each of the matrices

$$T_1(\theta) = \begin{pmatrix} \cos\theta & -\sin\theta \\ \sin\theta & \cos\theta \end{pmatrix},$$

$$T_2(\phi) = \begin{pmatrix} \cos 2\phi & \sin 2\phi \\ \sin 2\phi & -\cos 2\phi \end{pmatrix}.$$

Verify by matrix multiplication that

$$T_2(p)\, T_2(q) = T_1[2(p-q)],$$
$$T_2(p)\, T_1(q) = T_1(-q)\, T_2(p),$$

and interpret these results geometrically.   (N.)

19. The transformation $T:\mathbf{r} \to \mathbf{r}'$ of a three-dimensional space is given by $\mathbf{r}' = \mathbf{M}\mathbf{r}$ where

$$\mathbf{r} = \begin{pmatrix} x \\ y \\ z \end{pmatrix}, \quad \mathbf{r}' = \begin{pmatrix} x' \\ y' \\ z' \end{pmatrix} \quad \text{and} \quad \mathbf{M} = \begin{pmatrix} 1 & 2 & 3 \\ 2 & 0 & -2 \\ 3 & -2 & -7 \end{pmatrix}.$$

Show that T maps the whole space onto the plane Π whose equation is $x - 2y + z = 0$.
   Find the image under T of
   (i) the line $x = -y = \tfrac{1}{2}z$,
   (ii) the plane $x - y - z = 0$,
   (iii) the plane $x - z = 0$.

16 Further Calculus

16:1 TAYLOR'S SERIES

If it is required to approximate to the value of $f(x)$ when $x \approx a$ ($\neq 0$), it is usually convenient to employ Maclaurin's series in a different form, known as Taylor's series.

If
$$f(a+h) \equiv a_0 + a_1 h + a_2 h^2 + \ldots$$

(conventionally h is used as the variable), then, making the same assumptions as for Maclaurin's expansion (p. 297),

$$a_0 = f(a), \ a_1 = f'(a), \ \ldots, \ a_n = f^{(n)}(a)/n!, \ \ldots$$

$$\Rightarrow f(a+h) = f(a) + hf'(a) + \frac{h^2}{2!}f''(a) + \frac{h^3}{3!}f'''(a) + \ldots \qquad (16.1)$$

Example 1. Find an approximate value for $\tan 46°$.

$\tan 46° = \tan(\pi/4 + \pi/180)$ – remember that radian measure *must* be used in calculus work; putting $a = \pi/4$, $h = \pi/180$,

$$\tan 46° = \tan(\pi/4) + \tfrac{\pi}{180} \sec^2(\pi/4) + \tfrac{1}{2}(\tfrac{\pi}{180})^2 \cdot 2\sec^2(\pi/4)\tan(\pi/4) + \ldots$$
$$\approx 1 + \pi/90 + 2(\pi/180)^2 + \ldots$$
$$\approx 1 + 0{\cdot}0349 + 0{\cdot}0006 + \ldots$$
$$= 1{\cdot}0355 \text{ correct to 4 decimal places.}$$

Example 2. Taylor's theorem may be used to investigate limits of the form

$$\lim_{x \to a}\left[\frac{f(x)}{g(x)}\right], \text{ where } f(a) = 0 = g(a).$$

Writing $x = a + h$ and using Taylor's theorem gives

$$\lim_{x \to a}\left[\frac{f(x)}{g(x)}\right] = \lim_{h \to 0}\left[\frac{f(a+h)}{g(a+h)}\right] = \lim_{h \to 0}\left[\frac{hf'(a) + h^2\frac{f''(a)}{2!} + h^3\frac{f'''(a)}{3!} + \ldots}{hg'(a) + h^2\frac{g''(a)}{2!} + h^3\frac{g'''(a)}{3!} + \ldots}\right].$$

Clearly if $f^{(p)}(a)$, $g^{(q)}(a)$ are the first non-vanishing derivatives at $x = a$ of $f(x)$, $g(x)$ respectively, then the limit takes the value 0 if $p > q$, $f^{(p)}(a)/g^{(p)}(a)$ if $p = q$, $\pm \infty$ if $p < q$. This is the generalisation of l'Hôpital's rule.

For example, $\displaystyle\lim_{x \to a}\left[\frac{x^2 \ln x - a^2 \ln a}{x^2 - a^2}\right] = \lim_{x \to a}\left[\frac{2x \ln x + x}{2x}\right] = \ln a + \frac{1}{2}$

Taylor's series leads immediately to the Newton–Raphson method of approximation to a root of an equation (p. 365). For, if $f(a + h) = 0$ where h is small,

$$0 = f(a + h) \approx f(a) + hf'(a)$$

$$\Rightarrow h \approx -f(a)/f'(a).$$

16:2 FURTHER EXPANSIONS

For a complicated, or implicit, function, successive differentiation to obtain the coefficients in Maclaurin's expansion may be difficult or impossible. If an identity can be obtained involving perhaps y, dy/dx and d^2y/dx^2, successive differentiation of this identity may give the required results, as the following examples illustrate. Example 3 introduces an important result, of particular use in probability theory.

Example 1. Given that $y = \ln[x + \sqrt{(x^2 + 1)}]$, expand y in ascending powers of x as far as the term in x^5.

Writing y_1 for dy/dx, etc., so that $y_n = d^ny/dx^n$,

$$y_1 = [x + \sqrt{(x^2 + 1)}]^{-1}\left[1 + \frac{x}{\sqrt{(x^2 + 1)}}\right] = (x^2 + 1)^{-1/2} \Rightarrow y_1(0) = 1.$$

$$(1 + x^2)y_1{}^2 = 1 \Rightarrow 2xy_1^2 + (1 + x^2)2y_1y_2 = 0$$

$$\Rightarrow (1 + x^2)y_2 + xy_1 = 0 \text{ (since } y_1 \text{ is not in general zero)}$$

$$\Rightarrow y_2(0) = 0.$$

Differentiating and putting $x = 0$ successively,

$$(1 + x^2)y_3 + 3xy_2 + y_1 = 0 \Rightarrow y_3(0) = -1,$$

$$\Rightarrow (1 + x^2)y_4 + 5xy_3 + 4y_2 = 0 \Rightarrow y_4(0) = 0,$$

$$\Rightarrow (1 + x^2)y_5 + 7xy_4 + 9y_3 = 0 \Rightarrow y_5(0) = 9,$$

$$\Rightarrow y = x - \frac{1}{6}x^3 + \frac{3}{40}x^5 \ldots.$$

Example 2. Write down the expansion of $\ln[1 + (1/n)]$ in ascending powers of $1/n$ and show that, if n is a large positive number, $\left(1 + \dfrac{1}{n}\right)^n = e\left(1 - \dfrac{1}{2n} + \dfrac{11}{24n^2} - \dfrac{7}{16n^3}\right)$ approximately.

$$\ln\left(1 + \frac{1}{n}\right) = \frac{1}{n} - \frac{1}{2n^2} + \frac{1}{3n^3} - \ldots + \frac{(-1)^{r-1}}{rn^r} + \ldots$$

$$\Rightarrow n\ln\left(1 + \frac{1}{n}\right) = 1 - \frac{1}{2n} + \frac{1}{3n^2} - \frac{1}{4n^3} + O\left(\frac{1}{n^4}\right)$$

$$\Rightarrow \left(1 + \frac{1}{n}\right)^n = \exp\left[1 - \frac{1}{2n} + \frac{1}{3n^2} - \frac{1}{4n^3} + O\left(\frac{1}{n^4}\right)\right]$$

$$= e.\exp\left[-\frac{1}{2n} + \frac{1}{3n^2} - \frac{1}{4n^3} + O\left(\frac{1}{n^4}\right)\right]$$

$$= e\left[1 + \left(-\frac{1}{2n} + \frac{1}{3n^2} - \frac{1}{4n^3}\right) + \frac{1}{2!}\left(-\frac{1}{2n} + \frac{1}{3n^2}\right)^2\right.$$

$$\left. + \frac{1}{3!}\left(-\frac{1}{2n}\right)^3 + O\left(\frac{1}{n^4}\right)\right]$$

$$\Rightarrow \left(1 + \frac{1}{n}\right)^n = e\left(1 - \frac{1}{2n} + \frac{11}{24n^2} - \frac{7}{16n^3}\right) + O\left(\frac{1}{n^4}\right),$$

which implies the required result.

Example 3. Show that $\quad \lim\limits_{n \to \infty} \left(1 + \dfrac{x}{n}\right)^n = e^x.$

Here we use a theorem of pure mathematics which states that if

$$\lim_{t \to a} \{\ln f(t)\} = k, \quad \text{then} \quad \lim_{t \to a} \{f(t)\} = e^k.$$

In our case $\quad \lim\limits_{n \to \infty} \left[\ln\left(1 + \dfrac{x}{n}\right)^n\right] = \lim\limits_{n \to \infty} \left[n \ln\left(1 + \dfrac{x}{n}\right)\right]$

$$= \lim_{n \to \infty} \left(n\left[\frac{x}{n} + O\left(\frac{x^2}{n^2}\right)\right]\right) = x$$

$$\Rightarrow \lim_{n \to \infty} \left(1 + \frac{x}{n}\right)^n = e^x.$$

Similarly $\qquad\qquad\qquad\qquad \lim\limits_{n \to \infty} \left(1 + \dfrac{1}{n}\right)^{nx} = e^x.$

In particular $\qquad\qquad\qquad \lim\limits_{n \to \infty} \left(1 + \dfrac{1}{n}\right)^n = e.$

Exercise 16.2

1. If $y = \tan(e^x - 1)$, prove that

$$\frac{d^2 y}{dx^2} = \frac{dy}{dx}(1 + 2e^x y).$$

Obtain the Maclaurin expansion of y as far as the term in x^4 inclusive.   (C.)

2. If $y = \ln(\sec x + \tan x)$ prove that

$$e^y \frac{dy}{dx} = \left(\frac{dy}{dx}\right)^2 + \frac{d^2 y}{dx^2}.$$

Expand y in ascending powers of x as far as the term in x^3, and show that the coefficient of x^4 is zero.
(C.)

3. Prove that, if $t = \tan x$ and $n \geqslant 1$,

$$\frac{d}{dx}(t^n) = n(t^{n-1} + t^{n+1}).$$

Hence calculate the first five derivatives of $\tan x$ in terms of t and prove that

$$\tan x = x + \tfrac{1}{3}x^3 + \tfrac{2}{15}x^5 + \dots \quad$$ (O.C.)

4. If $y = \tan[\ln(1 + x)]$, prove that

$$(1 + x)\frac{dy}{dx} = 1 + y^2.$$

Find the value of y and its first four derivatives when $x = 0$, and hence or otherwise show that the power series for y begins with the terms

$$x - \tfrac{1}{2}x^2 + \tfrac{2}{3}x^3 - \tfrac{3}{4}x^4.$$ (O.C.)

5. If $y = \sqrt{(1 + \sin^2 x)}$, prove that

$$y\frac{d^2 y}{dx^2} + \left(\frac{dy}{dx}\right)^2 = 3 - 2y^2.$$

By repeated differentiation of this relation, obtain the Maclaurin expansion of y, neglecting x^5 and higher powers of x.

Verify this result by using the expansions of $\sin x$ and $(1 + u)^{1/2}$.   (O.C.)

6. Use Maclaurin's series to expand $\ln(1 + x)$, for $-1 < x < 1$, in ascending powers of x as far as the term in x^4.

If $y = \ln(1 + \sin x)$ prove that $\cos x \dfrac{d^2 y}{dx^2} + \dfrac{dy}{dx} = 0$.

Hence or otherwise prove that, neglecting powers of x higher than x^4, $\ln(1 + \sin x)$ differs from $\ln(1 + x)$ by $\frac{1}{6}(x^3 - x^4)$. (Any expansion used must be obtained as a Maclaurin series.)   (C.)

16:3 LEIBNIZ' THEOREM ON REPEATED DIFFERENTIATION OF A PRODUCT

As shown in the preceding section, in the course of obtaining an expansion it is necessary to differentiate a product repeatedly, either in the original function or in an identity. A useful result in this situation is Leibniz' theorem, which in some cases may enable the general term of an expansion to be obtained. The theorem states that if u and v are functions of x, then

$$(uv)_n = u_0 v_n + \binom{n}{1} u_1 v_{n-1} + \binom{n}{2} u_2 v_{n-2} + \cdots$$

$$+ \binom{n}{r} u_r v_{n-r} + \cdots + \binom{n}{n-1} u_{n-1} v_1 + u_n v_0. \qquad (16.2)$$

Here $\binom{n}{r}$ denotes the binomial coefficient $_nC_r = {}^nC_r = \dfrac{n!}{r!\,(n-r)!}$ and suffixes denote differentiation with respect to x,

e.g. $u_0 = u, \quad u_r v_{n-r} = \dfrac{d^r u}{dx^r} \cdot \dfrac{d^{n-r} v}{dx^{n-r}}.$

We prove this theorem by mathematical induction as follows:

Assuming the truth of the theorem for an integer $n = k$ and differentiating with respect to x, we obtain

$$(uv)_{k+1} = (u_0 v_{k+1} + u_1 v_k) + \binom{k}{1}(u_1 v_k + u_2 v_{k-1})$$

$$+ \binom{k}{2}(u_2 v_{k-1} + u_3 v_{k-2}) + \cdots$$

$$+ \binom{k}{r}(u_r v_{k+1-r} + u_{r+1} v_{k-r}) + \cdots$$

$$+ (u_k v_1 + u_{k+1} v_0).$$

The coefficient of the term involving $u_r v_{k+1-r}$ is

$$\binom{k}{r-1} + \binom{k}{r} = \frac{k!}{(r-1)!\,(k+1-r)!} + \frac{k!}{r!\,(k-r)!}$$

$$= \frac{(k+1)!}{r!\,(k+1-r)!} = \binom{k+1}{r}$$

$$\Rightarrow (uv)_{k+1} = u_0 v_{k+1} + \binom{k+1}{1} u_1 v_k$$

$$+ \binom{k+1}{2} u_2 v_{k-1} + \cdots$$

$$+ \binom{k+1}{r} u_r v_{k+1-r} + \cdots$$

$$+ \binom{k+1}{k} u_k v_1 + u_{k+1} v_0. \tag{16.3}$$

But equation (16.3) is merely the statement of the theorem with $k+1$ in place of k, and the theorem is true when $k = 1$ by virtue of the product rule. Hence it is true for $k = 1 + 1 = 2$ and therefore for $k = 2 + 1 = 3$ and so on for all $k \in \mathbb{N}$.

Example 1. Writing $u = x^2$, $v = e^{ax}$ in Leibniz' theorem gives, provided $n \geq 2$,

$$\frac{d^n}{dx^n}(x^2 e^{ax}) = x^2 \frac{d^n}{dx^n}(e^{ax}) + n.2x \frac{d^{n-1}(e^{ax})}{dx^{n-1}} + \tfrac{1}{2}n(n-1).2 \frac{d^{n-2}(e^{ax})}{dx^{n-2}}$$

$$= [a^2 x^n + 2nax + n(n-1)]\, a^{n-2} e^{ax}.$$

[The remaining terms vanish since $\dfrac{d^r(x^2)}{dx^r} = 0$ for $r > 2$.]

Example 2. If $y = (x^2 - 1)^n$, where $n \in \mathbb{N}$, prove that

$$(1-x^2)\frac{dy}{dx} + 2nxy = 0.$$

By differentiating this equation $(n+1)$ times and using Leibniz' theorem, or otherwise, show that the polynomial $p_n(x)$, defined by

$$p_n(x) = \frac{d^n}{dx^n}\{(x^2-1)^n\},$$

satisfies the equation

$$(1-x^2)\frac{d^2 p_n}{dx^2} - 2x\frac{dp_n}{dx} + n(n+1)p_n = 0.$$

Show also that

$$p_n(1) = (-1)^n p_n(-1) = 2^n n!$$

$$\ln y = n \ln(x^2 - 1) \Rightarrow \frac{1}{y}\frac{dy}{dx} = \frac{2nx}{x^2 - 1}$$

$$\Rightarrow (1-x^2)\frac{dy}{dx} + 2nxy = 0. \tag{1}$$

Differentiating $(n+1)$ times by using Leibniz' theorem we have

$$(1-x^2)\frac{d^{n+2}y}{dx^{n+2}} - 2(n+1)x\frac{d^{n+1}y}{dx^{n+1}} - (n+1)n\frac{d^n y}{dx^n}$$

$$+ 2n\left\{x\frac{d^{n+1}y}{dx^{n+1}} + (n+1)\frac{d^n y}{dx^n}\right\} = 0.$$

This reduces to the given equation on writing $p_n(x) = d^n/dx^n\{(x^2-1)^n\}$. To find $p_n(1)$ we need $d^n/dx^n[(x+1)^n(x-1)^n]$ when $x = 1$. Writing $u = (x+1)^n$, $v = (x-1)^n$ and noting that $d^r/dx^r[(x-1)^n] = n!(x-1)^{n-r}/(n-r)!$ vanishes when $x = 1$ except in the case $r = n$, we obtain the required value as

$$[(x+1)^n d^n/dx^n \{(x-1)^n\}]_{x=1} = 2^n.n!. \tag{2}$$

Similarly, to find $p_n(-1)$ we need $d^n/dx^n\{(x+1)^n (x-1)^n\}$ when $x = -1$ and this time the required value is

$$[(x-1)^n d^n/dx^n \{(x+1)^n\}]_{x=-1} = (-2)^n.n!. \tag{3}$$

Equations (2) and (3) are equivalent to the required results.

Example 3. If $y = f(x) = [x + \sqrt{(1+x^2)}]^{1/2}$, show that

$$(1+x^2)\frac{d^2 y}{dx^2} + x\frac{dy}{dx} - \frac{1}{4}y = 0.$$

Differentiate this equation n times by Leibniz' theorem and deduce that

$$f^{(n+2)}(0) = (\tfrac{1}{4} - n^2) f^{(n)}(0).$$

Hence find the expansion of y in ascending powers of x.

[Note that here we use $f^{(n)}(x)$ to denote $\dfrac{d^n}{dx^n}f(x).$]

$$\frac{dy}{dx} = \frac{1 + x/\sqrt{(1+x^2)}}{2[x + \sqrt{(1+x^2)}]^{1/2}} = \frac{y}{2\sqrt{(1+x^2)}} \tag{1}$$

$$\Rightarrow (1+x^2)\left(\frac{dy}{dx}\right)^2 = \frac{1}{4}y^2.$$

Differentiating and cancelling $2\,dy/dx$ gives

$$(1+x^2)\frac{d^2 y}{dx^2} + x\frac{dy}{dx} - \frac{1}{4}y = 0.$$

Leibniz' theorem gives

$$(1+x^2) f^{(n+2)}(x) + 2nx\, f^{(n+1)}(x) + n(n-1)\,f^{(n)}(x) + x\,f^{(n+1)}(x)$$
$$+ n\,f^{(n)}(x) - \tfrac{1}{4}f^{(n)}(x) = 0$$
$$\Rightarrow f^{(n+2)}(0) = (\tfrac{1}{4} - n^2)\,f^{(n)}(0). \tag{2}$$

But $f(0) = 1$ and equation (1) gives $f'(0) = \tfrac{1}{2}$.
    Then equation (2) gives

$$f^{2p}(0) = 1.\tfrac{1}{4}.(\tfrac{1}{4} - 2^2) \ldots \{\tfrac{1}{4} - (2p-2)^2\} = a_{2p},$$
$$f^{2p+1}(0) = \tfrac{1}{2}.(\tfrac{1}{4} - 1^2).(\tfrac{1}{4} - 3^2) \ldots \{\tfrac{1}{4} - (2p-1)^2\} = b_{2p+1}.$$

Hence

$$[x + \sqrt{(1+x^2)}]^{1/2} = \sum_{p=0}^{\infty} \frac{a_{2p}}{(2p)!} x^{2p} + \sum_{p=0}^{\infty} \frac{b_{2p+1}}{(2p+1)!} x^{2p+1}. \tag{3}$$

This result is valid for $|x| < 1$ since the radius of convergence of each of the series of equation (3) is unity. [See §17:2.]

Example 4. Maclaurin's theorem can be used, in conjunction with Leibniz' theorem, to obtain series solutions of certain differential equations subject to prescribed conditions when $x = 0$. In illustration we first derive the power series for $\sin x$ which we define as that solution of the equation

$$f''(x) + f(x) = 0 \tag{1}$$

for which $f(0) = 0$, $f'(0) = 1$.

Differentiating equation (1) n times and putting $x = 0$, we find

$$f^{(n+2)}(0) = -f^{(n)}(0).$$

It follows that

$$f^{(2r+1)}(0) = (-1)^r f'(0) = (-1)^r,$$
$$f^{(2r)}(0) = (-1)^{(r)} f(0) = 0.$$

The series expansion for $f(x)$ is therefore

$$x - \frac{x^3}{3!} + \frac{x^5}{5!} - \cdots + \frac{(-1)^r x^{2r+1}}{(2r+1)!} + \cdots.$$

Similarly we can derive the series for $\cos x$ when defined as that solution of equation (1) for which $f(0) = 1$, $f'(0) = 0$.

Example 5. We obtain that solution $y = f(x)$ of *Airy's equation* $d^2y/dx^2 - xy = 0$ for which $f(0) = 0$, $f'(0) = 1$.

Differentiating Airy's equation $(n+1)$ times by Leibniz' theorem, we find

$$f^{(n+3)}(x) - x f^{(n+1)}(x) - (n+1) f^n(x) = 0$$

so that

$$f^{(n+3)}(0) = (n+1) f^n(0), \quad (n \geqslant 0).$$

But, putting $x = 0$ in Airy's equation, we have $f''(0) = 0$. Hence

$$f^{(3r)}(0) = 0, \quad f^{(3r+2)}(0) = 0, \quad f^{(3r+1)}(0) = 2.5 \ldots (3r-1)$$

$$\Rightarrow f(x) = x + \frac{2x^4}{4!} + \frac{2.5 \, x^7}{7!} + \cdots + \frac{2.5 \ldots (3r-1)x^{3r+1}}{(3r+1)!} + \cdots.$$

Exercise 16.3

1. If $y = e^{ax} \cos bx$, show by induction that $d^n y/dx^n = r^n e^{ax} \cos(bx + n\theta)$ where $r^2 = a^2 + b^2$ and $\tan\theta = b/a$.

2. If $y = (x^3 - 3x^2)e^{2x}$, find $d^6 y/dx^6$.

3. If $x(1-x) y_2 + 2y = 0$, prove that

$$x(1-x) y_{n+2} + n(1-2x) y_{n+1} = (n+1)(n-2) y_n.$$

4. Prove that $\dfrac{d^{n+1}}{dx^{n+1}}(x\, y) = (n+1)\, y_n + x\, y_{n+1}$.

By taking $y = x^{n-1} e^{1/x}$ prove, by induction, that

$$\frac{d^n}{dx^n}(x^{n-1} e^{1/x}) = (-1)^n \frac{e^{1/x}}{x^{n+1}}.$$

5. If $y = \dfrac{\sin x}{1 - x^2}$, show that

(i) $(1 - x^2)\dfrac{d^2 y}{dx^2} - 4x\dfrac{dy}{dx} - (1 + x^2)y = 0$,

(ii) $y_{n+2}(0) - (n^2 + 3n + 1)y_n(0) - n(n-1)y_{n-2}(0) = 0$.

6. Find $d^6 y/dx^6$ for each of the following:

$$\text{(i) } y = x^5 \ln x, \quad \text{(ii) } y = e^{2x} \sin 2x.$$

7. If $x = \cos t$, $y = \cos 2pt$, show that

(i) $(1 - x^2)y_2 - xy_1 + 4p^2 y = 0$,

(ii) $(1 - x^2)y_{n+2} - (2n+1)xy_{n+1} + (4p^2 - n^2)y_n = 0$.

8. Determine the successive derivatives of $f(x)$ at $x = 0$, given that

$$\frac{df(x)}{dx} = x + f(x), \quad f(0) = 0,$$

and give the value of $f^{(n)}(0)$ for $n > 1$.
Hence show that the solution of the differential equation

$$\frac{dy}{dx} = x + y,$$

satisfying $y = 0$ when $x = 0$, is

$$y = e^x - 1 - x.$$

9. The *Bessel function of order zero* satisfies the equation

$$x \frac{d^2 J}{dx^2} + \frac{dJ}{dx} + xJ = 0,$$

together with the conditions that $J = 1$ when $x = 0$, and all its derivatives exist at $x = 0$.
Differentiate the equation n times, and show that

$$\frac{d^{n+1}J}{dx^{n+1}} = -\frac{n}{n+1}\frac{d^{n-1}J}{dx^{n-1}} \quad \text{when} \quad x = 0 \quad (n \geq 1).$$

Derive the Maclaurin series for $J(x)$.

16:4 INVERSE CIRCULAR FUNCTIONS

We use the notation $\operatorname{Sin}^{-1} x$ to denote "an angle whose sine is x". If $y = \operatorname{Sin}^{-1} x$, then $x = \sin y$ and the graph of $y = \operatorname{Sin}^{-1} x$ is shown in Fig. 16.1(i).
Clearly $\operatorname{Sin}^{-1} x$ is not a function of x, but we can define a function f with domain $-1 \leq x \leq 1$ and range $-\pi/2 \leq f(x) \leq \pi/2$. We shall write

$$f: x \to \sin^{-1} x,$$

where $f(x) = \sin^{-1} x$ is called the *principal value* of $\operatorname{Sin}^{-1} x$. The graph of $\sin^{-1} x$ is shown in Fig. 16.1(ii).
[The notation $\operatorname{Sin}^{-1} x$, with a capital S, for the general value and $\sin^{-1} x$, with a small s, for the principal value is not universally employed in books but we shall adhere to it for clarity.]
The graphs of $y = \operatorname{Cos}^{-1} x$ and $y = \cos^{-1} x$ are shown in Fig. 16.1(iii) and Fig. 16.1(iv). The principal value of $\operatorname{Cos}^{-1} x$ is the value of y in the range $0 \leq y \leq \pi$ for any given value of x.
The graphs of $y = \operatorname{Tan}^{-1} x$ and $y = \tan^{-1} x$ are shown in Fig. 16.1(v) and Fig. 16.1(vi). The principal value of $y = \operatorname{Tan}^{-1} x$ is the value of y in the range

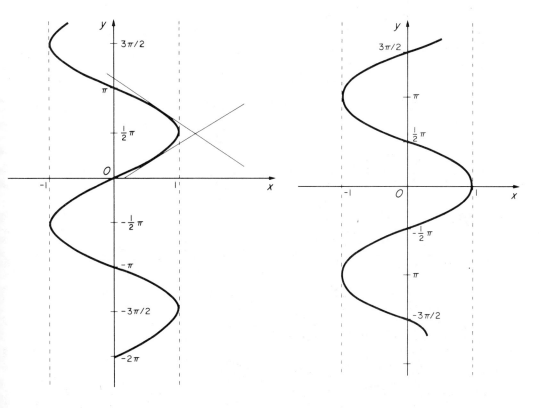

FIG. 16.1(i).   FIG. 16.1(iii).

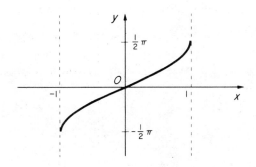

FIG. 16.1(ii).

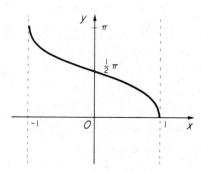

FIG. 16.1(iv).

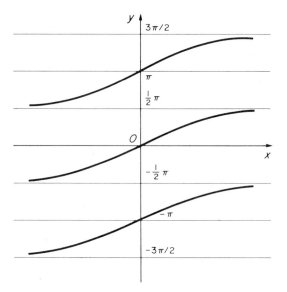

FIG. 16.1(v).

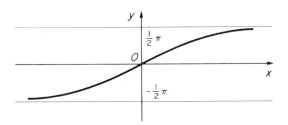

FIG. 16.1(vi).

$-\frac{1}{2}\pi \leqslant y \leqslant \frac{1}{2}\pi$ for any given value of x. The graphs show that $\operatorname{Sin}^{-1} x$ and $\operatorname{Cos}^{-1} x$ are defined for the domain $-1 \leqslant x \leqslant 1$ only but that $\operatorname{Tan}^{-1} x$ is defined $\forall\, x \in \mathbb{R}$. In some books $\sin^{-1} x$, $\tan^{-1} x$, etc., are denoted by $\arcsin x$, $\arctan x$, etc.

[Remember that the graph of $f^{-1}(x)$ can be obtained by reflection of the graph of $f(x)$ in the line $y = x$.]

Example 1. $\operatorname{Sin}^{-1}\left(\frac{1}{2}\right) = n\pi + (-1)^{n}\, \pi/6,$

$\qquad \sin^{-1}\left(\frac{1}{2}\right) = \pi/6,$

$\qquad \operatorname{Cos}^{-1}\left(\frac{1}{2}\right) = 2n\pi \pm \pi/3,$

$\qquad \cos^{-1}\left(\frac{1}{2}\right) = \pi/3,$

$\qquad \operatorname{Tan}^{-1}\left(\sqrt{3}\right) = n\pi + \pi/3,$

$\qquad \tan^{-1}\left(\sqrt{3}\right) = \pi/3,$

where $n \in \mathbb{Z}$.

Example 2. Solve the equation $\tan^{-1} 2x - \tan^{-1} x = \tan^{-1}\frac{1}{3}$, where $0 \leqslant 2x \leqslant \frac{1}{2}\pi$.

If $\tan^{-1} 2x = \alpha$, $\tan^{-1} x = \beta$ and $\tan^{-1}\frac{1}{3} = \gamma$, the given equation is equivalent to the equation *concerning angles*:

$$\alpha - \beta = \gamma \text{ where } \tan \alpha = 2x, \tan \beta = x, \text{ and } \tan \gamma = \tfrac{1}{3}$$

$$\Rightarrow \tan(\alpha - \beta) = \tfrac{1}{3}$$

$$\Leftrightarrow \frac{\tan \alpha - \tan \beta}{1 + \tan \alpha \tan \beta} = \frac{1}{3}$$

$$\Rightarrow \frac{x}{1 + 2x^2} = \frac{1}{3}$$

$$\Leftrightarrow 2x^2 - 3x + 1 = 0$$

$$\Leftrightarrow x = \tfrac{1}{2} \text{ or } x = 1.$$

16:5 THE DERIVATIVES OF THE INVERSE CIRCULAR FUNCTIONS

(i) If $y = \text{Sin}^{-1} x$, then $x = \sin y$ and

$$\frac{dx}{dy} = \cos y$$

$$\Rightarrow \frac{dy}{dx} = \frac{1}{\cos y} = \frac{1}{\pm \sqrt{(1 - x^2)}}.$$

Figure 16.1(i) shows the two-valued nature of the gradient,

$$\frac{1}{\cos y} > 0 \text{ for } -\tfrac{1}{2}\pi < y < \tfrac{1}{2}\pi$$

$$\Rightarrow \frac{d(\sin^{-1} x)}{dx} = \frac{1}{+\sqrt{(1 - x^2)}}. \tag{16.4}$$

(ii) If $y = \text{Cos}^{-1} x$, then $x = \cos y$ and

$$\frac{dx}{dy} = -\sin y$$

$$\Rightarrow \frac{dy}{dx} = -\frac{1}{\sin y} = \frac{-1}{\pm \sqrt{(1 - x^2)}}.$$

$$\frac{1}{\sin y} > 0 \text{ for } 0 < y < \pi$$

$$\Rightarrow \frac{d(\cos^{-1} x)}{dx} = \frac{1}{-\sqrt{(1 - x^2)}}. \tag{16.5}$$

(iii) If $y = \text{Tan}^{-1} x$, then $x = \tan y$ and

$$\frac{dx}{dy} = \sec^2 y$$

$$\Rightarrow \frac{dy}{dx} = \frac{1}{\sec^2 y} = \frac{1}{1 + x^2}.$$

Figure 16.1(v) confirms that dy/dx is always positive.

$$\frac{d}{dx}(\tan^{-1} x) = \frac{d(\text{Tan}^{-1} x)}{dx} = \frac{1}{1 + x^2}.$$ (16.6)

Application of the chain rule gives the results $(a > 0)$

$$\frac{d}{dx}\sin^{-1}\left(\frac{x}{a}\right) = \frac{1}{\sqrt{(a^2 - x^2)}},$$ (16.7)

$$\frac{d}{dx}\cos^{-1}\left(\frac{x}{a}\right) = -\frac{1}{\sqrt{(a^2 - x^2)}},$$ (16.8)

$$\frac{d}{dx}\tan^{-1}\left(\frac{x}{a}\right) = \frac{a}{x^2 + a^2}.$$ (16.9)

The inverse functions $\text{Cot}^{-1} x$, $\text{Sec}^{-1} x$, $\text{Cosec}^{-1} x$ have their principal values in the same ranges as $\tan^{-1} x$, $\cos^{-1} x$, $\sin^{-1} x$ respectively. The derivatives of these functions are best obtained from first principles as they arise.

Example 1. If $y = \text{Sin}^{-1}(\frac{1}{2}x)$, calculate the values of dy/dx in each of the cases (a) $y = -4\pi/3$, (b) $y = \frac{1}{3}\pi$, (c) $y = 11\pi/4$.

$$y = \text{Sin}^{-1}(\tfrac{1}{2}x) \Leftrightarrow x = 2 \sin y$$

$$\Rightarrow \frac{dx}{dy} = 2 \cos y \Leftrightarrow \frac{dy}{dx} = \frac{1}{2 \cos y}.$$

(a) When $y = -4\pi/3$, $\dfrac{dy}{dx} = -1.$

(b) When $y = \dfrac{1}{3}\pi$, $\dfrac{dy}{dx} = 1.$

(c) When $y = 11\pi/4$, $\dfrac{dy}{dx} = -\sqrt{2}/2.$

Example 2. If $y = e^{\sin^{-1} x}$, expand y in ascending powers of x as far as the term in x^4.

$$y(0) = e^0 = 1.$$

$$\ln y = \sin^{-1} x \Rightarrow \frac{y_1}{y} = \frac{1}{\sqrt{(1-x^2)}}$$

$$\Rightarrow (1-x^2)y_1^2 = y^2 \qquad\qquad \Rightarrow y_1(0) = 1$$

$$\Rightarrow (1-x^2)2y_1y_2 - 2xy_1^2 = 2yy_1$$

$$\Rightarrow (1-x^2)y_2 - xy_1 = y \qquad\qquad \Rightarrow y_2(0) = 1$$

$$\Rightarrow (1-x^2)y_3 - 3xy_2 - 2y_1 = 0 \qquad \Rightarrow y_3(0) = 2$$

$$\Rightarrow (1-x^2)y_4 - 5xy_3 - 5y_2 = 0 \qquad \Rightarrow y_4(0) = 5$$

$$\Rightarrow y = 1 + x + \tfrac{1}{2}x^2 + \tfrac{1}{3}x^3 + \tfrac{5}{24}x^4 + \dots.$$

Example 3. Expand $(\sin^{-1} x)^2$, where $-\pi/2 < \sin^{-1} x < \pi/2$, as a series of ascending powers of x giving the general term.

$$y = (\sin^{-1} x)^2$$

$$\Rightarrow \frac{dy}{dx} = \frac{2 \sin^{-1} x}{\sqrt{(1-x^2)}} \Rightarrow (1-x^2)y_1^2 = 4y$$

$$\Rightarrow (1-x^2)2y_1y_2 - 2xy_1^2 = 4y_1$$

$$\Rightarrow (1-x^2)y_2 - xy_1 = 2.$$

Differentiate n times by Leibniz' theorem,

$$(1 - x^2)y_{n+2} - 2nxy_{n+1} - n(n-1)y_n - xy_{n+1} - ny_n = 0.$$

Putting
$$x = 0, \quad y_{n+2}(0) = n^2 y_n(0).$$
$$y(0) = 0, \quad y_1(0) = 0, \quad y_2(0) = 2$$
$$\Rightarrow y_3(0) = y_5(0) = \ldots = y_{2n+1}(0) = 0$$

(verified by noting that y is an even function).

$$y_4(0) = 2^2.2, \, y_6(0) = 4^2.2^2.2, \ldots, y_{2n+2}(0) = 2^{2n+1}(n!)^2$$

$$\Rightarrow (\sin^{-1} x)^2 = x^2 + \sum_{r=1}^{\infty} 2^{2r+1}(r!)^2 x^{2r+2}/(2r+2)!$$

16:6 STANDARD INTEGRALS

$$\int \frac{1}{\sqrt{(a^2 - x^2)}} dx = \sin^{-1}\left(\frac{x}{a}\right) + C \qquad (a > 0), \qquad (16.10)$$

$$\int \frac{1}{a^2 + x^2} dx = \frac{1}{a} \tan^{-1}\left(\frac{x}{a}\right) + C \qquad (a > 0). \qquad (16.11)$$

For acute angles

$$\sin^{-1}\left(\frac{x}{a}\right) = \tfrac{1}{2}\pi - \cos^{-1}\left(\frac{x}{a}\right)$$

and for these angles it is therefore unnecessary to have a standard integral in terms of the inverse cosine, but care must be exercised in the use of these integrals since they refer only to the principal values of the inverse functions.

If $a > 0$, $b > 0$, extensions of the above results are

$$\int \frac{1}{\sqrt{(b^2 - a^2 x^2)}} dx = \frac{1}{a} \int \frac{1}{\sqrt{[(b^2/a^2) - x^2]}} dx = \frac{1}{a} \sin^{-1}\left[\frac{x}{b/a}\right] + C$$

$$= \frac{1}{a} \sin^{-1}\left(\frac{ax}{b}\right) + C, \qquad (16.10a)$$

$$\int \frac{1}{b^2 + a^2 x^2} dx = \frac{1}{a^2} \int \frac{1}{(b^2/a^2) + x^2} dx = \frac{1}{a^2(b/a)} \tan^{-1}\left(\frac{x}{b/a}\right) + C$$

$$= \frac{1}{ab} \tan^{-1}\left(\frac{ax}{b}\right) + C. \qquad (16.11a)$$

Example 1.
$$\int_0^{1/2} \frac{1}{\sqrt{(1 - 3x^2)}} dx = \frac{1}{\sqrt{3}} \int_0^{1/2} \frac{1}{\sqrt{(\frac{1}{3} - x^2)}} dx$$

$$= \left[\frac{1}{\sqrt{3}} \sin^{-1}(x\sqrt{3})\right]_0^{1/2} = \pi\sqrt{3}/9.$$

Example 2.
$$\int \frac{1}{\sqrt{(1-x-x^2)}}\,dx = \int \frac{1}{\sqrt{\{\frac{5}{4}-(x+\frac{1}{2})^2\}}}\,dx$$
$$= \sin^{-1}\{(x+\tfrac{1}{2})/\sqrt{5}/2\} + C = \sin^{-1}\{(2x+1)/\sqrt{5}\} + C.$$

Exampie 3.
$$\int \frac{1}{2x^2+4x+5}\,dx = \frac{1}{2}\int \frac{1}{(x+1)^2+\frac{1}{2}}\,dx$$
$$= \frac{1}{2}\frac{\sqrt{2}}{\sqrt{3}}\tan^{-1}\left\{\frac{(x+1)}{\sqrt{3}/\sqrt{2}}\right\} + C = \frac{1}{\sqrt{6}}\tan^{-1}\left\{\frac{(x+1)\sqrt{2}}{\sqrt{3}}\right\} + C.$$

Example 4.
$$\int_0^{a/2}\sin^{-1}\left(\frac{x}{a}\right)dx = \left[x\sin^{-1}\left(\frac{x}{a}\right)\right]_0^{a/2} - \int_0^{a/2}\frac{x\,dx}{\sqrt{(a^2-x^2)}}$$
$$= \left[x\sin^{-1}\left(\frac{x}{a}\right) + \sqrt{(a^2-x^2)}\right]_0^{a/2} = a\left(\frac{\pi}{12}+\frac{\sqrt{3}}{2}-1\right).$$

In this example $a > 0$.

Exercise 16.6

1. Sketch the graphs of $\mathrm{Sec}^{-1}x$, $\mathrm{Cosec}^{-1}x$ and $\mathrm{Cot}^{-1}x$.

2. Write down the principal values of each of the following
   (i) $\mathrm{Sin}^{-1}(\frac{1}{2})$, (ii) $\mathrm{Cos}^{-1}(-\sqrt{3}/2)$, (iii) $\mathrm{Tan}^{-1}(-1)$, (iv) $\mathrm{Cot}^{-1}(0)$,
   (v) $\mathrm{Cosec}^{-1}(-\sqrt{2})$, (vi) $\mathrm{Sec}^{-1}(-\sqrt{2})$.

3. In each of the following cases obtain the value of θ in the form of an inverse circular function and state the quadrant in which it lies.
   (i) $\theta = \tan^{-1}2 + \tan^{-1}1$, (ii) $\theta = \sin^{-1}(\frac{5}{13}) + \cos^{-1}(\frac{3}{5})$,
   (iii) $\theta = \tan^{-1}(-\frac{3}{4}) + \sin^{-1}(\frac{4}{5})$, (iv) $\theta = \cos^{-1}(-\frac{1}{2}) + \sin^{-1}(-\frac{1}{2})$,
   (v) $\theta = \tan^{-1}\frac{1}{2} - \tan^{-1}(-\frac{1}{3})$.

4. If $\mathrm{Tan}^{-1}x = n\pi + \alpha$ write down an expression for $\mathrm{Cot}^{-1}x$ and write down the values of $\sin\alpha$ and $\cos\alpha$.

5. Solve the equation $\tan^{-1}(4x/3) + \tan^{-1}x = \tan^{-1}7$ where both $\tan^{-1}(4x/3)$ and $\tan^{-1}x$ are acute.

6. If $\sin^{-1}x + \cos^{-1}2x = \frac{2}{3}\pi$ where both $\sin^{-1}x$ and $\cos^{-1}(2x)$ are acute, prove that $4x\sqrt{(1-x^2)} - x\sqrt{(1-4x^2)} + 1 = 0$.

7. If $\sin^{-1}x = -\alpha$ for $\alpha > 0$, write down the value of $\cos^{-1}x$.

8. Prove that $\tan^{-1}\left(\dfrac{a+x}{a-x}\right) - \tan^{-1}\left(\dfrac{x}{a}\right) = \dfrac{\pi}{4}$.

9. If $\mathrm{Sin}^{-1}x = n\pi + (-1)^n\alpha$ and $\mathrm{Cos}^{-1}x = 2n\pi \pm \beta$, prove that
   $$\tan(\alpha + 2\beta) = -x/\sqrt{(1-x^2)}.$$

10. Differentiate each of the following with respect to x:
    (i) $\sin^{-1}(2x)$, (ii) $\cos^{-1}(x/2)$, (iii) $\tan^{-1}\{(1+x)/(1-x)\}$,
    (iv) $\tan^{-1}\{(a+x)/(a-x)\}$, (v) $\sin^{-1}\left(\dfrac{1}{x}\right)$, (vi) $\tan^{-1}(\sin x)$, (vii) $\tan^{-1}(2\tan x)$,
    (viii) $\sin^{-1}\left\{\cos\left(x+\dfrac{1}{3}\pi\right)\right\}$, (ix) $\cos^{-1}(\sin 2x)$, (x) $\tan^{-1}(e^{2x}+1)$.

11. Write down the value of dy/dx at the point named for each of the following curves:
    (i) $y = \mathrm{Sin}^{-1}x$, $(-\frac{1}{2}, 7\pi/6)$, (ii) $y = \mathrm{Tan}^{-1}(x/4)$, $(4, 5\pi/4)$, (iii) $y = \mathrm{Sin}^{-1}(5x)$, $(-1/5, 5\pi/2)$,
    (iv) $y = \mathrm{Cos}^{-1}x$, $(-\frac{1}{2}, 2\pi/3)$.

12. Obtain each of the following derivatives:

$$\text{(i) } \frac{d}{dx}(\sec^{-1} x) \qquad \text{(ii) } \frac{d}{dx}(\operatorname{cosec}^{-1} x), \qquad \text{(iii) } \frac{d}{dx}(\cot^{-1} x).$$

13. Verify by using the tables or a calculator that the tangent to the curve $y = \operatorname{Tan}^{-1} x$ at the origin meets the curve again where $y \approx 2 \cdot 03$.

14. Find the equations of the tangents to the curve $y = \operatorname{Sin}^{-1} x$ at the points where $x = \frac{1}{2}$ for $0 < y < \pi$. If these tangents intersect at P and meet the x-axis at T and S respectively, calculate the area of the triangle PTS.

15. Calculate an approximate value for the small change in y which results from an increase in the value of x from $x = 1$ to $1 \cdot 05$ when $y = x \tan^{-1} x$.

16. If $y = \sin^{-1}\left(\dfrac{x}{a}\right)$, prove that $\dfrac{d^2 y}{dx^2} - x\left(\dfrac{dy}{dx}\right)^3 = 0$.

17. An aeroplane flying at 360 km h^{-1} in level flight and in a straight line at a height of 1 km above the ground is followed by a searchlight beam sited on the ground. The aeroplane passes directly over the searchlight. Calculate the angular speed and the angular acceleration of the beam at the instant when the aeroplane is directly overhead.

18. A bead P is threaded on to a straight wire and attached to a string which is threaded through a fixed ring O distant 1 m from the wire. The bead starts from rest at a point of the wire such that OP is perpendicular to the wire and it is caused to move along the wire with uniform acceleration 2 ms^{-2}, the string OP lengthening throughout the motion so as to remain taut. Calculate the angular speed (i.e. $d\theta/dt$ where $A\hat{O}P = \theta$) of the string (i) at time 1 s, (ii) when the bead has moved a distance 4 m.
Prove that the angular acceleration of the string is zero when $t^4 = \frac{1}{3}$.

19. Integrate $1/\sqrt{(x^2 - 1)}$ with respect to x by means of the substitutions $x = \sec \theta$, $\tan \frac{1}{2}\theta = t$.

20. Use the method of integration by parts to obtain each of the following integrals:

$$\text{(i) } \int \cos^{-1} x \, dx, \quad \text{(ii) } \int \tan^{-1} x \, dx, \quad \text{(iii) } \int \sec^{-1} x \, dx, \quad \text{(iv) } \int \cot^{-1} x \, dx.$$

21. Sketch the curve $y = a^3/(a^2 + x^2)$ for $a > 0$.
Calculate (i) the area of the region enclosed by the curve, the x-axis and the ordinates at $x = -a$ and $x = +a$;
(ii) the possible values of b in terms of a if the area of the region enclosed by the curve, the x-axis and the ordinates at $x = b$ and $x = 6b$ is $\frac{1}{4}\pi a^2$ units^2.

22. Find the volume of the solid of revolution generated by rotating about the x-axis the region defined in question 21 (i).

23. Evaluate each of the following definite integrals:

$$\text{(i) } \int_0^1 \frac{1}{\sqrt{(4 - x^2)}} \, dx, \qquad \text{(ii) } \int_0^3 \frac{1}{2x^2 + 6} \, dx, \qquad \text{(iii) } \int_{1/6}^{1/3} \frac{1}{\sqrt{(1 - 9x^2)}} \, dx,$$

$$\text{(iv) } \int_0^2 \frac{1}{x^2 + 2x + 2} \, dx, \qquad \text{(v) } \int_1^{1 \cdot 5} \frac{1}{\sqrt{(2x - x^2)}} \, dx.$$

24. Obtain each of the following integrals

$$\text{(i) } \int \frac{1}{x^2 + 4x + 5} \, dx, \qquad \text{(ii) } \int \frac{1}{2x^2 - 3x + 3} \, dx, \qquad \text{(iii) } \int \frac{1}{\sqrt{(1 - 2x - x^2)}} \, dx,$$

$$\text{(iv) } \int \frac{1}{\sqrt{(1 - 6x - 3x^2)}} \, dx, \qquad \text{(v) } \int \frac{1}{5x^2 + 10x + 8} \, dx.$$

25. Show that the area of the region enclosed by the ellipse $x^2/a^2 + y^2/b^2 = 1$ is πab.

26. Find the area of the region enclosed by the curve $y = \tan x$, the axis of y and the line $y = 1$.

27. Obtain $\int x \tan^{-1} x \, dx$.

28. Use partial fractions to integrate

$$\frac{5x^2 + 12}{x^4 - 16}$$

with respect to x.

29. Use the method of substitution to evaluate

$$\int_1^4 \frac{x}{1 + x^4} \, dx.$$

30. In the usual notation, an equation of motion for a particle moving in a straight line is

$$v = 4 \sqrt{(a^2 - s^2)}$$

where a is a positive constant. Calculate the time taken by the particle to move from $s = \frac{1}{2}a$ to $s = a$.

31. The curve $y = \sin^{-1} x$ intersects the curve $y = \cos^{-1} x$ at P and the latter curve intersects the x-axis at Q. Find the area of the region enclosed by the arcs OP and PQ and the x-axis. (O is the origin.)

16:7 THE HYPERBOLIC FUNCTIONS

At this stage we define certain combinations of exponential functions which are called *hyperbolic functions*.

Hyperbolic sine of $x = \sinh x = \dfrac{e^x - e^{-x}}{2} = x + \dfrac{x^3}{3!} + \dfrac{x^5}{5!} + \ldots ;$

hyperbolic cosine of $x = \cosh x = \dfrac{e^x + e^{-x}}{2} = 1 + \dfrac{x^2}{2!} + \dfrac{x^4}{4!} + \ldots ;$

$\tanh x = \dfrac{\sinh x}{\cosh x} ; \quad \coth x = \dfrac{1}{\tanh x} = \dfrac{\cosh x}{\sinh x} ;$

$\operatorname{cosech} x = \dfrac{1}{\sinh x} ; \quad \operatorname{sech} x = \dfrac{1}{\cosh x} .$

It is clear from the above definitions that these are all functions of x for $x \in \mathbb{R}$ (except $\coth x$ and $\operatorname{cosech} x$ at $x = 0$). The names of the new functions suggest an analogy with the circular functions. In this chapter we show that there is a close analogy between the properties of the hyperbolic functions and those of the circular functions. We discuss some of the properties of the hyperbolic functions below.

(a) $\hspace{4cm} \cosh^2 x - \sinh^2 x \equiv 1. \hspace{4cm}$ (16.12)

This result follows from the identity

$$\left(\frac{e^x + e^{-x}}{2} \right)^2 - \left(\frac{e^x - e^{-x}}{2} \right)^2 \equiv e^x \times e^{-x} \equiv 1$$

and is analogous to the identity $\cos^2 x + \sin^2 x \equiv 1$.

The point $(a\cosh t, b\sinh t)$ is a point *on one branch* of the hyperbola $x^2/a^2 - y^2/b^2 = 1$ for all values of t. Thus, in some respects, the hyperbolic functions $\sinh x$ and $\cosh x$

bear the same relation to the rectangular hyperbola $x^2 - y^2 = a^2$ as the circular functions bear to the circle $x^2 + y^2 = a^2$ and it is for this reason that they are so named.

(b)
$$\sinh x + \cosh x \equiv e^x. \tag{16.13}$$
$$\cosh x - \sinh x \equiv e^{-x}. \tag{16.14}$$

These results follow directly from the definitions; they are very useful in the processes of manipulating hyperbolic functions.

(c)
$$\frac{d}{dx}(\sinh x) = \cosh x. \tag{16.15}$$

$$\frac{d}{dx}(\cosh x) = \sinh x. \tag{16.16}$$

These results follow directly from the definitions.
(d) $y = \sinh x$ is an odd function of x.
  (i) There are no stationary values of y;
  (ii) when $x = 0$, $y = 0$;
  (iii) as $x \to +\infty$, $y \to +\infty$;
      as $x \to -\infty$, $y \to -\infty$.
(e) $y = \cosh x$ is an even function of x.
  (i) $y \geqslant 1 \ \forall x$;
  (ii) y has a stationary value, which is a minimum value, at $(0, 1)$;
  (iii) as $x \to +\infty$, $y \to +\infty$;
      as $x \to -\infty$, $y \to +\infty$.
(f) We define $\tanh x = \dfrac{\sinh x}{\cosh x} = \dfrac{e^x - e^{-x}}{e^x + e^{-x}} = \dfrac{1 - e^{-2x}}{1 + e^{-2x}}$.

$$y = \tanh x \Rightarrow \frac{dy}{dx} = \frac{\cosh^2 x - \sinh^2 x}{\cosh^2 x} = \frac{1}{\cosh^2 x} = \operatorname{sech}^2 x. \tag{16.17}$$

$y = \tanh x$ is an odd function of x,
  (i) $|y| < 1 \ \forall x$;
  (ii) as $x \to \pm\infty$, $y \to \pm 1$;
  (iii) there are no stationary values of y.
The graphs of $\sinh x$, $\cosh x$, $\tanh x$ are shown in Fig. 16.2.

Similarly
$$\frac{d}{dx}(\operatorname{sech} x) = -\operatorname{sech} x \tanh x, \tag{16.18}$$

$$\frac{d}{dx}(\operatorname{cosech} x) = -\operatorname{cosech} x \coth x, \tag{16.19}$$

$$\frac{d}{dx}(\coth x) = -\operatorname{cosech}^2 x. \tag{16.20}$$

(g)
$$\sinh (x + y) \equiv \sinh x \cosh y + \cosh x \sinh y. \tag{16.21}$$
$$\sinh (x - y) \equiv \sinh x \cosh y - \cosh x \sinh y. \tag{16.22}$$
$$\cosh (x + y) \equiv \cosh x \cosh y + \sinh x \sinh y. \tag{16.23}$$
$$\cosh (x - y) \equiv \cosh x \cosh y - \sinh x \sinh y. \tag{16.24}$$

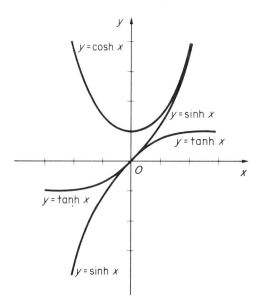

Fɪɢ. 16.2.

These results, which are in most respects analogous to the corresponding results for circular functions, differ in sign in the expanded form in (16.23) and (16.24).

The proof of (16.24) follows here: proofs for the other results follow similar lines.

$$\cosh (x - y) \equiv \tfrac{1}{2}(e^{x-y} + e^{-x+y})$$
$$\equiv \tfrac{1}{2}(e^x e^{-y} + e^{-x} e^y)$$
$$\equiv \tfrac{1}{2}\{(\sinh x + \cosh x)(\cosh y - \sinh y)$$
$$+ (\cosh x - \sinh x)(\sinh y + \cosh y)\}$$
$$\equiv \cosh x \cosh y - \sinh x \sinh y.$$

This kind of similarity persists for each of the formulae involving hyperbolic functions when compared with the corresponding formula for circular functions. The general rule, known as *Osborn's rule*, for changing a formula involving an algebraic relationship concerning circular functions (for *all* values of the variable) into the corresponding formula involving hyperbolic functions is as follows:

Replace each circular function by the corresponding hyperbolic function and change the sign of every product (or implied product) of two sines.

The justification for this rule will be given in §16:14.

Thus from $\sin x \sin y \equiv \tfrac{1}{2}\{\cos (x - y) - \cos (x + y)\}$ we have

$$\sinh x \sinh y \equiv \tfrac{1}{2}\{\cosh (x + y) - \cosh (x - y)\}.$$

Also from

$$\tan 2x \equiv \frac{2 \tan x}{1 - \tan^2 x}$$

we have

$$\tanh 2x \equiv \frac{2\tanh x}{1+\tanh^2 x},$$

because $\tan^2 x = \sin^2 x/\cos^2 x$ and so implies a product of two sines.

Proofs in particular cases of idèntities involving hyperbolic functions can be obtained by direct use of the definitions as illustrated in the proof of (16.24). The simpler results such as equations (16.21)–(16.24) can be used to prove the more complicated relations.

(h) $\int \sinh x \, dx = \cosh x + C.$ (16.25)

$\int \cosh x \, dx = \sinh x + C.$ (16.26)

Methods of integration, when the integrand involves hyperbolic functions, are similar, in most cases, to those used for integrands involving circular functions. Sometimes the substitution $e^x = u$ can be used to advantage as in example 5 below.

Example 1. Solve the equation $\tanh x + 4 \operatorname{sech} x = 4.$

The given equation can be written

$$\frac{e^x - e^{-x}}{e^x + e^{-x}} + \frac{8}{e^x + e^{-x}} = 4.$$

Multiplication by $e^x + e^{-x}$ ($\neq 0$) gives

$$3e^x - 8 + 5e^{-x} = 0.$$

Now multiplication throughout by e^x ($\neq 0$) gives

$$3e^{2x} - 8e^x + 5 = 0$$
$$\Leftrightarrow (3e^x - 5)(e^x - 1) = 0$$
$$\Leftrightarrow 3e^x = 5 \quad \text{or} \quad e^x = 1$$
$$\Leftrightarrow x = \ln(5/3) \quad \text{or} \quad 0.$$

Example 2. Prove that $\dfrac{1 + \sinh x + \cosh x}{1 - \sinh x - \cosh x} \equiv -\coth\left(\dfrac{x}{2}\right).$

The left-hand side $= \dfrac{1 + e^x}{1 - e^x}.$

Dividing numerator and denominator by $e^{x/2}$ gives

$$\text{left-hand side} \equiv \frac{e^{-x/2} + e^{x/2}}{e^{-x/2} - e^{x/2}} \equiv -\coth\left(\frac{x}{2}\right).$$

Example 3. Prove that $\cosh^2 x \cos^2 x - \sinh^2 x \sin^2 x \equiv \frac{1}{2}(1 + \cosh 2x \cos 2x).$

From equations (16.12) and (16.23),

$$\cosh^2 x \equiv \tfrac{1}{2}(\cosh 2x + 1),$$
$$\sinh^2 x \equiv \tfrac{1}{2}(\cosh 2x - 1).$$

But $\cos^2 x \equiv \tfrac{1}{2}(1 + \cos 2x), \quad \sin^2 x \equiv \tfrac{1}{2}(1 - \cos 2x)$

$$\Rightarrow \cosh^2 x \cos^2 x - \sinh^2 x \sin^2 x$$
$$\equiv \tfrac{1}{4}(\cosh 2x + 1)(1 + \cos 2x) - \tfrac{1}{4}(\cosh 2x - 1)(1 - \cos 2x)$$
$$\equiv \tfrac{1}{2}(1 + \cosh 2x \cos 2x).$$

Example 4. Show that the area of the region bounded by the x-axis, the curve $x^2/a^2 - y^2/b^2 = 1$ and the line OP where P is the point $(a \cosh t_1, b \sinh t_1)$ on the x-positive branch of the curve is $\left|\frac{1}{2}abt_1\right|$.

P_2 is a point $(a \cosh t_2, b \sinh t_2)$ where $t_2 < t_1$, on the x-positive, y-positive part of the curve (Fig. 16.3), N is the foot of the ordinate at P, A is the vertex of the hyperbola and the ordinate at P_2 meets Ox at M and OP at R.

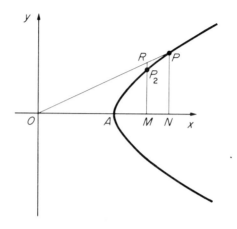

FIG. 16.3.

Then $P_2M = b \sinh t_2 = b \cosh t_2 \tanh t_2$ and $RM = b \cosh t_2 \tanh t_1$.

But $\tanh t_1 > \tanh t_2$ for $t_1 > t_2 > 0$

$$\Rightarrow P_2M < RM \text{ for all positions of } P_2 \text{ between } A \text{ and } P.$$

Hence the required area OPA = area of triangle OPN − area bounded by arc AP, PN, and the x-axis.
Area of $\triangle OPN = \frac{1}{2}ab \cosh t_1 \sinh t_1 = \frac{1}{4}ab \sinh 2t_1$.

$$\text{Area of } APN = \int_{a}^{a \cosh t_1} \frac{b}{a} \sqrt{(x^2 - a^2)}\, dx.$$

Put $x = a \cosh t$; then $dx/dt = a \sinh t$.

When $x = a \cosh t_1$, $t = t_1$; when $x = a$, $t = 0$

$$\Rightarrow \text{area of } APN = \int_{0}^{t_1} \frac{b}{a} a \sinh t\, a \sinh t\, dt$$

$$= ab \int_{0}^{t_1} \tfrac{1}{2}(\cosh 2t - 1)\, dt$$

$$= ab \left[\tfrac{1}{4} \sinh 2t - \tfrac{1}{2}t \right]_{0}^{t_1}$$

$$= \tfrac{1}{4}ab \sinh 2t_1 - \tfrac{1}{2}abt_1.$$

Thus the required area

$$OPA = \tfrac{1}{4}ab \sinh 2t_1 - (\tfrac{1}{4}ab \sinh 2t_1 - \tfrac{1}{2}abt_1) = \tfrac{1}{2}abt_1.$$

A similar result is obtained for the y-negative part of this branch of the curve where $t_1 < 0$. Hence the required area $= \frac{1}{2}|abt_1|$.

In the special case in which the curve is a rectangular hyperbola, [i.e. $a = b$], the area is $\left|\frac{1}{2}a^2 t_1\right|$ which is analogous to the sectorial area $\frac{1}{2}a^2\theta$ for a circle.

Example 5. $I = \int \operatorname{sech} x \, dx = \int \dfrac{2}{e^x + e^{-x}} \, dx.$

Putting $e^x = u$, so that $du/dx = e^x$, gives

$$I = \int \frac{2}{u^2 + 1} \, du = 2 \tan^{-1} u + C = 2 \tan^{-1}(e^x) + C.$$

Exercise 16.7

In questions 1–10 prove each of the following results for hyperbolic functions and check by Osborn's rule.

1. $\sinh 2x \equiv 2 \sinh x \cosh x$.

2. $\cosh 2x \equiv \cosh^2 x + \sinh^2 x = 2 \cosh^2 x - 1 = 1 + 2 \sinh^2 x$.

3. $\tanh 2x \equiv 2 \tanh x/(1 + \tanh^2 x)$.

4. $\sinh 3x \equiv 3 \sinh x + 4 \sinh^3 x$.

5. $\cosh 3x \equiv 4 \cosh^3 x - 3 \cosh x$.

6. $\sinh x + \sinh y \equiv 2 \sinh \dfrac{x+y}{2} \cosh \dfrac{x-y}{2}$.

7. $\cosh x \cosh y \equiv \frac{1}{2}\{\cosh (x + y) + \cosh (x - y)\}$.

8. $\tanh^2 x + \operatorname{sech}^2 x \equiv 1$.

9. $\tanh x \equiv \sqrt{\left(\dfrac{\cosh 2x - 1}{\cosh 2x + 1} \right)}$ for $x > 0$.

10. $\cosh^2 (x + y) - \cosh^2 (x - y) \equiv \sinh 2x \sinh 2y$.

11. If $\sinh x = 5/12$, calculate $\cosh x$ and $\tanh x$.

12. If $\tanh x = \frac{3}{4}$, calculate $\operatorname{sech} x$ and $\sinh x$.

13. Solve the equation $\sinh x + 3 \cosh x = 4 \cdot 5$.

14. Solve the equation $4 \tanh x = \coth x$.

15. Solve the equation $7 \operatorname{cosech} x + 2 \coth x = 6$.

16. Draw sketch graphs of (i) $\tanh x$, (ii) $\operatorname{cosech} x$, (iii) $\operatorname{sech} x$.

17. Differentiate each of the following:
(i) $\cosh (2x + 5)$,   (ii) $x \sinh x$,   (iii) $\sinh 2x \cosh 3x$,   (iv) $e^x \tanh x$,   (v) $\cosh (\ln x)$,   (vi) $(\operatorname{sech} x)/x$,   (vii) $\sinh (\sin x)$,   (viii) $\tan^{-1} (\sinh x)$,   (ix) $\ln \tanh 2x$,   (x) $\exp (\sinh x + \cosh x)$,   (xi) $\operatorname{cosec}^{-1} (\cosh x)$,   (xii) $\cosh^2 x - \sinh^2 x$.

18. Find the equation of the tangent and the equation of the normal to the curve $x = a \cosh t, y = b \sinh t$ at the point t_1 on the curve.

19. P is a point on the curve $y = c \cosh (x/c)$. [This curve is called the catenary.] PN is the ordinate at P and the tangent and normal at P meet the x-axis at T, G respectively. NZ is the perpendicular from N to PT and ZK is the perpendicular from Z to Ox. Find expressions in terms of the abscissa of P for PT, PG, TN and NG and prove that
(i) NZ is a constant length,
(ii) $ZK . PN = c^2$.

20. If $x = A \cosh nt + B \sinh nt$, where A, B, n are constants, show that

$$\frac{d^2 x}{dt^2} = n^2 x.$$

and find A and B given that $dx/dt = 4n$ when $t = 0$, and $x = 2$ when $t = 0$.

21. With the usual notation for the motion of a particle moving in a straight line,

$$x = 2 \cosh 4t + \sinh 4t.$$

Find the numerical values of the speed and the acceleration at time $t = 1$ each correct to 3 significant figures.

22. Find the minimum value of $13 \cosh x + 12 \sinh x$.

23. Find the stationary value of the function

$$y = (n^2 + 1)\cosh x + 2n \sinh x \text{ where } |n| \neq 1$$

and discuss the different cases $|n| > 1$ and $|n| < 1$. Sketch the curve in the cases
(i) $n = 2$, (ii) $n = -2$, (iii) $n = 1$, (iv) $n = -1$.

24. Obtain each of the following integrals:

(i) $\int \cosh 2x \, dx$, (ii) $\int \tanh x \, dx$, (iii) $\int \coth x \, dx$, (iv) $\int \operatorname{sech}^2 x \, dx$,
(v) $\int \operatorname{sech} x \tanh x \, dx$, (vi) $\int \operatorname{cosech} x \coth x \, dx$, (vii) $\int \cosh^2 x \, dx$,
(viii) $\int \cosh 2x \sinh 4x \, dx$, (ix) $\int x \sinh x \, dx$, (x) $\int e^x \cosh x \, dx$.

25. Evaluate each of the following integrals:

(i) $\displaystyle\int_0^1 x \cosh x \, dx$, (ii) $\displaystyle\int_0^1 \cosh^3 x \, dx$, (iii) $\displaystyle\int_0^{1/2} \cosh 2x \cosh 4x \, dx$,

(iv) $\displaystyle\int_0^2 \operatorname{sech} x \, dx$, (v) $\displaystyle\int_0^1 x \operatorname{sech}^2 x \, dx$.

26. Evaluate $\displaystyle\int_0^1 e^x \tanh x \, dx$ in terms of e.

27. Find the area of the region enclosed by the curves $y = \sinh x$, $y = \cosh x$, the y-axis and the ordinate $x = 1$.

28. In the usual notation for a particle moving in a straight line the speed of such a particle at time t is $5 \tanh 2t$. Calculate, correct to 3 significant figures,
(i) the distance moved by the particle in the first second,
(ii) the acceleration of the particle at time 1 second.

29. Calculate the volume formed when the region bounded by $y = \tanh x$, the x-axis and the ordinates $x = -1$ and $x = 1$ makes a complete revolution about the x-axis.

16:8 INVERSE HYPERBOLIC FUNCTIONS

By analogy with the inverse circular functions, $y = \sinh^{-1} x$ implies $x = \sinh y$ or "y is the quantity whose hyperbolic sine is x".

Graphs of $\sinh^{-1} x$ and $\operatorname{Cosh}^{-1} x$ (Fig. 16.4) show that $\operatorname{Cosh}^{-1} x$ is two-valued, but that $\sinh^{-1} x$ is a function of x, with domain $x \in \mathbb{R}$. $\operatorname{Tanh}^{-1} x$, $\coth^{-1} x$, $\operatorname{Sech}^{-1} x$ and $\operatorname{cosech}^{-1} x$ are defined in the same way. $\operatorname{Sech}^{-1} x$ is two-valued and the others are functions of x.

In some books the notation arsinh x, artanh x, etc., is used for $\sinh^{-1} x$, $\tanh^{-1} x$, etc.

The graphs of $\sinh^{-1} x$, $\operatorname{Cosh}^{-1} x$, $\tanh^{-1} x$ are shown in Fig. 16.4. Note that these graphs are obtained from those of Fig. 16.2 by reflection in the line $y = x$.

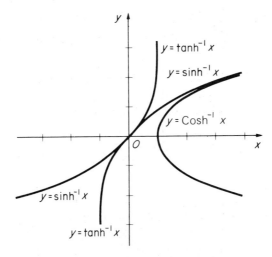

FIG. 16.4.

16:9 DERIVATIVES OF THE INVERSE HYPERBOLIC FUNCTIONS

If $y = \sinh^{-1} x$ so that $x = \sinh y$, then

$$\frac{dy}{dx} = \frac{1}{\cosh y} = \frac{1}{\sqrt{(\sinh^2 y + 1)}} = \frac{1}{\sqrt{(x^2 + 1)}}$$

the positive sign being taken for the radical because $\cosh y$ is positive for all values of y.

$$\Rightarrow \frac{d}{dx}(\sinh^{-1} x) = \frac{1}{\sqrt{(x^2 + 1)}}. \qquad (16.27)$$

Similarly $\dfrac{d}{dx}(\mathrm{Cosh}^{-1} x) = \dfrac{1}{\pm \sqrt{(x^2 - 1)}}$ where
$\mathrm{Cosh}^{-1} x$ is two-valued. If we define $y = \cosh^{-1} x$ as $y = \mathrm{Cosh}^{-1} x$ for $y \geqslant 0$, then

$$\frac{d}{dx}(\cosh^{-1} x) = \frac{1}{\sqrt{(x^2 - 1)}}. \qquad (16.28)$$

If $y = \tanh^{-1} x$ so that $x = \tanh y$, then

$$\frac{dy}{dx} = \frac{1}{\mathrm{sech}^2 y} = \frac{1}{1 - \tanh^2 y} = \frac{1}{1 - x^2}$$

$$\Rightarrow \frac{d}{dx}(\tanh^{-1} x) = \frac{1}{1 - x^2}. \qquad (16.29)$$

It follows from equations (16.27), (16.28), (16.29) that, for $a > 0$,

$$\frac{\mathrm{d}}{\mathrm{d}x}\sinh^{-1}\left(\frac{x}{a}\right) = \frac{1}{\sqrt{(x^2 + a^2)}}, \tag{16.30}$$

$$\frac{\mathrm{d}}{\mathrm{d}x}\cosh^{-1}\left(\frac{x}{a}\right) = \frac{1}{\sqrt{(x^2 - a^2)}}, \tag{16.31}$$

$$\frac{\mathrm{d}}{\mathrm{d}x}\tanh^{-1}\left(\frac{x}{a}\right) = \frac{a}{a^2 - x^2}. \tag{16.32}$$

Standard integrals

$$\int \frac{\mathrm{d}x}{\sqrt{(x^2 + a^2)}} = \sinh^{-1}\left(\frac{x}{a}\right) + C. \tag{16.33}$$

$$\int \frac{\mathrm{d}x}{\sqrt{(x^2 - a^2)}} = \cosh^{-1}\left(\frac{x}{a}\right) + C. \tag{16.34}$$

The integral

$$\int \frac{1}{a^2 - x^2}\,\mathrm{d}x$$

has already been considered (chapter 12) and it is more conveniently expressed in the form

$$\frac{1}{2a}\ln\left(\frac{a + x}{a - x}\right).$$

Each of the other integrals (16.33) and (16.34) expressed here in terms of "inverse hyperbolic functions" can be expressed in terms of natural logarithms, i.e. in terms of *inverse exponential functions*. Since the hyperbolic functions are very simple combinations of exponential functions this is a result which might well be expected. We examine this in detail in the next section.

16:10 LOGARITHMIC FORMS OF THE INVERSE HYPERBOLIC FUNCTIONS

(i) If $y = \sinh^{-1}(x/a)$ where $a > 0$, then

$$x = a \sinh y \tag{16.35}$$

$$\Rightarrow x^2 + a^2 = a^2 \sinh^2 y + a^2 = a^2 \cosh^2 y$$

$$\Rightarrow \sqrt{(x^2 + a^2)} = a \cosh y. \tag{16.36}$$

Here we retain the positive square root only since $a \cosh y > 0$. Addition of equations (16.35), (16.36) gives

$$x + \sqrt{(x^2 + a^2)} = a(\cosh y + \sinh y) = ae^y$$

$$\Rightarrow y = \sinh^{-1}\left(\frac{x}{a}\right) = \ln\left\{\frac{x + \sqrt{(x^2 + a^2)}}{a}\right\}. \tag{16.37}$$

(ii) If $y = \text{Cosh}^{-1}(x/a)$, where $a > 0$, then

$$x = a \cosh y \tag{16.38}$$

$$\Rightarrow x^2 - a^2 = a^2(\cosh^2 y - 1) = a^2 \sinh^2 y$$

$$\Rightarrow \pm \sqrt{(x^2 - a^2)} = a \sinh y. \tag{16.39}$$

Addition of equations (16.38), (16.39) gives

$$x \pm \sqrt{(x^2 - a^2)} = ae^y$$

$$\Rightarrow y = \text{Cosh}^{-1}\left(\frac{x}{a}\right) = \ln\left[\frac{x \pm \sqrt{(x^2 - a^2)}}{a}\right]. \tag{16.40}$$

Since

$$\frac{x - \sqrt{(x^2 - a^2)}}{a} = \frac{x^2 - (x^2 - a^2)}{a[x + \sqrt{(x^2 - a^2)}]} = \frac{a}{x + \sqrt{(x^2 - a^2)}},$$

we may rewrite equation (16.40) in the form

$$y = \text{Cosh}^{-1}\left(\frac{x}{a}\right) = \pm \ln\left[\frac{x + \sqrt{(x^2 - a^2)}}{a}\right]. \tag{16.41}$$

We take the positive sign in equation (16.41) to define the principal value of $\text{Cosh}^{-1}(x/a)$

$$\Rightarrow \cosh^{-1}\left(\frac{x}{a}\right) = +\ln\left[\frac{x + \sqrt{(x^2 - a^2)}}{a}\right]. \tag{16.41a}$$

(iii) If $y = \tanh^{-1}(x/a)$ where $a > 0$, then $x = a \tanh y$

$$\Rightarrow \frac{x}{a} = \frac{e^y - e^{-y}}{e^y + e^{-y}}$$

$$\Rightarrow e^{2y} = \frac{a + x}{a - x}$$

$$\Rightarrow y = \tanh^{-1}\left(\frac{x}{a}\right) = \tfrac{1}{2}\ln\left(\frac{a + x}{a - x}\right). \tag{16.42}$$

Example 1. $I = \displaystyle\int_0^a \sqrt{(x^2 + a^2)}\,dx.$

Put $x = a \sinh t$ so that $dx/dt = a \cosh t$.
When $x = a$, $\sinh t = 1$; when $x = 0$, $t = 0$.

Then
$$I = a^2 \int_0^{\sinh^{-1} 1} \cosh^2 t\,dt = a^2 \int_0^{\sinh^{-1} 1} \tfrac{1}{2}(\cosh 2t + 1)\,dt$$

$$= a^2\left[\tfrac{1}{4}\sinh 2t + \tfrac{1}{2}t\right]_0^{\sinh^{-1} 1} = a^2\left[\tfrac{1}{2}\sinh t \cosh t + \tfrac{1}{2}t\right]_0^{\sinh^{-1} 1}$$

When $\sinh t = 1$, $\cosh t = \sqrt{2}$ and $t = \ln(1 + \sqrt{2})$

$$\Rightarrow I = \tfrac{1}{2}a^2\{\sqrt{2} + \ln(1 + \sqrt{2})\}.$$

It is usual in numerical examples to use the logarithmic forms of the inverse functions, because these are the forms in which they are easily calculated with the help of a calculator. Most books of mathematical tables give values of natural logarithms but not of inverse hyperbolic functions.

Example 2. Solve the equation $3 \sinh^2 x - 2 \cosh x - 2 = 0$.
The given equation can be written

$$3(\cosh^2 x - 1) - 2 \cosh x - 2 = 0$$
$$\Leftrightarrow 3 \cosh^2 x - 2 \cosh x - 5 = 0$$
$$\Leftrightarrow (3 \cosh x - 5)(\cosh x + 1) = 0$$
$$\Leftrightarrow \cosh x = 5/3 \text{ since } \cosh x \geqslant 1$$
$$\Leftrightarrow x = \pm \ln \left\{ \frac{5}{3} + \sqrt{\left(\frac{25}{9} - 1 \right)} \right\} = \pm \ln 3.$$

16:11 METHODS OF INTEGRATION

Integrals of the types

$$\int \frac{1}{ax^2 + bx + c} \, dx, \quad \int \frac{1}{\sqrt{(ax^2 + bx + c)}} \, dx,$$

where a, b, c are constants, can be integrated with the help of one of the standard forms

$$\int \frac{1}{x^2 + a^2} \, dx = \frac{1}{a} \tan^{-1} \left(\frac{x}{a} \right) + C, \tag{16.43}$$

$$\int \frac{1}{x^2 - a^2} \, dx = \frac{1}{2a} \ln \left(\frac{x-a}{x+a} \right) + C \text{ when } |x| > a, \tag{16.44}$$

$$\int \frac{1}{a^2 - x^2} \, dx = \frac{1}{2a} \ln \left(\frac{a+x}{a-x} \right) + C \text{ when } |x| < a, \tag{16.45}$$

$$\int \frac{1}{\sqrt{(a^2 - x^2)}} \, dx = \sin^{-1} \left(\frac{x}{a} \right) + C, \tag{16.46}$$

$$\int \frac{1}{\sqrt{(a^2 + x^2)}} \, dx = \sinh^{-1} \left(\frac{x}{a} \right) + C, \tag{16.47}$$

$$\int \frac{1}{\sqrt{(x^2 - a^2)}} \, dx = \cosh^{-1} \left(\frac{x}{a} \right) + C. \tag{16.48}$$

Example 1. $\displaystyle \int \frac{1}{\sqrt{(3x^2 - 6x + 7)}} \, dx = \frac{1}{\sqrt{3}} \int \frac{1}{\sqrt{(x^2 - 2x + \frac{7}{3})}} \, dx$

$$= \frac{1}{\sqrt{3}} \int \frac{1}{\sqrt{\{(x-1)^2 + \frac{4}{3}\}}} \, dx = \frac{1}{\sqrt{3}} \sinh^{-1} \left\{ \frac{(x-1)\sqrt{3}}{2} \right\} + C.$$

Example 2. $\displaystyle\int \frac{1}{5x^2 + 2x + 6}\,dx = \frac{1}{5}\int \frac{1}{(x+\frac{1}{5})^2 + \frac{29}{25}}\,dx = \frac{1}{\sqrt{29}}\tan^{-1}\left(\frac{5x+1}{\sqrt{29}}\right) + C.$

It is important to notice that integrals of the form $\displaystyle\int \frac{1}{ax^2 + bx + c}\,dx,$ *where* $ax^2 + bx + c$ *can be resolved into rational factors, are best obtained by the method of partial fractions.*

Exercise 16.11

1. Show that $\tanh^{-1}\left(\dfrac{x^2-1}{x^2+1}\right) = \ln x.$

2. Solve the equation $\sinh^2 x - 4\cosh x + 5 = 0.$

3. Solve the equation $4\tanh^2 x - \operatorname{sech} x - 1 = 0.$

4. Solve the equations $\sinh x + \sinh y = 0\cdot 3,$
$$\cosh x + \cosh y = 2\cdot 7.$$
(Hint: Add and then subtract the equations.)

5. Express each of the following as logarithms

  (i) $\tanh^{-1} x,$ (ii) $\operatorname{sech}^{-1} x,$ (iii) $\operatorname{cosech}^{-1} x,$ (iv) $\sinh^{-1}\left(\dfrac{1}{x}\right),$ (v) $\cosh^{-1}(1+x^2).$

6. Differentiate with respect to x (i) $\coth^{-1} x,$ (ii) $\operatorname{sech}^{-1} x,$ (iii) $\operatorname{cosech}^{-1} x.$

7. Differentiate each of the following with respect to x:

  (i) $x\cosh^{-1} x,$ (ii) $\tanh^{-1}\left(\dfrac{x-a}{x+a}\right),$ (iii) $\sinh^{-1}(\tan x),$ (iv) $\cosh^{-1}\left(\dfrac{1}{x}\right),$ (v) $\cosh^{-1}(\sinh 2x).$

8. Use the method of substitution to evaluate each of the following correct to 3 significant figures:

  (i) $\displaystyle\int_0^3 \sqrt{(9+x^2)}\,dx,$ (ii) $\displaystyle\int_4^8 \sqrt{(x^2-16)}\,dx.$

9. Use the method of substitution to obtain the integrals

  (i) $\displaystyle\int \frac{x^2\,dx}{\sqrt{(1+x^2)}},$ (ii) $\displaystyle\int \frac{\sqrt{(x^2-1)}\,dx}{x}.$

10. Obtain each of the following integrals:

  (i) $\int \sinh^{-1} x\,dx,$ (ii) $\int \cosh^{-1} dx,$ (iii) $\int \tanh^{-1} x\,dx,$ (iv) $\int x\cosh^{-1} x\,dx.$

11. Find the area of the region enclosed by the curve $x^2/a^2 - y^2/b^2 = 1$ and the line $x = 2a,$ where $a > 0.$

12. Evaluate the following integrals:

  (i) $\displaystyle\int_2^3 \frac{dx}{\sqrt{(9+x^2)}},$ (ii) $\displaystyle\int_2^4 \frac{dx}{\sqrt{(4x^2-1)}},$ (iii) $\displaystyle\int_1^2 \frac{dx}{\sqrt{(x^2+2x+2)}},$

  (iv) $\displaystyle\int_1^2 \frac{dx}{\sqrt{(2x^2+4x-1)}},$ (v) $\displaystyle\int_a^b \frac{dx}{\sqrt{\{(x+a)^2+b^2\}}}.$

13. Find the area of the region between the x-axis, the lines $x = -a,$ $x = +a$ and the curve $y\sqrt{(a^2+x^2)} = a^2,$ where $a > 0.$

14. Prove that, for the curve $y = \ln \cosh x$,

$$x = \tfrac{1}{2} \ln \left\{ \left(1 + \frac{dy}{dx}\right) \middle/ \left(1 - \frac{dy}{dx}\right) \right\}.$$

Obtain each of the following integrals:

15. $\displaystyle \int \frac{dx}{\sqrt{(x^2 - 2x + 3)}},$ **16.** $\displaystyle \int \frac{dx}{\sqrt{(1 - 4x - x^2)}},$ **17.** $\displaystyle \int \frac{dx}{\sqrt{(x^2 - 2x - 5)}},$

18. $\displaystyle \int \frac{dx}{2x^2 + 5x + 1},$ **19.** $\displaystyle \int \frac{dx}{3 - x - x^2},$ **20.** $\displaystyle \int \frac{dx}{x^2 + 2x + 5},$

21. $\displaystyle \int \frac{dx}{\sqrt{(1 - 4x - 2x^2)}},$ **22.** $\displaystyle \int \frac{dx}{\sqrt{(3x^2 - 6x - 5)}},$

23. $\displaystyle \int \frac{dx}{5 - 4x - 2x^2},$ **24.** $\displaystyle \int \frac{dx}{\sqrt{(1 + x + x^2)}}.$

16:12 THE INTEGRALS $\displaystyle \int \frac{px + q}{ax^2 + bx + c}\,dx$ AND $\displaystyle \int \frac{px + q}{\sqrt{(ax^2 + bx + c)}}\,dx$

The first of these integral types can be changed to a sum of two integrals, one of which is of

the form $\displaystyle k_1 \int \frac{f'(x)}{f(x)}\,dx$ and the other of the form $\displaystyle k_2 \int \frac{1}{ax^2 + bx + c}\,dx,$

where k_1 and k_2 are constants. The second can be changed to a sum of two integrals one of which is of

the form $\displaystyle k_3 \int \frac{f'(x)}{\sqrt{\{f(x)\}}}\,dx$

and the other of the form $\displaystyle k_4 \int \frac{1}{\sqrt{(ax^2 + bx + c)}}\,dx.$

Example 1. $\displaystyle \int \frac{(3x + 7)}{2x^2 + 4x + 1}\,dx = \frac{3}{4} \int \frac{(4x + 4)}{2x^2 + 4x + 1}\,dx + \int \frac{4}{2x^2 + 4x + 1}\,dx$

$\displaystyle = \frac{3}{4} \int \frac{(4x + 4)}{2x^2 + 4x + 1}\,dx + 2 \int \frac{1}{x^2 + 2x + \frac{1}{2}}\,dx$

$\displaystyle = \frac{3}{4} \int \frac{(4x + 4)}{2x^2 + 4x + 1}\,dx + 2 \int \frac{1}{(x + 1)^2 - \frac{1}{2}}\,dx$

$\displaystyle = \frac{3}{4} \ln (2x^2 + 4x + 1) + \sqrt{2} \ln \left\{ \frac{x + 1 - 1/\sqrt{2}}{x + 1 + 1/\sqrt{2}} \right\} + C$

$\displaystyle = \frac{3}{4} \ln (2x^2 + 4x + 1) + \sqrt{2} \ln \left\{ \frac{2x + 2 - \sqrt{2}}{2x + 2 + \sqrt{2}} \right\} + C.$

Example 2. $\displaystyle\int \frac{(x+1)}{\sqrt{(3x^2-x-4)}}dx = \frac{1}{6}\int\frac{(6x-1)}{\sqrt{(3x^2-x-4)}}dx + \frac{7}{6}\int\frac{1}{\sqrt{(3x^2-x-4)}}dx$

$$= \frac{1}{6}\int\frac{(6x-1)}{\sqrt{(3x^2-x-4)}}dx + \frac{7}{6\sqrt{3}}\int\frac{1}{\sqrt{\left(x^2-\frac{x}{3}-\frac{4}{3}\right)}}dx$$

$$= \frac{1}{6}\int\frac{(6x-1)}{\sqrt{(3x^2-x-4)}}dx + \frac{7}{6\sqrt{3}}\int\frac{1}{\sqrt{\left\{\left(x-\frac{1}{6}\right)^2-\frac{47}{36}\right\}}}dx$$

$$= \frac{1}{3}\sqrt{(3x^2-x-4)} + \frac{7}{6\sqrt{3}}\cosh^{-1}\left(\frac{6x-1}{\sqrt{47}}\right)+C.$$

Exercise 16.12

Obtain each of the following integrals:

1. $\displaystyle\int\frac{(x+2)\,dx}{x^2+5x+1}$,

2. $\displaystyle\int\frac{(2x+3)\,dx}{2x^2+x+3}$,

3. $\displaystyle\int\frac{(3x+2)\,dx}{\sqrt{(x^2+x+1)}}$,

4. $\displaystyle\int\frac{5x\,dx}{\sqrt{(x^2-3x-1)}}$,

5. $\displaystyle\int\frac{(3x+4)\,dx}{2x^2+x+5}$,

6. $\displaystyle\int\frac{x\,dx}{4x^2+2x-3}$,

7. $\displaystyle\int\frac{(x+1)\,dx}{\sqrt{(1-3x-x^2)}}$,

8. $\displaystyle\int\frac{(5x+2)\,dx}{1+4x-x^2}$,

9. $\displaystyle\int\frac{(x+2)\,dx}{\sqrt{(x^2+6x+4)}}$,

10. $\displaystyle\int\frac{(3x^2+5x+6)\,dx}{x^2+x+1}$.

16:13 SUMMARY OF STANDARD INTEGRALS AND METHODS OF INTEGRATION

Constants of integration have been omitted in the following list; in all cases where the logarithm of a function occurs it is implied that reference is to the positive values of the function only; it is also implied that reference is made only to those values of x for which the integrand exists.

(1) $\displaystyle\int x^n\,dx = \frac{x^{n+1}}{n+1}, \quad (n \ne -1).$ (2) $\displaystyle\int\frac{1}{x}\,dx = \ln x.$

(3) $\displaystyle\int \sin x\,dx = -\cos x.$ (4) $\displaystyle\int \cos x\,dx = \sin x.$

(5) $\displaystyle\int \sec^2 x\,dx = \tan x.$ (6) $\displaystyle\int \csc^2 x\,dx = -\cot x.$

(7) $\displaystyle\int \tan x\,dx = -\ln\cos x = \ln\sec x.$ (8) $\displaystyle\int \cot x\,dx = \ln\sin x.$

(9) $\displaystyle\int \sec x\,dx = \ln(\sec x + \tan x) = \ln\tan\left(\frac{\pi}{4}+\frac{x}{2}\right).$

(10) $\displaystyle\int \csc x\,dx = -\ln(\csc x + \cot x) = \ln\tan\left(\frac{x}{2}\right).$

(11) $\int \sin^2 x \, dx$ and $\int \cos^2 x \, dx$ are obtained by using the transformations $\sin^2 x = \frac{1}{2}(1 - \cos 2x)$ and $\cos^2 x = \frac{1}{2}(1 + \cos 2x)$ respectively.

(12) Integrals of the type $\int \sin mx \cos nx \, dx$ are obtained by using the product-sum transformations.

Example. $\int \sin 3x \cos 4x \, dx = \frac{1}{2}\int (\sin 7x - \sin x) \, dx$

$$= \frac{1}{2} \cos x - \frac{1}{14} \cos 7x.$$

---

(13) $\int \sin^m x \cos^n x \, dx$ where $m, n \in \mathbb{N}$.

(i) m and n even. Use 11, 12.

Example. $\int \sin^2 x \cos^4 x \, dx = \frac{1}{8}\int (1 - \cos 2x)(1 + \cos 2x)^2 \, dx$

$$= \frac{1}{8}\int (1 - \cos 2x)[1 + 2\cos 2x + \frac{1}{2}(1 + \cos 4x)] \, dx$$
$$= \frac{1}{8}\int [\frac{3}{2} + \frac{1}{2}\cos 2x - (1 + \cos 4x) + \frac{1}{2}\cos 4x - \frac{1}{4}(\cos 6x + \cos 2x)] \, dx$$
$$= \frac{1}{32}\int (2 + \cos 2x - 2\cos 4x - \cos 6x) \, dx$$
$$= \frac{1}{32}(2x + \frac{1}{2}\sin 2x - \frac{1}{2}\sin 4x - \frac{1}{6}\sin 6x).$$

(The techniques explained in §13:1 may also be used to express $\sin^m x \cos^n x$ in terms of sines and cosines of multiples of x.)

(ii) m even, n odd. Work in terms of $\sin x$, for example,

$$\int \sin^4 x \cos^7 x \, dx = \int \sin^4 x \cos^6 x \cos x \, dx;$$
$$\text{put } \sin x = s \Rightarrow \cos x \, dx = ds,$$
$$\int \sin^4 x \cos^7 x \, dx = \int s^4(1 - s^2)^3 \, ds = \int (s^4 - 3s^6 + 3s^8 - s^{10}) \, ds$$
$$= \frac{1}{5} \sin^5 x - \frac{3}{7} \sin^7 x + \frac{1}{3} \sin^9 x - \frac{1}{11} \sin^{11} x.$$

(iii) m and n odd. Work in terms of $\sin x$ or $\cos x$.

(14) Trigonometric functions containing only the product $\sin x \cos x$ and/or *even* powers of $\sin x$, $\cos x$ [other than 13(i) above]. Use the substitution $\tan x = t \Rightarrow \sec^2 x \, dx = dt$. For example

$$\int \frac{1}{5 \sin^2 x - 2 \cos^2 x} \, dx = \int \frac{\sec^2 x}{5 \tan^2 x - 2} \, dx = \int \frac{1}{5t^2 - 2} \, dt$$
$$= \left(\frac{\sqrt{2}}{4\sqrt{5}}\right) \ln \left[\frac{\sqrt{5}\tan x - \sqrt{2}}{\sqrt{5}\tan x + \sqrt{2}}\right].$$

(15) Trigonometric functions for which no other method is available. Use the substitution

$$\tan (x/2) = t \Rightarrow dx = \frac{2dt}{1 + t^2}.$$

Example 1. $\displaystyle\int\frac{1}{4+5\sin x}\,dx = \int\frac{1}{(2t^2+5t+2)}\,dt = \int\left\{\frac{2}{3(2t+1)} - \frac{1}{3(t+2)}\right\}dt$

$$= \frac{1}{3}\ln\left[\frac{1+2t}{2+t}\right] = \frac{1}{3}\ln\left[\frac{1+2\tan\left(\frac{1}{2}x\right)}{2+\tan\left(\frac{1}{2}x\right)}\right].$$

Example 2. $\displaystyle\int_0^{\pi/2}\frac{1}{(2+\cos x)}\,dx = 2\int_0^1\frac{1}{(t^2+3)}\,dt = \left[\frac{2}{\sqrt3}\tan^{-1}\left(\frac{t}{\sqrt3}\right)\right]_0^1 = \frac{\pi}{3\sqrt3}.$

---

(16) $\displaystyle\int\frac{1}{\sqrt{(a^2-x^2)}}\,dx = \sin^{-1}\left(\frac{x}{a}\right), \qquad (a>0).$

(17) $\displaystyle\int\frac{1}{x^2+a^2}\,dx = \frac{1}{a}\tan^{-1}\left(\frac{x}{a}\right).$

(18) $\displaystyle\int e^{kx}\,dx = \frac{1}{k}e^{kx}.$

(19) $\displaystyle\int\frac{1}{ax+b}\,dx = \frac{1}{a}\ln(ax+b).$

(20) $\displaystyle\int\frac{1}{(ax+b)^n}\,dx = \frac{-1}{a(n-1)(ax+b)^{n-1}}, \qquad (n \neq 1).$

(21) $\displaystyle\int\frac{f'(x)}{f(x)}\,dx = \ln f(x).$

Example $\displaystyle\int\frac{\cos x\,dx}{3+4\sin x} = \int\frac{\frac{d(\sin x)}{dx}}{3+4\sin x}\,dx = \frac{1}{4}\ln(3+4\sin x).$

---

(22) $\int \sinh x\,dx = \cosh x.$

(23) $\int \cosh x\,dx = \sinh x.$

Integrals 24–27 follow directly by integration as the reverse of differentiation.

(24) $\int \tanh x\,dx = \ln\cosh x.$

(25) $\int \coth x\,dx = \ln\sinh x.$

(26) $\int \operatorname{sech}^2 x\,dx = \tanh x.$

(27) $\int \operatorname{cosech}^2 x\,dx = -\coth x.$

(28) $\displaystyle\int\frac{1}{\sqrt{(x^2+a^2)}}\,dx = \sinh^{-1}\left(\frac{x}{a}\right)$ or $\ln\{x+\sqrt{(x^2+a^2)}\}$

Note $\sinh^{-1}\left(\dfrac{x}{a}\right) = \ln\left\{\dfrac{x+\sqrt{(x^2+a^2)}}{a}\right\} = \ln\{x+\sqrt{(x^2+a^2)}\} + \ln\left(\dfrac{1}{a}\right)$ and the last term forms part of the constant of integration.

(29) $\int \dfrac{1}{\sqrt{(x^2-a^2)}}\,dx = \cosh^{-1}\left(\dfrac{x}{a}\right)$ or $\ln\{x+\sqrt{(x^2-a^2)}\}$.

(30) $\int \dfrac{1}{x^2-a^2}\,dx = \dfrac{1}{2a}\ln\left(\dfrac{x-a}{x+a}\right)$.

(31) $\int \dfrac{1}{a^2-x^2}\,dx = \dfrac{1}{2a}\ln\left(\dfrac{a+x}{a-x}\right)$.

(32) *Surds.*   No general rules can be laid down when the integrand involves surds. However, we give below some examples which are intended to serve as a guide.

Example 1. If the integrand involves $\sqrt{(ax+b)}$, then the substitution $ax+b = u^2$ may reduce the integral to one of the types considered earlier, e.g.

$$\int\left\{\dfrac{1+\sqrt{(ax+b)}}{1-\sqrt{(ax+b)}}\right\}dx = \dfrac{2}{a}\int\dfrac{u(1+u)\,du}{(1-u)} = \dfrac{2}{a}\int\left(-u-2+\dfrac{2}{1-u}\right)du$$

$$= \dfrac{-u^2-4u-4\ln(1-u)}{a}$$

$$= \dfrac{-ax-b-4\sqrt{(ax+b)}-4\ln\{(1-\sqrt{(ax+b)}\}}{a}.$$

Example 2. $\displaystyle\int_{1/2}^{1}\dfrac{1}{x\sqrt{(5x^2-4x+1)}}\,dx.$

Putting $x = 1/u$ transforms this integral into

$$\int_{1}^{2}\dfrac{1}{\sqrt{(5-4u+u^2)}}\,du = \int_{1}^{2}\dfrac{1}{\sqrt{[1+(u-2)^2]}}\,du = \left[\sinh^{-1}(u-2)\right]_{1}^{2} = -\sinh^{-1}(-1)$$

$$= \sinh^{-1}1 = \ln(1+\sqrt{2}).$$

(33) *Hyperbolic functions.* Integration of a polynomial in $\sinh x$, $\cosh x$ or $\tanh x$ can usually be evaluated by methods akin to those used for similar trigonometric functions. In other cases, however, this method may not be suitable and the substitution $e^x = u$ is appropriate. It is particularly advisable that results should be checked by differentiation.

Example 1. $\int \sinh^4 x\,dx = \frac{1}{8}\int(\cosh 4x - 4\cosh 2x + 3)\,dx$
$$= \tfrac{1}{32}(\sinh 4x - 8\sinh 2x + 12x).$$

Example 2. $\int \cosh^5 x\,dx = \int(1+\sinh^2 x)^2\,d(\sinh x)$
$$= \sinh x + \tfrac{2}{3}\sinh^3 x + \tfrac{1}{5}\sinh^5 x.$$

Example 3. $\displaystyle\int \operatorname{cosech} x\,dx = \int\dfrac{2\,dx}{e^x-e^{-x}} = 2\int\dfrac{d(e^x)}{e^{2x}-1} = \ln\left(\dfrac{e^x-1}{e^x+1}\right)$

$$= \ln\tanh\dfrac{x}{2}.$$

Systematic integration

(i) Whenever possible an integral is obtained directly from one of the standard forms quoted above.

(ii) If the integrand is a quotient, it should be examined to find out whether

(a) it can be integrated directly as a logarithm, or

(b) the denominator factorises so that the method of partial fractions can be used, or,

(c) it is one of the forms discussed in § 16:12 of this chapter.

(iii) The possibility of simplifying the integration by the method of substitution should be considered.

(iv) The possibility of using the method of integration by parts should be considered.

The student will find that, with experience, he can proceed directly to the most favourable method of integration in particular cases, but in the early stages a systematic search for the method is desirable.

Exercise 16.13

In questions 1–54 integrate the given expressions with respect to x.

1. $x \sec^2 x$.

2. $\sec x \tan^2 x$.

3. $\ln x$.

4. $x \ln x$.

5. $\tanh^{-1} x$.

6. $\sinh^{-1} (x/a)$.

7. $\sinh^{-1} (a/x)$.

8. $(x-1)^2 \ln x$.

9. $(1+x^2) \tan^{-1} x$.

10. $\dfrac{x \sin^{-1} x}{\sqrt{(1-x^2)}}$.

11. $x(\ln x)^2$.

12. $x \sin^2 x \cos x$.

13. $\sec^{-1}\left(\dfrac{x}{a}\right)$.

14. $\dfrac{x^2}{\sqrt{(a^2-x^2)}}$.

15. $\sin x \ln (1 - a \cos x)$.

16. $x \sin^{-1} x$.

17. $x e^{-x^2}$.

18. $\sec^2 x \tan x$.

19. $\sec^4 x$.

20. $x^{-1} \ln x$.

21. $\cos^3 x \sin^4 x$.

22. $\dfrac{x^3}{1-x^4}$.

23. $\dfrac{x}{9+x^4}$.

24. $x \sqrt{(a^2+x^2)}$.

25. $\dfrac{e^x}{5+3e^x}$.

26. $x^3 \sqrt{(a^2-x^2)}$.

27. $\dfrac{\sin x}{16+9 \cos^2 x}$.

28. $\dfrac{e^{ax}}{\sqrt{(b^2+e^{2ax})}}$.

29. $\sqrt{\left[\dfrac{x}{c-x}\right]}$.

30. $\dfrac{e^{3x}}{1+e^{6x}}$.

31. $\sinh^n x \cosh^3 x$.

32. $\dfrac{\sin x}{a+b \cos x}$.

33. $\dfrac{1}{2+\cos^2 x}$.

34. $\dfrac{1}{5 \cosh x + 4 \sinh x}$.

35. $\sec^{3/2} x \tan^3 x$.

36. $x \sqrt{\left[\dfrac{a+x}{a-x}\right]}$.

37. $\dfrac{(1+x^3)}{x^2(1-x)}$.

38. $\dfrac{1}{x(1+x^5)}$.

39. $\dfrac{1}{1+\sin x}$.

40. $\dfrac{\cos x}{3+\cos x}$.

41. $\dfrac{1}{1+8 \sin^2 x}$.

42. $\dfrac{\sqrt{(x-1)}}{x}$.

43. $\dfrac{1}{(1-\sqrt{x})\sqrt{x}}$.

44. $\dfrac{5 \cos x}{2 \cos x + \sin x + 2}$.

45. $\dfrac{1}{8-x^{3/2}}$.

46. $\dfrac{x+1}{x\sqrt{(x-1)}}.$ **47.** $\dfrac{4x+5}{\sqrt{(2x^2+7x+3)}}.$ **48.** $\dfrac{\sin x}{5-3\sin^2 x}.$

49. $\dfrac{3+\sin x}{1+\cos x}.$ **50.** $\dfrac{1}{(1-x)\sqrt{(x+1)}}.$ **51.** $\dfrac{1}{1+\sin x+\cos x}.$

52. $\dfrac{\cos x-\sin x}{\cos x+\sin x}.$ **53.** $(x+1)\sqrt{(x^2+x+1)}.$ **54.** $\operatorname{sech}^4 x.$

In questions 55–86 evaluate the given definite integrals.

55. $\displaystyle\int_1^2 \dfrac{1}{e^x-1}\,dx.$ **56.** $\displaystyle\int_0^{\pi/2} \dfrac{\sin x\cos x\,dx}{a\cos^2 x+b\sin^2 x}.$ **57.** $\displaystyle\int_0^1 2^{3x}\,dx.$

58. $\displaystyle\int_0^{\pi/4} \sin x\ln(\cos x)\,dx.$ **59.** $\displaystyle\int_0^{\pi/4} \theta\sec^2\theta\,d\theta.$ **60.** $\displaystyle\int_0^{\pi/4} x\tan^2 x\,dx.$

61. $\displaystyle\int_0^{\pi/2} \sin\theta\ln(1+\sin\theta)\,d\theta.$ **62.** $\displaystyle\int_{1/2}^1 \sin^{-1}(\sqrt{x})\,dx.$ **63.** $\displaystyle\int_0^{\pi/2} x\sin^2 x\,dx.$

64. $\displaystyle\int_0^{\pi/2} \dfrac{1}{3\sin x+2}\,dx.$ **65.** $\displaystyle\int_0^{\pi/2} \dfrac{1}{4-\cos^2\theta}\,d\theta.$ **66.** $\displaystyle\int_1^2 \dfrac{(x-1)\,dx}{\sqrt{(8+2x-x^2)}}.$

67. $\displaystyle\int_1^\infty \dfrac{1}{x\sqrt{(25x^2-1)}}\,dx.$ **68.** $\displaystyle\int_0^{\pi/2} \dfrac{1}{3\sin\theta+\cos\theta+1}\,d\theta.$ **69.** $\displaystyle\int_0^{\pi/6} \dfrac{1}{\cos\theta+\cos^3\theta}\,d\theta.$

70. $\displaystyle\int_0^1 \dfrac{1}{(1+x)\sqrt{(1-x^2)}}\,dx.$ **71.** $\displaystyle\int_0^\pi \dfrac{\sin 2x\,dx}{2-\cos x}.$ **72.** $\displaystyle\int_0^{\pi/4} \dfrac{2\,dx}{3\sin 2x+4\cos 2x}.$

73. $\displaystyle\int_0^{\pi/2} \dfrac{\sin 2x\,dx}{1+\cos^2 x}.$ **74.** $\displaystyle\int_0^{\pi/2} \dfrac{1}{1-2a\cos\theta+a^2}\,d\theta,\ |a|\neq 1.$

75. $\displaystyle\int_0^\infty \dfrac{1}{1-2x\cos\alpha+x^2}\,dx,$ when $0<\alpha\leqslant\tfrac12\pi.$ **76.** $\displaystyle\int_0^{\pi/2} \dfrac{\sin 3x}{1+\cos^2 x}\,dx.$

77. $\displaystyle\int_{1/8}^{1/3} \dfrac{1}{\sqrt{x}\sqrt{(1+x)}}\,dx.$ **78.** $\displaystyle\int_0^{\pi/2} \dfrac{\cos x\,dx}{7+\cos 2x}.$ **79.** $\displaystyle\int_0^\pi \dfrac{1}{2+\cos x}\,dx.$

80. $\displaystyle\int_1^2 \left(\dfrac{x-1}{2-x}\right)^{1/2}\,dx.$ **81.** $\displaystyle\int_0^1 x^2\sin^{-1}x\,dx.$ **82.** $\displaystyle\int_0^{\pi/4} \dfrac{x\,dx}{1+\cos 2x}.$

83. $\displaystyle\int_{1/2}^1 \dfrac{1}{1+x^3}\,dx.$ **84.** $\displaystyle\int_0^{1/2} \left(\dfrac{x^2+x+1}{x^2-x+1}\right)\,dx.$ **85.** $\displaystyle\int_0^{1/2} x\ln(1-x^2)\,dx.$

86. $\displaystyle\int_0^1 x^2\tan^{-1}x\,dx.$

87. Show that

$$\frac{x^5}{(x+1)(x-2)} = x^3 + x^2 + 3x + 5 + \frac{1}{3(x+1)} + \frac{32}{3(x-2)}.$$

Hence integrate this function.

88. Using the identity $x^4 + x^2 + 1 \equiv (x^2 + 1)^2 - x^2$, show that

$$\int_0^1 \frac{x^2\,dx}{x^4 + x^2 + 1} = \frac{\pi}{4\sqrt{3}} - \frac{1}{4}\ln 3.$$

16:14 EXPONENTIAL VALUES OF SINE AND COSINE

Since, for real values of θ,

$$e^{i\theta} = \cos\theta + i\sin\theta,$$
$$e^{-i\theta} = \cos\theta - i\sin\theta$$
$$\Rightarrow \cos\theta = \tfrac{1}{2}(e^{i\theta} + e^{-i\theta}),$$

$$\sin\theta = \frac{1}{2i}(e^{i\theta} - e^{-i\theta})$$

(16.49)

for real values of θ. These values of $\sin\theta$ and $\cos\theta$ are analogous to the definitions $\sinh x = \tfrac{1}{2}(e^x - e^{-x})$ and $\cosh x = \tfrac{1}{2}(e^x + e^{-x})$ for real values of x.

The reason for the close relation between the properties of the two sets of functions is now clear.

We *define* $\sin z$ and $\cos z$ for complex values of z by the equations

$$\sin z = \frac{e^{iz} - e^{-iz}}{2i}, \quad \cos z = \frac{e^{iz} + e^{-iz}}{2}$$

and we similarly extend the definition of $\sinh x$ and $\cosh x$ to include

$$\sinh z = \tfrac{1}{2}(e^z - e^{-z}), \quad \cosh z = \tfrac{1}{2}(e^z + e^{-z}).$$

These definitions include the special cases (in which z is purely real or purely imaginary) considered earlier.

Now we have

$$\cosh iz = \cos z, \quad \sinh iz = i\sin z \qquad (16.50)$$

and it can be proved that

$$\cos iz = \cosh z, \quad \sin iz = i\sinh z. \qquad (16.51)$$

Formulae which we have established for both the circular and hyperbolic functions of real variables can now be extended to the corresponding functions of complex variables and can be shown to be true by simple processes of algebra. Formulae for hyperbolic functions can be obtained from the corresponding formulae for circular functions by replacing cos by cosh and sin by i sinh. This is equivalent to the rule given earlier whereby the formulae for hyperbolic functions were obtained from the corresponding formulae for circular functions by changing the sign in front of every product of two sines.

Example 1. Prove that

$$\cos(z+w) \equiv \cos z \cos w - \sin z \sin w,$$

where z and w are complex variables, and deduce the corresponding formula for $\cosh(z+w)$.

$$\cos(z+w) \equiv \tfrac{1}{2}\{e^{i(z+w)} + e^{-i(z+w)}\} \equiv \tfrac{1}{2}(e^{iz}e^{iw} + e^{-iz}e^{-iw})$$
$$\equiv \tfrac{1}{2}[(\cos z + i \sin z)(\cos w + i \sin w) + (\cos z - i \sin z)(\cos w - i \sin w)]$$
$$\Rightarrow \cos(z+w) \equiv \cos z \cos w - \sin z \sin w.$$

By a similar algebraic process we could obtain

$$\cos i(z+w) \equiv \cos iz \cos iw - \sin iz \sin iw$$
$$\Rightarrow \cosh(z+w) \equiv \cosh z \cosh w - i^2 \sinh z \sinh w$$
$$\equiv \cosh z \cosh w + \sinh z \sinh w.$$

Example 2. Evaluate $\mathrm{Cos}^{-1} 1{\cdot}5$.

If $\mathrm{Cos}^{-1} 1{\cdot}5 = x + iy$, where x and y are real, then

$$\cos(x+iy) = \cos x \cos iy - \sin x \sin iy = 1{\cdot}5$$
$$\Rightarrow \cos x \cosh y - i \sin x \sinh y = 1{\cdot}5.$$

Equating real and imaginary parts

$$\cos x \cosh y = 1{\cdot}5, \tag{1}$$
$$\sin x \sinh y = 0. \tag{2}$$

Hence from equation (2),

$$\sin x = 0 \Rightarrow x = n\pi, \quad \text{where } n \in \mathbb{Z},$$
$$\text{or } \sinh y = 0 \Rightarrow y = 0.$$

But $y = 0$ involves $\cos x = 1{\cdot}5$ from equation (1), which is contrary to the hypothesis that x is real. Also $x = n\pi$ when n is an odd integer involves $\cosh y = -1{\cdot}5$, which is a contrary to the hypothesis that y is real,

$$\Rightarrow x = 2m\pi \quad (m \in \mathbb{Z})$$

and

$$y = \mathrm{Cosh}^{-1} 1{\cdot}5 = \pm \ln\left(\frac{3+\sqrt{5}}{2}\right)$$
$$\Rightarrow \mathrm{Cos}^{-1} 1{\cdot}5 = 2m\pi \pm i \ln\left(\frac{3+\sqrt{5}}{2}\right) \quad (m \in \mathbb{Z}).$$

Example 3. Find the real and imaginary parts of $\tan(x+iy)$.

If

$$u + iv = \tan(x+iy) \tag{1}$$
$$\Leftrightarrow u - iv = \tan(x-iy). \tag{2}$$

Adding (1) and (2)
$$2u = \tan(x+iy) + \tan(x-iy)$$
$$= \frac{\sin(x+iy)\cos(x-iy) + \sin(x-iy)\cos(x+iy)}{\cos(x+iy)\cos(x-iy)}$$
$$= \frac{\sin 2x}{\frac{1}{2}(\cos 2x + \cos 2iy)}$$
$$= \frac{2\sin 2x}{\cos 2x + \cosh 2y}$$
$$\Rightarrow u = \frac{\sin 2x}{\cos 2x + \cosh 2y}.$$

Also, subtracting (2) from (1), $2iv = \tan(x+iy) - \tan(x-iy)$
$$\Rightarrow v = \frac{\sinh 2y}{\cos 2x + \cosh 2y}.$$

This example is an illustration of an important technique which is used to find the real and imaginary parts of $f(z)$ where $z = x + iy$.

For, if
$$w = u + iv = f(z),$$
then
$$w^* = u - iv = [f(z)]^*$$
and so
$$2u = f(z) + [f(z)]^*, \qquad 2iv = f(z) - [f(z)]^*.$$

It should be observed that $[f(z)]^* = f(z^*)$ only if the numerical coefficients in $f(z)$ are all real.

Exercise 16.14

1. Express $\sin(x-iy) + \cos(x+iy)$ in the form $a + ib$.

2. Express (i) $\sinh(1-i)$, (ii) $\cosh(1-i)$ in the form $a + ib$.

3. Use the exponential form of $\sin z$ and $\cos z$ to prove each of the following formulae and, in each case, write down the corresponding formula for hyperbolic functions:

(i) $\sin 2z = 2\sin z \cos z$, (ii) $\sin 3z = 3\sin z - 4\sin^3 z$,

(iii) $\cos z + \cos w = 2\cos\frac{1}{2}(z+w)\cos\frac{1}{2}(z-w)$,

(iv) $\tan 3z = (3\tan z - \tan^3 z)/(1 - 3\tan^2 z)$,

(v) $\tan^2 z = \dfrac{1 - \cos 2z}{1 + \cos 2z}$.

4. Find the real and imaginary parts of $\tanh(x+iy)$.

5. Show that all solutions of the equation
$$\sin z = 2i\cos z$$
are given by $z = (\frac{1}{2} \pm n)\pi + \frac{1}{2}i \ln 3$ where n is zero or any positive integer.   (L.)

6. Evaluate each of the following in the form $a + ib$ where a and b are real numbers:

(i) $\text{Sin}^{-1} 2$, (ii) $\text{Cos}^{-1} 3$, (iii) $\text{Sec}^{-1}(-\frac{1}{2})$.

7. Find the polar equation of the curve in the Argand diagram described by the point z when it varies so that ze^z is real.

Sketch that part of the curve which is such that $\arg(ze^z) = 0$ and $-\pi < \arg z < \pi$ indicating the asymptotes.   (L.)

8. If $\tan(\frac{1}{4}\pi + iv) = re^{i\theta}$ where v, r and θ are all real, prove that $r = 1$, $\tan\theta = \sinh 2v$ and $\tanh\theta = \tan\frac{1}{2}\theta$.   (L.)

9. Find the general solution of the equation $\sinh z = 2\cosh z$, where z is complex.   (L.)

10. If x, y, u, v are real and $\cosh(x+iy) = \tan(u+iv)$ prove that

$$\cosh 2x + \cos 2y = 2\left\{ \frac{\cosh 2v - \cos 2u}{\cosh 2v + \cos 2u} \right\}.$$

(L.)

16:15 ELEMENTARY PARTIAL DIFFERENTIATION

If $f(x, y)$ is a function of the two *independent* variables x and y, we define the partial derivatives of $f(x, y)$ with respect to x and y as

$$\lim_{h \to 0} \frac{f(x+h,\, y) - f(x, y)}{h} \quad \text{and} \quad \lim_{k \to 0} \frac{f(x,\, y+k) - f(x, y)}{k}$$

respectively when these limits exist. The partial derivative of f with respect to x is effectively the derivative of f with respect to x when y is treated as a constant. This partial derivative we denote by $\partial f/\partial x$ or f_x. Similarly the partial derivative of f with respect to y is the derivative of f with respect to y when x is treated as a constant. We denote this partial derivative by $\partial f/\partial y$ or f_y.

Example 1. If $f = xe^y \ln(2x+3y) + \sin(x^2 y)$, then

$$\frac{\partial f}{\partial x} = e^y \ln(2x+3y) + \frac{2xe^y}{2x+3y} + 2xy \cos(x^2 y),$$

$$\frac{\partial f}{\partial y} = xe^y \ln(2x+3y) + \frac{3xe^y}{2x+3y} + x^2 \cos(x^2 y).$$

Example 2. If $z = F(ax+by) + G(x^m y^p)$, where a, b, m, p are constants, find $\partial z/\partial x$, $\partial z/\partial y$.

Here $F(ax+b)$ is the function $F(u)$ of the function $u = ax+by$, $G(x^m y^p)$ is the function $G(v)$ of the function $v = x^m y^p$.

Then $\partial z/\partial x = F'(ax+by)\,(\partial(ax+by)/\partial x) + G'(x^m y^p)\,(\partial(x^m y^p)/\partial x)$ where $F'(ax+by)$ denotes the function $F'(u)$ of the function $u = ax+by$. In fact we may consider $F'(ax+by)$ as the derivative of $F(ax+by)$, where $ax+by$ is regarded as a single variable

$$\Rightarrow \frac{\partial z}{\partial x} = aF'(ax+by) + mx^{m-1} y^p G'(x^m y^p).$$

Similarly

$$\frac{\partial z}{\partial y} = bF'(ax+by) + px^m y^{p-1} G'(x^m y^p).$$

Example 3. If $u = \sin^{-1}(y/2x)$, find $\partial u/\partial x$ and $\partial u/\partial x$ and *verify* that

$$x\frac{\partial u}{\partial x} + y\frac{\partial u}{\partial y} = 0.$$

By the chain rule

$$\frac{\partial u}{\partial x} = \frac{1}{\sqrt{\{1 - (y/2x)^2\}}} \frac{\partial}{\partial x}\left(\frac{y}{2x}\right)$$

$$= \frac{-y}{2x^2\sqrt{\{1-(y/2x)^2\}}} = \frac{-y}{x\sqrt{(4x^2-y^2)}}.$$

Similarly

$$\frac{\partial u}{\partial y} = \frac{1}{\sqrt{\{1-(y/2x)^2\}}} \frac{\partial}{\partial y}\left(\frac{y}{2x}\right)$$

$$= \frac{1}{2x\sqrt{\{1-(y/2x)^2\}}} = \frac{1}{\sqrt{(4x^2-y^2)}}.$$

It follows at once by substitution that

$$x\frac{\partial u}{\partial x} + y\frac{\partial u}{\partial y} = \frac{-y}{\sqrt{(4x^2-y^2)}} + \frac{y}{\sqrt{(4x^2-y^2)}} = 0.$$

Since $\partial f/\partial x$ is a function of x and y, we may differentiate it with respect to x giving

$$\frac{\partial}{\partial x}\left(\frac{\partial f}{\partial x}\right),$$

the result being denoted by $\partial^2 f/\partial x^2$ or f_{xx}. Similarly,

$$\frac{\partial}{\partial y}\left(\frac{\partial f}{\partial y}\right)$$

is denoted by $\partial^2 f/\partial y^2$ or f_{yy}. In addition we have the "mixed" derivatives

$$\frac{\partial}{\partial y}\left(\frac{\partial f}{\partial x}\right) \quad \text{and} \quad \frac{\partial}{\partial x}\left(\frac{\partial f}{\partial y}\right).$$

Except in special circumstances, rarely if ever encountered in physical problems, the order of differentiation is immaterial and we denote either of these derivatives by $\partial^2 f/\partial x\,\partial y$ or f_{xy}. Similarly by $\partial^n f/(\partial x^p\,\partial y^q)$, where $p+q=n$, we denote the result of q successive differentiations of f with respect to y followed by p successive differentiations with respect to x.

Example 1. Show that $\Phi = e^{-x^2/y}$ satisfies the equation

$$\frac{\partial^2\Phi}{\partial x^2} = 4\frac{\partial\Phi}{\partial y} - \frac{2\Phi}{y}.$$

$$\frac{\partial\Phi}{\partial x} = -\frac{2xe^{-x^2/y}}{y} = -\frac{2x\Phi}{y}$$

$$\Rightarrow \frac{\partial^2\Phi}{\partial x^2} = -\frac{2\Phi}{y} - \frac{2x}{y}\frac{\partial\Phi}{\partial x} = -\frac{2\Phi}{y} + \frac{4x^2\Phi}{y^2}.$$

$$\frac{\partial\Phi}{\partial y} = \frac{x^2\,e^{-x^2/y}}{y^2} = \frac{x^2\Phi}{y^2}$$

$$\Rightarrow \frac{\partial^2\Phi}{\partial x^2} - 4\frac{\partial\Phi}{\partial y} + \frac{2\Phi}{y} = -\frac{2\Phi}{y} + \frac{4x^2\Phi}{y^2} - \frac{4x^2\Phi}{y^2} + \frac{2\Phi}{y} = 0$$

$$\Leftrightarrow \frac{\partial^2\Phi}{\partial x^2} = 4\frac{\partial\Phi}{\partial y} - \frac{2\Phi}{y}.$$

Example 2. *Verify* that

$$V = f(x+ct) + g(x-ct),$$

where f and g are arbitrary differentiable functions and c is constant, satisfies the partial differential equation (the one-dimensional wave equation)

$$\frac{\partial^2 V}{\partial t^2} = c^2 \frac{\partial^2 V}{\partial x^2}.$$

If $V = 0$ at $x = 0$ for all t, show that $g(u) = -f(-u)$. If also $V = 0$ at $x = l$ for all t show that $f(u + l) = f(u - l)$. Deduce that $f(u)$ is periodic with period $2l$.

$$\frac{\partial V}{\partial t} = f'(x + ct) \frac{\partial(x + ct)}{\partial t} + g'(x - ct) \frac{\partial(x - ct)}{\partial t}$$

$$= cf'(x + ct) - cg'(x - ct).$$

$$\frac{\partial^2 V}{\partial t^2} = cf''(x + ct) \frac{\partial(x + ct)}{\partial t} - cg''(x - ct) \frac{\partial(x - ct)}{\partial t}$$

$$= c^2 [f''(x + ct) + g''(x - ct)].$$

Similarly $\partial^2 V/\partial x^2 = f''(x + ct) + g''(x - ct)$ and the required result follows. If $V = 0$, at $x = 0$ for all t, then

$$f(ct) + g(-ct) = 0 \text{ for all } t.$$
$$\Rightarrow g(u) = -f(-u) \tag{1}$$

on writing $ct = -u$.
 If $V = 0$ at $x = l$ for all t, then

$$f(l + ct) + g(l - ct) = 0 \text{ for all } t$$
$$\Rightarrow f(l + u) + g(l - u) = 0 \text{ for all } u.$$

But, from (1), $g(l - u) = -f(u - l)$, and so

$$f(u + l) = f(u - l) \tag{2}$$

as required. Finally, writing $u = v + l$, we have

$$f(v + 2l) = f(v)$$

indicating that $f(v)$ is periodic with period $2l$.

Example 3. If $\psi = F(y + x) + G(y + 2x)$, where F and G are arbitrary functions, satisfies the equation

$$a \frac{\partial^2 \psi}{\partial x^2} - 3 \frac{\partial^2 \psi}{\partial x \partial y} + b \frac{\partial^2 \psi}{\partial y^2} = 0,$$

find the values of the constants a and b.

 Substitution gives

$$(a - 3 + b) F''(y + x) + (4a - 6 + b) G''(y + 2x) = 0. \tag{1}$$

Therefore since F and G are arbitrary functions the coefficients of F″ and G″ in equation (1) must vanish identically

$$\Rightarrow a - 3 + b = 0, \quad 4a - 6 + b = 0$$
$$\Rightarrow a = 1, b = 2.$$

Example 4. If the arbitrary function $z = g(x + my)$, where m is a numerical constant, satisfies the partial differential equation

$$6 \frac{\partial^2 z}{\partial x^2} + \frac{\partial^2 z}{\partial x \partial y} - \frac{\partial^2 z}{\partial y^2} = 0,$$

find the two possible values of m.

Making the given substitution we find

$$(6 + m - m^2)\, g''(x + my) = 0.$$

Since g is arbitrary, we must have

$$6 + m - m^2 = 0.$$

This quadratic equation has roots $m = -2, 3$ and these are the possible values of m.

This example illustrates a technique for solving certain linear partial differential equations with constant coefficients. In fact the general solution of the given equation is

$$z = g_1(x - 2y) + g_2(x + 3y),$$

where g_1 and g_2 are arbitrary functions.

Example 5. If z is defined implicitly in terms of the independent variables x and y by the relation

$$xy = f(x + z), \tag{1}$$

prove that

$$x \frac{\partial z}{\partial x} - y \frac{\partial z}{\partial y} + x = 0, \tag{2}$$

$$x^2 \frac{\partial^2 z}{\partial x^2} = y^2 \frac{\partial^2 z}{\partial y^2}. \tag{3}$$

Differentiating the given relation (1) partially with respect to x we find, using the chain rule, that

$$y = f'(x + z) \frac{\partial}{\partial x}(x + z) = \left(1 + \frac{\partial z}{\partial x}\right) f'(x + z)$$

$$\Rightarrow \frac{\partial z}{\partial x} = \frac{y - f'(x + z)}{f'(x + z)}. \tag{4}$$

Similarly

$$x = f'(x + z) \frac{\partial z}{\partial y}$$

$$\Rightarrow \frac{\partial z}{\partial y} = \frac{x}{f'(x + z)}. \tag{5}$$

Elimination of $f'(x + z)$ from equations (4) and (5) leads at once to equation (2).

To derive equation (3) we differentiate equation (2) first partially with respect to x and second partially with respect to y obtaining respectively

$$x \frac{\partial^2 z}{\partial x^2} + \frac{\partial z}{\partial x} - y \frac{\partial^2 z}{\partial x \partial y} + 1 = 0, \tag{6}$$

$$x \frac{\partial^2 z}{\partial x \partial y} - y \frac{\partial^2 z}{\partial y^2} - \frac{\partial z}{\partial y} = 0. \tag{7}$$

Then $x \times$ equation (6) $- y \times$ equation (7) and use of equation (2) leads to the required result. [Note that in this case we do not have to find $\partial^2 z/\partial x^2$, $\partial^2 z/\partial y^2$ explicitly in terms of the derivatives of f.]

If $f(x_1, x_2, \ldots, x_n)$ is a function of the n *independent* variables $x_1, x_2, \ldots, x_n$, we define the partial derivative of f with respect to x_r, $\partial f/\partial x_r$, $(r = 1, 2, \ldots, n)$ as the derivative of f with respect to x_r when the remaining $n - 1$ variables are treated as constants. Similarly, we define partial derivatives of higher orders.

Example 1. Find the relation between the constants a_1, a_2, a_3, p, c, if the (three-dimensional) wave equation

$$\frac{\partial^2 \Phi}{\partial x_1^2} + \frac{\partial^2 \Phi}{\partial x_2^2} + \frac{\partial^2 \Phi}{\partial x_3^2} = \frac{1}{c^2} \frac{\partial^2 \Phi}{\partial t^2}$$

is satisfied by $\Phi = \sin(x_1/a_1)\sin(x_2/a_2)\sin(x_3/a_3)\cos pt$.

$$\frac{\partial^2\Phi}{\partial x_1^2} = -\frac{1}{a_1^2}\sin\left(\frac{x_1}{a_1}\right)\sin\left(\frac{x_2}{a_2}\right)\sin\left(\frac{x_3}{a_3}\right)\cos pt = \frac{-\Phi}{a_1^2}.$$

Similarly for

$$\frac{\partial^2\Phi}{\partial x_2^2}, \quad \frac{\partial^2\Phi}{\partial x_3^2}.$$

Also

$$\frac{1}{c^2}\frac{\partial^2\Phi}{\partial t^2} = -\frac{p^2}{c^2}\sin\left(\frac{x_1}{a_1}\right)\sin\left(\frac{x_2}{a_2}\right)\sin\left(\frac{x_3}{a_3}\right)\cos pt = -\frac{p^2\Phi}{c^2}.$$

Therefore the required condition is

$$\left(\frac{p^2}{c^2} - \frac{1}{a_1^2} - \frac{1}{a_2^2} - \frac{1}{a_3^2}\right)\Phi = 0$$

$$\Rightarrow \frac{1}{a_1^2} + \frac{1}{a_2^2} + \frac{1}{a_3^2} = \frac{p^2}{c^2}.$$

Example 2. Show that $V = 1/r$, where $r^2 = (x-a)^2 + (y-b)^2 + (z-c)^2$ and a, b, c are constants, satisfies Laplace's equation

$$\frac{\partial^2 V}{\partial x^2} + \frac{\partial^2 V}{\partial y^2} + \frac{\partial^2 V}{\partial z^2} = 0.$$

$$\frac{\partial V}{\partial x} = -\frac{1}{r^2}\frac{\partial r}{\partial x}, \quad \frac{\partial^2 V}{\partial x^2} = \frac{2}{r^3}\left(\frac{\partial r}{\partial x}\right)^2 - \frac{1}{r^2}\frac{\partial^2 r}{\partial x^2}.$$

But

$$2r\frac{\partial r}{\partial x} = 2(x-a) \quad \text{and} \quad r\frac{\partial^2 r}{\partial x^2} + \left(\frac{\partial r}{\partial x}\right)^2 = 1$$

$$\Rightarrow \frac{\partial r}{\partial x} = \frac{x-a}{r}, \quad \frac{\partial^2 r}{\partial x^2} = \frac{1}{r} - \frac{(x-a)^2}{r^3}$$

$$\Rightarrow \frac{\partial^2 V}{\partial x^2} = \frac{2(x-a)^2}{r^5} - \frac{1}{r^3} + \frac{(x-a)^2}{r^5} = \frac{3(x-a)^2}{r^5} - \frac{1}{r^3}.$$

Therefore by symmetry

$$\frac{\partial^2 V}{\partial x^2} + \frac{\partial^2 V}{\partial y^2} + \frac{\partial^2 V}{\partial z^2} = \frac{3\left[(x-a)^2 + (y-b)^2 + (z-c)^2\right]}{r^5} - \frac{3}{r^3} = 0,$$

since $(x-a)^2 + (y-b)^2 + (z-c)^2 = r^2$.

Exercise 16.15

1. Show that $u = x^2 \sin(\ln y)$ satisfies

$$2y^2\frac{\partial^2 u}{\partial y^2} + 2y\frac{\partial u}{\partial y} + x\frac{\partial u}{\partial x} = 0.$$

2. If $V = (Ar^n + Br^{-n})\sin n\theta$, where A, B, n are constants, prove that

$$r\frac{\partial}{\partial r}\left(r\frac{\partial V}{\partial r}\right) + \frac{\partial^2 V}{\partial \theta^2} = 0.$$

3. (i) If $r\Phi = \sin pr \cos cpt$, where p and c are constants, show that

$$\frac{\partial^2 \Phi}{\partial r^2} + \frac{2}{r}\frac{\partial \Phi}{\partial r} = \frac{1}{c^2}\frac{\partial^2 \Phi}{\partial t^2}.$$

(ii) If $z = f(x - 2y) + g(3x + y)$, where f and g are arbitrary functions and if

$$\frac{\partial^2 z}{\partial x^2} + a\frac{\partial^2 z}{\partial x \partial y} + b\frac{\partial^2 z}{\partial y^2} = 0,$$

find a, b.

4. (i) Show that $\Phi = e^{-2x}\sin x \sin t$ satisfies the equation

$$\frac{\partial^2 \Phi}{\partial x^2} + 4\frac{\partial \Phi}{\partial x} = 5\frac{\partial^2 \Phi}{\partial t^2}.$$

(ii) If $u = \ln(1 + xy^2)$, show that

$$2x\frac{\partial u}{\partial x} = y\frac{\partial u}{\partial y} \quad \text{and} \quad 2\frac{\partial^2 u}{\partial x^2} + y^3\frac{\partial^2 u}{\partial x \partial y} = 0.$$

5. (i) If $z = \ln r$, where $r^2 = x^2 + y^2$, prove that

$$\frac{\partial^2 z}{\partial x^2} + \frac{\partial^2 z}{\partial y^2} = 0.$$

(ii) If $z = f(x + ay) + \Phi(x - ay) - \dfrac{x}{2a^2}\cos(x + ay)$, where f and Φ are arbitrary functions, prove that

$$a^2\frac{\partial^2 z}{\partial x^2} - \frac{\partial^2 z}{\partial y^2} = \sin(x + ay).$$

6. If $r^2 = x^2 + y^2$, prove that

$$\frac{\partial r}{\partial x} = \frac{x}{r} \quad \text{and} \quad \frac{\partial^2 r}{\partial x^2} = \frac{y^2}{r^3}.$$

If $V = e^{k(r-x)}$, where k is constant, prove that

(i)
$$\left(\frac{\partial V}{\partial x}\right)^2 + \left(\frac{\partial V}{\partial y}\right)^2 + 2Vk\frac{\partial V}{\partial x} = 0,$$

(ii)
$$\frac{\partial^2 V}{\partial x^2} + \frac{\partial^2 V}{\partial y^2} + 2k\frac{\partial V}{\partial x} = \frac{Vk}{r}.$$

7. If $r^2 = x^2 + y^2$ and $u = \cos kr$, where k is constant, prove that

(i)
$$\frac{\partial r}{\partial x} = \frac{x}{r}, \qquad \text{(ii) } y\frac{\partial u}{\partial x} = x\frac{\partial u}{\partial y},$$

(iii)
$$\frac{\partial^2 u}{\partial x^2} + \frac{\partial^2 u}{\partial y^2} + k^2 u = -\frac{k}{r}\sin kr.$$

8. (i) Find for what values of n the function $v = r^n(3\cos^2\theta - 1)$ satisfies the equation

$$\frac{\partial}{\partial r}\left(r^2\frac{\partial v}{\partial r}\right) + \frac{1}{\sin\theta}\frac{\partial}{\partial\theta}\left(\sin\theta\frac{\partial v}{\partial\theta}\right) = 0.$$

(ii) Find k so that $U = t^k \exp(-r^2/4t)$ satisfies the equation

$$\frac{1}{r^2}\frac{\partial}{\partial r}\left(r^2\frac{\partial U}{\partial r}\right) = \frac{\partial U}{\partial t}.$$

9. Show that $u = e^{-3y}\cos 4y \cos 5x$ satisfies the equation

$$\frac{\partial^2 u}{\partial x^2} = \frac{\partial^2 u}{\partial y^2} + 6\frac{\partial u}{\partial y}.$$

10. If $z = y^2 \, f\left(\dfrac{x}{y}\right)$, prove that

$$x^2 \frac{\partial^2 z}{\partial x^2} + 2xy \frac{\partial^2 z}{\partial x \, \partial y} + y^2 \frac{\partial^2 z}{\partial y^2} = 2z.$$

11. Show that the numerical values of a and b (other than zero) can be found so that $u = \exp(-ax + bt^2)$ satisfies the equation

$$\frac{\partial^2 u}{\partial t^2} - 4t \frac{\partial u}{\partial t} = 100 \frac{\partial^2 u}{\partial x^2}$$

for all values of x and t, and find a and b.

16:16 DIFFERENTIALS

If $y = f(x)$ and $f(x)$ is differentiable, i.e. $f'(x)$ exists, then corresponding to a small increment δx in x, the definition of $f'(x)$ implies that the increment δy in y is given by

$$\delta y = [f'(x) + \eta] \, \delta x,$$

where $\eta \to 0$ as $\delta x \to 0$

$$\Rightarrow \delta y \approx f'(x) \, \delta x, \tag{16.52}$$

a result of importance in the discussion of small errors and small increments. This result can be extended to functions of several *independent* variables; e.g. if $y = f(u, v, w)$, then

$$\delta y \approx \frac{\partial f}{\partial u} \delta u + \frac{\partial f}{\partial v} \delta v + \frac{\partial f}{\partial w} \delta w.$$

Example 1. The length of the side a of a triangle ABC is calculated from measurements of the sides b, c and the angle A, using the cosine formula $a^2 = b^2 + c^2 - 2bc \cos A$. Find the approximate error δa in a due to small errors δb, δc, δA in b, c, A respectively.

$$a^2 = b^2 + c^2 - 2bc \cos A$$

$$\Rightarrow 2a\delta a = 2b\delta b + 2c\delta c - 2[\delta b.c \cos A + \delta c.b \cos A - bc \sin A.\delta A]$$

$$\Rightarrow \delta a = \frac{1}{a}\left[(b - c \cos A)\delta b + (c - b \cos A)\delta c + bc \sin A.\delta A\right].$$

Example 2. The range R of a projectile fired with velocity V at an angle of elevation α is $(V^2 \sin 2\alpha)/g$. If V and g are fixed show that the error in elevation required to make the shot fall short by the small quantity a at the maximum range D is $\pm (a/2D)^{1/2}$ radians.

R is clearly a maximum, D, when $\alpha = \frac{1}{4}\pi$ and $D = V^2/g$. Using Taylor's series,

$$\delta R = \frac{2V^2 \cos 2\alpha \delta \alpha}{g} - \frac{4V^2 \sin 2\alpha (\delta \alpha)^2}{2g} + O\{(\delta \alpha)^3\}.$$

Therefore at $\alpha = \frac{1}{4}\pi$ and for $\delta R = -a$ we must have $-a = -2V^2(\delta\alpha)^2/g$ which gives the required result.

We now define the *differential* of y, written dy, by the equation

$$dy = f'(x)\delta x, \tag{16.53}$$

so that in Fig. 16.5, if $PB = \delta x$, then $QB = \delta y$, and if $PB = dx$, $AB = dy$; but notice that, as compared with δx, there is no requirement for dx to be *small*. For any B' on PB,

$$PB' = dx \Rightarrow A'B' = dy.$$

Since in result (16.53), $f'(x)$ is the coefficient of dx, $f'(x)$ is sometimes called the *differential coefficient* of $f(x)$.

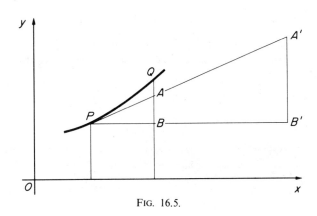

FIG. 16.5.

Example (i) $d(\cos ax) = -a \sin ax\, dx.$

(ii) $xe^{x^2} dx = d(\tfrac{1}{2}e^{x^2}).$

(iii) $\dfrac{dx}{a^2 + x^2} = \dfrac{1}{a}d\left[\tan^{-1}\left(\dfrac{x}{a}\right)\right].$

(iv) $\dfrac{dx}{x} = d(\ln x).$

(v) $d(uv) = u\, dv + v\, du.$

(vi) $d\left(\dfrac{u}{v}\right) = \dfrac{v\, du - u\, dv}{v^2}.$

(vii) $\dfrac{dv}{v^2} = -d\left(\dfrac{1}{v}\right).$

Miscellaneous Exercise 16

1. Find a pair of real values (x, y) satisfying the equations

$$12(\cosh x - \cosh y) = 5,$$
$$12(\sinh x - \sinh y) = 7.$$ (N.)

2. If x is real, prove that

$$1 - x^2 \leqslant \frac{1}{1+x^2} \leqslant 1 - x^2 + x^4,$$

and hence show that, if $x > 0$,

$$x - \tfrac{1}{3}x^3 < \tan^{-1} x < x - \tfrac{1}{3}x^3 + \tfrac{1}{5}x^5.$$

A chord of length 2c divides a circle into two segments. The height of one of them is h. Prove that the length S of its arc is given by

$$S = 2\left(\frac{c^2}{h}+h\right)\tan^{-1}\left(\frac{h}{c}\right).$$

Deduce that, if $a = h/c$,

$$1+\frac{2a^2}{3}-\frac{a^4}{3} < \frac{S}{2c} < 1+\frac{2a^2}{3}-\frac{2a^4}{15}+\frac{a^6}{5}. \quad \text{(N.)}$$

3. Find the area of the finite region enclosed between the two curves $y = c\cosh(x/c)$, $y = c\sinh(x/c)$ and the ordinates $x = 0$, $x = h > 0$.

Find the volume V_1 swept out when this region revolves about the axis of x.

Show that the volume V_2 swept out when this region revolves about the axis of y is given by

$$V_2 = 2\pi c^3\left\{1-\left(\frac{h}{c}+1\right)e^{-h/c}\right\}.$$

4. If $y = \sqrt{(1-x^2)}\sin^{-1}x$, find dy/dx. Express $\left(1-\frac{dy}{dx}\right)\left(\frac{1}{x}-x\right)$ in terms of y. (N.)

5. Differentiate the functions

(i) $2x\tan^{-1}x - \ln(1+x^2)$,　(ii) $\sin^{-1}\sqrt{(x-1)}$.

State the restrictions on the value of x for the second of these functions. (N.)

6. Prove that

$$\int_0^1 \frac{x^2+6}{(x^2+4)(x^2+9)}\,dx = \frac{\pi}{20}. \quad \text{(N.)}$$

7. If $y = 2/(\sinh x)$, prove that

$$1+(dy/dx)^2 = \tfrac14(y^2+2)^2. \quad \text{(N.)}$$

8. Find the real solutions of the simultaneous equations

$$\cosh x\cosh y = 2,\quad \sinh x\sinh y = 1. \quad \text{(N.)}$$

9. Prove that

$$\int_0^1 \frac{x\,dx}{1+x^4} = \frac{\pi}{8}. \quad \text{(N.)}$$

10. Make in one diagram rough sketches of the curves

$$y = 12e^{-x} \text{ and } y = 25\sinh x.$$

Determine the abscissa of their point of intersection. Show that the area of the closed region bounded by the curves and the y-axis is 2 square units. (N.)

11. Find the value of dy/dx in terms of t at a point of the curve whose parametric equations are

$$x = a(t-\tanh t),\ y = a\operatorname{sech} t.$$

Prove that

$$a\frac{d^2y}{dx^2} = \frac{\cosh^3 t}{\sinh^4 t}. \quad \text{(N.)}$$

12. Prove that

$$\text{(i)}\int_1^2 \frac{1}{\sqrt{(1+6x-3x^2)}}\,dx = \frac{\pi}{3\sqrt3},$$

$$(ii) \quad \int_0^{\pi/4} \cos^3 \theta \, d\theta = \frac{5\sqrt{2}}{12}. \tag{N.}$$

13. Starting from the definitions of $\sinh x$ and $\cosh x$, prove that
(a) $\cosh^2 x - \sinh^2 x = 1$,
(b) $\operatorname{arsinh} x = \ln\{x + \sqrt{(x^2 + 1)}\}$.
Solve the equation

$$4 \cosh^2 x = 1 + 7 \sinh x,$$

giving the roots in logarithmic form.
Find the area of the finite region bounded by the curve $y = \operatorname{arsinh} x$, the x-axis and the line $x = 1$.
(L.)

14. (i) Sketch the graphs of e^x, $\sinh x$, $\cosh x$ and $\operatorname{arcosh} x$ in the same diagram showing their correct relative positions.
(ii) Find, as a logarithm to base e, the real root of the equation

$$2 \sinh x - \cosh x = 1.$$

(iii) Evaluate $\displaystyle\int_0^1 \operatorname{arsinh} x \, dx.$
(L.)

15. If $y = (\operatorname{arsinh} x)^2$, prove that

$$(1 + x^2)\frac{d^2 y}{dx^2} + x\frac{dy}{dx} = 2.$$

Prove that the Maclaurin expansion for y is

$$\sum_{n=1}^{\infty} \frac{(-1)^{n-1} 2^{2n-1}[(n-1)!]^2}{(2n)!} x^{2n}. \tag{L.}$$

16. (i) Solve the simultaneous equations

$$\cosh x + \cosh y = 4,$$
$$\sinh x + \sinh y = 2,$$

expressing your answers in terms of logarithms.

(ii) Evaluate $\displaystyle\int_1^2 \operatorname{arsinh}\left(\frac{1}{x}\right) dx.$
(L.)

17. Solve, for real x, y, the simultaneous equations

$$\cosh x \cosh y = 4, \quad \sinh x \sinh y = 1,$$

expressing your answers in terms of logarithms.
(L.)

18. (i) Prove that, if $V = (1 - 2xy + y^2)^{-1/2}$,
(a) $\dfrac{\partial V}{\partial x} + \dfrac{\partial V}{\partial y} = xV^3$,

(b) $x\dfrac{\partial V}{\partial x} - y\dfrac{\partial V}{\partial y} = y^2 V^3$.

(ii) The base radius r and the semi-vertical angle α of a right circular cone are measured and the volume calculated from the formula $V = \frac{1}{3}\pi r^3 \cot \alpha$. Measured values are $r = 8$ cm, $\alpha = 45°$, and these measurements are liable to errors of ± 0.04 cm and $\pm 0.25°$ respectively. Show that the greatest percentage error that can occur in the calculated volume is a little less than $2\frac{1}{2}\%$.
(O.C.)

19. Express $\operatorname{Cos}^{-1} 2$ and $\operatorname{Cosh}^{-1}\frac{1}{2}$ in the form $a + ib$ where a and b are real.

20. If

$$2 \sinh x + 3 \cosh x = 3 + k,$$

where k is small, prove that $x = \frac{1}{2}k$ and

$$x = \frac{k}{2} - \frac{3k^2}{16}$$

are successive approximations to a real root of this equation.

21. If $u + iv = \cosh(x + iy)$, where u, v, x, y are real, express each of u and v as a function of x and y. Show that

$$\frac{\partial u}{\partial x} = \frac{\partial v}{\partial y}, \quad \frac{\partial u}{\partial y} = -\frac{\partial v}{\partial x}.$$

If x is held constant while y varies, show that the locus of the point with rectangular cartesian coordinates (u, v) is an ellipse. If y is held constant while x varies, show that the point (u, v) describes a hyperbola, which intersects the ellipse at right angles.

22. Prove that

$$\sum_{n=1}^{\infty} \frac{\sin nx}{n!} = e^{\cos x} \sin(\sin x).$$

23. If $x + y + z = f(x - z)$, prove that

$$\frac{\partial z}{\partial x} - 2\frac{\partial z}{\partial y} = 1.$$

24. If $z = x^n f(y/x)$ prove that

$$x\frac{\partial z}{\partial x} + y\frac{\partial z}{\partial y} = nz.$$

If $V = z \ln r$, where $r^2 = x^2 + y^2$, prove that

$$x\frac{\partial V}{\partial x} + y\frac{\partial V}{\partial y} = nV + z.$$

25. Show that $u = Ae^{mx}\cos(\omega t + mx) + Be^{-mx}\cos(\omega t - mx)$ is a solution of $\frac{\partial^2 u}{\partial x^2} = 2\frac{\partial u}{\partial t}$, where A, B, m and ω are constants provided that $m^2 = \omega$.

Find the values of the constants, given the conditions (i) $m > 0$, (ii) u remains finite as $x \to \infty$, (iii) $u = \cos t$ when $x = 0$.

26. The coordinates (x, y) of a point P are expressible in terms of real variables u and v by the formula

$$x + iy = (u + iv)^2.$$

Prove that the locus of P is a parabola (i) when u varies and v is constant, and also (ii) when v varies and u is constant. Prove also that all the parabolas have a common focus and a common axis.   (O.C.)

27. Prove that, if u, v, x, y are real numbers connected by the relation

$$u + iv = e^{x + iy},$$

then

$$u = e^x \cos y, \quad v = e^x \sin y.$$

The variable point (x, y) describes, in a counter-clockwise sense, the boundary of the semi-infinite strip defined by the three straight lines

$$y = \tfrac{1}{4}\pi, \quad x \leqslant 0; \quad x = 0; \quad y = \tfrac{1}{2}\pi, \quad x \leqslant 0.$$

Illustrate the corresponding locus of the point (u, v) by a sketch, giving a careful explanation.   (O.C.)

28. Find the nth differential coefficients of the following functions:

(i) $x^3 \ln x$, $n > 3$, (ii) $\cos^3 2x$, (iii) $\ln\left(\dfrac{1-x}{1+x}\right)$, (iv) $x^2 \cos 2x$.

29. Prove that $\tan x = x + \dfrac{x^3}{3} + \dfrac{2x^5}{15} + \dfrac{17x^7}{315} + \ldots$ (x being an acute angle).

(i) Hence prove that

$$\ln \sec x = \frac{x^2}{2} + \frac{x^4}{12} + \frac{x^6}{45} + \ldots$$

When $x = 0.1$, show that $1 + \ln \sec x - \cosh x$ is approximately 4.2×10^{-6}.

(ii) Deduce the expansion of $\tan^2 x$ as far as the term in x^6. Hence show that $\sec^2 x$ differs from $27/(3 - x^2)^3$ by about $x^6/135$ when x is small.

30. (a) Evaluate $\displaystyle\int_0^\pi \frac{1 + \cos x}{5 - 3\cos x}\,dx.$

(b) Transform $\displaystyle\int_1^2 \frac{\ln x}{1 + x^2}\,dx$ by the substitution $x = \dfrac{1}{t}$.

Deduce the value of $\displaystyle\int_{1/2}^2 \frac{\ln x}{1 + x^2}\,dx.$ (N.)

31. Denoting by $\tan^{-1} x$ the principal value of the inverse tangent, prove that $\tan^{-1} x > x - \tfrac{1}{3}x^3$ for all values of $x > 0$.

32. Find the integral $\displaystyle\int \frac{1}{(x - 3)\sqrt{(2x^2 - 12x + 19)}}\,dx.$ (N.)

33. Differentiate

$$\tan^{-1}\left(\frac{1}{1-x}\right) - \tan^{-1}\left(\frac{1}{1+x}\right), \quad \sin^{-1}\left\{\frac{2x}{\sqrt{(4 + x^4)}}\right\}.$$

Explain why the two answers are numerically equal.

If $y = A + B\sin^{-1} x + (\sin^{-1} x)^2$, where A and B are constants, verify that

$$(1 - x^2)\frac{d^2y}{dx^2} - x\frac{dy}{dx} = 2.$$ (O.C.)

34. Prove that

$$\frac{d}{dx}\left\{\tan^{-1}\left(\frac{\sin x}{\cos 2x}\right)\right\} = \frac{3\cos x - \cos 3x}{2(1 - \sin x \sin 3x)}.$$ (O.C.)

35. If $y = \sqrt{\dfrac{1 - x^2}{1 + x^2}}$, prove that $\dfrac{dy}{dx} = -\sqrt{\dfrac{1 - y^4}{1 - x^4}}.$

By differentiation, or otherwise, prove that

$$\int \frac{1}{\sqrt{(\cos 2x - \cos 4x)}}\,dx = -\frac{1}{\sqrt{6}}\sinh^{-1}\sqrt{\frac{\sin 3x}{\sin^3 x}}.$$ (O.C.)

36. Evaluate the integrals

(i) $\displaystyle\int \frac{1}{\sqrt{(x^2 - 4x + 3)}}\,dx;$ (ii) $\displaystyle\int \frac{1}{x(x^2 - 4x + 3)}\,dx;$ (iii) $\displaystyle\int \sin^4 x\,dx.$

By using the substitution $x + 1 = 1/y$, or otherwise, prove that

$$\int_1^3 \frac{1}{(x + 1)\sqrt{(4x - 3 - x^2)}}\,dx = \frac{\pi}{2\sqrt{2}}.$$ (O.C.)

37. Evaluate the integrals

$$\text{(i)} \int \frac{x\,dx}{(x+1)^2\,(x^2+4)}; \quad \text{(ii)} \int_0^{\pi} \frac{1}{4-3\cos x}\,dx; \quad \text{(iii)} \int_1^2 \frac{1}{x\sqrt{(x^2+4)}}\,dx. \tag{L.}$$

38. Evaluate the integrals

$$\text{(i)} \int \left(\frac{1+x}{1-x}\right)^{1/2}\,dx, \quad \text{(ii)} \int x\tan^{-1}dx. \tag{L.}$$

39. Evaluate

$$\text{(i)} \int_0^1 \frac{1}{(x+1)(x^2+2x+2)}\,dx, \quad \text{(ii)} \int_0^{\pi/4} x^2 \sin 2x\,dx.$$

$$\text{Find} \int \frac{\sin\theta}{1+\sin\theta}\,d\theta. \tag{L.}$$

40. Evaluate (i) $\int \dfrac{(2x^3+7)\,dx}{(2x+1)(x^2+2)}$, (ii) $\int \dfrac{1}{x\sqrt{(1+2x-x^2)}}\,dx.$

$$\text{Show that} \int_0^{\pi/2} \frac{1}{3+5\cos\theta}\,d\theta = \frac{1}{4}\ln 3. \tag{L.}$$

41. Evaluate the integrals

$$\text{(i)} \int e^x \cos x\,dx, \quad \text{(ii)} \int \frac{(x+2)\,dx}{\sqrt{(x^2+2x-3)}}. \tag{L.}$$

42. Integrate $\dfrac{1}{\sinh x + 2\cosh x}.$ \hfill (L.)

43. Find indefinite integrals of

$$\text{(i)} \frac{x+1}{x(x^2+4)}, \quad \text{(ii)} \frac{x}{\sqrt{(4x-x^2)}}, \quad \text{(iii)} x(\ln x)^2. \tag{L.}$$

44. Show that

$$\int_0^{\pi/4} \frac{2}{3\sin 2x + 4\cos 2x}\,dx = \frac{1}{5}\ln 6. \tag{L.}$$

45. Evaluate

$$\text{(a)} \int_1^4 \frac{3x-1}{\sqrt{(x^2-2x+10)}}\,dx, \quad \text{(b)} \int \frac{e^{3x}}{1+e^{6x}}\,dx. \tag{L.}$$

46. (i) Evaluate the indefinite integral $\int x^n \ln x\,dx,$

(a) when $n \neq -1$, (b) when $n = -1$.
Hence or otherwise evaluate

$$\int_1^e \frac{(x+1)^3}{x^2}\,\ln x\,dx.$$

(ii) Show that $\int \sqrt{(x^2 + a^2)}\,dx = \frac{1}{2}x\sqrt{(x^2 + a^2)} + \frac{a^2}{2}\sinh^{-1}(x/a)$ and evaluate

$$\int_1^4 \sqrt{(x^2 - 2x + 17)}\,dx.$$

(L.)

47. Given that $y_n(x) = e^{x^2}\dfrac{d^n}{dx^n}e^{-x^2}$, where n is a positive integer, prove that

$$y_{n+2} + 2xy_{n+1} + 2(n+1)y_n = 0.$$

Prove also that

$$\frac{dy_n(x)}{dx} = -2ny_{n-1}(x).$$

Deduce that $y_n(x)$ is a polynomial in x of degree n.   (O.C.)

17 Convergence of Sequences and Series: Further Inequalities

17:1 CONVERGENCE OF SEQUENCES

We consider a sequence $u_1, u_2, u_3, \ldots, u_n, \ldots$ defined $\forall n \in \mathbb{N}$ so that $u_n = f(n)$. Examples of sequences are the set $\mathbb{N}$ of positive integers

$$1, 2, 3, \ldots, n, \ldots,$$

and the terms of a geometric progression

$$a, ar, ar^2, \ldots, ar^{n-1}, \ldots.$$

Here we concern ourselves with the meaning of the term *limit* applied to a sequence u_n as $n \to +\infty$. Later, §17:2, we consider the convergence of series.

We say that the sequence u_n converges to the finite limit l if

$$|u_n - l| < \varepsilon, \tag{17.1}$$

however small the assigned positive number ε may be, for all n exceeding some number $n_0(\varepsilon)$, that is a number n_0 which depends on ε. We express this in the form

$$\lim_{n \to \infty} u_n = l. \tag{17.1a}$$

If a sequence does not converge, then it may:

(i) tend to $+\infty$, i.e. exceed any assigned positive number N for all n sufficiently large [by "for all n sufficiently large" we mean for all n exceeding some number n_1 which depends on N]; or

(ii) tend to $-\infty$, i.e. be less than any assigned negative number $-N$, $(N > 0)$, for all n sufficiently large; or

(iii) oscillate, that is, does not tend to a limit or to $+\infty$ or to $-\infty$. A sequence oscillates boundedly if $\exists M > 0$ such that $|u_n| < M \, \forall n$. Otherwise the sequence oscillates unboundedly.

Example 1. The sequence $(n + 1)/n$ converges to the limit 1.

Example 2. The sequence n^2 diverges to ∞.

Example 3. The sequence $-n^4$ diverges to $-\infty$.

546

Example 4. The sequence $\cos n\pi$ oscillates boundedly (between the values -1 and 1).

Example 5. The sequence $\{1 + 2(-1)^n\}n$ oscillates unboundedly.

A sequence u_n is said to be *monotonic increasing* when $u_{n+1} \geqslant u_n$ for $n \in \mathbb{N}$, and to be *strictly monotonic increasing* if $u_{n+1} > u_n$. Similarly, monotonic decreasing and strictly monotonic decreasing sequences are those for which $u_{n+1} \leqslant u_n, u_{n+1} < u_n$ respectively.

It is proved in books on analysis that if a sequence is monotonic increasing, then it either tends to a finite limit or diverges to $+\infty$. In particular, if this sequence is bounded above, i.e. $\exists K : u_n \leqslant K \ \forall \, n \in \mathbb{N}$, then the sequence tends to a finite limit which cannot exceed K. Similarly a monotonic decreasing sequence tends to a finite limit or diverges to $-\infty$, and must tend to a finite limit if it is bounded below.

Example 1. The sequence $1 - (1/n)$ is strictly monotonic increasing, bounded above (every term is less than 1), and tends to the limit 1.

Example 2. The sequence $-n$ is strictly monotonic decreasing and diverges to $-\infty$.

Example 3. *The limit of $u_n = x^n$ for $x \in \mathbb{R}$.*

Here we illustrate the importance in analysis of considering all the cases which arise.
The results for the cases $x = 0, \pm 1$ are obvious (trivial) thus:
  (i) When $x = 0$, $x^n = 0$ for all n and u_n converges to zero.
  (ii) When $x = 1$, $x^n = 1$ for all n and u_n converges to 1.
  (iii) When $x = -1$, $x^n = (-1)^n$ and u_n oscillates boundedly between -1 and 1.
Now we consider the general cases when $x > 0$.
  (iv) If $x > 1$, then $u_{n+1} = xu_n > u_n$ and so u_n increases monotonically. Hence $u_n \to +\infty$ or to a finite limit $l > 1$. Suppose that this latter result holds so that $\lim\limits_{n \to \infty} x^n = l$. Then clearly

$$xl = x \lim_{n \to \infty} x^n = \lim_{n \to \infty} x^{n+1} = l$$

by definition. Therefore $xl = l$. But $l > 1$ and so $x = 1$ contrary to the fact that $x > 1$. Hence our assumption that the sequence converges gives rise to a contradiction and so must be false. Therefore $\lim\limits_{n \to \infty} x^n \to +\infty$ for $x > 1$.

Note: The above illustrates a standard technique of proof in mathematics. Thus: *If two and only two mutually exclusive conclusions A, B are possible and the assumption that A is true leads to a contradiction then B must be true.*

  (v) If $0 < x < 1$, then $x = 1/y$, where $y > 1$. Then by (iv) above $y^n \to +\infty$ as $n \to +\infty$ and so x^n converges to 0.
  (vi) If $-1 < x < 0$, then $x^n = (-1)^n |x|^n$ and, by (v), $|x|^n \to 0 \Rightarrow x^n \to 0$.
  (vii) If $x < -1$, then $x^n = (-1)^n |x|^n$ and, by (iv), $|x|^n \to \infty$. Hence, in this case x^n oscillates unboundedly. [We use the results of this example in §17:2.]

---

*Example 4. *The general principle of convergence.*
A necessary and sufficient condition that the sequence u_n converges is that, given $\varepsilon > 0$, $\exists \, N(\varepsilon) \in \mathbb{N}$:

$$|u_{n_1} - u_{n_2}| < \varepsilon \quad \forall \, n_1, n_2 \in \mathbb{N} > N. \tag{1}$$

If the sequence converges to l, then, by the definition of convergence, given $\varepsilon' > 0$, $\exists N(\varepsilon')$:

$$|u_n - l| < \varepsilon'$$

for $n > N(\varepsilon')$. Hence, if $n_1, n_2 > N$,

$$|u_{n_1} - l| < \varepsilon', \quad |u_{n_2} - l| < \varepsilon'$$
$$\Rightarrow |u_{n_1} - u_{n_2}| \leqslant |u_{n_1} - l| + |u_{n_2} - l| < 2\varepsilon'$$

which, on writing $\varepsilon' = \frac{1}{2}\varepsilon$ implies that condition (1) is necessary.

It is proved in books on analysis that condition (1) is also sufficient for convergence of the sequence.

---

*Example 5. $a_n \in \mathbb{R}^+$ satisfies the relationship

$$a_n = \frac{1}{2}\left(a_{n-1} + \frac{a^2}{a_{n-1}}\right) \tag{1}$$

$\forall n \in \mathbb{N} \ (\neq 1)$ and $a > 0$. Prove that, if $a_1 > 0$,

(i) $a_n \geqslant a$ for $n \geqslant 2$,
(ii) $a_{n-1} \geqslant a_n$ for $n \geqslant 3$,
(iii) $a_n - a \leqslant \frac{1}{2}(a_{n-1} - a)$ for $n \geqslant 3$.

Hence show that a_n tends to the limit a as n tends to infinity.

Since $a_1 > 0$, it follows from equation (1) that $a_2 > 0$. Clearly by mathematical induction $a_n > 0$ for $n \geqslant 1$.

(i) From equation (1),

$$a_n - a = \frac{(a - a_{n-1})^2}{2a_{n-1}} \geqslant 0 \quad \text{for} \quad n \geqslant 2. \tag{2}$$

Hence $a_n \geqslant a$ for $n \geqslant 2$.

(ii) From equation (1),

$$a_n - a_{n-1} = \frac{a^2 - a_{n-1}^2}{2a_{n-1}} \leqslant 0 \quad \text{for} \quad n \geqslant 3$$

from result (i).

(iii) Also equation (2) can be written

$$a_n - a = \frac{1}{2}(a_{n-1} - a)\left(1 - \frac{a}{a_{n-1}}\right)$$

and since $a/a_{n-1} \geqslant 0$ this implies that $a_n - a \leqslant \frac{1}{2}(a_{n-1} - a)$ for $n \geqslant 3$. It follows that

$$(a_n - a) \leqslant (\tfrac{1}{2})^{n-2}(a_2 - a_1)$$

and by the results of Example 3 above $|a_n - a| < \varepsilon$ for n sufficiently large. Therefore $a_n \to a$ as $n \to \infty$.

Note: We could derive this result from (i), (ii) by noting that a_n is a decreasing sequence of positive terms bounded below and we derive the limit l by writing $a_{n-1} = a_n = l$ in (1) and solving the resulting quadratic equation, taking the positive root.

Exercise 17.1

1. Consider the behaviour as $n \to \infty$ of the sequences whose nth terms are the following:

(i) $\dfrac{n}{n^2 + 1}$; (ii) $\dfrac{n^2 + 1}{2n}$; (iii) $\cos(n\pi/3)$; (iv) $n\cos(n\pi/3)$; (v) $n^r x^n$ where $r \in \mathbb{N}$.

2. (i) Give an example of a sequence $s_1, s_2, s_3, \ldots, s_n, \ldots \in \mathbb{R}$ which does not have a limit, and justify your answer.

(ii) Sketch roughly the graphs of $\sin^3 x$ and $\sin(x^3)$. Show that if $P_1, P_2, \ldots, P_n, \ldots$ are the successive positive zeros of $\sin(x^3)$, then

$$P_n - P_{n-1} \to 0 \quad \text{as} \quad n \to \infty.$$

3. If $n \in \mathbb{N}$ and

$$x_n = \frac{1}{n} + \frac{1}{n+1} + \ldots + \frac{1}{2n},$$

show that

$$x_{n+1} < x_n \quad \text{and} \quad \tfrac{1}{2} < x_n \leqslant \tfrac{3}{2}.$$

Deduce that x_n tends to a limit as n tends to infinity.

4. If $a > 0$, discuss the limits of

$$\frac{a^n - 1}{a^n + 1}$$

as $n \in \mathbb{N}$ tends to infinity.

5. The numbers $x_n (n \in \mathbb{N})$ satisfy the recurrence relation

$$x_{n+1} = \frac{x_n^2 + x_n + 1}{x_n + 2}.$$

If $x_1 > -2$, show that $(x_n - x_{n+1})$, $(x_n - 1)$, $(x_{n+1} - 1)$ all have the same sign when $n > 1$.
If $x_1 > -2$, show that $x_n \to 1$ as $n \to \infty$.

6. The sequence $a_0, a_1, a_2, \ldots$ is defined by

$$a_0 = a \neq 0, \quad a_{n+1} = \tfrac{1}{2}\left(a_n + \frac{k}{a_n}\right), \quad n = 0, 1, 2, \ldots,$$

where k is positive.
Show that

$$a_{n+1}^2 - k \geqslant 0.$$

Show also that if $a > 0$, then
(i) $a_{n+1} \leqslant a_n$ for $n \geqslant 1$,
(ii) the sequence is bounded below.
Determine the limit of the sequence in this case.
Show that if $a < 0$, then the sequence is still convergent but to a different limit, and state this limit.

(N.)

7. A sequence $\{x_n\}$ is defined by

$$x_1 = a, \quad x_{n+1} = 3 - \frac{4}{x_n + 2}.$$

Prove that, for $a = 1$, $0 < x_n < 2$, and that x_n tends to a limit as $n \to \infty$. Find the value of this limit.
Investigate the behaviour of x_n as $n \to \infty$ in the case $a = 3$.   (N.)

17:2 CONVERGENCE OF SERIES

We now give an elementary discussion of the convergence of series.
A series is made up of terms $u_1, u_2, u_3, \ldots, u_n, \ldots$ so that

$$s_n = u_1 + u_2 + u_3 + \ldots + u_n$$

is the sum of the first n terms. If the sequence s_n tends to a finite limit s as $n \to \infty$, i.e.

$\lim\limits_{n \to \infty} s_n = s$, the infinite series is *convergent* and its *sum* is s. If $s_n \to \pm \infty$ as $n \to \infty$, the series is *divergent*; if s_n neither tends to a limit nor diverges to $\pm \infty$, the series *oscillates*. (See §17:1.)

Example. Consider the geometric series

$$s_n = a + ar + ar^2 + \ldots + ar^{n-1} = a\frac{(1 - r^n)}{(1 - r)},$$

if $r \neq 1$. Without loss of generality we take $a > 0$.

We determine $\lim\limits_{n\to\infty} s_n$ by using the results of Example 3, p. 547.

If $|r| < 1$, $\lim\limits_{n\to\infty} s_n = a/(1-r)$ and the series converges to this value.

If $r = 1$, $s_n = na$ and the series diverges.

If $r = -1$, $s_{2n} = 0$, $s_{2n+1} = a$ and the series oscillates between 0 and a.

If $r > 1$, s_n increases indefinitely with n and the series diverges.

If $r < -1$, s_n oscillates unboundedly, i.e. $|s_n|$ increases indefinitely.

We conclude that the geometric series converges if and only if $|r| < 1$.

A first test for the convergence of any infinite series is obtained by direct application of the definition of convergence. If a number s exists such that $|s - s_n|$ can be made as small as we please by taking n large enough, the series converges to the sum s; this test applies whether the terms are positive or negative.

A second test is derived by comparison with the geometric series $\sum\limits_{p=1}^{\infty} ar^{p-1}$. We need consider only positive values of r. If a series of positive terms $\sum\limits_{p=1}^{\infty} u_p$ is such that (apart from a finite number of terms at the beginning) its terms are each less than the corresponding terms of a convergent geometric series then, by the definition of convergence given above, $\sum\limits_{p=1}^{\infty} u_p$ must converge also. Similarly, if the terms of the series $\sum\limits_{p=1}^{\infty} u_p$ are correspondingly greater than those of a divergent geometric series, then the series $\sum\limits_{p=1}^{\infty} u_p$ diverges. This leads to the *ratio test*:

"An infinite series, whose terms u_n are such that

$$\lim_{n\to\infty} \left| \frac{u_{n+1}}{u_n} \right| = l,$$

is *convergent* if $l < 1$, is *divergent* if $l > 1$."

If $l = 1$ the series may converge or diverge; more elaborate tests are required in this case. Usually, if $l = 1$ and the series converges, it does so very slowly and a large number of terms must be taken to give a good approximation to the sum; such series are of little value for numerical computation. The ratio test will not find the sum of a series; it can only indicate convergence or divergence. It should be noted that, if $\sum\limits_{p=1}^{\infty} u_p$ converges, then $u_n \to 0$ as $n \to \infty$. (See Example 3, p. 551.)

A series which contains positive and negative terms is *absolutely convergent* if $\sum\limits_{1}^{\infty} |u_p|$ is convergent, i.e. if the sum of the moduli of the terms converges. If a series is absolutely convergent, then it is convergent. If a series with positive and negative terms is convergent but the series $\sum\limits_{1}^{\infty} |u_p|$ is divergent, $\sum\limits_{1}^{\infty} u_p$ is said to be *conditionally convergent*.

Example 1. The series for e^x is convergent for all values of x.

$$e^x = \sum_0^\infty \frac{x^p}{p!} \quad \text{so that} \quad \frac{u_{n+1}}{u_n} = \frac{x}{n}.$$

Therefore $\lim_{n \to \infty} \left| \frac{u_{n+1}}{u_n} \right| = \lim_{n \to \infty} \frac{|x|}{n} = 0 \quad$ for all x.

Therefore by the ratio test the series is convergent.

Example 2. The series $\sum_1^\infty \frac{1}{p(p+1)}$ converges to the sum 1.

Since

$$\frac{1}{p(p+1)} = \frac{1}{p} - \frac{1}{p+1},$$

$$S_n = \sum_1^n \frac{1}{p(p+1)} = \left(\frac{1}{1} - \frac{1}{2} \right) + \left(\frac{1}{2} - \frac{1}{3} \right) + \ldots + \left(\frac{1}{n} - \frac{1}{n+1} \right)$$

$$= 1 - \frac{1}{n+1}.$$

Therefore

$$\lim_{n \to \infty} s_n = \lim_{n \to \infty} \left(1 - \frac{1}{n+1} \right) = 1.$$

Therefore directly from the definition of convergence the series converges to the sum 1.

Example 3. Show that the condition $a_n \to 0$ is necessary but not sufficient for the convergence of the series $\sum_1^\infty a_n$.

If the series converges, then, by the general principle of convergence, given any $\varepsilon > 0$, $\exists N(\varepsilon) \in \mathbb{R}$ such that

$$|s_n - s_m| < \varepsilon \quad \forall m, n > N(\varepsilon)$$

and in particular $|s_n - s_{n-1}| = |a_n| < \varepsilon$ and so a necessary condition for convergence is that $a_n \to 0$ as $n \to \infty$. That the condition is not sufficient for convergence is clearly shown by the *counter-example*

$$s_k = \sum_{n=1}^k n^{-1/2} \tag{1}$$

in which the nth term tends to zero as $n \to \infty$ but, since every term on the right-hand side of (1) except the last exceeds $n^{-1/2}$,

$$s_k > k \cdot k^{-1/2} = k^{1/2} \quad \text{for} \quad k > 1. \tag{2}$$

Inequality (2) implies that, given any positive p, $s_{p^2} > p$ and hence the series $\sum_1^\infty n^{-1/2}$ diverges although the nth term tends to zero.

Many of the infinite series which occur in mathematics and physics consist of terms $u_n(x)$ which are functions of a variable x. In such cases, if a series converges, its sum $s(x)$ will in general be a function of x also. A problem of major importance, as in the derivation of Maclaurin's series, is to determine whether $\Sigma u'_n(x) = s'(x)$, i.e. whether such an infinite series may be differentiated term by term. We *quote* results which cover the most important type of series, the power series

$$f(x) = \sum_0^\infty a_p x^p = a_0 + a_1 x + a_2 x^2 + \ldots + a_n x^n + \ldots.$$

Application of the tests for convergence usually shows that this series is absolutely convergent when $|x| < R$ where R is a positive number (which may be infinite) called the "radius of convergence" of the series. In this case two theorems of pure mathematics state that

(i) $f'(x) = \sum_0^\infty pa_p x^{p-1},$

(ii) $\int_0^x f(t)\,dt = \sum_0^\infty a_p \frac{x^{p+1}}{(p+1)},$

whenever $|x| < R$. This means that a power series may be differentiated or integrated term by term when $|x| < R$.

Example 1. If $|x| < 1$, then the geometric progression $\sum_0^\infty x^p$ converges

$$\Rightarrow (1-x)^{-1} = \sum_0^\infty x^p \quad \text{for} \quad |x| < 1.$$

Differentiating, $(1-x)^{-2} = \sum_0^\infty px^{p-1}$ for $|x| < 1$.

By the ratio test this last series is absoutely convergent for $|x| < 1$ and so, differentiating again,

$$2(1-x)^{-3} = \sum_0^\infty p(p-1)x^{p-2}, \quad \text{etc.}$$

Example 2. If $|x| < 1$, then $1 - x^2 + x^4 - \ldots + (-1)^n x^{2n} + \ldots = (1+x^2)^{-1}$. Integrating gives

$$\int_0^x \frac{dt}{1+t^2} = \tan^{-1} x = x - \frac{x^3}{3} + \frac{x^5}{5} - \ldots + \frac{(-1)^n x^{2n+1}}{(2n+1)} + \ldots$$

$$\text{for} \quad |x| < 1.$$

Finally we give a test for convergence known as the *integral test*. "If $f(x)$ is a function of x which is positive and continuous for all $x > 1$ and steadily decreases as x increases, then

$$\sum_{r=1}^\infty f(r) \quad \text{and} \quad \int_1^\infty f(x)\,dx$$

both converge or both diverge."

(For a detailed discussion of the convergence of integrals, see §19:2.)

Since $f(r) > f(x) > f(r+1)$ for $r < x < r+1$, where r is positive integer,

$$\int_r^{r+1} f(r)\,dx > \int_r^{r+1} f(x)\,dx > \int_r^{r+1} f(r+1)\,dx$$

$$\Leftrightarrow f(r) > \int_r^{r+1} f(x)\,dx > f(r+1).$$

Therefore, by addition,

$$\sum_{1}^{n-1} f(r) > \int_{1}^{n} f(x)\,dx > \sum_{1}^{n-1} f(r+1) = \sum_{2}^{n} f(r).$$

Letting $n \to \infty$ we have the required result.

This result is illustrated in Fig. 17.1, where [the sum of the areas of the larger rectangles] exceeds [the area between the curve, the x-axis and the ordinates $x = 1$ and $x = n$] which exceeds [the sum of the areas of the smaller rectangles].

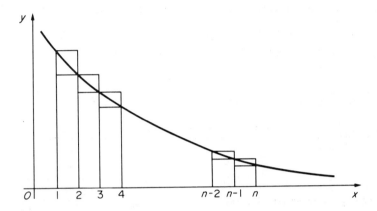

FIG. 17.1.

We note that the sequence u_n, where

$$u_n = \sum_{r=1}^{n-1} f(r) - \int_{1}^{n} f(x)\,dx,$$

is an increasing sequence. But

$$u_n - f(1) + f(n) = \sum_{r=2}^{n} f(r) - \int_{1}^{n} f(x)\,dx < 0.$$

Hence u_n is bounded above and therefore tends to a limit. (See p. 547.)

Example 1. $\sum_{1}^{\infty} \dfrac{1}{n^{\alpha}}$ converges if $\displaystyle\int_{1}^{\infty} \dfrac{1}{x^{\alpha}}\,dx$ converges. This integral takes the value

$$\lim_{N \to \infty} \left(\frac{N^{1-\alpha}-1}{1-\alpha} \right) \text{ if } \alpha \neq 1, \qquad \lim_{N \to \infty} (\ln N) \text{ if } \alpha = 1.$$

Clearly the intergral is only convergent if $\alpha > 1$. Hence

$$\sum_{1}^{\infty} \frac{1}{n^{\alpha}} \text{ is convergent if and only if } \alpha > 1. \text{ In particular the series } \sum_{1}^{\infty} \frac{1}{n} \text{ diverges.}$$

Further, by the above result, it follows that

$$\sum_{r=1}^{n-1} \frac{1}{r} - \int_1^n \frac{1}{x} \, dx = \sum_{r=1}^{n-1} \frac{1}{r} - \ln n$$

tends to a finite limit. In fact we can write

$$\lim_{n \to \infty} \left\{ \sum_{r=1}^{n-1} \frac{1}{r} - \ln n \right\} = \gamma$$

where γ is a finite constant, called *Euler's constant*.

---

*Example 2. The convergence of the series with positive and negative terms

$$\sum_1^\infty \frac{(-1)^{n-1}}{n} = 1 - \frac{1}{2} + \frac{1}{3} - \frac{1}{4} + \frac{1}{5} - \cdots$$

is investigated as follows.

Since $s_{2m} = \left(1 - \frac{1}{2}\right) + \left(\frac{1}{3} - \frac{1}{4}\right) + \left(\frac{1}{5} - \frac{1}{6}\right) + \cdots + \left(\frac{1}{2m-1} - \frac{1}{2m}\right)$ and

$s_{2m+2} = s_{2m} + \left(\frac{1}{2m+1} - \frac{1}{2m+2}\right)$, the sequence of "partial" sums $s_2, s_4, s_6, \ldots, s_{2m}, \ldots$ is an increasing sequence of numbers. We may also write

$$s_{2m} = 1 - \left(\frac{1}{2} - \frac{1}{3}\right) - \left(\frac{1}{4} - \frac{1}{5}\right) - \cdots - \left(\frac{1}{2m-2} - \frac{1}{2m-1}\right) - \frac{1}{2m} < 1.$$

This is true for all values of m. Therefore any member of the sequence $s_2, s_4, s_6, \ldots, s_{2m}, \ldots$ is greater than its predecessors but is less than unity. This is sufficient to prove that $\lim_{m \to \infty} s_{2m} = l_1$, where $0 < l_1 \leq 1$. (See p. 547). Similarly we can prove that s_{2m+1}, the sum of an odd number of terms, forms a decreasing sequence and also tends to a limit l_2, where $0 < l_2 \leq 1$. Finally, we prove that these two limits are the same.

$$s_{2m+1} = s_{2m} + \frac{1}{2m+1}$$

$$\Rightarrow \lim_{m \to \infty} (s_{2m+1} - s_{2m}) = \lim_{m \to \infty} \frac{1}{2m+1} = 0$$

$$\Rightarrow l_1 = l_2.$$

The reader should note that it is difficult to give general rules for handling conditionally convergent series. In fact, by taking the terms in a different order the series can be made to converge to any value.

Example 3. Discuss the convergence of the series

(i) $\sum_{n=2}^\infty \frac{n}{\sqrt{(n^3 - 4)}}$,   (ii) $\sum_{n=1}^\infty \frac{\sqrt{n}}{n^2 - 2}$,   (iii) $\sum_{n=1}^\infty \frac{(\frac{1}{2}n)^n}{n!}$,   (iv) $\sum_{n=1}^\infty \frac{1}{n} \sin \frac{2n\pi}{3}$.

(i) Each term exceeds the corresponding term of the series $\Sigma n^{-1/2}$. This latter series diverges by the integral test and so the given series diverges. [It is easy to show that, if Σv_n is a series of positive terms and $\lim_{n \to \infty} \frac{u_n}{v_n}$ is a finite non-zero number, then the series Σu_n and Σv_n both converge or both diverge.]

(ii) The nth term is comparable with $n^{-3/2}$ and $\Sigma n^{-3/2}$ converges. Hence the given series converges.

(iii) The ratio test gives

$$\frac{u_{n+1}}{u_n} = \frac{1}{2}\left(1 + \frac{1}{n}\right)^n.$$

As $n \to \infty$, $\left(1 + \dfrac{1}{n}\right)^n \to e$ (see p. 497), and so $u_{n+1}/u_n \to e/2 > 1$. By the ratio test, the series diverges.

(iv) The series is

$$\frac{\sqrt{3}}{2}\left(\frac{1}{1} - \frac{1}{2} + \frac{1}{4} - \frac{1}{5} + \frac{1}{7} - \frac{1}{8} + \frac{1}{10} - \frac{1}{11} + \cdots\right).$$

It is easy to show, by the method used in Example 2 above, that this series converges.

Exercise 17.2

In questions 1–6 test the following series (Σu_n), with u_n as stated, for convergence or divergence:

1. $\dfrac{x^n}{n^2}$.

2. $\dfrac{\sin nx}{n(n+1)}$.

3. $\dfrac{n}{(n+1)(n+2)}$.

4. $n(n+1)\left(\dfrac{x}{a}\right)^n$, $a > 0$.

5. $n!x^n$.

6. $\dfrac{1}{n \ln n}$.

7. If $t_n = \displaystyle\sum_{r=1}^n \frac{1}{r(r+1)}$ and $s_n = \displaystyle\sum_{r=1}^n \frac{1}{r(r+2)}$, prove that $\displaystyle\lim_{n \to \infty} \frac{s_n}{t_n} = \frac{3}{4}$.

8. Find the real values of x for which the following series are *convergent*:

(i) $x - \dfrac{x^3}{3} + \dfrac{x^5}{5} - \dfrac{x^7}{7} + \cdots$,

(ii) $1 - \dfrac{3x^2}{2!} + \dfrac{5x^4}{4!} - \dfrac{7x^6}{6!} + \cdots$,

(iii) $\displaystyle\sum_{n=1}^\infty n^{-1}(x/2)^n$, (iv) $\displaystyle\sum_{n=1}^\infty \frac{x^n}{\sqrt{n}}$, (v) $\displaystyle\sum_{n=1}^\infty 2^{-n} \cos nx$.

9. Determine whether the following series are convergent or divergent:

(i) $\displaystyle\sum_{n=0}^\infty \frac{2n-1}{n^2-2}$, (ii) $\displaystyle\sum_{n=0}^\infty \frac{n+1}{n^3-2}$,

(iii) $\displaystyle\sum_{n=1}^\infty \frac{2n-1}{3n+2}$, (iv) $\displaystyle\sum_{n=1}^\infty \frac{n+3}{2n^2+5}$, (v) $\displaystyle\sum_{n=1}^\infty n \cdot 3^{-n}$,

(vi) $\displaystyle\sum_{n=1}^\infty \frac{(-1)^n}{n^2}$, (vii) $\displaystyle\sum_{n=1}^\infty \frac{(-1)^n}{\sqrt{n}}$.

10. Prove that, $\forall n \in \mathbb{N}$,

$$(1 - x)(1 + x + x^2 + \cdots + x^{n-1}) = 1 - x^n,$$

and sum the series

$$\sum_{n=1}^\infty \frac{1 + 2 + 2^2 + \cdots + 2^{n-1}}{n!}.$$ (L.)

11. Find

(i) $\displaystyle\lim_{x \to 0} \{x^{1/x}\}$, (ii) $\displaystyle\lim_{x \to 0} \left\{\left(\frac{1+x}{1-x}\right)^{1/x}\right\}$.

12. Evaluate the limits of the following expressions (a and b being positive):

(i) $\sqrt[n]{(a^n + b^n)}$, as $n \to \infty$;

(ii) $\dfrac{1}{x^2} - \text{cosec}^2\, x$, as $x \to 0$;

(iii) $\dfrac{\ln b - \ln x}{b - x}$, as $x \to b$;

(iv) $(1-x)^{-3} \int_1^x \ln(3t - 3t^2 + t^3)dt$, as $x \to 1$;

(v) $\dfrac{5^x - 4^x}{3^x - 2^x}$, as $x \to 0$;

(vi) $xe^{-n^2 x}$ as $x \to \infty$;

(vii) $x \sin(\pi/x)$ as $x \to \infty$;

(viii) $\dfrac{x^2 + 2\cos x - 2}{x^4}$ as $x \to 0$.

13. Determine the constants a and b so that $(1 + a\cos 2x + b\cos 4x)/x^4$ may have a finite limit as $x \to 0$, and find the value of the limit.

14. Find

(i) $\lim\limits_{n \to \infty} \left(1 + \dfrac{2}{n}\right)^{1/n}$,   (ii) $\lim\limits_{x \to 0} \dfrac{x \sin^2 x}{\sin 2x - 2\sin x}$,

(iii) $\lim\limits_{x \to \pi/2} \dfrac{\cos x}{\cos 3x}$,   (iv) $\lim\limits_{n \to \infty} n(\sqrt{(n^2 + 1)} - n)$,

(v) $\lim\limits_{x \to 0} \dfrac{1 - 2^x}{\sin x}$,   (vi) $\lim\limits_{x \to \infty} (3x^2 - 1)e^{-x}$,

(vii) $\lim\limits_{x \to 0} \dfrac{x \sinh x}{1 - \cos x}$,   (viii) $\lim\limits_{x \to \infty} \left(\dfrac{x+2}{x+1}\right)^x$.

15. Find the sum to infinity of the series in which the nth terms are

(a) $x^n/(2n)!$,   (b) x^n/n,   (c) $\exp(nx)/2^n$,

stating the set of values of x for which each series converges.   (L.)

16. (i) Find constants A, B, C such that

$$2n^2 + 3 \equiv An(n-1) + Bn + C.$$

Deduce that the sum of the series $\displaystyle\sum_{n=1}^{\infty} \dfrac{2n^2 + 3}{n!}$ is $7e - 3$.

(ii) Prove that each of the following series is convergent:

(a) $\displaystyle\sum_{n=1}^{\infty} \dfrac{2^n - 1}{3^n + 1}$,   (b) $\displaystyle\sum_{n=1}^{\infty} \dfrac{3n + 2}{n(n+1)(n+2)}$.   (L.)

17:3 THE CALCULUS APPLIED TO INEQUALITIES

In §7:10 we considered simple algebraic inequalities. We now consider the application of differentiation to inequalities using geometrical ideas. If the derivative $f'(x)$ of a function $f(x)$ is continuous and $f'(x) > 0$ in the interval $a \leqslant x \leqslant b$, then $f(x)$ is increasing at every point of the interval. In this case $f(x)$ is a *strictly monotonic increasing* function of x. This is represented graphically by the fact that the curve $y = f(x)$ rises throughout the whole interval, Fig. 17.2(a).

If $f'(x) \geqslant 0$ in the interval $a \leqslant x \leqslant b$, i.e. $f'(x)$ may vanish at some points, $f(x)$ is said to be *monotonic increasing*, Fig. 17.2(b). In this case the curve $y = f(x)$ never descends at any point. If x_1 and x_2 are any two points of the interval such that $x_1 < x_2$, the ordinate at Q_2, Fig. 17.2(b), must be at least as high as the ordinate at Q_1. In particular the curve neither falls below its starting level, $f(a)$, nor rises above its finishing level, $f(b)$. All these results are embodied in the formal statement:

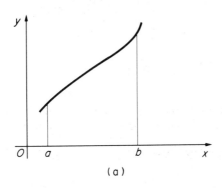

(a)

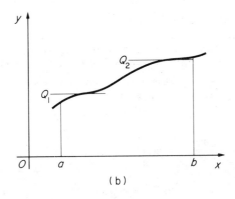

(b)

FIG. 17.2.

"If $f'(x) \geqslant 0$ for all x in the interval $a \leqslant x \leqslant b$, and $a < x_1 < x_2 < b$, then

$$f(a) \leqslant f(x_1) \leqslant f(x_2) \leqslant f(b)."$$

In the latter series of inequalities the "equal" signs do *not* hold for a *strictly* monotonic increasing function. The corresponding results if $f'(x) \leqslant 0$ for all x in the range $a \leqslant x \leqslant b$ are

$$f(a) \geqslant f(x_1) \geqslant f(x_2) \geqslant f(b),$$

and may be derived from the previous case by considering the function $-f(x)$ which is an increasing function.

Example 1. If $f(x) = x - \ln(1 + x)$,

$$f'(x) = 1 - \frac{1}{1+x} = \frac{x}{1+x} \geqslant 0 \text{ for all } x \geqslant 0$$

$$\Rightarrow f(x) \text{ never falls below its starting value } f(0)$$
$$\Rightarrow x - \ln(1+x) \geqslant f(0) = 0, \text{ for } x \geqslant 0.$$

Similarly
$$f'(x) = \frac{x}{1+x} \leqslant 0 \text{ for } -1 \leqslant x \leqslant 0$$

$$\Rightarrow f(x) \text{ never falls below its finishing value } f(0)$$
$$\Rightarrow x - \ln(1+x) \geqslant 0 \text{ for } -1 \leqslant x \leqslant 0.$$

Example 2. Show that $1 - \dfrac{x^2}{2} \leqslant \cos x \leqslant 1 - \dfrac{x^2}{2} + \dfrac{x^4}{24}$.

Consider
$$f(x) = \cos x - 1 + \frac{x^2}{2}.$$

$$f'(x) = x - \sin x, \quad f''(x) = 1 - \cos x$$
$$\Rightarrow f''(x) \geqslant 0 \text{ for all } x \text{ and, since } f'(0) = 0, f'(x) \geqslant 0 \text{ for } x \geqslant 0.$$
$$\text{But } f(0) = 0$$
$$\Rightarrow f(x) \geqslant 0 \text{ for } x \geqslant 0$$
$$\Rightarrow 1 - \frac{x^2}{2} \leqslant \cos x \text{ for } x \geqslant 0.$$

Similarly considering
$$g(x) = 1 - \frac{x^2}{2} + \frac{x^4}{24} - \cos x,$$

$$g^{(4)}(x) = 1 - \cos x \geqslant 0 \quad \text{and, since} \quad g'''(0) = 0, \quad g''(x) \geqslant 0 \text{ for } x \geqslant 0.$$

Proceeding in this way we find $g(x) \geqslant 0$ which is the other part of the result for $x \geqslant 0$. Since we are dealing with even functions the result holds for negative values of x also.

Example 3. Prove that $x > \dfrac{3 \sin x}{2 + \cos x}$ for all values of $x > 0$.

If
$$f(x) \equiv x - \frac{3 \sin x}{2 + \cos x}, \quad \text{then}$$

$$f'(x) \equiv 1 - \frac{3 \cos x (2 + \cos x) + \sin x (3 \sin x)}{(2 + \cos x)^2}$$

$$\equiv \frac{1 - 2 \cos x + \cos^2 x}{(2 + \cos x)^2}$$

$$\equiv \left(\frac{1 - \cos x}{2 + \cos x} \right)^2.$$

Hence $f'(x)$ is finite and $\geqslant 0$ for all values of x. Thus, except at the points where $x = 2n\pi$, where $f'(x) = 0$, and which are points of inflexion, the function increases throughout the range.
But when $x = 0$, $f(x) = 0$

$$\Rightarrow f(x) > 0 \text{ for all values of } x > 0.$$

In some cases the theory of maxima and minima may be of value. Thus, if the greatest and least values of $f(x)$ are M, m respectively in the range $a \leqslant x \leqslant b$, then

$$m \leqslant f(x) \leqslant M \quad \text{for} \quad a \leqslant x \leqslant b.$$

Example 1. Prove by differentiation, or otherwise, that

$$xy \leqslant e^{x-1} + y \ln y$$

for all real x and all positive y. When does the sign of equality hold?

Consider the function $f(x) = xy - e^{x-1}$, regarded as a function of x. Then

$$f'(x) = y - e^{x-1}, \quad f''(x) = -e^{x-1} < 0 \,\forall\, x,$$

so that $f(x)$ has a maximum when

$$x = 1 + \ln y. \tag{1}$$

This maximum value is $y \ln y$. Hence

$$xy - e^{x-1} \leqslant y \ln y$$

which is equivalent to the required result. Equality holds only when x is given by equation (1).

Example 2. By considering the stationary value of the function $x - \ln x$, show that, for $x > 0$.

$$\ln x \leqslant x - 1.$$

Deduce that, if $a_1, a_2, \ldots, a_n$ are positive numbers and $A = \dfrac{1}{n}\sum_1^n a_r$, then

$$\sum_{r=1}^{n} \ln\left(\frac{a_r}{A}\right) \leqslant \left(\sum_{r=1}^{n} \frac{a_r}{A}\right) - n = 0.$$

Hence deduce that

$$\frac{a_1 + a_2 + \ldots + a_n}{n} \geqslant (a_1 a_2 \ldots a_n)^{1/n}.$$

Use this result to prove that if u, v, w are positive quantities and $u + v + w = 1$, then

$$\frac{1}{u^2} + \frac{1}{v^2} + \frac{1}{w^2} \geqslant 27.$$

If $f(x) = x - \ln x$, then $f'(x) = (x-1)/x$ and clearly $f(x)$ has a minimum, unity, when $x = 1$. Therefore $x - \ln x \geqslant 1$, i.e. $\ln x \leqslant x - 1$ for $x > 0$.

Using this result for each of the quantities $a_1/A, a_2/A, \ldots$, where $A = \left(\sum_{r=1}^{n} a_r\right)\Big/ n$, and adding we find

$$\sum_{r=1}^{n} \ln\left(\frac{a_r}{A}\right) \leqslant \sum_{r=1}^{n}\left(\frac{a_r}{A} - 1\right) = \left(\sum_{r=1}^{n} \frac{a_r}{A}\right) - n = \frac{nA}{A} - n = 0.$$

It follows that

$$\ln\left(\frac{a_1 a_2 \ldots a_n}{A^n}\right) \leqslant 0$$

$$\Leftrightarrow a_1 a_2 \ldots a_n \leqslant A^n$$

$$\Leftrightarrow (a_1 a_2 \ldots a_n)^{1/n} \leqslant \frac{a_1 + a_2 + \ldots + a_n}{n}. \tag{1}$$

The quantities on the left-hand and right-hand sides of equation (1) are called the geometric mean and the arithmetic mean respectively of the n positive numbers a_r. Hence the geometric mean of n positive numbers cannot exceed their arithmetic mean.

Now

$$\frac{1}{u^2} + \frac{1}{v^2} + \frac{1}{w^2} = \frac{u^2v^2 + v^2w^2 + w^2u^2}{u^2v^2w^2},$$

and by the above result

$$u^2v^2 + v^2w^2 + w^2u^2 \geqslant 3(u^4v^4w^4)^{1/3}.$$

Also

$$1 = (u + v + w) \geqslant 3(uvw)^{1/3}$$

$$\Rightarrow \frac{1}{u^2} + \frac{1}{v^2} + \frac{1}{w^2} \geqslant \frac{3}{(uvw)^{2/3}} \geqslant \frac{3}{3^{-2}} = 27.$$

17:4 FURTHER ALGEBRAIC INEQUALITIES

Equation (1) of Example 2 on p. 559 above may be established by algebraic methods based on a fundamental inequality which we now establish.

$$(a - b)^2 \geqslant 0$$

$$\Leftrightarrow a^2 - 2ab + b^2 \geqslant 0$$

$$\Leftrightarrow a^2 + b^2 \geqslant 2ab. \tag{17.2}$$

From (17.2),

$$a^2 + b^2 + 2ab \geqslant 4ab$$

$$\Leftrightarrow (a + b)^2 \geqslant 4ab. \tag{17.3}$$

From (17.3), *if a and b are positive,*

$$a + b \geqslant 2\sqrt{(ab)}$$

$$\Leftrightarrow \tfrac{1}{2}(a + b) \geqslant \sqrt{(ab)}, \tag{17.4}$$

the equality occurring when $a = b$.

The arithmetic mean of two positive numbers is greater than their geometric mean, unless the two numbers are equal, in which case the means are equal.

In the following examples we approach similar problems from first principles and by the same methods.

Example 1. Prove that $(a^2 + b^2)(a^4 + b^4) \geqslant (a^3 + b^3)^2$.

$$(a^2 + b^2)(a^4 + b^4) = a^6 + b^6 + a^2b^4 + b^2a^4$$

$$= (a^3 + b^3)^2 - 2a^3b^3 + a^2b^4 + b^2a^4$$

$$= (a^3 + b^3)^2 + a^2b^2(a - b)^2.$$

But

$$a^2b^2(a - b)^2 \geqslant 0$$

$$\Rightarrow (a^2 + b^2)(a^4 + b^4) \geqslant (a^3 + b^3)^2.$$

Example 2. If x and y are positive, prove that

$$\frac{1}{x^2} + \frac{1}{y^2} \geqslant \frac{8}{(x + y)^2}.$$

$$\frac{1}{x^2} + \frac{1}{y^2} - \frac{8}{(x + y)^2} = \frac{x^4 + x^2y^2 + 2x^3y + x^2y^2 + y^4 + 2xy^3 - 8x^2y^2}{(x + y)^2x^2y^2}$$

$$= \frac{x^4 + y^4 - 6x^2y^2 + 2x^3y + 2xy^3}{(x + y)^2x^2y^2}$$

$$= \frac{(x^2 - y^2)^2 + 2xy(x - y)^2}{(x + y)^2x^2y^2}.$$

Each term of the numerator is positive or zero and the denominator is positive

$$\Rightarrow \frac{1}{x^2} + \frac{1}{y^2} - \frac{8}{(x+y)^2} \geq 0$$

$$\Rightarrow \frac{1}{x^2} + \frac{1}{y^2} \geq \frac{8}{(x+y)^2}.$$

Example 3. Prove that

$$a^4 + b^4 + c^4 \geq a^2 b^2 + b^2 c^2 + c^2 a^2 \geq abc(a+b+c).$$

From (17.2),
$$a^4 + b^4 \geq 2a^2 b^2,$$
$$b^4 + c^4 \geq 2b^2 c^2,$$
$$c^4 + a^4 \geq 2c^2 a^2.$$

Adding

$$2(a^4 + b^4 + c^4) \geq 2(a^2 b^2 + b^2 c^2 + c^2 a^2)$$
$$\Rightarrow a^4 + b^4 + c^4 \geq a^2 b^2 + b^2 c^2 + c^2 a^2.$$

Also from (17.2),

$$a^2(b^2 + c^2) \geq 2a^2 bc,$$
$$b^2(c^2 + a^2) \geq 2b^2 ac,$$
$$c^2(a^2 + b^2) \geq 2c^2 ab.$$

Adding
$$2\Sigma a^2 b^2 \geq 2abc\Sigma a$$
$$\Leftrightarrow \Sigma a^2 b^2 \geq abc\Sigma a.$$

Example 4. Prove that, if x, y, z are any positive numbers, $(y+z)(z+x)(x+y) \geq 8xyz$. Prove also that, if a, b, c are any three positive numbers such that each is less than the sum of the other two,

$$(b+c-a)(c+a-b)(a+b-c) \leq abc.$$

Since $y+z \geq 2\sqrt{(yz)}$ with two similar equations, multiplication gives the desired result at once. The second inequality follows on writing

$$y+z = a, \quad z+x = b, \quad x+y = c.$$

Exercise 17.4

1. Prove that

(i) $a(1-a) \leq \frac{1}{4}$ for all values of a,   (ii) if $b \leq c \leq 1$, then $b(1-c) \leq \frac{1}{4}$,

(iii) $xe^{1-x} \leq 1$ for all values of x.

2. If $0 < x < 1$, prove that

$$x + \frac{x^2}{2} + \frac{x^3}{3} + \ldots + \frac{x^n}{n} < \ln\left(\frac{1}{1-x}\right) < x + \frac{x^2}{2} + \ldots + \frac{x^{n-1}}{n-1} + \frac{x^n}{n(1-x)}.$$

3. Prove that $2^n > n^2$ for $n > 4$.

4. Show that, if $x > 0$,

$$x - \tfrac{1}{3}x^3 + \tfrac{1}{5}x^5 > \tan^{-1} x > x - \tfrac{1}{3}x^3.$$

5. Show that, if $x > 0$, $\ln (1 + x) > x/(1 + x)$. Show also that, if $x > 0$,

$$x - \tfrac{1}{2}x^2 < \ln (1 + x) < x - \tfrac{1}{2}x^2 + \tfrac{1}{3}x^3.$$

6. If n is a positive integer and x is a positive variable, prove by differentiation that

$$\frac{(n + 1 + x)^{n+1}}{(n + x)^n}$$

is an increasing function of x.

7. Show that, if $x \in \mathbb{R}$, $2x^2 + 6x + 9 > 0$.

Hence, or otherwise, solve the inequation

$$\frac{(x + 1)(x + 3)}{x} > \frac{(x + 6)}{3}.$$

8. Prove that if $x > 0$,

$$\text{(i)} \ \ln (1 + x) > x - \tfrac{1}{2}x^2, \qquad \text{(ii)} \ x > \frac{5 \sin x}{4 + \cos x}.$$

9. Prove that for all real angles α, β

$$|\sin (\alpha + \beta)| \leqslant |\sin \alpha| + |\sin \beta|.$$

10. Find three real linear factors of $2x^3 - 3x^2 - 3x + 2$.

Hence solve, for $x \in \mathbb{R}$,

$$\text{(i)} \ 2x^3 - 3x^2 - 3x + 2 \geqslant 0, \qquad \text{(ii)} \ 2e^{3x} - 3e^{2x} - 3e^x + 2 \geqslant 0.$$

11. Prove that if a, b, c, d are positive constants such that $bc > ad$, the expression

$$\frac{a \cos x + b \sin x}{c \cos x + d \sin x}$$

increases throughout the range $x = 0$ to $x = \tfrac{1}{2}\pi$.

12. Prove that $\forall a$, b,

$$\sin a \sin b \leqslant \sin^2 \tfrac{1}{2}(a + b).$$

Show further that, if a, b, c and d all lie between 0 and π, then

$$\sin a \sin b \sin c \sin d \leqslant \left[\sin \tfrac{1}{4}(a + b + c + d) \right]^4$$

and, by writing $d = \tfrac{1}{3}(a + b + c)$, deduce that

$$\sin a \sin b \sin c \leqslant \left[\sin \tfrac{1}{3}(a + b + c) \right]^3.$$

Show precisely where the restrictions on a, b, c and d are used in the proof of the last two inequalities.

13. If $pq > 0$, prove that $p/q > 0$. Find

(i) the set of·values of x for which $\dfrac{1 - 4x}{2x - 3} > 0$,

(ii) the set of values of x for which

$$\frac{2x}{x - 1} + \frac{x - 5}{x - 2} > 3.$$

14. Prove that the value of

$$3 \sin x - \sin 3x$$

steadily increases as x increases from $-\frac{1}{2}\pi$ to $\frac{1}{2}\pi$.

Deduce that $\tan 3x \, \cot x$ steadily increases as x increases from 0 to $\pi/6$.

15. Prove that, for $x > 0$, each of the functions

$$\ln\left(1 + \frac{1}{x}\right) - \frac{2}{2x+1}$$

and

$$\frac{1}{\sqrt{\{x(x+1)\}}} - \ln\left(1 + \frac{1}{x}\right)$$

decreases as x increases. State the limit of each of these functions as $x \to \infty$.

Deduce that $\dfrac{2}{2x+1} < \ln\left(1 + \dfrac{1}{x}\right) < \dfrac{1}{\sqrt{\{x(x+1)\}}}$ for $x > 0$.

16. Sketch the curve $y = x^3 e^{-x}$, indicating the points of inflexion. Find the equations of the tangents to the curve that pass through the origin.

Show that, for $a > 6$,

$$5 < \int_0^a x^3 e^{-x} dx < 6. \tag{L.}$$

17. Defining $\ln t$ for $t > 1$ as $\displaystyle\int_1^t \frac{dx}{x}$, prove that

$$1 - \frac{1}{t} < \ln t < t - 1. \tag{L.}$$

17:5 LINEAR DIFFERENCE EQUATIONS

A typical *difference equation*, sometimes called a *recurrence relation*, can be written in the form

$$a_r u_{n+r} + a_{r-1} u_{n+r-1} + \ldots + a_1 u_{n+1} + a_0 u_n = f(n) \tag{17.5}$$

which relates members of the sequence of values $u_1, u_2, \ldots, u_n, \ldots$. In this equation the coefficients a_r may be functions of n, and n and r take only positive integral values, i.e. $n, r \in \mathbb{N}$. The value of r determines the *order* of the difference equation.

Physical problems such as those concerning transmission lines, beams supported at a number of points along their length, and problems in which the coefficients of a series need to be determined, often furnish equations such as (17.5). The solution of (17.5) consists of finding a formula giving u_n in terms of n.

Equation (17.5) is linear in u_n and bears a close analogy with linear differential equations (see §18:8). Here we consider only those difference equations whose solutions are obtained by methods analogous to those used to solve differential equations. First, we note that the solution of (17.5) consists of a "complementary function", which contains arbitrary constants, to which is added a "particular solution". The complementary function satisfies the reduced equation (17.5) with $f(n) = 0$ and the "particular solution", usually obtained by a trial method, satisfies the full equation (17.5).

If the coefficients a_r in equation (17.5) do not involve n we use the methods and results of § 18:8 as a guide in solving the difference equation with constant coefficients. We obtain the "complementary function", which is a solution of the (reduced) difference equation

$$a_r u_{n+r} + a_{r-1} u_{n+r-1} + \ldots + a_1 u_{n+1} + a_0 u_n = 0, \tag{17.6}$$

by assuming a trial solution $u_n = z^n$. The complementary function is, in fact, $\sum_{s=1}^{r} C_s \alpha_s^n$ where the α_s are the roots of the "auxiliary equation"

$$a_r z^r + a_{r-1} z^{r-1} + \ldots + a_1 z + a_0 = 0,$$

provided that these roots are all distinct. Examples illustrate the procedure if the roots of the auxiliary equation are not all real and distinct.

If the solution contains q arbitrary constants and q terms of the sequence u_1, $u_2, \ldots$ are prescribed, e.g. in physical problems by specific "boundary conditions", then the arbitrary constants can be determined. The technique of solution is illustrated in the following examples, where the various possibilities are exhibited both in the complementary function and in the particular solution. The methods we illustrate in the following examples are all analogous to those used in solving differential equations.

Example 1. Find the general solution of the difference equation

$$u_{n+1} - 3u_n = 1. \tag{1}$$

The auxiliary equation is $z - 3 = 0$. Hence the complementary function is $C\,3^n$.
The corresponding differential equation $dy/dx - 3y = 1$ has the particular integral $y = \frac{1}{3}$. Hence we try $u_n = k$, a constant, as the particular solution of equation (1). Direct substitution gives $k - 3k = 1 \Leftrightarrow k = -\frac{1}{2}$, and hence the required general solution of equation (1) is

$$u_n = C\,3^n - \tfrac{1}{2}.$$

Example 2. Find the solution of the difference equation

$$u_{n+2} - 3u_{n+1} - 4u_n = 6n + 19 \tag{1}$$

for which $u_1 = 1$, $u_2 = 10$.

The auxiliary equation is $z^2 - 3z - 4 = 0$ with roots 4 and -1. Hence the complementary function is $C_1 4^n + C_2(-1)^n$.
The corresponding differential equation $d^2y/dx^2 - 3\,dy/dx - 4y = 6x + 19$ has a particular integral of the form $px + q$ where p and q are constants. Accordingly we seek a particular solution of equation (1) in the form $u_n = an + b$, where a and b are constants. Substitution gives

$$a(n+2) + b - 3\{a(n+1)+b\} - 4(an+b) = 6n + 19$$
$$\Leftrightarrow -6na - (a+6b) = 6n + 19$$
$$\Leftrightarrow -6a = 6, \quad -(a+6b) = 19$$
$$\Leftrightarrow a = -1, \quad b = -3.$$

Hence the particular solution is $-(n+3)$ and the general solution is

$$u_n = C_1 4^n + C_2(-1)^n - n - 3.$$

The conditions $u_1 = 1$, $u_2 = 10$ give

$$4C_1 - C_2 - 1 - 3 = 1, \quad 16C_1 + C_2 - 2 - 3 = 10$$

$\Leftrightarrow C_1 = 1, C_2 = -1$. Hence the required solution is

$$u_n = 4^n + (-1)^{n+1} - n - 3.$$

Example 3. If $a > 1$, show that the general solution of the equation

$$u_{n+2} - 2au_{n+1} + u_n = 0 \tag{1}$$

can be expressed in the form

$$u_n = C_1 \cosh n\theta + C_2 \sinh n\theta,$$

where $\theta = \ln\{a + \sqrt{(a^2 - 1)}\}$.

Equation (1) can be written

$$u_{n+2} - 2\cosh\theta . u_{n+1} + u_n = 0,$$

where $\cosh\theta = a$, i.e. $\theta = \ln\{a + \sqrt{(a^2 - 1)}\}$,

$$\Rightarrow u_{n+2} - (e^\theta + e^{-\theta})u_{n+1} + u_n = 0.$$

The auxiliary equation $z^2 - (e^\theta + e^{-\theta})z + 1 = 0$ has roots e^θ, $e^{-\theta}$. Hence the solution can be written

$$u_n = A e^{n\theta} + B e^{-n\theta}$$

or, writing

$$A = \tfrac{1}{2}(C_1 + C_2), \quad B = \tfrac{1}{2}(C_1 - C_2),$$
$$u_n = C_1 \cosh n\theta + C_2 \sinh n\theta.$$

Similarly the solution of the equation

$$u_{n+2} - 2au_{n+1} + u_n = 0,$$

where $|a| < 1$, can be expressed in the form

$$u_n = C_1 \cos n\theta + C_2 \sin n\theta,$$

where $\cos\theta = a$.

Example 4. Solve the equation

$$u_{n+2} - 2u_{n+1} + 2u_n = 0.$$

The auxiliary equation has roots $1 \pm i$, i.e. $\sqrt{2}e^{i\pi/4}$ and $\sqrt{2}e^{-i\pi/4}$. Hence the general solution is

$$u_n = 2^{n/2}(Ae^{ni\pi/4} + Be^{-ni\pi/4})$$
$$\Rightarrow u_n = 2^{n/2}\{C_1 \cos(n\pi/4) + C_2 \sin(n\pi/4)\}.$$

Example 5. Find the general solution of the equation

$$u_{n+3} + u_{n+2} - 2u_{n+1} + 12u_n = a$$

where a is a constant.

The auxiliary equation is

$$z^3 + z^2 - 2z + 12 = (z + 3)(z^2 - 2z + 4) = 0,$$

with roots $-3, 1 \pm i\sqrt{3} = 2e^{\pm i\pi/3}$. Hence the complementary function is

$$C_1(-3)^n + 2^n\{C_2 \cos(n\pi/3) + C_3 \sin(n\pi/3)\}.$$

The particular solution is found by trial (assuming λa, where λ is a constant, as a particular solution) to be $a/12$. Hence the general solution required is

$$C_1(-3)^n + 2^n\{C_2 \cos (n\pi/3) + C_3 \sin (n\pi/3)\} + a/12.$$

Example 6. Find the solution of

$$u_{n+1} - ku_n = k^n \tag{1}$$

given that $u_1 = 1$.

The complementary function is Ck^n and hence the particular solution cannot be of the form λk^n, where λ is a constant. In fact the corresponding differential equation $dy/dx - ky = e^{kx}$ has a particular integral of the form pxe^{kx}, where p is constant. Accordingly we seek a particular solution of equation (1) in the form $u_n = \lambda nk^n$. Substitution gives

$$\lambda\{(n+1)k^{n+1} - k^{n+1}n\} = k^n$$

$\Rightarrow \lambda = 1/k$. Hence a particular solution is nk^{n-1} and the general solution of equation (1) is

$$u_n = Ck^n + nk^{n-1}.$$

Since $u_1 = 1$, $C = 0$ and hence the required solution is

$$u_n = nk^{n-1}.$$

Example 7. Find the general solution of the equation

$$u_{n+2} - 4u_{n+1} + 4u_n = 2^n.$$

The auxiliary equation $z^2 - 4z + 4 = 0$ has roots 2, 2 and hence the complementary function is $(C_1 + C_2 n)2^n$.

Since 2^n and $n2^n$ occur in the complementary function, we use $\lambda n^2 2^n$ as a trial function for the particular solution. Then substitution gives

$$\lambda\{(n+2)^2 2^{n+2} - 4(n+1)^2 2^{n+1} + 4n^2 2^n\} = 2^n$$

$\Rightarrow \lambda = \frac{1}{8}$. Hence the general solution required is

$$u_n = (C_1 + C_2 n + n^2/8)2^n.$$

Example 8. The Fibonacci sequence u_n satisfies the recurrence relation

$$u_{r+2} = u_{r+1} + u_r \tag{1}$$

with

$$u_0 = 0 \quad \text{and} \quad u_1 = 1.$$

Find an expression for u_r, and hence or otherwise prove that

$$v_r = \frac{u_{r+1}}{u_r}$$

converges to the limit $\frac{1}{2}(1 + \sqrt{5})$ as $r \to \infty$.

The auxiliary equation is

$$z^2 - z - 1 = 0$$

with solutions $z = (1 \pm \sqrt{5})/2$. Therefore

$$u_r = A\left(\frac{1+\sqrt{5}}{2}\right)^n + B\left(\frac{1-\sqrt{5}}{2}\right)^n.$$

The conditions $u_0 = 0$, $u_1 = 1$

$$\Rightarrow A = 1/\sqrt{5} = B$$

$$\Rightarrow u_r = \frac{1}{\sqrt{5}} \left\{ \left(\frac{1+\sqrt{5}}{2} \right)^n - \left(\frac{1-\sqrt{5}}{2} \right)^n \right\}.$$

By definition

$$v_r = \frac{\{(1+\sqrt{5})^{r+1} - (1-\sqrt{5})^{r+1}\}}{2\{(1+\sqrt{5})^r - (1-\sqrt{5})^r\}}$$

$$= \frac{1}{2}(1+\sqrt{5})(1-x^{r+1})/(1-x^r),$$

where
$$x = (1-\sqrt{5})/(1+\sqrt{5}) = -(6-2\sqrt{5})/4.$$

Since
$$|x| < 1, \lim_{r \to \infty} x^r = 0 \text{ and so } \lim_{r \to \infty} v_r = \tfrac{1}{2}(1+\sqrt{5}).$$

Generating function of a power series

If the coefficients $u_0, u_1, \ldots$ of a convergent power series satisfy a recurrence relation (say)

$$au_{n+2} + bu_{n+1} + cu_n = 0 \quad \text{for} \quad n \in \{0, \mathbb{N}\},$$

let
$$S \equiv u_0 + u_1 x + u_2 x^2 + \ldots.$$

Then
$$aS \equiv au_0 + au_1 x + au_2 x^2 + au_3 x^3 + \ldots,$$
$$bxS \equiv \qquad bu_0 x + bu_1 x^2 + bu_2 x^3 + \ldots$$
$$cx^2 S \equiv \qquad\qquad cu_0 x^2 + cu_1 x^3 + \ldots.$$

$$\Rightarrow (a + bx + cx^2)S = au_0 + (au_1 + bu_0)x \qquad \text{(on addition)}$$

$$\Leftrightarrow S \equiv \frac{au_0 + (au_1 + bu_0)x}{a + bx + cx^2}.$$

This expression is called the *generating function* of the power series.

Exercise 17.5

Solve the difference equations in questions 1–12,

1. $u_{n+1} = \lambda u_n$.

2. $u_{n+2} - 3u_{n+1} + 2u_n = 0$, given that $u_1 = 3$, $u_2 = 5$.

3. $15u_{n+2} - 47u_{n+1} + 28u_n = 0$.   **4.** $4u_{n+2} - 12u_{n+1} + u_n = 0$.

5. $u_{n+3} - 2u_{n+2} - u_{n+1} + 2u_n = 0$, given that $u_1 = \frac{1}{2}$, $u_2 = 1$, $u_3 = 2$.

6. $u_{n+2} - 2u_{n+1} + 2u_n = 0$, given that $u_1 = 1$, $u_2 = 2$.

7. $u_{n+2} - 4u_{n+1} + 4u_n = 0$, given that $u_1 = 0$, $u_2 = 1$.

8. $u_{n+2} + 8u_{n+1} - 9u_n = 8.3^{n+1}$, given that $u_1 = -6$, $u_2 = 90$.

9. $u_{n+2} - 7u_{n+1} + 12u_n = 6n + 1$, given that $u_1 = 3$, $u_2 = 10$.

10. $2u_{n+2} - 5u_{n+1} + 2u_n = 1$, given that $u_1 = 1$ and u_n is finite as $n \to \infty$.

11. $u_{n+2} - (a+b)u_{n+1} + abu_n = 0$, given that $u_1 = p$, $u_N = 0$. Consider the cases $a \neq b$, $a = b$ separately. (Assume $a \neq 0$.)

12. $u_{n+2} - 8u_{n+1} + 16u_n = 4^n$, given that $u_1 = \frac{1}{8}$, $u_2 = 2$.

13. Verify that the recurrence relation

$$u_n = (n-1)(u_{n-1} + u_{n-2})$$

is satisfied by $u_n = n!$

By putting $u_n = n!v_n$, and then $v_n - v_{n-1} = w_n$, or otherwise, find (in the form of a finite series) the solution of the recurrence relation for which $u_1 = 0$, $u_2 = 1$.

14. The sequence $\{u_n\}$ is given by the relation

$$u_{n+1} = (2u_n)^{1/2} \text{ for } n \geqslant 1 \text{ and } u_1 = 1.$$

Prove that $u_n < u_{n+1} < 2$. Deduce that $\{u_n\}$ converges and find its limit.

Verify your result by taking logarithms, writing $v_n = \log u_n$ and solving the difference equation for v_n.

15. The sequence $\{y_r\}$ of real numbers is defined by the relation $y_r^4 \cdot y_{r+2} = y_{r+1}^5$, where $y_0 = 1$ and $y_1 = 2$. By putting $y_r = 2^{x_r}$, or otherwise, find an expression for y_r.

16. Show that, if the sequence u_n satisfies the recurrence formula

$$au_{n+2} + bu_{n+1} + cu_n = 0 \quad (n \geqslant 0)$$

then u_n is the coefficient of x^n in the expansion of a function of the form

$$\frac{px - q}{cx^2 + bx + a}.$$

Prove that, if

$$p = q \text{ and } s_n = \sum_0^n u_r, \text{ then } as_2 + bs_1 + cs_0 = 0$$

and s_n satisfies the same recurrence formula as u_n.

Prove that, if

$$p \neq q \text{ and } v_n = \sum_0^n u_r \left(\frac{q}{p}\right)^r,$$

then

$$ap^2v_{n+2} + bpqv_{n+1} + cq^2v_n = 0$$

for $n \geqslant 0$.   (O.C.)

17. The general term of a sequence is u_n. The relation

$$u_{n+2} - 2au_{n+1} + a^2u_n = 0$$

is satisfied when $n \geqslant 0$. Prove that

$$\sum_{n=0}^{\infty} u_n x^n = \frac{u_0 + (u_1 - 2au_0)x}{(1 - ax)^2}$$

when $|ax| < 1$.

Hence, or otherwise, prove that

$$u_n = u_0(1 - n)a^n + u_1 na^{n-1}.$$

Prove that, if $v_0 = 2$, $v_1 = a + 1$, and the relation

$$v_{n+2} - 2av_{n+1} + a^2v_n = (a - 1)^2$$

is satisfied when $n \geqslant 0$, then

$$v_n = a^n + 1.$$ (O.C.)

17:6 CONVERGENCE OF ITERATIVE PROCESSES

In §12:9 we considered the Newton–Raphson method for approximation to a root a of the equation $f(x) = 0$. We now consider the convergence of this process. If x is an approximation to the root a of $f(x) = 0$ then, if $x + e = a$, by Taylor's theorem

$$0 = f(x + e) = f(x) + ef'(x) + O(e^2).$$

Hence an approximate value for e is $-f(x)/f'(x)$. Thus we regain the Newton–Raphson iteration formula

$$x_{n+1} = x_n - \frac{f(x_n)}{f'(x_n)}. \qquad (17.7)$$

To investigate the convergence of the Newton–Raphson process we write $x_n = a - e_n$. Then

$$f(x_n) = f(a - e_n) = f(a) - e_n f'(a) + \tfrac{1}{2} e_n^2 f''(a) + O(e_n^3),$$
$$f'(x_n) = f'(a - e_n) = f'(a) - e_n f''(a) + O(e_n^2).$$

Then equation (17.7) can be written

$$e_{n+1} = e_n + \frac{f(x_n)}{f'(x_n)}$$

$$\Rightarrow e_{n+1} = e_n + \frac{f(a - e_n)}{f'(a - e_n)}$$

$$\Rightarrow e_{n+1} = e_n + \frac{f(a) - e_n f'(a) + \tfrac{1}{2} e_n^2 f''(a) + O(e_n^3)}{f'(a) - e_n f''(a) + O(e_n^2)}.$$

Assuming that e_n is small and remembering that $f(a) = 0$, we deduce

$$e_{n+1} = e_n - e_n \left[1 - \tfrac{1}{2} e_n \frac{f''(a)}{f'(a)} + O(e_n^2) \right] \left[1 - e_n \frac{f''(a)}{f'(a)} + O(e_n^2) \right]^{-1}$$

$$\Rightarrow e_{n+1} = e_n - e_n \left[1 + \tfrac{1}{2} e_n \frac{f''(a)}{f'(a)} + O(e_n^2) \right]$$

$$\Rightarrow e_{n+1} = -\frac{f''(a)}{2f'(a)} e_n^2 + O(e_n^3)$$

$$\approx -\mu_n e_n^2,$$

where $\mu_n = f''(a)/\{2f'(a)\}$. This shows that the method is a *second-order process* because of the second-degree factor e_n^2. This means that the errors decrease rapidly for small e_n. Furthermore, $|\mu_n|$ need not be less than 1 for convergence.

The solution of the equation $\phi(x) = x$

Many functions $f(x)$ are such that the equation $f(x) = 0$ can easily be written in the form

$$\phi(x) = x, \qquad (17.8)$$

where $\phi(x)$ is a well-behaved function in the neighbourhood of the root $x = \alpha$. The determination of α is equivalent to finding the intersection of the line $y = x$ and the curve $y = \phi(x)$.

Equation (17.8) can often be solved by the iteration formula

$$x_{n+1} = \phi(x_n). \tag{17.9}$$

This process produces a series of points on a diagram either by a step-like path or a spiral path as illustrated in Figs. 17.3(i), (ii).

A_n is (x_n, x_n); P_n is (x_n, y_n), $y_n = \phi(x_n)$;

A_{n+1} is (x_{n+1}, x_{n+1}), $x_{n+1} = y_n$; P_{n+1} is (x_{n+1}, y_{n+1}), $y_{n+1} = \phi(x_{n+1})$;

etc. Clearly the procedure is convergent if the steps or the spiral approach the root. [Figure 17.3 (i) illustrates convergent cases whereas Fig. 17.3(ii) illustrates divergent cases.] The method will be convergent if the slope of the curve $y = \phi(x)$ at $x = \alpha$ is numerically less than unity. To make the process approximate to the root when $|\phi'(\alpha)| > 1$ we would have to use the recurrence relation

$$x_n = \phi(x_{n+1}),$$

a relation which is inconvenient for practical use. Hence we deduce that the iteration is convergent if $|\phi'(\alpha)| < 1$. This condition is a sufficient condition, but the iteration may possibly converge if $|\phi'(\alpha)| = 1$, though it will converge very slowly.

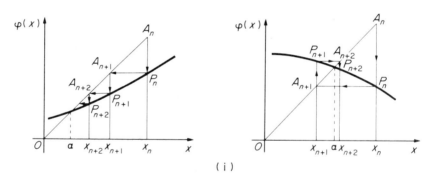

(i)

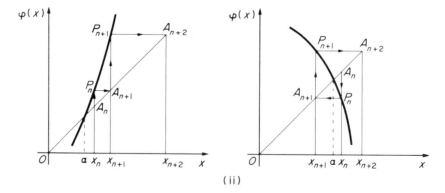

(ii)

Fig. 17.3.

The relation between consecutive errors in this method is

$$\alpha - e_{n+1} = \phi(\alpha - e_n) = \phi(\alpha) - e_n\phi'(\alpha) + O(e_n^2),$$

where $x_n = \alpha - e_n$

$$\Rightarrow e_{n+1} = e_n\phi'(\alpha) + O(e_n^2),$$

which shows that the iteration is a *linear or first-order process* which converges more quickly the smaller the value of $\phi'(\alpha)$.

Since any iterative process provides a sequence of values $x_1, x_2, \ldots$ in which successive terms are related, usually, by an explicit formula, any of the tests for convergence of a sequence (not of a series) mentioned in §17:1 can be applied.

Example. Find the real root of the equation

$$f(x) = x^5 + 5x - 1 = 0.$$

This equation has one positive root which lies in the interval $(0, 1)$ and no negative roots, since $f(0 \cdot 0) < 0$, $f(1) > 0$. Writing the equation in the form

$$x = \tfrac{1}{5} - \tfrac{1}{5}x^5$$

we determine the root from the iterative relationship

$$x_{n+1} = \tfrac{1}{5} - \tfrac{1}{5}x_n^5,$$

with $x_1 = 0 \cdot 5$. We find

x	$1 - x^5$	$0 \cdot 2(1 - x^5)$
0·5	0·968 75	0·193 750
0·193 750	0·999 727 0	0·199 945 4
0·199 945 4	0·999 680 4	0·199 936 1
0·199 936 1	0·999 680 5	0·199 936 1

Further iterations do not affect the seventh place of decimals, so that the root, correct to seven places, is 0·199 936 1.

In some circumstances the approximate solution of an equation is wanted, not numerically, but in terms of some parameter in the coefficients. If this is small it is possible to solve the equation by obtaining the root as a power series in this parameter. Expansions by methods already explained (the binomial, exponential, sine series, or Taylor's or Maclaurin's theorems), or expansions by successive approximations may be used. The successive approximations in this case will include successively higher powers of the small parameter.

We illustrate the techniques in the following examples, the results of which can be checked by use of the Newton–Raphson method.

Example 1. If p/m is small find the numerically small root of the equation

$$x^3 - mx + p = 0.$$

We first write the equation in the form

$$x = \frac{p}{m} + \frac{x^3}{m}.$$

This is of the form $x = f(x)$ where the right-hand side is small. We may therefore substitute the approximate value $x_0 = 0$

$$\Rightarrow x_1 = \frac{p}{m}.$$

Substituting again,

$$x_2 = \frac{p}{m} + \frac{1}{m} \cdot \frac{p^3}{m^3} = \frac{p}{m}\left(1 + \frac{p^2}{m^3}\right)$$

$$\Rightarrow x_3 = \frac{p}{m} + \frac{1}{m} \cdot \frac{p^3}{m^3}\left(1 + \frac{p^2}{m^3}\right)^3$$

$$\Rightarrow x_3 = \frac{p}{m} + \frac{p^3}{m^4}\left(1 + \frac{3p^2}{m^3}\right) + O\left(\frac{p^7}{m^{10}}\right)$$

$$\Rightarrow x_3 = \frac{p}{m}\left(1 + \frac{p^2}{m^3} + \frac{3p^4}{m^6}\right).$$

Notice that each approximation adds another power of p^2/m^3 inside the bracket. No advantage is gained by expanding $(1 + p^2/m^3)^3$ to include more terms.

Example 2. Find, as far as the term in ε^3, the solution, in ascending powers of ε, of the equation $\theta = \alpha + \varepsilon \sin \theta$, where ε is small.

$$\theta_0 = \alpha$$
$$\Rightarrow \theta_1 = \alpha + \varepsilon \sin \theta_0 = \alpha + \varepsilon \sin \alpha$$
$$\Rightarrow \theta_2 = \alpha + \varepsilon \sin \theta_1 = \alpha + \varepsilon \sin (\alpha + \varepsilon \sin \alpha)$$
$$\Rightarrow \theta_2 = \alpha + \varepsilon[\sin \alpha \cos (\varepsilon \sin \alpha) + \cos \alpha \sin (\varepsilon \sin \alpha)].$$

Retaining only terms of $O(\varepsilon)$ in the bracket, so that $\cos (\varepsilon \sin \alpha) = 1$ and $\sin (\varepsilon \sin \alpha) = \varepsilon \sin \alpha$, we have

$$\theta_2 = \alpha + \varepsilon \sin \alpha + \varepsilon^2 \sin \alpha \cos \alpha$$
$$\Rightarrow \theta_3 = \alpha + \varepsilon[\sin \alpha \cos (\varepsilon \sin \alpha + \varepsilon^2 \sin \alpha \cos \alpha) + \cos \alpha \sin (\varepsilon \sin \alpha + \varepsilon^2 \sin \alpha \cos \alpha)]$$
$$\Rightarrow \theta_3 = \alpha + \varepsilon \sin \alpha + \varepsilon^2 \sin \alpha \cos \alpha + \tfrac{1}{2}\varepsilon^3 \sin \alpha(3 \cos^2 \alpha - 1) + O(\varepsilon^4).$$

Note once more that we derive *one* additional term at each stage.

Exercise 17.6

1. Find graphically the number of roots of the equation

$$4x^2 - 1 = \tan x$$

in the range $-1 \cdot 5 < x < 1 \cdot 5$.
Taking $x = 0 \cdot 65$ as the initial approximation to one root, obtain a second approximation
(a) by the Newton–Raphson method,
(b) by writing the equation in the form $x = \tfrac{1}{2}\sqrt{(1 + \tan x)}$, and using an iterative method.
Give your answers to three decimal places.   (L.)

2. Show graphically, or otherwise, that the iterative process

$$x_{n+1} = 2/(x_n^2 + 3)$$

converges, from any initial value, to a root of the equation

$$x^3 + 3x - 2 = 0.$$

Using this or any other suitable process, calculate the root correct to three significant figures.

3. A sequence is defined by $x_{n+1} = F(x_n)$, $x_0 = a$. State a sufficient condition for the convergence of the sequence to a root α of the equation $x = F(x)$.

The positive root α of the equation $x^2 - x - 1 = 0$ is to be calculated by iteration using the relation

$$x_{n+1} = (x_n + 1)/x_n$$

with $x_0 = 2$. By sketching the curve $y = (x + 1)/x$, the line $y = x$ and a polygon with vertices at the points (x_0, x_0), (x_0, x_1), (x_1, x_1), (x_1, x_2), . . . , examine graphically the behaviour of the sequence.   (L.)

4. Explain what is meant by linear convergence of an iteration. Assuming that the sequence defined by the formula

$$x_{n+1} = \frac{1 + ax_n}{a(1 + x_n)},$$

where a is positive, is convergent, prove that the convergence is linear.
   Prove that the Newton–Raphson method for the solution of equations has quadratic convergence. (L.)

5. Show that the equation

$$2 \sinh x + 3 \cosh x = 3 + k,$$

where k is small and positive, has two real roots. Show also that, if k^3 is neglected, the positive root is approximately

$$\frac{k}{2} - \frac{3k^2}{16}. \qquad \text{(L.)}$$

6. Find an approximation, as far as the term in ε^2, to the solution near $x = a$ of the equation

$$x = a + \varepsilon f(x),$$

where ε is small when compared with a and neither $f(x)$ nor any of its derivatives is large near $x = a$.

17:7 CONTINUITY OF A FUNCTION

   In §6:3 we mentioned continuity as a significant property of the graph of a function, and hence of the function; we said that the "natural" requirement for a function to be continuous throughout its domain was that it must be possible to draw its entire graph without lifting pencil from paper. If there was a "break" in the graph this corresponded to a discontinuity of the function. In fact this intuitive explanation of continuity provides a sound basis for an analytic definition.
   We say that the function f with domain Df is continuous at $x = \xi$ if

$$(1) \quad \xi \in Df \qquad\qquad ,$$

$$(2) \quad \lim_{x \to \xi -} f(x) = f(\xi) = \lim_{x \to \xi +} f(x).$$

Example 1. (a) Consider $f: x \to -x \quad (x < 0)$
$$\qquad\qquad\qquad x \to x \quad (x \geqslant 0),$$
or $x \to |x|$,   at $x = 0$.

Here $\lim_{x \to 0 -} f(x) = f(0) = \lim_{x \to 0+} f(x) = 0$
$$\Leftrightarrow f \text{ is continuous at } x = 0, \text{ see Fig. 17.4(i).}$$

(b) Consider $g: x \to \dfrac{3}{x - 1}$ at $x = 1$.

$$1 \notin Dg \Rightarrow g \text{ is discontinuous at } x = 1, \text{ see Fig. 17.4 (ii).}$$

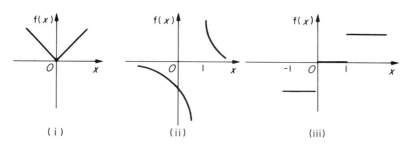

FIG. 17.4.

(c) Consider $h:x \to [x]$ at $x = 0$.
(Remember $[x]$ is the integer part of x so that $[2\cdot5] = 2$, $[-0\cdot4] = -1$.)

$$\lim_{x\to 0-} h(x) = -1, \text{ but } h(0) = \lim_{x\to 0+} = 0$$

$\Rightarrow h$ is discontinuous at $x = 0$, see Fig. 17.4 (iii).

(d) Consider $q_1: x \to \dfrac{\sin x}{x} \quad$ for $\quad x \neq 0$,

$\qquad\qquad x \to 0 \quad$ for $\quad x = 0$.

$$\lim_{x\to 0-} q_1(x) = \lim_{x\to 0+} q_1(x) = 1 \text{ (see §4:6)}$$

But $\quad q_1(0) = 0 \Rightarrow q_1$ is discontinuous at $x = 0$.

(e) Consider $q_2: x \to \dfrac{\sin x}{x} \quad$ for $\quad x \neq 0$,

$\qquad\qquad x \to 1$ for $x = 0$.

$$\lim_{x\to 0-} q_1(x) = \lim_{x\to 0+} q_2(x) = q_2(0) = 1 \Rightarrow q_2 \text{ is continuous at } x = 0.$$

Example 2. Consider $p:x \to \sqrt{(x-2)}$ at $x = 2$.

Here we cannot discuss $\lim\limits_{x\to 2-} p(x)$ since $x \in Dp \Leftrightarrow x \geqslant 2$. It is natural to modify our requirements for continuity at an end point of an interval within a domain, so that if $x \in Df \Leftrightarrow x:a \leqslant x \leqslant b$ we say that f is continuous at $x = a \Leftrightarrow \lim\limits_{x\to a+} f(x) = f(a)$ and f is continuous at $x = b \Leftrightarrow \lim\limits_{x\to b-} f(x) = f(b)$.

In our case $\lim\limits_{x\to 2+} p(x) = p(2) = 0 \Rightarrow p$ is continuous at $x = 2$.

Differentiability of a function

It might seem that if a function is continuous at $x = \xi$ it will have a derivative there; but Fig. 17.4(i) indicates that though continuity of a function is certainly a *necessary* condition for its differentiability, it is not a sufficient one. In this example $\lim\limits_{x\to 0-} f'(x) = -1$ and $\lim\limits_{x\to 0+} f(x) = 1$. The necessary and sufficient condition for a function F to be differentiable at $x = \xi$ is that F' is continuous at $x = \xi$.

Example 1. $r{:}x \rightarrow 3x^{2/3} + 1$.

        $r'{:}x \rightarrow 2x^{-1/3}$.   [Fig. 17.5(i)]

The function r is continuous at $x = 0$; but since r' is discontinuous at $x = 0$, r is not differentiable there. Note that although $r'(0) \neq 0$, $r(x)$ has a minimum value at $x = 0$; maximum or minimum values of $f(x)$ occur *either* at points where $f'(x) = 0$ *or* where $f'(x)$ has a discontinuity.

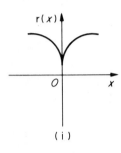
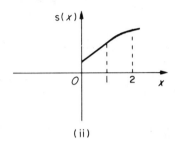

               (i)   (ii)

FIG. 17.5.

Example 2. $s{:}x \rightarrow 1 + 2x$, $(0 \leqslant x \leqslant 1)$

            $x \rightarrow 4x - x^2$. $(1 < x \leqslant 2)$

    $s'{:}x \rightarrow 2$ $(0 \leqslant x \leqslant 1)$

            $x \rightarrow 4 - 2x$ $(1 < x \leqslant 2)$, see Fig. 17.5(ii).

$$\lim_{x \rightarrow 1-} s'(x) = s'(1) = \lim_{x \rightarrow 1+} s'(x) = 2$$

$$\Rightarrow s' \text{ is continuous at } x = 1 \Leftrightarrow s \text{ is differentiable at } x = 1.$$

Exerise 17.7

1. $f(x)$ is a real function defined over $\mathbb{R}$; it is also odd. State, for each of the following statements, whether or not it is necessarily true:

    (i) $f(x)$ is continuous for all x;

    (ii) $f(x)$ is continuous in some neighbourhood of $x = 0$;

    (iii) $f(x)$ is periodic.

For each false statement, give a specific counter-example.   (O.C.S.M.P.)

2. Sketch, for $|x| < 3$, the graph of the function $f(x)$ defined by

$$f(x) = \frac{1}{x} + 1 \quad\quad (-3 < x \leqslant -1)$$
$$= \tfrac{1}{4}x^3 \quad\quad\quad (-1 < x \leqslant 2)$$
$$= 3x - 4 \quad\quad (2 < x < 3).$$

If $f(x)$ has an inverse function in the interval $|x| < 3$, sketch it.

    Is $f(x)$ differentiable (i) at $x = 2$, (ii) at $x = -1$? (Justify your answers by arguments from first principles.) Sketch the graph of $f'(x)$, for the points at which $f'(x)$ exists.   (O.C.S.M.P.)

3. Given any real-valued function f of a real variable x, a new function g is defined by setting $g(x) = f(x)$ when $f(x) \geqslant 0$ and $g(x) = -f(x)$ when $f(x) < 0$. Sketch the graph of g for the case when f is defined in the interval $-2 \leqslant x \leqslant 5$ by

$$f(x) = -1, \qquad -2 \leqslant x < 0,$$
$$= 1, \qquad 0 \leqslant x \leqslant 1,$$
$$= 2 - x, \qquad 1 \leqslant x \leqslant 3,$$
$$= -1, \qquad 3 \leqslant x \leqslant 5.$$

State for each of the following assertions whether it is true or false. Give a proof or counter-example as appropriate.

(i) If, for any function f, the function g is defined as above, then when g is continuous at a point f must also be continuous at that point.

(ii) If *any* real-valued function of a real variable is continuous at a point, it is also differentiable at that point.
(O.C.S.M.P.)

4. State carefully what is meant by the statement that 'f(x) is differentiable at $x = a$'. Use your definition to evaluate the derivative of $g(x)$ at $x = 1$, where

$$g(x) = x - 1, \qquad x \leqslant 1$$
$$= x^2 - x, \qquad x > 1.$$
(O.C.S.M.P.)

Miscellaneous Exercise 17

1. (i) The terms of a sequence $u_1, u_2, u_3, \ldots,$ satisfy the relation

$$19u_{n+1} = u_n{}^3 + 30 \qquad (n \geqslant 1)$$

and $u_1 = 1$. Prove the following results:
(a) $u_n > 1$ for $n > 1$,
(b) $u_n < 2$,
(c) $u_{n+1} > u_n$.
Assuming that the sequence u_n tends to a finite limit l as $n \to \infty$, find l.
(ii) For each of the following series, find the set of values of x for which it is convergent:

(a) $\displaystyle\sum_{n=1}^{\infty} \frac{n^2 x^{2n}}{3^n},$
(b) $\displaystyle\sum_{n=1}^{\infty} \frac{1}{(1 + 2x^2)^n}.$
(L.)

2. (i) Show that, when n is greater than 2^{2k},

$$\left(1 + \frac{1}{2} + \frac{1}{3} + \frac{1}{4} + \ldots + \frac{1}{n}\right) > k + 1.$$

(ii) If $a_n = (4n - 1)/(2n + 2)$, show that a_n tends to a limit as n tends to infinity. Find the smallest integer n such that a_n differs from the limit by less than 10^{-6}.
(iii) Find the limit as $n \to \infty$ of

(a) ne^{-n},   (b) $\{\sqrt{(n^2 + n)} - \sqrt{(n^2 - n)}\}.$
(L.)

3. Find the sets of values of x for which the following series are convergent:

(a) $\displaystyle\sum_{n=1}^{\infty} \frac{1}{(2 - x)^n},$
(b) $\displaystyle\sum_{n=1}^{\infty} (1 + \ln x)^n,$
(c) $\displaystyle\sum_{n=1}^{\infty} \frac{x^n}{n},$

(d) $\displaystyle\sum_{n=1}^{\infty} n(4x^2)^n,$
(e) $\displaystyle\sum_{n=1}^{\infty} \frac{(n + 3)x^n}{(n + 1)(n + 2)},$
(f) $\displaystyle\sum_{n=1}^{\infty} \frac{\cosh nx}{2^n}.$
(L.)

4. Find the sum of each of the series

$$\text{(a)} \sum_{n=1}^{\infty} \frac{1}{2^{2n-1}}, \quad \text{(b)} \sum_{n=1}^{\infty} \frac{1}{2^n n!}, \quad \text{(c)} \sum_{n=1}^{\infty} \frac{1}{n2^n}, \quad \text{(d)} \sum_{n=1}^{\infty} \frac{n}{2^{3n-2}}. \qquad \text{(L.)}$$

5. Decide, stating reasons, whether $\sum_{r=1}^{\infty} \sin(\theta^r)$ and $\sum_{r=1}^{\infty} \cos(\theta^r)$ are convergent when $0 < \theta < 1$. (N.)

6. Show that the sum of the series

$$1 + 3(\tfrac{1}{2}) + 5(\tfrac{1}{2})^2 + \ldots + (2n-1)(\tfrac{1}{2})^{n-1}$$

is $6 - R_n$, where

$$R_n = \frac{(2n+3)}{2^{n-1}}.$$

Show that, for $n \geqslant 4$,

$$\frac{R_{n+1}}{R_n} \leqslant \frac{13}{22},$$

and deduce that, for $n \geqslant 4$,

$$R_n \leqslant \frac{11}{8}\left(\frac{13}{22}\right)^{n-4}. \qquad \text{(N.)}$$

7. Given that

$$f(x) = x - \ln(1+x)$$

for all x greater than -1, determine the values of x for which $f'(x) > 0$. Show that for $x > 0$, $\ln(1+) < x$. State whether this inequality is true for $-1 < x < 0$, and illustrate the inequality by sketch graphs of $y = \ln(1+x)$ and $y = x$.

Investigate the convergence of $\sum_{n=0}^{\infty} u_n$ in each of the following cases:

$$\text{(a) } u_n = \ln(1 + x^n), \quad x \geqslant 0, \qquad \text{(b) } u_n = \sin\frac{\pi}{2^n}, \qquad \text{(c) } u_n = \cos\frac{\pi}{2^n}.$$

8. Show that, for $x > 0$,

$$x > \sin x > x - \tfrac{1}{6}x^3. \qquad \text{(L.)}$$

9. Using the same axes, sketch the graphs of the functions $\sinh x$, $\cosh x$ and $\tanh x$. Prove independently that for $x > 0$

$$\cosh x > \sinh x > x > \tanh x.$$

State without proof the corresponding inequalities for $x < 0$. (L.)

10. (i) Prove the inequalities
(a) $e^x \geqslant x + 1$, for all values of x,
(b) $x - 1 \geqslant \ln x$, for $x > 0$.
(ii) If the sum of the positive numbers a, b, c is 3, find the range of the possible values of $(a^2 + b^2 + c^2)$. (L.)

11. If a and b are positive and n is a positive integer, show by induction, or otherwise, that

$$\left(\frac{a+b}{2}\right)^n \leqslant \frac{a^n + b^n}{2}.$$

Hence, or otherwise, show that, for all values of c,

$$(a+b+c)^4 \leqslant 27(a^4 + b^4 + c^4). \qquad \text{(L.)}$$

12. (i) If a, b are positive numbers, prove that

$$(1-a)(1-b) > 1 - a - b.$$

Deduce, or prove otherwise, that if a, b, c are positive, then

$$(1 - a)(1 - b)(1 - c) > 1 - a - b - c$$

when one at least of a, b, c is less than unity.   (O.C.)

(ii) Show that $x(x - 1)(x - 2)(x - 3)$ differs by a constant from a perfect square. Determine the range of values of x for which

$$64x(x - 1)(x - 2)(x - 3) < 17.$$

13. Show that there is no value of x for which

$$|x| + |x - 2| = x^2 - 2x + 4$$

and sketch on the same axes the graphs of

(i) $y = |x| + |x - 2|$,   (ii) $y = x^2 - 2x + 4$.

Shade the region satisfying the four inequalities

(a) $y > 2$, (b) $y < x^2 - 2x + 4$, (c) $y > 2x - 2$ and (d) $y > 2 - 2x$.   (N.)

14. By considering the graphs of $x^2 + y^2 - x = 0$ and $x^2 + y^2 - 1 = 0$, or otherwise, prove that if $x^2 + y^2 < x$, then $x^2 + y^2 < 1$.

15. (i) Find the set of values of x for which

$$\left|\frac{x - 3}{x + 1}\right| < 2.$$

(ii) If x, y and z are positive, show that

(a) $2(x^3 + y^3) \geq (x^2 + y^2)(x + y)$,

(b) $3(x^3 + y^3 + z^3) \geq (x^2 + y^2 + z^2)(x + y + z)$.

16. Given that

$$A = \int_{0 \cdot 6}^{0 \cdot 8} \frac{1}{\sqrt{(1 - x^2)}}\,dx, \qquad B = \int_{0 \cdot 6}^{0 \cdot 8} \sqrt{\left(\frac{x}{1 - x^2}\right)}\,dx$$

and

$$C = \int_{0 \cdot 6}^{0 \cdot 8} \frac{x}{\sqrt{(1 - x^2)}}\,dx,$$

show, without integrating, that

$$\sqrt{(0 \cdot 8)}A > B > \frac{1}{\sqrt{(0 \cdot 8)}}C.$$

Evaluate A and C by integration as accurately as your tables permit and use Simpson's rule with three ordinates to obtain a value for B. Use the inequalities as a rough check on your results.   (L.)

17. Determine the sets of values of x for which the following series are convergent:

(a) $\displaystyle\sum_{n=0}^{\infty} \frac{1}{(1 + x)^n}$,   (b) $\displaystyle\sum_{n=1}^{\infty} \frac{(x - 2)^n}{n}$,   (c) $\displaystyle\sum_{n=0}^{\infty} \frac{x^n}{1 + x^{2n}}$,

(d) $\displaystyle\sum_{n=1}^{\infty} \left(\frac{2x}{1 + x^2}\right)^n$,   (e) $\displaystyle\sum_{r=1}^{\infty} r^b(x + 1)^r$,   (f) $\displaystyle\sum_{n=1}^{\infty} x^n e^{-nx}$.   (L.)

18. Solve the difference equation $u_{n+2} - (k + 2)u_{n+1} + 2ku_n = 0$, where k is a positive constant, distinguishing between the cases where $k = 2$ and $k \neq 2$.

Find the particular solution, when $k = 2$, $u_1 = 0$, $u_2 = 4$.

19. Solve the difference equations:

(i) $u_{n+3} - 2u_{n+2} - u_{n+1} + 2u_n = 0$, given that $u_1 = 0$, $u_2 = 2$ and u_n remains finite as $n \to \infty$.

(ii) $u_{n+3} - 4u_{n+2} + 5u_{n+1} - 2u_n = 0$, given that $u_1 = 1$, $u_2 = 0$, $u_3 = -5$.

(iii) $u_{n+2} - u_{n+1} + u_n = 0$, given that $u_1 = \frac{1}{2}(1 + \sqrt{3})$, $u_2 = \frac{1}{2}(-1 + \sqrt{3})$.

20. Two sequences $u_0, u_1, \ldots$ and $v_0, v_1, \ldots$ are such that

$$u_0 = 0, \quad u_1 = 3, \quad u_n = 3v_n + 2v_{n-1}, \quad v_n = u_n + u_{n-1}.$$

Express u_n in terms of n, and prove that $\lim_{n \to \infty} (u_n - 2v_n) = 0$.   (L.)

21. Verify that the recurrence relation

$$u_n = u_{n-1} + 2u_{n-2} + 1$$

is satisfied by

$$u_n = A\alpha^n + B\beta^n + \gamma,$$

where A and B are arbitrary constants, if α, β and γ are suitably chosen non-zero constants. Find the values of α, β and γ.

If $u_1 = 1$ and $u_2 = 2$, find the values of A and B.   (O.C.)

22. Discuss the convergence of the infinite geometric series Σar^n and deduce that, if $\lim_{n \to \infty} u_n^{1/n} < 1$, a series Σu_n of positive terms is convergent.

If an infinite series $u_1 + u_2 + \ldots + u_n + \ldots$ is such that

$$u_{2m-1} = a^m, \quad u_{2m} = k^m a^m,$$

where $1 < k < \frac{1}{a}$, find $\lim_{n \to \infty} \frac{u_{n+1}}{u_n}$ and $\lim_{n \to \infty} u_n^{1/n}$ for the cases $n = 2m$, $n = 2m+1$ and test the series for convergence.

(ii) Test for convergence the series for which the nth terms are

(a) $(-1)^{n-1}\left(\sec\frac{\alpha}{n} - 1\right)$.   (b) $\frac{x^n}{x^{n-1}+n}$, $(x > 0)$.   (L.)

23. (i) Find the two conditions that the sequence $a_1, a_2, \ldots a_n, \ldots$, where

$$a_n = pn^2 - qn + r,$$

p, q, r being real constants, is such that $a_{n+1} \geq a_n$ for all n.

(ii) The sequence $b_1, b_2, \ldots b_n, \ldots$ is defined by

$$b_n = \left(1 + \frac{1}{n}\right)\sin\frac{n\pi}{2}.$$

Determine the behaviour of its sub-sequences for the separate cases n odd and n even and find the limits of the sub-sequences.

(iii) (a) Find

$$\lim_{x \to 0} (1 - 2x)^{3/x}.$$

(b) Find, when y decreases through positive values to zero,

$$\lim\left(\frac{y^2 \ln y}{\sin y}\right).$$ (L.)

24. A sequence satisfies the relation

$$u_n = u_{n-1} + u_{n-2} \quad (n \geq 2).$$

Prove that

(i) $u_n^2 - u_{n-1} u_{n+1} = (-1)^{n-1} (u_1^2 - u_0 u_2),$

(ii) $u_1 u_2 + u_2 u_3 + \ldots + u_{2n} u_{2n+1} = u_{2n+1}^2 - u_1^2 \quad (n \geq 1).$ (L.)

25. Discuss the existence of the derivative at $x = 0$ of the function

$$f : x \rightarrow x|x|,$$

showing that you have considered both positive and negative values as x tends to 0.

State with reasons which of the following are true and which are false:

(i) f is an even function;

(ii) f is continuous at $x = 0$;

(iii) f is differentiable at $x = 0$.

Sketch a graph of the function f and find the derived function f'.

Is f' differentiable at $x = 0$?   (O.C.S.M.P.)

18 Differential Equations

18:1 FIRST-ORDER DIFFERENTIAL EQUATIONS WITH VARIABLES SEPARABLE

We start with a reminder of the method of solution introduced in § 12:7.
If a differential equation can be expressed in the form

$$\frac{dy}{dx} = f(x)g(y), \qquad (18.1)$$

where $f(x)$ is a function of x only and $g(y)$ is a function of y only, it can be written

$$\frac{dy}{g(y)} = f(x)\,dx. \qquad (18.2)$$

[Note that we now use differentials, see §16:16.]
In this case the variables are said to be *separable* and integration gives

$$\int \frac{dy}{g(y)} = \int f(x)\,dx + C. \qquad (18.3)$$

In general, the integrals in (18.3) give a relationship between x, y which involves an arbitrary constant. This is the general solution of the differential equation; it may not be possible to express y as an explicit function of x and C.

Example 1. Solve the differential equation $\dfrac{dy}{dx} = \dfrac{1+y^2}{1+x^2}$.

This equation can be written

$$\frac{dy}{1+y^2} = \frac{dx}{1+x^2}.$$

Integration gives

$$\int \frac{1}{1+y^2}\,dy = \int \frac{1}{1+x^2}\,dx + C$$

$$\Leftrightarrow \tan^{-1}y = \tan^{-1}x + C$$

$$\Leftrightarrow \tan^{-1}y - \tan^{-1}x = \tan^{-1}\left(\frac{y-x}{1+xy}\right) = C = \tan^{-1}A, \text{ say.}$$

Hence the general solution can be written

$$y - x = A(1+xy) \quad \text{or} \quad y = (x+A)/(1-Ax).$$

581

Example 2. $y(1+x^2)\dfrac{dy}{dx}+1 = y^2$.

This equation can be written

$$(1 + x^2)\frac{dy}{dx} = \frac{y^2 - 1}{y}$$

$$\Leftrightarrow \int \frac{y\,dy}{y^2 - 1} = \int \frac{1}{1+x^2}\,dx$$

$$\Leftrightarrow \tfrac{1}{2}\ln(y^2 - 1) + \ln C = \tan^{-1} x$$

$$\Rightarrow x = \tan[\ln\{C\sqrt{(y^2 - 1)}\}].$$

(The solution can alternatively be expressed in the form

$$y^2 = 1 + Ae^{2\tan^{-1}x}.)$$

18:2 HOMOGENEOUS DIFFERENTIAL EQUATIONS

A differential equation of the form

$$\phi(x, y)\frac{dy}{dx} + \psi(x, y) = 0, \tag{18.4}$$

where $\phi(x, y)$ and $\psi(x, y)$ are each *homogeneous* in x, y and of the same degree p [i.e. $\phi(\lambda x, \lambda y) = \lambda^p \phi(x, y)$ and $\psi(\lambda x, \lambda y) = \lambda^p \psi(x, y)$], is called a homogeneous differential equation and can be expressed in the form

$$\frac{dy}{dx} = f\left(\frac{y}{x}\right). \tag{18.5}$$

Writing $y = xv$, where v is a function of x only and is to be determined, equation (18.5) becomes

$$x\frac{dv}{dx} + v = f(v)$$

$$\Leftrightarrow x\frac{dv}{dx} = f(v) - v.$$

In this equation the variables are separable and the solution can be found from

$$\int \frac{1}{f(v) - v}\,dv = \int \frac{1}{x}\,dx + C, \tag{18.6}$$

where, after integration, v must be replaced by y/x.

Example 1. $x(x+y)\dfrac{dy}{dx} = x^2 + y^2$, $y = 0$ when $x = 1$.

This equation is homogeneous and can be expressed in the form

$$\frac{dy}{dx} = \frac{1 + \left(\dfrac{y}{x}\right)^2}{1 + \left(\dfrac{y}{x}\right)}.$$

The substitution $y = xv$ gives

$$v + x\frac{dv}{dx} = \frac{1 + v^2}{1 + v}$$

$$\Leftrightarrow x\frac{dv}{dx} = \frac{1 - v}{1 + v}.$$

Separation of the variables and integration gives

$$\int\frac{1 + v}{1 - v}dv = \int\frac{1}{x}dx + C$$

$$\Leftrightarrow -2\ln(1 - v) - v = \ln x + C$$

$$\Rightarrow v = \ln\left\{\frac{A}{x(1 - v)^2}\right\}.$$

Hence, writing $v = y/x$, the general solution is

$$y = x\ln\left\{\frac{Ax}{(x - y)^2}\right\}.$$

Since $y = 0$ when $x = 1$, $A = 1$ and the particular solution required is

$$y = x\ln\left\{\frac{x}{(x - y)^2}\right\}.$$

Example 2. $x\cos\left(\dfrac{y}{x}\right).\dfrac{dy}{dx} - \left\{x\sin\left(\dfrac{y}{x}\right) + y\cos\left(\dfrac{y}{x}\right)\right\} = 0.$

Writing $y = xv$ gives

$$x\cos v\left(v + x\frac{dv}{dx}\right) = x(\sin v + v\cos v)$$

$$\Leftrightarrow x\frac{dv}{dx} = \tan v.$$

Separation and integration gives $\int\cot v\,dv = \int\frac{1}{x}dx + C$

$$\Leftrightarrow \ln\sin v = \ln x + C.$$

Hence the solution can be written

$$\sin\left(\frac{y}{x}\right) = Ax.$$

Equations reducible to the homogeneous form. The differential equation

$$\frac{dy}{dx} = \frac{(a_1x + b_1y + c_1)^r}{(a_2x + b_2y + c_2)^r}, \tag{18.7}$$

where $a_1, b_1, c_1, a_2, b_2, c_2, r$ are constants, can usually be reduced to the homogeneous form by changing the independent and dependent variables to X, Y respectively where $x = X + h$, $y = Y + k$ and choosing h, k, so that

$$a_1h + b_1k + c_1 = 0,$$
$$a_2h + b_2k + c_2 = 0. \tag{18.8}$$

The equation (18.7) then becomes

$$\frac{dY}{dX} = \frac{(a_1 X + b_1 Y)^r}{(a_2 X + b_2 Y)^r}$$

which can be reduced to separable form by the substitution $Y = XV$.

Effectively equation (18.7) is made homogeneous by a translation of axes in the plane Oxy, the origin being moved to the point (h, k) which is the intersection of the straight lines

$$a_1 x + b_1 y + c_1 = 0, \quad a_2 x + b_2 y + c_2 = 0.$$

An exception occurs when these two lines are parallel and equations (18.8) have no solution. In this case, however, $a_2 = f a_1, b_2 = f b_1$, where f is a constant, and equation (18.7) can be reduced by the substitution

$$z = a_1 x + b_1 y \qquad (18.9)$$

to a separable equation in x and z.

Example 1.   Solve the equation $\dfrac{dy}{dx} = \dfrac{4x - 3y - 1}{3x + 4y - 7}$.

The intersection of the lines

$$4x - 3y - 1 = 0, \quad 3x + 4y - 7 = 0$$

is the point $(1, 1)$. Writing $x = X + 1, y = Y + 1$, gives

$$\frac{dY}{dX} = \frac{4X - 3Y}{3X + 4Y}.$$

Then the substitution $Y = XV$ gives

$$V + X\frac{dV}{dX} = \frac{4 - 3V}{3 + 4V}$$

$$\Leftrightarrow X\frac{dV}{dX} = \frac{4 - 6V - 4V^2}{3 + 4V}.$$

Separation and integration gives

$$\int \frac{(3 + 4V)dV}{2(2 - 3V - 2V^2)} = \int \frac{1}{X}dX + C$$

$$\Leftrightarrow -\tfrac{1}{2}\ln(2 - 3V - 2V^2) = \ln X + C$$

$$\Rightarrow (2 - 3V - 2V^2)X^2 = A$$

or, putting $V = Y/X$,

$$(2X + Y)(X - 2Y) = A.$$

Hence the general solution is

$$(2x + y - 3)(x - 2y + 1) = A.$$

Example 2. $\dfrac{dy}{dx} = \dfrac{x - y - 1}{y - x - 1}$.

Since the lines $x - y - 1 = 0, y - x - 1 = 0$ are parallel we make the substitution $z = x - y$. Then

$$\frac{dz}{dx} = 1 - \frac{dy}{dx}$$

and the equation becomes

$$1 - \frac{dz}{dx} = \frac{z-1}{-z-1}$$

$$\Leftrightarrow \frac{dz}{dx} = \frac{2z}{z+1}.$$

Separation and integration gives

$$\int \left(1 + \frac{1}{z}\right) dz = 2 \int 1 \, dx + C$$

$$\Leftrightarrow z + \ln z = 2x + C$$

$$\Leftrightarrow \ln (x - y) - x - y = C.$$

_____

Exercise 18.2

1. (i) Solve the equation $(1 - x^2)\dfrac{dy}{dx} = xy(1 + y^2)$, given that $y = 1$ when $x = 0$.

   (ii) By the substitution $z = x + y$, transform the equation

   $$\frac{dy}{dx} = \frac{1 + 2x + 2y}{1 - 2x - 2y}$$

   into an equation involving z and x only. Hence, or otherwise, solve the given equation.

2. Find y as a function of x when

   $$\frac{1}{y}\frac{dy}{dx} - x = xy$$

   and $y = 1$ when $x = 0$. Show that y becomes infinite when $x^2 = \ln 4$.

3. By means of the substitution $x - y = z$, solve the differential equation

   $$\frac{dy}{dx} = \sin (x - y),$$

   given that $y = 0$ when $x = 0$.

   Solve the differential equations:

4. $\dfrac{dy}{dx} + \dfrac{2x(x + y)}{x^2 + y^2} = 0,$ 5. $xy\dfrac{dy}{dx} = x^2 + y^2.$

6. $(5x - 2y)\dfrac{dy}{dx} - y = 2x.$ 7. $\dfrac{dy}{dx} = e^{-y/x} + y/x.$

8. $x\dfrac{dy}{dx} = y + \sqrt{(x^2 + y^2)}.$ 9. $\dfrac{dy}{dx} = \dfrac{x + 3y}{3x + y}$ given $y = 0$ when $x = 1.$

10. $\dfrac{dy}{dx} = \dfrac{3x^2 - 3xy + y^2}{2x^2 + 3xy}$ given that $y = -2$ when $x = 1.$

11. $\dfrac{dy}{dx} = \dfrac{4x + 2y + 2}{y - x - 2}.$ 12. $\dfrac{dy}{dx} = \dfrac{x - y}{x - 8y + 7}.$

13. $\dfrac{dy}{dx} = \dfrac{x + 2y - 1}{2x + 4y + 3}.$ 14. $\dfrac{dy}{dx} = \dfrac{2x - y}{x + 2y - 5}.$ 15. $xy\dfrac{dy}{dx} = y^2 + x^2 e^{y/x}.$

18:3 FIRST-ORDER LINEAR EQUATIONS: BERNOULLI'S EQUATION

The equation
$$\frac{dy}{dx} + Py = Q, \tag{18.10}$$

where P and Q are functions of x only, is a *linear* differential equation of the first order. The equation is linear because it involves y and its derivatives only in the first degree.

If $Q = 0$, equation (18.10) is separable and can be expressed in the form
$$\frac{dy}{y} + P\,dx = 0$$

which has solution
$$ye^{\int P\,dx} = C. \tag{18.11}$$

Differentiating the left-hand side of equation (18.11) gives
$$\frac{d}{dx}\left(ye^{\int P\,dx}\right) = \left(\frac{dy}{dx} + Py\right)e^{\int P\,dx},$$

i.e. multiplication of the left-hand side of equation (18.10) by $e^{\int P\,dx}$ makes that left-hand side the derivative of the product $ye^{\int P\,dx}$. In this case equation (18.10) can be written
$$\frac{d}{dx}\left(ye^{\int P\,dx}\right) = Qe^{\int P\,dx}$$

which has solution
$$ye^{\int P\,dx} = \int Qe^{\int P\,dx}\,dx + C$$

or
$$y = e^{-\int P\,dx}\int Qe^{\int P\,dx}\,dx + Ce^{-\int P\,dx}. \tag{18.12}$$

The factor $e^{\int P\,dx}$, which makes the left-hand side of equation (18.10) the derivative of a product, is called an *integrating factor*.

Example 1. Use the formula for differentiating a product to find an expression for the 'integrating factor' of a linear first order differential equation.

Let u be the integrating factor. Then
$$u\frac{dy}{dx} + uPy = \frac{d}{dx}(uy) = u\frac{dy}{dx} + y\frac{du}{dx}.$$

Hence we require
$$\frac{du}{dx} = uP \Leftrightarrow \int\frac{du}{u} = \int P\,dx$$

$$\Leftrightarrow \ln u = \int P\,dx \Leftrightarrow u = e^{\int P\,dx}.$$

If the whole equation is multiplied by u, then
$$u\frac{dy}{dx} + uPy = \frac{d}{dx}(uy) = uQ \quad \Leftrightarrow uy = \int uQ\,dx + C$$

which is equivalent to equation (18.12).

Example 2. $\dfrac{dy}{dx} + \dfrac{3}{x} y = 8x^4$, where $y = 0$ when $x = 1$.

Here, $P = 3/x$ and the integrating factor is

$$e^{\int (3/x)dx} = e^{3 \ln x} = e^{\ln x^3} = x^3,$$

on using the relation $e^{\ln f(x)} = f(x)$. Multiplied by this integrating factor, the given equation takes the form

$$x^3 \frac{dy}{dx} + 3x^2 y = 8x^7. \tag{1}$$

The left-hand side of equation (1) *must be* $\dfrac{d}{dx}(x^3 y)$ and hence (1) can be written

$$\frac{d}{dx}(x^3 y) = 8x^7.$$

Integrating:

$$x^3 y = x^8 + C.$$

Hence the *general* solution is

$$y = x^5 + C/x^3.$$

The particular solution for which $y = 0$ when $x = 1$ is obtained by writing $C = -1$.

Example 3. $\sin x \dfrac{dy}{dx} + 2y \cos x = \cos x.$ (1)

Before equation (1) can be expressed in the form (18.10) it must be divided by $\sin x$ giving

$$\frac{dy}{dx} + 2y \cot x = \cot x. \tag{2}$$

The integrating factor is

$$e^{\int 2 \cot x \, dx} = e^{2 \ln \sin x} = e^{\ln \sin^2 x} = \sin^2 x.$$

After multiplication by this integrating factor equation (2) becomes

$$\sin^2 x \frac{dy}{dx} + 2y \sin x \cos x = \sin x \cos x. \tag{3}$$

Equation (3) can be written

$$\frac{d}{dx}(y \sin^2 x) = \sin x \cos x$$

which integrates to give

$$y \sin^2 x = \tfrac{1}{2} \sin^2 x + C$$
$$\Leftrightarrow y = \tfrac{1}{2} + C \operatorname{cosec}^2 x.$$

In simple cases such as this the integrating factor may sometimes be apparent by inspection.

Example 4. $(1 + 3x)\dfrac{dy}{dx} + (3 - 9x)y = 3.$

After division by $(1 + 3x)$ and multiplication by the integrating factor

$$\exp \left[\int \frac{(3 - 9x)}{(1 + 3x)} dx \right] = \exp\{2 \ln(1 + 3x) - 3x\} = (1 + 3x)^2 e^{-3x},$$

the equation can be written

$$\frac{d}{dx}\{(1+3x)^2 e^{-3x}y\} = 3(1+3x)e^{-3x}.$$

This integrates to give

$$(1+3x)^2 e^{-3x}y = -(3x+2)e^{-3x} + C$$
$$\Leftrightarrow y = (Ce^{3x} - 3x - 2)/(1+3x)^2.$$

BERNOULLI'S EQUATION

$$\frac{dy}{dx} + Py = Qy^n, \qquad (18.13)$$

where $n(\neq 1)$ is constant and, as before, P and Q are functions of x, can be reduced to linear form on dividing by y^n and substituting $z = y^{1-n}$. For, since

$$\frac{dz}{dx} = \frac{(1-n)}{y^n}\frac{dy}{dx},$$

equation (18.13) becomes

$$\frac{dz}{dx} + (1-n)Pz = (1-n)Q, \qquad (18.14)$$

which is linear. If $n = 1$, equation (18.13) is separable.

Example 1. $\dfrac{dy}{dx} = y \cot x + y^3 \operatorname{cosec} x.$ \hfill (1)

Dividing by y^3 and rearrangement gives

$$\frac{1}{y^3}\frac{dy}{dx} - \frac{1}{y^2}\cot x = \operatorname{cosec} x. \qquad (2)$$

The substitution $z = \dfrac{1}{y^2}$, so that $\dfrac{dz}{dx} = -\dfrac{2}{y^3}\dfrac{dy}{dx}$, transforms equation (2) into

$$\frac{dz}{dx} + 2z \cot x = -2 \operatorname{cosec} x.$$

Multiplication by the integrating factor $\sin^2 x$ gives

$$\frac{d}{dx}(z \sin^2 x) = -2 \sin x$$
$$\Leftrightarrow z \sin^2 x = 2 \cos x + C$$
$$\Leftrightarrow y^2 = \sin^2 x/(2 \cos x + C).$$

Example 2. $x(x^2 - 1)\dfrac{dy}{dx} - y = x^3 y^2$, where $y = -\frac{1}{2}$ when $x = 2$.

The equation can be written

$$x(x^2-1)\frac{1}{y^2}\frac{dy}{dx} - \frac{1}{y} = x^3.$$

Writing $z = \dfrac{1}{y}$, so that $\dfrac{dz}{dx} = -\dfrac{1}{y^2}\dfrac{dy}{dx}$, and dividing by $x(x^2 - 1)$ gives

$$\frac{dz}{dx} + \frac{z}{x(x^2 - 1)} = \frac{-x^2}{(x^2 - 1)}.$$

The integrating factor is

$$\exp\left\{\int \frac{1}{x(x^2 - 1)}\,dx\right\} = \exp\left\{\int\left(-\frac{1}{x} + \frac{1}{2(x - 1)} + \frac{1}{2(x + 1)}\right)dx\right\}$$
$$= \exp\left\{-\ln x + \tfrac{1}{2}\ln(x - 1) + \tfrac{1}{2}\ln(x + 1)\right\}$$
$$= \frac{\sqrt{(x^2 - 1)}}{x}.$$

Multiplication by this integrating factor gives

$$\frac{d}{dx}\left\{\frac{z\sqrt{(x^2 - 1)}}{x}\right\} = -\frac{x}{\sqrt{(x^2 - 1)}}$$

$$\Leftrightarrow \frac{z\sqrt{(x^2 - 1)}}{x} = -\sqrt{(x^2 - 1)} + C$$

$$\Leftrightarrow \frac{1}{y} = -x + \frac{Cx}{\sqrt{(x^2 - 1)}}.$$

Since $y = -\tfrac{1}{2}$ when $x = 2$, the particular solution required is given when $C = 0$, i.e. $y = -1/x$.

Exercise 18.3

Solve the differential equations 1–10.

1. $\dfrac{dy}{dx} + \dfrac{y}{x} = e^x.$

2. $\dfrac{dy}{dx} + y\cot x = e^x.$

3. $(1 + x^2)\dfrac{dy}{dx} + xy = x.$

4. $\dfrac{dy}{dx} + \dfrac{y}{x - 1} = e^x.$

5. $\dfrac{dy}{dx} + ay = b.$

6. $\dfrac{dy}{dx} + y = \cos x.$

7. $\dfrac{dr}{d\theta} + r\tan\theta = \cos\theta.$

8. $\sin x\,\dfrac{dy}{dx} - y = \sin^2 x.$

9. $\cos x\,\dfrac{dy}{dx} - y = \cos x.$

10. $\dfrac{dy}{dx} = \dfrac{y + x - 1}{x}.$

11. Make the substitution $y = \dfrac{1}{z}$ and hence obtain the complete solution of the equation $\dfrac{dy}{dx} - y\cot x = y^2\operatorname{cosec} x.$

12. Solve the equation $\dfrac{dy}{dx} + \dfrac{y}{x} = x^2 y^3$ by means of the substitution $z = 1/y^2.$

13. With the usual notation for the motion of a particle moving in a straight line, the equation of motion of such a particle is $\dfrac{dv}{dt} = -2v + t.$ If $v = 1$ when $t = 0$, calculate v when $t = 1$ and if $s = 0$ when $t = 0$ calculate s when $t = 1$.

14. The current i in an electric circuit of resistance R and self-induction L satisfies the equation

$$L\frac{di}{dt} + Ri = 40\sin 100\,t.$$

If $i = 0$ when $t = 0$ and if $L = 0.20$ and $R = 20$, find i when $t = 0.01$.

15. Solve the differential equation

$$x\frac{dy}{dx} + 3y = \frac{a^3}{x^2}\cos\left(\frac{x}{a}\right)$$

with the condition $y = 0$ when $x = \pi a/2$.

Show that $|y| \leqslant \dfrac{2a^4}{x^3}$ for $x > 0$.   (N.)

16. (a) Solve the differential equation

$$\frac{dy}{dx} + 2y = e^{-2x}$$

(b) By means of the substitution $y = xz$, where z is a function of x, reduce the differential equation

$$x\frac{dy}{dx} - y = \tfrac{1}{4}x^2 - y^2$$

to a differential equation involving z and x only. Hence solve it, given that $y = 0$ when $x = \ln 2$.   (N.)

17. Solve the equation $x\dfrac{dy}{dx} + 2y = x^3$, given that $y = 0$ when $x = 1$.   (L.)

18. Solve the differential equation $x\dfrac{dy}{dx} - y = 3x^4$ given that $y = -1$ when $x = 1$.   (L.)

19. Solve the equation $\dfrac{dy}{dx} - y\cot x = \tan^2 x$ given that $y = 0$ when $x = \pi/4$.   (N.)

20. Find the solution of the equation

$$\left(1 - \frac{dy}{dx}\right)\sin x = y\cos x$$

for which $y = \sqrt{2} - 1$ when $x = \pi/4$.   (O.C.)

18:4 APPLICATIONS

Problems which lead to differential equations are so numerous that it is impossible to give a comprehensive list. One important group concerns the growth of a population of living organisms; in such a population the number of new organisms appearing is proportional to the number already present. Similarly, radioactive disintegration of an element is a phenomenon which occurs at random in a collection of radioactive atoms; the greater the number of atoms present the greater—proportionally—is the number which disintegrate. It follows that the rate of disintegration—or disappearance—of the substance is proportional to the amount present at any instant. In a chemical reaction the rate at which the reaction takes place is governed by the concentrations of the various substances involved. This type of growth problem usually leads to the appearance of exponential functions.

Conduction of heat along a rod is another physical problem which gives rise to a differential equation. The rate at which heat is conducted across a section of the rod is proportional to the temperature gradient, $d\theta/dx$, say, at that point. If the temperature falls steeply, $d\theta/dx$ is large and heat is conducted rapidly. Diffusion of one gas into another (or the spreading of a smell across a room), when it takes place in one dimension, e.g. down a narrow tube, is governed by an ordinary differential equation;

the rate at which a gas diffuses is proportional to the gradient of its concentration, dc/dx, down the tube. Diffusion and conduction of heat are closely similar processes and give rise to important partial differential equations when they take place in two or three dimensions.

The examples which follow together with Exercise 18.4 give other simple examples of the occurrence of differential equations in a variety of problems.

Example 1. The rate of decay of a substance is kx, where x is the amount of the substance remaining and k is a constant. Show that the half-life (the time in which the amount is halved) of the substance is $\tau = (\ln 2)/k$. Find also the time required for five-eighths of the substance to disintegrate, giving the result in terms of the half-life τ.

The rate of decay of the substance is $-dx/dt$. Hence

$$\frac{dx}{dt} = -kx.$$

The general solution of this differential equation is

$$x = Ae^{-kt}.$$

If the initial quantity is $x = x_0$ then $A = x_0$, and

$$x = x_0 e^{-kt}.$$

The half-life, τ, is the time that has elapsed when half the original amount remains, i.e.

$$\tfrac{1}{2}x_0 = x_0 e^{-k\tau} \Leftrightarrow \tfrac{1}{2} = e^{-k\tau}$$
$$\Leftrightarrow k\tau = \ln 2 \Leftrightarrow \tau = (\ln 2)/k.$$

In the second situation the amount remaining is $\tfrac{3}{8}x_0$ and the required time is T where

$$\tfrac{3}{8}x_0 = x_0 e^{-kT} \Leftrightarrow kT = \ln \tfrac{8}{3}$$
$$\Leftrightarrow \frac{T}{\tau} = \frac{\ln \tfrac{8}{3}}{\ln 2}.$$

Example 2. A tank contains V cm^3 of brine in which there is S g of salt. Water is pumped into the tank at the rate of u cm^3 s^{-1} and brine is pumped out at the rate of v cm^3 s^{-1}. Assuming perfect mixing throughout, formulate and solve the differential equation for the amount s g of salt remaining in the tank after t s. Consider also the special case $u = v$.

The volume of liquid in the tank alters with time, unless $u = v$, so we suppose that this volume is y cm^3 at time t. The brine which is removed takes away some of the salt, which is not replaced. Assuming perfect mixing the concentration of salt at time t is (s/y) g cm^{-3}, and so salt is removed at a rate (vs/y) g s^{-1}

$$\Rightarrow \frac{ds}{dt} = -\frac{vs}{y}. \tag{1}$$

The changes in volume are given by

$$\frac{dy}{dt} = u - v. \tag{2}$$

From (2) we see that $y = A + (u-v)t$, and the initial condition that $y = V$ when $t = 0$ gives

$$y = V + (u-v)t. \tag{3}$$

Equations (1) and (3) give

$$\frac{ds}{dt} = -\frac{vs}{V + (u-v)t} \quad \text{or} \quad \frac{ds}{s} = -\frac{v\,dt}{V + (u-v)t}$$

$$\Leftrightarrow \ln s = -\frac{v}{u-v}\ln\{V + (u-v)t\} + B.$$

Initially, $s = S$, $t = 0$ and therefore

$$\ln S = -\frac{v}{u-v}\ln V + B$$

$$\Rightarrow \ln\left(\frac{s}{S}\right) = \frac{v}{u-v}\ln\left\{\frac{V}{V+(u-v)t}\right\}$$

$$\Leftrightarrow s = S\left\{\frac{V}{V+(u-v)t}\right\}^{v/(u-v)}.$$

In the special case $u = v$ equation (2) gives $y = V$, i.e. y is constant. Hence equation (1) becomes

$$\frac{ds}{dt} = -\frac{vs}{V}, \quad s = Ce^{-vt/V},$$

where C is an arbitrary constant. From the initial conditions $C = S$ and so

$$s = Se^{-vt/V}.$$

Example 3. In a chemical reaction in which a compound X is formed from a compound Y and other agents the masses of X and Y present at time t are x and y respectively. The sum of the two masses is constant and at any time the rate at which x is increasing is proportional to the product of the two masses at that time. Show that the equation governing the reaction is of the form

$$\frac{dx}{dt} = kx(a-x)$$

and interpret the constant a. If $x = 0 \cdot 1a$ at time $t = 0$ find, in terms of k and a, the time at which the mass of Y is $0 \cdot 1a$.

The information given can be expressed in the two equations

$$\frac{dx}{dt} = k_1 xy, \qquad x + y = \text{constant},$$

where k_1 is a constant of proportionality. A comparison of these equations and the given differential equation shows that

$$a = x + y = \text{constant}, \qquad k_1 = k$$

$$\Rightarrow \frac{dx}{dt} = kx(a-x).$$

We separate the variables and obtain

$$\frac{dx}{x(a-x)} = \left(\frac{1}{x} + \frac{1}{a-x}\right)\frac{dx}{a} = k\,dt$$

$$\Leftrightarrow \ln x - \ln(a-x) = akt + A.$$

Since $x = 0 \cdot 1a$ initially,

$$\ln(0 \cdot 1a) - \ln(0 \cdot 9a) = \ln(1/9) = A$$

$$\Rightarrow \ln\left(\frac{x}{a-x}\right) = \ln\left(\frac{x}{y}\right) = akt + \ln\left(\frac{1}{9}\right)$$

$$\Leftrightarrow \frac{9x}{y} = e^{akt} \qquad \text{or} \qquad 9(a-y) = ye^{akt}.$$

When $y = 0 \cdot 1a$,

$$9(0 \cdot 9a) = 0 \cdot 1\,ae^{akt} \Leftrightarrow 81 = e^{akt}$$

$$\Leftrightarrow \ln 81 = 4\ln 3 = akt \Leftrightarrow t = (4\ln 3)/(ak).$$

Exercise 18.4

1. The pressures at height h and at ground level in the atmosphere are p and p_0 respectively and v and v_0 are the corresponding volumes per unit mass. Given that, if δp is the infinitesimal increase in pressure due to an increase δh in height,

$$\delta p = -(1/v)\delta h,$$

show that, if the adiabatic relation $pv^\gamma = p_0 v_0^\gamma$ holds, then

$$dh/dp = -v_0(p_0/p)^{1/\gamma}.$$

Solve this differential equation to find an expression for the pressure at height h.

2. A chemical compound C is formed from two compounds A and B. If at time t the compound C has mass x, and a and b are the masses of A and B, the rate of increase of C's mass is given by the differential equation

$$\frac{dx}{dt} = kab \qquad (k \text{ a positive constant}),$$

where $a = a_0 - \alpha x$, $b = b_0 - \beta x$. The quantities a_0 and b_0 are the initial masses of A and B, and α, β are two related positive constants. Find the time taken for the compound A to become half-exhausted, being given that this occurs before B is completely exhausted.

3. A chemical compound X is formed from two compounds Y and Z. The masses of the compounds at time t are respectively x, y, z, and initially $x = 0$, $y = a$, $z = b (a > b)$. The process is such that $x + y$ and $x + z$ are both constant, and $dx/dt = kyz$, where k is a positive constant. Show that

$$x = \frac{ab(1-e^{-u})}{a-be^{-u}},$$

where $u = k(a-b)t$.

4. In a chemical reaction between two substances X and Y, the respective concentrations are x and y at time t. It is known that the rate of change of $\ln x$ is equal to the excess of the concentration of Y over that of X, while the rate of change of $\ln y$ equals the sum of the concentrations of X and Y.
Show that

$$\frac{dy}{dx} = \frac{y(y+x)}{x(y-x)}.$$

If at some stage of the reaction both concentrations become equal to unity, prove that

$$xy = \exp\{(y-x)/x\}.$$

5. The velocity of a reversible chemical reaction is given by

$$\frac{dx}{dt} = k_1(3-x)(4-x) - k_2 x(1+x),$$

where k_1 and k_2 are positive constants. If $x = 0$ at $t = 0$, show, graphically or otherwise, that x increases steadily up to an equilibrium value which is between 0 and 3.
If $k_1 = 3$, $k_2 = 1$, solve the differential equation, and show that x reaches nine-tenths of its equilibrium value in a time $(3 \ln 2)/14 = 0.149$ approximately.

6. In a certain type of chemical reaction the amount x of a substance which has been transformed in time t is such that the rate of increase of x is proportional to the amount of substance that has not yet been transformed. If the original amount of the substance is 100 g and 60 g has been transformed after 2 min, find how much will be transformed after 6 min.

7. A vessel A contains a mixture of 70% of water and 30% of another liquid L. A second vessel B contains water only and the total amount of liquid in each vessel is the same. At time $t = 0$ the vessels are connected, allowing a flow of liquid from each vessel to the other, the total amount of liquid in each vessel remaining unaltered. The net flow of liquid L is from that vessel in which the concentration of L is greater, and the rate of flow is proportional to the difference between the concentrations in the two vessels. Prove that x, the percentage of liquid L present in A at time t, is given by a differential equation of the form

$$\frac{dx}{dt} = -k(x-15)$$

where k is a positive constant.

Integrate this equation, expressing t as a function of x. If $x = 25$ when $t = 5$, find to the nearest tenth of a unit, the value of t when $x = 18$.

8. The temperature θ of a cooling liquid is known to decrease at a rate proportional to $(\theta - \alpha)$, where α is the constant temperature of the surrounding medium. Show that $\theta - \alpha$ must be proportional to e^{-kt}, where t is the time and k is a positive constant.

If the constant temperature of the surrounding medium is $15°$ and the temperature of the liquid falls from $60°$ to $45°$ in 4 minutes, find

(i) the temperature after a further 4 minutes,

(ii) the time in which the temperature falls from $45°$ to $30°$.

9. The natural rate of growth of a colony of micro-organisms in a liquid is k_1 organisms per organism per minute. If the organisms are removed at the rate of k_2 organisms per minute, write down a differential equation to describe the variation in size of the colony.

Initially the colony contains N organisms. Find the number of organisms after t minutes and prove that the colony will be removed in a finite time provided that $N < k_2/k_1$.

10. At time t a population of organisms has x members. The rate of increase due to reproduction is $a + bt$ per cent and the rate of decrease due to mortality $a' + b't$ per cent, where a, a', b, b' are positive constants. Construct a differential equation connecting x with t, and solve it, given that initially $x = x_0$. If $a > a'$ and $b < b'$, show that x will at first increase but will ultimately tend to zero. Find also the maximum value of x.

11. A flask contains a growing bacterial culture and food which is being consumed by the culture. Each unit of food consumed results in an increase of three-quarters of a unit in the culture. At time t the flask contains x units of culture and y units of food, and the time-rate of increase of x is equal to xy; initially $x = 3$, $y = 100$. Construct a differential equation for dx/dt, and hence determine x as a function of the time.

12. A population of insects is placed in an experimental environment and allowed to grow for several days. Its net time-rate of increase is proportional to its size, x insects, at any time t days, and its increases during the fourth day and the fifth day are estimated (from counts of newly hatched and dead insects) as 3566 and 4143 insects respectively.

By constructing and solving a suitable differential equation, determine the initial size and initial time-rate of increase of the population.

13. The output, N articles per day, of a machine slows down in such a way that the rate of decrease of N is proportional to the product of N and the total time t that the machine has been in use. Express this statement as a differential equation and solve it for N in terms of t and any necessary constants.

Initially the output was 1000 articles per day but after 50 days it has dropped to 950 articles per day. Calculate how much longer the machine will be kept in use if it is to be discarded as soon as its output falls to 500 articles per day.

14. A radioactive substance disintegrates in accordance with the equation $dm/dt = -km$, where m is the mass remaining at time t and k is a constant. Prove that $m = m_0 e^{-kt}$, where m_0 is the initial mass.

A mixture of two radioactive substances is known to consist initially of 300 mg of one substance and 100 mg of the other. After one week the total mass of the mixture is found to be 200 mg and after one further week the total mass is 112 mg. It is required to deduce the values of the constants k_1 and k_2 which give the rates of disintegration of the two substances respectively in accordance with the above equation, the unit of time being one week. By using the substitutions $x_1 = e^{-k_1}$, $x_2 = e^{-k_2}$ show that the data allow of two different solutions. In order to select the correct solution a further measurement of the total mass is made three weeks after the start, and the total mass is then found to be 70·4 mg. Deduce the values of k_1 and k_2.

18:5 EQUATIONS OF HIGHER ORDERS

The general solution of a differential equation of the second order contains two arbitrary constants.

An equation of the form

$$\frac{d^n y}{dx^n} = f(x)$$

can be solved by n direct integrations, each integration introducing one arbitrary constant. The final solution is of the form $y = F(x)$ where $F(x)$ contains n arbitrary constants.

Equations of the form $\dfrac{d^2 y}{dx^2} = f(y)$ are transformed by the substitution $\dfrac{dy}{dx} = p$ to a first-order equation in p and y with variables separable.

Example 1. The equation of a simple harmonic motion is

$$\frac{d^2 y}{dt^2} = -\omega^2 y,$$

where ω is constant.

Put $\dfrac{dy}{dt} = v$. Then

$$\frac{d^2 y}{dt^2} = \frac{dv}{dt} = \frac{dv}{dy}\frac{dy}{dt} = v\frac{dv}{dy}$$

$$\Rightarrow v\frac{dv}{dy} = -\omega^2 y$$

$$\Leftrightarrow \int v\, dv = -\int \omega^2 y\, dy$$

$$\Leftrightarrow \frac{v^2}{2} = \frac{a^2 \omega^2}{2} - \frac{\omega^2 y^2}{2},$$

where a is constant and $a^2\omega^2/2$ is the constant of integration,

$$\Leftrightarrow v = \frac{dy}{dt} = \pm\omega\sqrt{(a^2 - y^2)}.$$

$$\Leftrightarrow \int \frac{1}{\sqrt{(a^2 - y^2)}}\, dy = \pm \int \omega\, dt$$

$$\Leftrightarrow \sin^{-1}\left(\frac{y}{a}\right) = \pm\omega t + b \quad \text{where } b \text{ is constant}$$

$$\Leftrightarrow y = a\sin(\pm\omega t + b)$$

$$\Leftrightarrow y = a\sin\omega t \cos b + a\cos\omega t \sin b$$

$$\text{or } y = a\cos\omega t \sin b - a\sin\omega t \cos b$$

and in either case since a and b are arbitrary constants we may write the solution

$$y = A\cos\omega t + B\sin\omega t,$$

where A and B are arbitrary.

———————————————

*Example 2. Solve the equation $\dfrac{d^2y}{dx^2} = e^{2y}$.

Put $\dfrac{dy}{dx} = p$.   Then

$$\frac{d^2y}{dx^2} = \frac{d\left(\dfrac{dy}{dx}\right)}{dy} \cdot \frac{dy}{dx} = p\frac{dp}{dy}.$$

The differential equation is transformed to

$$p\frac{dp}{dy} = e^{2y}$$

$$\Leftrightarrow \int p\, dp = \int e^{2y} dy$$

$$\Leftrightarrow \frac{p^2}{2} = \frac{e^{2y}}{2} + \frac{a^2}{2},$$

where a is an arbitrary constant,

$$\Leftrightarrow \frac{dy}{dx} = \pm \sqrt{(e^{2y} + a^2)}$$

$$\Leftrightarrow \int \frac{1}{\sqrt{(e^{2y} + a^2)}} dy = \pm \int dx.$$

Put $e^{2y} = t$ so that $\dfrac{dt}{dy} = 2e^{2y} = 2t$. Then

$$\int \frac{1}{2t\sqrt{(t + a^2)}} dt = \pm \int dx.$$

Put $(t + a^2) = u^2$ so that $dt/du = 2u$. Then

$$\int \frac{2u\,du}{2(u^2 - a^2)u} = \pm \int 1\, dx$$

$$\Leftrightarrow x = \pm \frac{1}{2a} \ln\left(\frac{u - a}{u + a}\right) + b,$$

where b is an arbitrary constant,

$$\Rightarrow x = \frac{1}{2a} \ln\left\{\frac{\sqrt{(e^{2y} + a^2)} - a}{\sqrt{(e^{2y} + a^2)} + a}\right\} + b.$$

The ambiguous $\pm$ sign is omitted from the final statement of the solution because the range of possible values of the arbitrary a includes negative values as well as the corresponding positive values and therefore the solution as stated includes all possible solutions.

Exercise 18.5

1. Solve the equation $\dfrac{d^2y}{dx^2} + 4y = 0$ given that $y = 1$ when $x = \pi/4$ and $y = 2$ when $x = \tfrac{1}{2}\pi$.

2. Solve the equation $\dfrac{d^2y}{dx^2} + 9y = 0$ given that when $x = 0$, $y = 2$, and $\dfrac{dy}{dx} = 0$.

3. Solve the equation $\dfrac{d^2y}{dx^2} + 64y = 0$ given that $y = 0$ and $\dfrac{dy}{dx} = 1$ when $x = \tfrac{1}{2}\pi$.

4. Solve $\dfrac{d^2y}{dx^2} + 9y = 0$ given that $y = \sqrt{2}$ when $x = \pi/12$ and $\dfrac{dy}{dx} = 1$ when $x = \pi/6$.

5. Solve the equation $\dfrac{d^2y}{dx^2} = n^2 y \quad (n \neq 0)$.

6. Solve the equation $\dfrac{d^2y}{dx^2} - y = 0$ given that when $x = 0$, $y = 4$ and $\dfrac{dy}{dx} = 2$.

7. Solve the equation $\dfrac{d^2y}{dx^2} + \sin x = 0$.

*8. Solve the equation $y^3 \dfrac{d^2y}{dx^2} = 1$ given that when $x = 0$, $y = 1$ and $\dfrac{dy}{dx} = 1$.

18:6 LINEAR EQUATIONS OF THE SECOND ORDER WITH CONSTANT COEFFICIENTS

We consider the equation

$$a\frac{d^2y}{dx^2} + b\frac{dy}{dx} + cy = P, \tag{18.15}$$

where a, b and c are constants and P is a function of x.

(i) Put $y = u + v$ where u and v are functions of x. Then

$$\frac{dy}{dx} = \frac{du}{dx} + \frac{dv}{dx}, \quad \frac{d^2y}{dx^2} = \frac{d^2u}{dx^2} + \frac{d^2v}{dx^2}.$$

The equation transforms to

$$\left(a\frac{d^2u}{dx^2} + b\frac{du}{dx} + cu\right) + \left(a\frac{d^2v}{dx^2} + b\frac{dv}{dx} + cv\right) = P$$

and it is therefore satisfied by the solution $y = u + v$ if

$$(1) \quad a\frac{d^2u}{dx^2} + b\frac{du}{dx} + cu = 0,$$

$$(2) \quad a\frac{d^2v}{dx^2} + b\frac{dv}{dx} + cv = P.$$

Thus, if $y = u$ is a solution of the original equation with the right hand side replaced by zero and $y = v$ is a solution of the original equation, then

$$y = u + v$$

is a solution of the original equation.

If, further, $y = u$ is the *general* solution of the equation

$$a\frac{d^2y}{dx^2} + b\frac{dy}{dx} + cy = 0 \tag{18.16}$$

and therefore contains two arbitrary constants, and $y = v$ is a particular solution of

$$a\frac{d^2 y}{dx^2} + b\frac{dy}{dx} + cy = P,$$

then $y = u + v$ is the general solution of the equation

$$a\frac{d^2 y}{dx^2} + b\frac{dy}{dx} + cy = P.$$

Of these two parts of the solution, u is called the *complementary function* (C.F.) and v is called the *particular integral* (P.I.).

(ii) If $y = u_1$ and $y = u_2$ are particular solutions of

$$a\frac{d^2 y}{dx^2} + b\frac{dy}{dx} + cy = 0,$$

then $y = Au_1 + Bu_2$ is the general solution, for

(1) it contains two arbitrary constants,

(2) $a\dfrac{d^2}{dx^2}(Au_1 + Bu_2) + b\dfrac{d}{dx}(Au_1 + Bu_2) + c(Au_1 + Bu_2)$

$$= A\left(a\frac{d^2 u_1}{dx^2} + b\frac{du_1}{dx} + cu_1\right) + B\left(a\frac{d^2 u_2}{dx^2} + b\frac{du_2}{dx} + cu_2\right) = 0.$$

We require, therefore, to find two particular solutions u_1 and u_2 of the equation

$$a\frac{d^2 y}{dx^2} + b\frac{dy}{dx} + cy = 0$$

in order to write down the complementary function as $Au_1 + Bu_2$.

18:7 THE COMPLEMENTARY FUNCTION

In the equation

$$a\frac{d^2 y}{dx^2} + b\frac{dy}{dx} + cy = 0, \tag{18.16}$$

we try as a solution $y = e^{mx}$, where m is a constant to be determined, so that $\dfrac{dy}{dx} = me^{mx}$

and $\dfrac{d^2 y}{dx^2} = m^2 e^{mx}$. Then

$$am^2 e^{mx} + bm e^{mx} + c e^{mx} = 0$$
$$\Leftrightarrow e^{mx}(am^2 + bm + c) = 0.$$

Hence $y = e^{m_1 x}$ and $y = e^{m_2 x}$, where m_1, m_2 are the roots of

$$am^2 + bm + c = 0,$$

are solutions of

$$a\frac{d^2 y}{dx^2} + b\frac{dy}{dx} + cy = 0.$$

(The equation in m is called the *auxiliary equation*.) The complementary function is therefore

$$Ae^{m_1 x} + Be^{m_2 x} \text{ provided that } m_1 \neq m_2. \tag{18.17}$$

Special cases arise when $m_1 = m_2$, and when m_1 and m_2 are complex. These two cases are considered separately below.

(i) If $m_1 = m_2$, then

$$\frac{b}{a} = -2m_1; \quad \frac{c}{a} = m_1^2,$$

and the differential equation (18.16) can be written

$$\frac{d^2 y}{dx^2} - 2m_1\frac{dy}{dx} + m_1^2 y = 0.$$

Put $y = ze^{m_1 x}$ where z is a function of x. Then

$$\frac{dy}{dx} = e^{m_1 x}\frac{dz}{dx} + m_1 ze^{m_1 x}, \quad \frac{d^2 y}{dx^2} = e^{m_1 x}\frac{d^2 z}{dx^2} + 2m_1 e^{m_1 x}\frac{dz}{dx} + m_1^2 e^{m_1 x} z$$

and the equation transforms to

$$e^{m_1 x}\left(\frac{d^2 z}{dx^2} + 2m_1\frac{dz}{dx} + m_1^2 z - 2m_1\frac{dz}{dx} - 2m_1^2 z + m_1^2 z\right) = 0,$$

$$\Leftrightarrow e^{m_1 x}\frac{d^2 z}{dx^2} = 0 \quad \Leftrightarrow \frac{d^2 z}{dx^2} = 0$$

$$\Leftrightarrow z = Ax + B, \text{ where } A \text{ and } B \text{ are constants.}$$

Hence in this case the C.F. is

$$e^{m_1 x} z = e^{m_1 x}(Ax + B). \tag{18.18}$$

(ii) If m_1 and m_2 are complex, then, assuming that a, b, c are all real, we can write $m_1 = \alpha + i\beta$, $m_2 = \alpha - i\beta$. The C.F. is

$$Ae^{\alpha x + i\beta x} + Be^{\alpha x - i\beta x} = e^{\alpha x}(Ae^{i\beta x} + Be^{-i\beta x}).$$

But $Ae^{i\beta x} = A(\cos\beta x + i\sin\beta x)$ and $Be^{-i\beta x} = B(\cos\beta x - i\sin\beta x)$.

Hence the C.F. is $\qquad e^{\alpha x}(C\cos\beta x + D\sin\beta x) \tag{18.19}$

where C and D are constants.

This result may also be obtained by the method used in (i), i.e. with the substitution $y = e^{\alpha x} z$, reducing the equation to the form

$$\frac{d^2 z}{dx^2} + \beta^2 z = 0.$$

Example 1. *Roots of the auxiliary equation are real and different.*

$$2\frac{d^2y}{dx^2} - 5\frac{dy}{dx} + 2y = 0.$$

The auxiliary equation is

$$2m^2 - 5m + 2 = 0$$
$$\Leftrightarrow (2m - 1)(m - 2) = 0$$
$$\Leftrightarrow m = \tfrac{1}{2} \text{ or } 2.$$

Hence the general solution is $y = Ae^{x/2} + Be^{2x}$.

Example 2. *Roots of the auxiliary equation are equal.*

$$\frac{d^2y}{dx^2} + 4\frac{dy}{dx} + 4y = 0.$$

The auxiliary equation is

$$m^2 + 4m + 4 = 0$$
$$\Leftrightarrow m = -2 \text{ (a repeated root).}$$

Hence the general solution is $y = (Ax + B)e^{-2x}$.

Example 3. *Roots of the auxiliary equation are complex.*

$$\frac{d^2y}{dx^2} - 2\frac{dy}{dx} + 2y = 0.$$

The auxiliary equation is

$$m^2 - 2m + 2 = 0$$
$$\Leftrightarrow m = 1 \pm i.$$

Hence the general solution is $y = e^x(A\cos x + B\sin x)$.

Example 4. *The linear equation is of higher order than the second.*

The methods used below are justified by similar reasoning to that used above for equations of the second order.

$$\frac{d^3y}{dx^3} - y = 0.$$

The auxiliary equation is

$$m^3 - 1 = 0.$$

The solutions of this equation are $m = 1, \omega, \omega^2$, i.e. $1, \tfrac{1}{2}(-1 \pm i\sqrt{3})$. The general solution of the differential equation is therefore

$$y = Ae^x + e^{-x/2}\left\{B\cos\left(\frac{\sqrt{3}}{2}x\right) + C\sin\left(\frac{\sqrt{3}}{2}x\right)\right\}.$$

Example 5. $\dfrac{d^3y}{dx^3} + 2\dfrac{d^2y}{dx^2} + \dfrac{dy}{dx} = 0.$

The auxiliary equation is

$$m^3 + 2m^2 + m = 0.$$

The solution of this equation is $m = 0, -1, -1$. The general solution of the differential equation is therefore

$$y = A + e^{-x}(Bx + C).$$

Summary. If $f(m) = 0$ is the auxiliary equation of the linear differential equation, then

(i) for every real root m_r which is not a repeated root, the C.F. contains a term $A_r e^{m_r x}$;

(ii) for every double root m_p the C.F. contains a term $e^{m_p x}(A_p x + B_p)$; [if there are n roots each equal to m_s the C.F. contains a term $e^{m_s x} g(x)$ where $g(x)$ is a polynomial involving n arbitrary constants and of degree $n-1$ in x];

(iii) for every pair of conjugate complex roots $\alpha \pm i\beta$ the C.F. contains a term $e^{\alpha x}(A \cos \beta x + B \sin \beta x)$.

Exercise 18.7

Obtain the general solutions of the following equations:

1. $2\dfrac{d^2 y}{dx^2} - 5\dfrac{dy}{dx} - 3y = 0$ 2. $4\dfrac{d^2 y}{dx^2} - \dfrac{dy}{dx} - 5y = 0$.

3. $\dfrac{d^2 y}{dx^2} - 9y = 0$.   4. $\dfrac{d^2 y}{dx^2} + 5\dfrac{dy}{dx} = 0$.

5. $\dfrac{d^2 y}{dx^2} - 4\dfrac{dy}{dx} + 4y = 0$.   6. $2\dfrac{d^2 y}{dx^2} + 2\dfrac{dy}{dx} + y = 0$.

7. $\dfrac{d^2 y}{dx^2} + 3\dfrac{dy}{dx} + 4y = 0$.   8. $\dfrac{d^2 y}{dx^2} + 5\dfrac{dy}{dx} + y = 0$.

9. $4\dfrac{d^2 y}{dx^2} + 12\dfrac{dy}{dx} + 9y = 0$.   10. $\dfrac{d^3 y}{dx^3} + \dfrac{d^2 y}{dx^2} = 0$.

18:8 THE PARTICULAR INTEGRAL

We have to find any particular solution of the equation

$$a\frac{d^2 y}{dx^2} + b\frac{dy}{dx} + cy = P \tag{18.15}$$

where P is a function of x. In special cases we can state intuitively the nature of the function which is a particular integral, and the intuitive statement can then be confirmed by substitution into the equation.

(i) If P is a polynomial function of x of degree n, there is a solution of the form $y = f(x)$ where $f(x)$ is a polynomial of degree n.

(ii) If P is a function of the form $g \cos nx + h \sin nx$, there is a solution of the form $y = k \cos nx + l \sin nx$.

(iii) If P is of the form pe^{nx} there is a solution of the form $y = qe^{nx}$.

(iv) If P is a function of x which is the sum of some or all of the functions enumerated in (i) to (iii), there is a solution which is also the sum of such functions.

(v) If the C.F. already contains terms of the types suggested for the particular integral, it is possible to obtain particular integrals in case (i) of the form $y = kx^{p+1}$, where the term of highest degree in the C.F. involves x^p, and in case (ii) of the form $kx \cos nx + lx \sin nx$, and in case (iii) of the form $qx\, e^{nx}$ unless there is a similar term in the C.F. in which case the P.I. is of the form $qx^2\, e^{nx}$, etc.

Example 1. $\dfrac{d^2 y}{dx^2} - 4\dfrac{dy}{dx} + 3y = x^3.$

The auxiliary equation is $m^2 - 4m + 3 = 0$ with roots $m = 3, 1$. Hence the C.F. is

$$Ae^{3x} + Be^x.$$

Try $y = ax^3 + bx^2 + cx + d$ as the P.I. in the differential equation.

Then $\dfrac{dy}{dx} = 3ax^2 + 2bx + c, \qquad \dfrac{d^2 y}{dx^2} = 6ax + 2b$

$$\Rightarrow 6ax + 2b - 12ax^2 - 8bx - 4c + 3ax^3 + 3bx^2 + 3cx + 3d \equiv x^3$$

$$\Leftrightarrow 3ax^3 - (12a - 3b)x^2 + (6a - 8b + 3c)x + (2b - 4c + 3d) \equiv x^3$$

Equating coefficients of powers of x we find

$$a = \tfrac{1}{3},$$
$$12a - 3b = 0 \Rightarrow b = \tfrac{4}{3},$$
$$6a - 8b + 3c = 0 \Rightarrow c = \tfrac{26}{9},$$
$$2b - 4c + 3d = 0 \Rightarrow d = \tfrac{80}{27}.$$

Hence the P.I. is

$$\frac{1}{3}x^3 + \frac{4}{3}x^2 + \frac{26x}{9} + \frac{80}{27}$$

and the general solution of the equation is

$$y = Ae^{3x} + Be^x + \frac{1}{3}x^3 + \frac{4}{3}x^2 + \frac{26x}{9} + \frac{80}{27}.$$

Example 2. $\dfrac{d^2 y}{dx^2} + 2\dfrac{dy}{dx} + 5y = \sin 2x.$

The auxiliary equation is $m^2 + 2m + 5 = 0$ with roots $m = -1 \pm 2i$. Hence the C.F. is

$$e^{-x}(A \cos 2x + B \sin 2x).$$

Try $y = a \sin 2x + b \cos 2x$ as the P.I. in the differential equation. Then

$$\frac{dy}{dx} = 2a \cos 2x - 2b \sin 2x, \qquad \frac{d^2 y}{dx^2} = -4a \sin 2x - 4b \cos 2x$$

$$\Rightarrow -4a \sin 2x - 4b \cos 2x + 4a \cos 2x - 4b \sin 2x + 5a \sin 2x + 5b \cos 2x \equiv \sin 2x$$

$$\Rightarrow (a - 4b) \sin 2x + (4a + b) \cos 2x \equiv \sin 2x.$$

Equating coefficients of $\sin 2x$ and $\cos 2x$,

$$a - 4b = 1,$$
$$4a + b = 0$$
$$\Rightarrow a = \tfrac{1}{17}, \quad b = -\tfrac{4}{17}.$$

Hence a particular integral is $\tfrac{1}{17} \sin 2x - \tfrac{4}{17} \cos 2x$, and the general solution is

$$y = e^{-x}(A \cos 2x + B \sin 2x) + \tfrac{1}{17} \sin 2x - \tfrac{4}{17} \cos 2x.$$

Example 3. $\dfrac{d^2 y}{dx^2} + 4\dfrac{dy}{dx} + 4y = e^x.$

The auxiliary equation is $m^2 + 4m + 4 = 0$ which has $m = -2$ as a repeated root. Hence the C.F. is

$$e^{-2x}(Ax + B).$$

Try $y = ae^x$ as the P.I. in the differential equation. Then

$$\frac{dy}{dx} = ae^x, \quad \frac{d^2 y}{dx^2} = ae^x$$

$$\Rightarrow ae^x + 4ae^x + 4ae^x \equiv e^x$$

$$\Leftrightarrow a = 1/9.$$

Hence a particular integral is $y = e^x/9$ and the general solution is

$$y = e^{-2x}(Ax + B) + e^x/9.$$

Example 4. $\dfrac{d^2 y}{dx^2} - 3\dfrac{dy}{dx} + 2y = 2e^x.$

The auxiliary equation is $m^2 - 3m + 2 = 0$ with roots $m = 2, 1$. Hence the C.F. is

$$Ae^{2x} + Be^x.$$

If the substitution $y = ae^x$ is made, $\dfrac{d^2 y}{dx^2} - 3\dfrac{dy}{dx} + 2y$ is identically zero. In this case we try $y = axe^x$ as the P.I. Then

$$\frac{dy}{dx} = ae^x + axe^x, \quad \frac{d^2 y}{dx^2} = 2ae^x + axe^x$$

$$\Rightarrow 2ae^x + axe^x - 3ae^x - 3axe^x + 2axe^x \equiv 2e^x$$

$$\Rightarrow -ae^x \equiv 2e^x$$

$$\Leftrightarrow a = -2.$$

Hence a particular integral is $-2xe^x$, and the general solution is

$$y = Ae^{2x} + Be^x - 2xe^x.$$

If Axe^x had also been a term of the C.F., the trial substitution to make in order to find the P.I. would have been $y = ax^2 e^x$.

Example 5. $\dfrac{d^2 y}{dx^2} + n^2 y = \sin nx.$

The C.F. is $A \sin nx + B \cos nx$.
Because $A \sin nx$ is a term of the C.F. we try $y = x(a \sin nx + b \cos nx)$ as the P.I. Then

$$\frac{dy}{dx} = (a \sin nx + b \cos nx) + x(an \cos nx - bn \sin nx),$$

$$\frac{d^2 y}{dx^2} = 2n(a \cos nx - b \sin nx) + x(-an^2 \sin nx - bn^2 \cos nx)$$

$$\Rightarrow 2an \cos nx - 2bn \sin nx - an^2 x \sin nx - bn^2 x \cos nx + an^2 x \sin nx + bn^2 x \cos nx \equiv \sin nx$$

$$\Leftrightarrow 2an \cos nx - 2bn \sin nx \equiv \sin nx$$

$$\Rightarrow a = 0, \quad b = -\frac{1}{2n}.$$

Hence a particular integral is $-\dfrac{x}{2n}\cos nx$ and the general solution is

$$y = A \sin nx + B \cos nx - \frac{x}{2n}\cos nx.$$

It must be emphasized that the analysis of methods for finding the P.I. of a linear equation of the second order with constant coefficients which we have just made is concerned only with special cases. In the equation

$$a\frac{d^2 y}{dx^2} + b\frac{dy}{dx} + cy = P,$$

P is any function of x. Those functions which we have considered here are the ones occurring most frequently.

Exercise 18.8

Obtain the general solution of each of the differential equations in questions 1–20.

1. $\dfrac{d^2 y}{dx^2} + 5\dfrac{dy}{dx} + 6y = 4.$

2. $\dfrac{d^2 y}{dx^2} + 9y = 1.$

3. $\dfrac{d^2 y}{dx^2} - \dfrac{dy}{dx} - 2y = \sin x.$

4. $\dfrac{d^2 y}{dx^2} - 2\dfrac{dy}{dx} + y = 2e^x.$

5. $\dfrac{d^2 y}{dx^2} + 3\dfrac{dy}{dx} + 2y = e^{-2x}.$

6. $3\dfrac{d^2 y}{dx^2} + 2\dfrac{dy}{dx} + y = \cos 2x.$

7. $\dfrac{d^2 y}{dx^2} - 3\dfrac{dy}{dx} + 3y = 2.$

8. $\dfrac{d^2 y}{dx^2} - 4\dfrac{dy}{dx} + 5y = x^2 + x + 1.$

9. $\dfrac{d^2 y}{dx^2} - \dfrac{dy}{dx} - 2y = 1 + x.$

10. $\dfrac{d^2 y}{dx^2} - 4\dfrac{dy}{dx} + 4y = x + e^x.$

11. $\dfrac{d^2 y}{dx^2} + 4\dfrac{dy}{dx} + 5y = \sin 2x.$

12. $\dfrac{d^2 y}{dx^2} - 5\dfrac{dy}{dx} = 4.$

13. $\dfrac{d^2 y}{dx^2} + 2\dfrac{dy}{dx} + 2y = \sin 2x + \cos x.$

14. $\dfrac{d^3 y}{dx^3} + \dfrac{d^2 y}{dx^2} = 1 + e^{-x}.$

15. $\dfrac{d^2 y}{dx^2} - 6\dfrac{dy}{dx} + 9y = e^{3x}.$

16. $\dfrac{d^2 y}{dx^2} - 3\dfrac{dy}{dx} + 2y = \sin 2x.$

17. $p^2\dfrac{d^2 y}{dx^2} - p\dfrac{dy}{dx} = a(1 + e^x), \quad p \neq 0 \text{ or } 1.$

18. $\dfrac{d^4 y}{dx^4} - 16y = x.$

19. $\dfrac{d^2 x}{dt^2} - 2\dfrac{dx}{dt} - 3x = \cos(2t + \pi/4).$

20. $\dfrac{d^2 x}{dt^2} - 2\dfrac{dx}{dt} + 2x = \cos(2t + \pi/3).$

21. The equation of motion for the small oscillations of a pendulum swinging in a resisting medium is

$$\frac{d^2\theta}{dt^2} + k\frac{d\theta}{dt} + n^2\theta = 0$$

where θ is the angular displacement of the pendulum from its mean position and k and n are constants. Show that if $k^2 > 4n$ and if the pendulum starts from rest at $\theta = \alpha$, then

$$\theta = \frac{\alpha}{m_2 - m_1}(m_2 e^{m_1 t} - m_1 e^{m_2 t})$$

where m_1 and m_2 are the roots of $m^2 + km + n^2 = 0$.

22. Solve the equation

$$\frac{d^2 y}{dt^2} + 3\frac{dy}{dt} + 2y = 10\cos t$$

given that $y = 1$ and $\dfrac{dy}{dt} = 0$ when $t = 0$.

23. Solve the equation

$$\frac{d^2 x}{dt^2} - 4x = 5e^{3t}$$

given that $x = -2$ and $\dfrac{dx}{dt} = -3$ when $t = 0$. Find the value of t for which $\dfrac{dx}{dt} = 0$.   (O.C.)

24. In the usual notation for a particle moving in a straight line, the equation of motion of a particle is

$$\frac{d^2 s}{dt^2} + 5\frac{ds}{dt} + 6s = ae^{-t}$$

where $a > 0$. If $s = 0$ and $v = 0$ when $t = 0$, prove that s is positive for all values of t and if $a = 100$ calculate each of s and v, correct to three significant figures, when $t = 1$.

25. If the equation of motion for a particle moving in a straight line is

$$\frac{d^2 s}{dt^2} + 2\frac{ds}{dt} + 5s = 10a\sin t,$$

prove that for large values of t, and whatever the initial conditions,

$$s \approx \sqrt{5}a\sin(t - \alpha) \text{ where } \tan\alpha = \tfrac{1}{2}.$$

26. If the equation of motion for a particle moving in a straight line is

$$\frac{d^2 s}{dt^2} + 3\frac{ds}{dt} + 2s = 16e^{-3t},$$

and $s = 4$, $v = -15$ when $t = 0$, calculate the minimum value of s.

27. If $\dfrac{d^2 y}{dx^2} + n^2 y = kx$, when n and k are real constants, and $y = 0$ when $x = 0$ and when $x = a$, find the value of y when $x = a/2$.   (L.)

28. Solve the equation

$$\frac{d^2 y}{dt^2} + 6\frac{dy}{dt} + 25y = 6\sin t,$$

given that when $t = 0$, $y = 0$ and $dy/dt = 0$.   (L.)

29. Solve the equation

$$\frac{d^2 y}{d\theta^2} + 9y = 8\sin\theta$$

given that $y = dy/d\theta = 0$ when $\theta = \pi/2$.   (L.)

30. A particle moves on a fixed straight line through a fixed point O and its equation of motion is

$$\frac{d^2x}{dt^2} + 2k\frac{dx}{dt} + n^2x = \cos nt,$$

x being its displacement from O at time t, $(n > k)$. Find x in terms of t, given that $x = \dfrac{1}{2k^2}$

and $\dfrac{dx}{dt} = 0$ when $t = 0$. Find the speed at time t if $n = 2k$.   (L.)

18:9 THE MEAN-VALUE THEOREM AND LINEAR APPROXIMATIONS

If a function $f(x)$ has a derivative at all points within the interval $a < x < b$, its graph is a smooth curve starting at height $f(a)$ and finishing at height $f(b)$. It seems clear from a figure that the tangent to the curve is parallel to the chord joining the points $[a, f(a)]$, $[b, f(b)]$ for at least one point inside the interval, (see Fig. 18.1.) In fact

$$f'(\xi_1) = f'(\xi_2) = \frac{f(b) - f(a)}{b - a},   \tag{18.20}$$

where $a < \xi_1 < \xi_2 < b$. This is the *mean value theorem*.

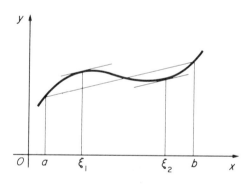

FIG. 18.1.

If $f(a) = f(b) = 0$ this result means that there is at least one value $x = \xi (a < \xi < b)$ for which the derivative vanishes; or we may say that at least one zero of $f'(x)$ separates two zeros of $f(x)$. In this form the theorem is known as *Rolle's theorem*. This result is one of such importance in calculus that a proof which simply appeals to a figure is not strictly sufficient. Usually in books on analysis a proof of Rolle's theorem is given first and the mean value theorem is then deduced. An alternative form of equation (18.20) is obtained by writing $b = a + h$ to give

$$f(a + h) = f(a) + hf'(a + \theta h),   \tag{18.21}$$

where $0 < \theta < 1$. Reference to Fig. 18.1 suggests that, although the actual value of θ is not known, this value depends not only on the function $f(x)$ but also upon the values of a and b (or h).

Many functions commonly occurring in physics and engineering arise from differential equations; when numerical values are required in a particular problem, the value of some function $y = f(x)$ and one or more of its derivatives may be known for a single value of x only. To estimate values of the function at neighbouring points we can use equation (18.21). Suppose that $f''(x)$ exists for points of the curve near $x = a$, then

$$f(a + h) = f(a) + hf'(a + \theta_1 h) = f(a) + h[f'(a) + \theta_1 hf''(a + \theta_2 h)],$$

where $0 < \theta_2 < \theta_1 < 1$,

$$\Rightarrow f(a + h) = f(a) + hf'(a) + O(h^2). \qquad (18.22)$$

Provided h is small we may neglect the term $O(h^2)$ in equation (18.22) and obtain an estimate for $f(a + h)$. This is equivalent to replacing the curve by its tangent at P, see Fig. 18.2.

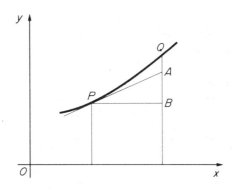

FIG. 18.2.

Note that equation (18.22) is a particular case of Taylor's series, p. 495, and can, in fact, be used to prove Taylor's theorem formally.

In many cases the value of a function, $f(x)$, may be known at a number of neighbouring points whereas $f'(x)$ is unknown, e.g. the values $f(x)$ at P and Q in Fig. 18.2 may be known. To estimate the values of $f(x)$ for points lying between P and Q we replace the curve by chord PQ and estimate $f(x)$ from points on this line. (Figure 18.2 suggests that, in general, the errors will be of the same order as that made by using A instead of Q.) The more closely the chord approaches the curve the more accurate the estimates will be, but, if the points P, Q are widely separated or the gradient of the curve is altering rapidly in the neighbourhood, the estimates are unlikely to be accurate. The commonest use of this method, which is a form of *linear interpolation*, occurs with tabulated functions such as logarithms, sines, cosines, etc. In the tables the values of a function are given at convenient intervals over a range of the independent variable. For example, five-figure sine tables do not list the value of $\sin 35° 16' 20''$ but an estimate of this can be obtained by adding one-third of the difference between $\sin 35° 16'$ ($0·577\,38$) and $\sin 35° 17'$ ($0·577\,61$) to the former, giving $0·577\,46$. It should be noted that the accuracy of a result derived by linear interpolation cannot exceed the accuracy of the tables from which it is derived. Figures 18.1 and 18.2 also indicate that the errors made in estimates using linear interpolation lie on the concave side of the curve, whereas those made using the mean value theorem lie on the convex side of the curve. If it is possible to use both methods, then a very good approximation may be found by averaging. (It is assumed that the points P and Q of Fig. 18.2 are close together.)

18:10 STEP-BY-STEP METHODS FOR APPROXIMATE SOLUTION OF FIRST-ORDER DIFFERENTIAL EQUATIONS

Most first-order differential equations can be expressed in the form

$$\frac{dy}{dx} = f(x, y). \tag{18.23}$$

We have seen how to obtain exact solutions of such an equation when $f(x, y)$ takes one of certain particular forms; but in many cases no exact solution of the equation is possible. An approximation to the solution curve, often called the *integral curve*, of such an equation through a given point $P_0(x_0, y_0)$ can be obtained by a "step-by-step" method, giving an open polygon $P_0 P_1 P_2 \ldots$, of which the vertices approximate to the points on the solution curve with the same x-coordinates, Fig. 18.3. The simplest of such methods is as follows.

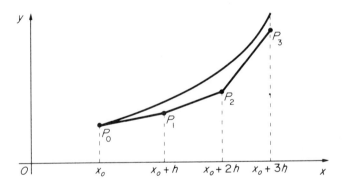

FIG. 18.3.

The differential equation gives the gradient $f(x_0, y_0)$ of the solution curve at P_0, and we take this to be the gradient of $P_0 P_1$. Hence the coordinates (x_1, y_1) of P_1 are given by

$$x_1 = x_0 + h, \quad y_1 = y_0 + hf(x_0, y_0),$$

where h is the chosen step-length. Similarly, $P_2(x_2, y_2)$ is given by

$$x_2 = x_1 + h, \quad y_2 = y_1 + hf(x_1, y_1),$$

and $P_{r+1}(x_{r+1}, y_{r+1})$ by

$$x_{r+1} = x_r + h, \quad y_{r+1} = y_r + hf(x_r, y_r).$$

The process clearly fails if there is a point P_r at which $f(x_r, y_r)$ is undefined.

Example. Obtain an approximation to the solution curve for

$$\frac{dy}{dx} = x^2 - y^2$$

through $(0, 0)$ using steps of 0.2.

x_{r+1}	0·2	0·4	0·6	0·8	1·0
$x_r^2 - y_r^2$	0	0·04	0·1599	0·3584	0·6275
$h(x_r^2 - y_r^2)$	0	0·008	0·0320	0·0717	0·1255
$y_{r+1} = y_r + h(x_r^2 - y_r^2)$	0	0·0080	0·0400	0·1117	0·2372

A more accurate step-by-step method

An improvement on the above method relies on the use of the mean value theorem [stating that, for a continuous curve through $P_r(x_r, y_r)$ and $P_{r+1}(x_{r+1}, y_{r+1})$, the gradient at some point $x = \xi$ where $x_r < \xi < x_{r+1}$ is equal to the gradient of $P_r P_{r+1}$].

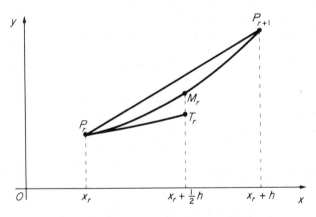

FIG. 18.4.

We assume that $\xi = x_r + \tfrac{1}{2}h$, so that in Fig. 18.4 the gradient of $P_r P_{r+1}$ would equal the gradient of the curve at M_r. However, we cannot find this from the differential equation since we do not know the y-coordinate of M_r; we therefore use the coordinates of T_r, the point on the tangent at P_r for which $x = x_r + \tfrac{1}{2}h$, to find the gradient of the solution curve through T_r. To summarise, we are now making two assumptions:

  (i) the gradient of the solution curve through P_r, at the point M_r where $x = x_r + \tfrac{1}{2}h$, is equal to the gradient of the chord $P_r P_{r+1}$;
 (ii) the gradient at T_r of the solution curve through T_r is equal to the gradient at M_r of the solution curve through M_r. The point T_r has coordinates

$[x_r + \frac{1}{2}h, y_r + \frac{1}{2}hf(x_r, y_r)]$; thus the gradient of the solution curve through T_r is $f(T_r)$ and this is taken to be the gradient of $P_r P_{r+1}$, thus enabling us to make the step from P_r to P_{r+1}.

Example. We will apply this improved method to the previous example.

$$\frac{dy}{dx} = x^2 - y^2 = f(x, y).$$

	x_{r+1}	0·2	0·4	0·6	0·8	1·0
T_r	$x_r + \frac{1}{2}h$	0·1	0·3	0·5	0·7	0·9
	$y_r + \frac{1}{2}hf(x_r, y_r)$	0	0·0060	0·0360	0·1052	0·2268
	$f(T_r)$	0·01	0·0900	0·2487	0·4789	0·7586
$y_r + hf(T_r) = y_{r+1}$		0·002	0·0110	0·0697	0·1655	0·3172

Exercise 18.10

In questions 1–3 use a step-by-step method and compare your results with the exact solution.

1. If $\dfrac{dy}{dx} + x = xy^2$, find y when $x = 1$, given that $y = 2$ when $x = 0$.

2. If $(1 + x)\dfrac{dy}{dx} - xy = x$, find y when $x = 0·6$, given that $y = 0$ when $x = 0$.

3. If $(1 + x^2)\dfrac{dy}{dx} = x(1 - y^2)$, find y when $x = 2$, given that $y = 0$ when $x = 1$.

4. If $\dfrac{dy}{dx} = \sqrt{(x + y)}$, and $y = 0·25$ when $x = 0$, find approximate values of y for the range 0·0 (0·1) 0·3 of x.

5. If $\dfrac{dy}{dx} = x + y$, and $y = -1$ when $x = 0$, find approximate values of y for the range 0 (0·25) 1 of x.

6. Use linear interpolation to find an estimate for the value of ln 1·714 given that, to five decimal places,

$$\ln 1·71 = 0·53649, \ln 1·72 = 0·54232.$$

7. Justify graphically the approximation

$$y_{n+1} - y_n \approx h\left(\frac{dy}{dx}\right)_{x = x_n},$$

where y_n and y_{n+1} are the ordinates at $x = x_n$ and $x = x_n + h$ respectively.
Use this step-by-step method, with $h = 0·5$, to find approximations to the values of y at $x = 0·5$ and $x = 1·0$ given that

$$\frac{dy}{dx} = y^2 \text{ and } y(0) = 0·2.$$

Obtain the exact solution of this differential equation and find the relative errors in your approximate values. Explain why the relative errors in the values found by the step-by-step method will increase as x increases.   (L.)

8. By means of a sketch, or otherwise, indicate the reason for the validity of the approximation

$$f'(x_0) \approx \frac{f(x_0 + h) - f(x_0 - h)}{2h}$$

for small h.

If $f(x) = y$, then y satisfies the differential equation

$$x\frac{dy}{dx} = 2x - y + 1,$$

with $f(1) = 0$ and $f(1 \cdot 2) = 0 \cdot 5333$.

Use the above approximation (with a step length $h = 0 \cdot 2$) to determine $f(1 \cdot 8)$, working to three decimal places.

[Tabulate enough of your calculations to make your method clear.]   (N.)

9. The following method is used to find approximately the integral curve through a given point of the differential equation

$$\frac{dy}{dx} = f(x, y).$$

If $x = nh$, $y = y_n$ is a point on this curve and if $f_n = f(nh, y_n)$, then at $x = (n+1)h$ the value y_{n+1} of y is given approximately by

$$y_{n+1} = y_n + hf\{(n+\tfrac{1}{2})h, y_n + \tfrac{1}{2}hf_n\}.$$

Use this method with $h = 0 \cdot 1$ to find the approximate values of y at $x = 0 \cdot 1$ and $x = 0 \cdot 2$ on the integral curve through the origin of the differential equation

$$\frac{dy}{dx} = 1 + x - y^2.$$

Work to three decimal places.   (L.)

18:11 VECTOR DIFFERENTIAL EQUATIONS

We suppose that the components of a vector $\mathbf{a}$ are functions of a single scalar variable, t say, so that

$$\mathbf{a} = a_1(t)\mathbf{i} + a_2(t)\mathbf{j} + a_3(t)\mathbf{k}.$$

We *define*

$$\frac{d\mathbf{a}}{dt} = a_1'(t)\mathbf{i} + a_2'(t)\mathbf{j} + a_3'(t)\mathbf{k}.\qquad(18.24)$$

Note that this definition is essentially dependent on $\mathbf{i}, \mathbf{j}, \mathbf{k}$ being (unit) vectors in *fixed* directions.

The following results are easily obtained from applications of the definition of equation (18.24) and the product rule for differentiation.

(1) If λ, $\mathbf{a}$ both depend on t and $\mathbf{b} = \lambda\mathbf{a}$, then

$$\frac{d\mathbf{b}}{dt} = \frac{d\lambda}{dt}\mathbf{a} + \lambda\frac{d\mathbf{a}}{dt}.$$

(2) If $\mathbf{a}, \mathbf{b}$ depend upon t and $u = \mathbf{a}.\mathbf{b}$, then

$$\frac{du}{dt} = \frac{d\mathbf{a}}{dt}.\mathbf{b} + \mathbf{a}.\frac{d\mathbf{b}}{dt}.$$

(3) If $\mathbf{c} = \mathbf{a} \times \mathbf{b}$, then

$$\frac{d\mathbf{c}}{dt} = \frac{d\mathbf{a}}{dt} \times \mathbf{b} + \mathbf{a} \times \frac{d\mathbf{b}}{dt}.$$

If $\mathbf{b} = \dfrac{d\mathbf{a}}{dt}$ then we *define* $\mathbf{a} = \int \mathbf{b}\, dt$. Clearly $\mathbf{a}$ will be indefinite to the extent of a constant (arbitrary) vector.

The solution of a vector differential equation can be obtained by resolution with components parallel to *fixed* cartesian axes, as illustrated in the following examples.

Example 1. Solve the differential equation $\dfrac{d\mathbf{v}}{dt} = -k\mathbf{v}$, where k is constant, given that $\mathbf{v} = \mathbf{v}_0$ when $t = 0$.

The cartesian resolutes of this equation are (taking $\mathbf{v} = v_1\mathbf{i} + v_2\mathbf{j} + v_3\mathbf{k}$)

$$\frac{dv_1}{dt} = -kv_1, \quad \frac{dv_2}{dt} = -kv_2, \quad \frac{dv_3}{dt} = -kv_3,$$

with solutions

$$v_1 = A_1 e^{-kt}, \quad v_2 = A_2 e^{-kt}, \quad v_3 = A_3 e^{-kt},$$

where A_1, A_2, A_3 are arbitrary constants. Therefore

$$\mathbf{v} = \mathbf{A} e^{-kt},$$

where $\mathbf{A}$ is an *arbitrary vector*. The initial condition gives $\mathbf{A} = \mathbf{v}_0$, so the required solution is

$$\mathbf{v} = \mathbf{v}_0 e^{-kt}.$$

Example 2. The general solution of the differential equation

$$\ddot{\mathbf{r}} = -n^2 \mathbf{r}$$

where n is constant, is, by solution of the component equations

$$\ddot{x} = -n^2 x \text{ etc,}$$

$$\mathbf{r} = \mathbf{A} \cos nt + \mathbf{B} \sin nt,$$

where $\mathbf{A}, \mathbf{B}$ are constant vectors.

Example 3. The position vector $\mathbf{r}$ of a particle moving in a plane is given at any instant by

$$\frac{d\mathbf{r}}{dt} = \frac{k\mathbf{r}}{(\mathbf{a}.\mathbf{r})}, \tag{1}$$

where $\mathbf{a}$ and k are constant, $\mathbf{a} \neq \mathbf{0}$ and $k > 0$. Form the scalar product of each side of this equation with $\mathbf{a}$ and solve the resulting differential equation for $\mathbf{a}.\mathbf{r}$. Deduce that, if $\mathbf{r} = \mathbf{a}$ when $t = 0$, then

$$\frac{d\mathbf{r}}{dt} = \frac{k\mathbf{r}}{\mathbf{a}^2 + kt}. \tag{2}$$

Hence find $\mathbf{r}$ at time t.   (L.)

Taking scalar product of the given differential equation with $\mathbf{a}$, we obtain

$$\mathbf{a}.\frac{d\mathbf{r}}{dt} = \frac{k\mathbf{a}.\mathbf{r}}{\mathbf{a}.\mathbf{r}} = k.$$

But, since **a** is constant,

$$\frac{d}{dt}(\mathbf{a} . \mathbf{r}) = \mathbf{a} . \frac{d\mathbf{r}}{dt}$$

$$\Rightarrow \frac{d}{dt}(\mathbf{a} . \mathbf{r}) = k$$

$$\Rightarrow \mathbf{a} . \mathbf{r} = kt + C.$$

$$\mathbf{r} = \mathbf{a} \quad \text{when} \quad t = 0 \Rightarrow C = \mathbf{a}^2.$$

Therefore equation (1) can be written in the form (2).
The resolutes of (2), e.g.

$$\frac{dx}{dt} = \frac{kx}{\mathbf{a}^2 + kt}, \text{ etc.,}$$

integrate (on separation) to give

$$x = A_1(\mathbf{a}^2 + kt), \text{ etc.}$$
$$\Rightarrow \mathbf{r} = \mathbf{A}(\mathbf{a}^2 + kt).$$

The initial condition gives $\mathbf{A} = \mathbf{a}/\mathbf{a}^2$.

$$\Rightarrow \mathbf{r} = \mathbf{a}(\mathbf{a}^2 + kt)/\mathbf{a}^2.$$

Example 4. By resolution into components and use of the results of § 18:7 we obtain the following:

(i) If $\ddot{\mathbf{r}} - (\alpha + \beta)\dot{\mathbf{r}} + \alpha\beta\mathbf{r} = 0$, where α, β are real and distinct, then

$$\mathbf{r} = \mathbf{a}e^{\alpha t} + \mathbf{b}e^{\beta t}$$

(ii) If $\ddot{\mathbf{r}} + 2\alpha\dot{\mathbf{r}} + \alpha^2\mathbf{r} = 0$, where α is a real constant, then

$$\mathbf{r} = (\mathbf{a} + \mathbf{b}t)e^{-\alpha t}.$$

(iii) If $\ddot{\mathbf{r}} + 2k\dot{\mathbf{r}} + (k^2 + \omega^2)\mathbf{r} = 0$, where k, ω are real constants, then

$$\mathbf{r} = e^{-kt}(\mathbf{a} \cos \omega t + \mathbf{b} \sin \omega t).$$

In each case **a**, **b** are arbitrary vectors.

(iv) The solution of the equation

$$\ddot{\mathbf{r}} + 2k\dot{\mathbf{r}} + (k^2 + \omega^2)\mathbf{r} = \mathbf{h}f(t),$$

where **h** is a constant vector and $f(t)$ a prescribed function of t, is the sum of the complementary function $\mathbf{r}_c$, and the particular integral $\mathbf{r}_p$, where $\mathbf{r}_c$ is given by (iii) above, and

$$\mathbf{r}_p = \mathbf{h}\phi(t).$$

The scalar function $\phi(t)$ is the particular integral of the equation

$$\ddot{\phi} + 2k\dot{\phi} + (k^2 + \omega^2)\phi = f(t).$$

When $f(t)$ is known, this equation may be solved by the methods of § 18:8.

Exercise 18.11

1. Solve the vector differential equation

$$\frac{d^2\mathbf{r}}{dt^2} - 5\frac{d\mathbf{r}}{dt} + 4\mathbf{r} = 0,$$

given that $\mathbf{r} = -3\mathbf{i}$, $d\mathbf{r}/dt = 6\mathbf{j}$ when $t = 0$.

(L.)

2. Solve the vector differential equation

$$\frac{d^2\mathbf{r}}{dt^2} + \lambda \frac{d\mathbf{r}}{dt} = \mathbf{a},$$

where $\mathbf{a}$ is a constant vector and λ is a non-zero constant scalar. (L.)

3. (i) Find the velocity $\mathbf{v}$ and acceleration $\mathbf{f}$ of a particle which moves so that at time t its position vector is

$$\mathbf{r} = \mathbf{i} \sin \omega t + \mathbf{j} \cos \omega t + \mathbf{k} t,$$

where ω is constant.
Find also the times at which

$$(\mathbf{r} \times \mathbf{f}) . \mathbf{j} = 0.$$

(ii) Show that a vector function $\mathbf{v}(t)$ of the variable t will have a constant magnitude if and only if $\mathbf{v} . \dfrac{d\mathbf{v}}{dt} = 0$. (L.)

4. Solve the equations
   (i) $\ddot{\mathbf{r}} + 5\dot{\mathbf{r}} + 6\mathbf{r} = \mathbf{0}$, given that $\mathbf{r} = \mathbf{i}$, $\dot{\mathbf{r}} = \mathbf{j}$, at $t = 0$,
   (ii) $\ddot{\mathbf{r}} + k\dot{\mathbf{r}} = \mathbf{a}$, where $\mathbf{a}$ is a constant vector, k is constant and $\mathbf{r} = \mathbf{0}$, $\dot{\mathbf{r}} = \mathbf{V}$ at $t = 0$,
   (iii) $\ddot{\mathbf{r}} + 2n\dot{\mathbf{r}} + n^2\mathbf{r} = \mathbf{0}$, where n is constant, given that $\mathbf{r} = \mathbf{0}$, $\dot{\mathbf{r}} = \mathbf{V}$ at $t = 0$,
   (iv) $\ddot{\mathbf{r}} + \omega^2\mathbf{r} = \mathbf{a} \cos pt$, where $\mathbf{a}$ is a constant vector and ω, p are positive constants.

5. (i) The position vector $\mathbf{r}$ of a point P at time t satisfies the equation

$$\frac{d\mathbf{r}}{dt} = \boldsymbol{\omega} \times \mathbf{r},$$

where $\boldsymbol{\omega}$ is a constant non-zero vector. Show that $|\mathbf{r}|$ is constant and that P lies in a fixed plane.
Deduce that P describes a circle with constant speed.
(ii) A particle Q moves so that its position vector $\mathbf{r}$ satisfies the equation

$$\frac{d^2\mathbf{r}}{dt^2} = -n^2\mathbf{r},$$

where $n \neq 0$. Show that Q describes a closed plane curve in time $2\pi/n$. (L.)

Miscellaneous Exercise 18

1. (i) Solve the differential equation

$$dy/dx + y = x,$$

given that $y = 1$ when $x = 0$.
(ii) A body is placed in a room which is kept at a constant temperature. The temperature of the body falls at a rate $k\theta°$C per minute, where k is a constant and θ is the difference between the temperature of the body and that of the room at time t. Express this information in the form of a differential equation and hence show that $\theta = \theta_0 e^{-kt}$, where θ_0 is the temperature difference at time $t = 0$.
   The temperature of the body falls $5°$C in the first minute and $4°$C in the second minute. Show that the fall of temperature in the third minute is $3\cdot2°$C. (L.)

2. Integrate the equations

$$\text{(i)} \quad \frac{dy}{dx} = \frac{x+y-1}{2x},$$

$$\text{(ii)} \quad \frac{d^2y}{dx^2} - \frac{dy}{dx} - 2y = \sin x. \qquad \text{(N.)}$$

3. (a) Solve the differential equation

$$\frac{dy}{dx} - x \operatorname{cosec}^2 y = 3 \operatorname{cosec}^2 y.$$

(b) If

$$x\frac{dy}{dx} + (x-1)y = x^2 e^{2x},$$

find the value of y in terms of x, given that $y = 1$ when $x = 3$.   (N.)

4. Gas is allowed to expand in a cylinder, maintained at a constant absolute temperature T, and does work on a piston sliding in the cylinder. For any small increase in the volume of the gas from V to $V + \delta V$, the work done by the gas is approximately $P\delta V$, where P is the pressure of the gas at volume V. If P, V and T satisfy the equation

$$PV = 85\,000\,T,$$

show that the work W done by the gas in expanding to volume V satisfies the differential equation

$$V\frac{dW}{dV} = 85\,000\,T.$$

If $T = 300$ and P changes from 2000 to 1030, find the percentage increase in the volume of the gas, and calculate the work done by the gas during the change.   (N).

5. The rate of decay at any instant of a radioactive substance is proportional to the amount of substance remaining at that instant. If the initial amount of the substance is A and the amount remaining after time t is x, prove that

$$x = Ae^{-kt}$$

where k is constant.

If the amount remaining is reduced from $\frac{1}{2}A$ to $\frac{1}{3}A$ in 8 hours, prove that the initial amount of the substance was halved in about 13·7 hours.   (N.)

6. Integrate the equation $\dfrac{d^2 y}{dx^2} + 2\dfrac{dy}{dx} + 5y = x^2$.   (N.)

7. (a) If $x(x+1)\dfrac{dy}{dx} = y(y+1)$ and $y = 2$ when $x = 1$, find y when $x = 2$.

(b) Solve completely the differential equation

$$\frac{d^2 x}{dt^2} + 5\frac{dx}{dt} + 6x = 2e^{-t}.$$

If $x = 0$ and $\dfrac{dx}{dt} = 1$ when $t = 0$, show that the maximum value of x is $\frac{1}{4}$.   (N.)

8. Integrate the equation

$$\frac{d^2 y}{dx^2} - 3\frac{dy}{dx} + 2y = 1 + \cosh x.$$ (N.)

9. During a fermentation process the rate of decomposition of substance at any instant is related to the amounts of yg of substance and xg of active ferment by the law

$$\frac{dy}{dt} = -0.25\,xy.$$

The value of x time t is $4/(1+t)^2$; when $t = 0$, $y = 10$. Express $\dfrac{dy}{dt}$ in terms of y and t, and hence determine y as a function of t.

Prove that the amount of substance ultimately remaining is approximately 3·7 g.   (N.)

10. (a) Solve the differential equation

$$x\frac{dy}{dx} + 3y = \sqrt{(1+3x^3)},$$

given that $y = 1$ when $x = 1$.

(b) If $e^{-x^2}\dfrac{dy}{dx} = x(y+2)^2$, find the value of y in terms of x, given that $y = 0$ when $x = 0$.   (N.)

11. A vessel containing V cm^3 of liquid is in the form of a cone of semi-vertical angle $30°$ with vertex downwards and axis vertical. At time $t = 0$ a small plug is removed from the bottom of the vessel and liquid flows out at a rate $k\sqrt{y}$ cm^3 per minute, k being a constant and y cm the depth of the liquid in the vessel at time t minutes. Find t as a function of y.

When $t = t_1$ the vessel contains a volume $\frac{1}{2}V$ cm^3 of liquid; when $t = t_2$, it is empty. Calculate t_1/t_2.

12. Solve the differential equation

$$\frac{d^2 y}{dx^2} + 2k\frac{dy}{dx} + (1+k^2)y = \cos x, \quad 0 < k < 1.$$ (N.)

13. Integrate the equations

$$\text{(i)} \quad \frac{d^3 y}{dx^3} - y = x^4 + e^{2x},$$

$$\text{(ii)} \quad (1+x^2)\frac{dy}{dx} + xy = 2x.$$ (N.)

14. (i) Find the general solution of the equation

$$(x^2 - 1)\frac{dy}{dx} + 2y = x + 1.$$

(ii) A point starts from rest at the origin and its abscissa at time t satisfies the equation

$$\frac{d^2 x}{dt^2} + 4\frac{dx}{dt} + 3x = 6ae^{-2t}$$

where $a > 0$. Determine the greatest value of x.   (N.)

15. The horizontal cross-section of a tank has a constant area A m^2. Water is poured into the tank at the rate of K m^{3}s^{-1}. At the same time water flows out through a hole in the bottom at the rate of $H\sqrt{y}$ m^{3}s^{-1}, where y m is the depth of the water. Show that at time t seconds

$$A\frac{dy}{dt} = K - H\sqrt{y}.$$

By putting $y = x^2$ find the general solution of this equation when H, K are constants. If $y = 0$ at $t = 0$, deduce that

$$\frac{H^2 t}{2AK} = \ln\left(\frac{K}{K - H\sqrt{y}}\right) - \frac{H\sqrt{y}}{K}.$$

If $A = 1000$, $H = 100$, $K = 600$, show that when $y = 25$ the value of t is 115 approximately. Take $\ln 6$ as $1\cdot7918$.

16. A particle moving along a straight line OX is at a distance x from O at time t, and its speed is given by

$$8t^2\frac{dx}{dt} = (1 - t^2)x^2.$$

If $x = 4$ when $t = 1$, prove that $x = 8t/(1 + t^2)$.
Find (i) the speed when $t = 0$,
       (ii) the maximum distance from O,
       (iii) the maximum speed *towards* O for positive values of t.

17. At time t after a battery has been switched on, the current x in an electric circuit satisfies the differential equation

$$L\frac{dx}{dt} + Rx = E,$$

where L, R and E are positive constants. Show that the substitution $x = y + (E/R)$ reduces this equation to

$$L\frac{dy}{dt} + Ry = 0.$$

Solve the latter equation and deduce x as a function of t, given that $x = 0$ when $t = 0$.
The total charge q that has crossed any section of the circuit in time t is given by

$$q = \int_0^t x\,dt.$$

By integrating the original differential equation with respect to t, or otherwise, show that

$$q = \frac{E}{R}t - \frac{L}{R}x. \qquad \text{(N.)}$$

18. (i) Find the general solution of the equation

$$\frac{dy}{dx} + y\cot x = x\cot x.$$

(ii) The abscissa x of a point moving along the x-axis satisfies the equation

$$\frac{d^2x}{dt^2} + 2n\frac{dx}{dt} + 2n^2x = n^2a(\cos nt - 2\sin nt).$$

Initially $x = 0$ and $\dfrac{dx}{dt} = na$. Show that when the point first returns to the origin, its speed is $na(1 - e^{-\pi/2})$.
$$\text{(N.)}$$

19. (a) Solve the equation

$$(1 + x)^2\left(x\frac{dy}{dx} - 2y\right) = x^3.$$

If $y = 1$ when $x = 1$, find y when $x = 2$.
(b) The abscissa of a point P moving along the x-axis satisfies the equation

$$4\frac{d^4x}{dt^4} + \frac{d^2x}{dt^2} = 9\sin t + 12\cos t.$$

When $t = 0$, $x = 9$, $\dfrac{dx}{dt} = 3$, $\dfrac{d^2x}{dt^2} = -4$, $\dfrac{d^3x}{dt^3} = -3$. Prove that P oscillates between the origin and the point
$x = 10$. $\qquad\qquad\qquad\qquad\qquad\qquad\qquad\qquad\qquad\qquad\qquad\qquad\qquad$ (N.)

20. (a) If $(x^2 - 1)\dfrac{dy}{dx} + 2y = 0$, find the value of y in terms of x, given that $y = 3$ when $x = 2$.
(b) Solve the differential equation

$$x\frac{dy}{dx} + 2y = \frac{2\sin x}{x\cos^3 x}. \qquad \text{(N.)}$$

21. In a certain chemical reaction the rate of decomposition of a substance at any time t is proportional to the amount m that remains. In any small change, the percentage increase in the pressure p inside the vessel in which the reaction takes place is proportional to the percentage decrease in the amount m of the substance.
Express the above statements in calculus notation using a and b as the respective constants of proportionality.
If m and p have the values m_0, p_0 when $t = 0$, find

(i) m in terms of m_0, a and t,

(ii) p in terms of p_0, m_0, b and m,
and hence obtain p in terms of p_0, a, b and t. $\qquad\qquad\qquad\qquad\qquad\qquad$ (N.)

22. Integrate the equations

(i) $x(1+x)\dfrac{dy}{dx}+y^2(1-y)=0,$

(ii) $\dfrac{d^2y}{dx^2}+2\dfrac{dy}{dx}+5y=3-2e^{-x}+\cos 2x.$ (N.)

23. (a) If $y=Ax\sin x+Bx\cos x$, where A and B are constants, obtain a linear differential equation independent of A and B which is satisfied by y.

(b) Solve the differential equations

(i) $\dfrac{d^2y}{dx^2}+3\dfrac{dy}{dx}+2y=x,$

(ii) $(x-x^2)\dfrac{dy}{dx}+y=x^2+2x$

giving the solution of (ii) for which $y=0$ when $x=-1$. (N.)

24. (a) Solve the equation

$$\frac{dy}{dx}+\frac{y(x+2y)}{x(y+2x)}=0.$$

(b) The coordinates of a point moving in the $x-y$ plane satisfy the equations

$$\frac{dx}{dt}-\omega y=\omega a,\qquad \frac{dy}{dt}+4\omega x=4\omega a.$$

Given that $x=2a$ and $y=a$ when $t=0$, show that the path is an ellipse of area $4\pi a^2$. (N.)

25. (i) Find the general solution of the differential equation

$$\frac{dy}{dx}=2x(1-y),\qquad (y<1).$$

(ii) Solve the differential equation

$$\frac{d^2y}{dx^2}+\frac{dy}{dx}-6y=e^{2x}.$$ (L.)

26. (i) Solve the differential equation

$$\frac{d^2x}{dt^2}+6\frac{dx}{dt}+9x=4e^t,$$

given that $x=1$ and $\dfrac{dx}{dt}=-\dfrac{3}{2}$ when $t=0$.

(ii) Find the general solution of the differential equation

$$x\frac{dy}{dx}-3y=x^4\ln x.$$ (L.)

27. (a) In the differential equation

$$(x+y)\frac{dy}{dx}=x^2+xy+x+1$$

change the dependent variable from y to z, where $z=x+y$. Deduce the general solution of the given equation.

(b) The normal at the point $P(x,y)$ on a curve meets the x-axis at Q, and N is the foot of the ordinate of P. If

$$NQ=\frac{x(1+y^2)}{1+x^2},$$

find the equation of the curve, given that it passes through the point $(3,1)$. (N.)

28. (i) Find the function y of x satisfying the equation

$$(x+1)\frac{dy}{dx} - y(y+1)(x+2) = 0$$

and such that $y = \frac{1}{2}$ when $x = 0$.

(ii) By means of the substitution $x = e^t$ find the function y of x satisfying the equation

$$x^2\frac{d^2y}{dx^2} + x\frac{dy}{dx} + y = 0$$

and such that $y = 1$ and $\frac{dy}{dx} = 0$ when $x = 1$.   (L.)

29. The normal to a plane curve at a variable point P meets the axes of x and y at Q and R respectively. The orthogonal projection of PQ on the x-axis is equal to that of PR on the y-axis. Find the equation of the curve, given that it passes through $(1, 1)$ but not through $(0, 0)$.   (L.)

30. Solve the differential equations

(i) $x(x-y)dy/dx + x^2 + y^2 = 0$,

(ii) $dy/dx + y \tan x = \sec x$,

(iii) $d^2y/dx^2 + 2dy/dx + 5y = x + 1$.   (L.)

31. (i) Solve the differential equation

$$\frac{d^2y}{dx^2} - 4y = 0.$$

given that $y = 3$ and $dy/dx = 2$ when $x = 0$.

(ii) At time t the speed v of a particle satisfies the differential equation $\frac{dv}{dt} + 2v = t$. Given that $v = 0$ when $t = 0$, solve this differential equation and hence find, at time $t = 2$, the distance of the particle from its position at $t = 0$.   (L.)

32. (i) Find the general solution of the differential equation

$$\frac{dx}{dt} + 2x = e^{-2t}\left(t^3 + \frac{1}{t}\right) \qquad (t > 0).$$

(ii) If

$$\frac{d^2y}{dx^2} + 9y = 0,$$

find the general solution for y in terms of x. If $y = 2$ when $x = 0$ and $y = -4$ when $x = \frac{1}{2}\pi$, find the value of dy/dx when $x = \pi$.   (L.)

33. Solve the differential equations

(i) $\dfrac{dy}{dx} = \dfrac{2x+3y}{3y-2x}$;

(ii) $\dfrac{dy}{dx} + y = xy^3$;

(iii) $\dfrac{dy}{dx} + 2y = 3e^{-x}$, given that $y = 0$ when $x = 0$.   (L.)

34. Solve the differential equations

(i) $x\,dy/dx + 2y = \cos x$;

(ii) $d^2y/dt^2 + 10\,dy/dt + 29y = 0$, given that $y = 0$ and $dy/dt = 4$ when $t = 0$.   (L.)

35. (i) Find y in terms of x given that

$$x\frac{dy}{dx} + 3y = 3, \quad \text{and} \quad y = 3 \quad \text{when } x = 1.$$

(ii) Solve

$$\frac{d^2y}{dx^2} + 4\frac{dy}{dx} + 3y = \sin x + e^x + 3.$$ (L.)

36. Solve the differential equations

(i) $x \sin y \frac{dy}{dx} - \cos y = 1$, given that $y = 0$ when $x = 1$;

(ii) $\frac{d^2y}{dx^2} + 4\frac{dy}{dx} + 8y = \cosh x.$ (L.)

37. (i) Obtain the solution of the equation

$$(1 - x^2)\frac{dy}{dx} = 1 + \cos 2y$$

for which $y = \frac{1}{4}\pi$ when $x = 0$.

(ii) By substituting $t = 3x - y - 3$, or otherwise, solve the equation

$$\frac{dy}{dx} = \frac{3x - y + 3}{3x - y - 1}.$$ (L.)

38. (i) Find y in terms of x if

$$\sin x \frac{dy}{dx} - 2y \cos x = 3 \sin x.$$

(ii) Solve

$$\frac{dy}{dx} = \frac{x + 3y}{3x + y}$$

given $y = 0$ when $x = 1$. (L.)

39. (i) Solve the differential equation

$$x^2\frac{dy}{dx} - xy = 1,$$

given that $y = 2$ when $x = 1$.

(ii) Solve the differential equation

$$\frac{d^2x}{dt^2} + 2\frac{dx}{dt} + 10x = 5 + e^t.$$ (L.)

40. (i) By use of the substitution $z = 1/y^2$, find the general solution of the differential equation

$$\frac{dy}{dx} + \frac{y}{x} = x^2y^3.$$

(ii) Obtain the general solution of the differential equation

$$\frac{d^2y}{dx^2} + 2\frac{dy}{dx} + (1 + k^2)y = 0,$$

where k is a positive constant. Show that, if $y = 0$ at $x = 0$ and $dy/dx + 2y = 0$ at $x = 1$, and y is not identically zero, then k must be a root of the equation $\sin k + k \cos k = 0$. (L.)

41. If $\frac{d^2y}{dx^2} + 4\frac{dy}{dx} + 3y = 0$ and $y = 0$, $\frac{dy}{dx} = 2$ at $x = 0$, find the maximum value of y. (L.)

42. Solve the equation

$$\frac{d^2y}{dx^2} - 7\frac{dy}{dx} + 6y = 36x,$$

given that $y = 0$ and $\frac{dy}{dx} = 4$ when $x = 0$.

Show that, for this solution, $\dfrac{d^2y}{dx^2}$ is zero for $x = \frac{1}{5}\ln\frac{2}{9}$. (L.)

43. Solve the equation $\dfrac{d^2y}{dx^2} + 4y = 12\cos 4x$, given that $y = 1$ when $x = \pi/4$ and $y = -3$ when $x = \pi/2$.
Hence, or otherwise, show that y is a maximum when $x = n\pi \pm \pi/6$, where n is an integer. (L.)

44. Solve the equations

(i) $x\dfrac{dy}{dx} + \dfrac{y}{x-1} = x(x-1),$

(ii) $x(x-1)\cos y\,\dfrac{dy}{dx} + \sin y = x(x-1)^2.$ (L.)

45. Solve the equation

$$x\frac{dy}{dx} = y + \sqrt{(x^2 - y^2)}, \text{ by writing } y = vx, \text{ or otherwise.}$$ (L.)

46. A heavy particle B is attached to one end of a light inextensible string AB of length c. The string rests on a smooth horizontal plane, with A at the origin and B at the point $(0, c)$, the axes of coordinates being taken to lie in the plane. The end A of the string is now moved along the x-axis, and the particle moves in such a way that the string remains taut and AB always touches at B the curve traced by B. Show that if (x, y) are the coordinates of B,

$$(c^2 - y^2)^{1/2}\,(dy/dx) + y = 0.$$

Deduce that the equation of the curve is

$$x + (c^2 - y^2)^{1/2} - c\ln\left\{\frac{c + (c^2 - y^2)^{1/2}}{y}\right\} = 0.$$ (N.)

47. A body of mass m moves in a straight line under a constant propelling force and a resistance mkv, where v is its speed and k is a constant. The body starts from rest at time $t = 0$ and the speed tends to a limiting value V as t increases. Write down differential equations connecting (i) v with t, (ii) v with the distance covered. Solve these equations.
Show that the average speed during the interval from the start to the instant when $v = \frac{1}{2}V$ is

$$V\left(1 - \frac{1}{2\ln 2}\right).$$

48. In a certain chemical reaction the amount x of one substance at time t is related to the velocity of the reaction dx/dt by the equation

$$\frac{dx}{dt} = k(a-x)(2a-x)$$

where a and k are constants, and $x = 0$ when $t = 0$. If $x = 2.0$ when $t = 1$ and $x = 2.8$ when $t = 3$, show that $a = 3$ and find (i) the value of k, (ii) the value of x when $t = 2$. (N.)

49. At time t a population of organisms has x members. The rate of increase due to reproduction is $a + bt$ per cent and the rate of decrease due to mortality $a' + b't$ per cent, where a, a', b, b' are positive constants. Construct a differential equation connecting x with t, and solve it, given that initially $x = x_0$. If $a > a'$ and $b < b'$, show that x will at first increase but will ultimately tend to zero. Find also the maximum value of x. (N.)

50. (a) The normal at any point of a curve passes through the point $(1, 1)$. Express this condition in the form of a differential equation, and hence find the equation of the family of curves which satisfy the condition.
(b) Given that y satisfies the differential equation

$$\frac{dy}{dx} + 2y\tan x = \sin x,$$

and that $y = 1$ when $x = \pi/3$, express dy/dx in terms of x. (N.)

51. (a) The curve $y = f(x)$ passes through the point $(3, 1)$, and its gradient at the point (x, y) is given by the differential equation

$$\frac{dy}{dx} = \tfrac{1}{2}\left(1 - \frac{y}{x}\right).$$

Find $f(x)$.

(b) By substituting $y = z^{-1/2}$ or otherwise, solve the differential equation

$$\frac{dy}{dx} = y - xe^{-2x}y^3.$$

(N.)

52. A particle moves so that its position vector at time t satisfies the differential equation

$$\frac{d^2\mathbf{r}}{dt^2} = \mathbf{E} + \frac{d\mathbf{r}}{dt} \times \mathbf{B},$$

where $\mathbf{E}$, $\mathbf{B}$ are constant vectors. At time $t = 0$, $\mathbf{r} = \mathbf{0}$ and $d\mathbf{r}/dt = \mathbf{V}$. Show that

(a) $$\frac{d\mathbf{r}}{dt} = \mathbf{E}t + \mathbf{r} \times \mathbf{B} + \mathbf{V},$$

(b) $$\left(\frac{d\mathbf{r}}{dt}\right)^2 = \mathbf{V}^2 + 2\mathbf{E}.\mathbf{r}.$$

If, further, $\mathbf{E}$ is parallel to $\mathbf{V}$ but perpendicular to $\mathbf{B}$, show that

$$\frac{d\mathbf{r}}{dt}.\mathbf{B} = 0.$$

(L.)

19 Definite Integrals: Further Applications of Integration

19:1 PROPERTIES OF DEFINITE INTEGRALS

In most cases $\int_a^b f(x)\,dx$ is found by obtaining the indefinite integral,

$$\int f(x)\,dx = \phi(x) + C \text{ say,}$$

and then using the definition

$$\int_a^b f(x)\,dx = \phi(b) - \phi(a).$$

There are cases, however, in which, because the indefinite integral is difficult to obtain or even unobtainable in terms of functions we have so far defined, it is desirable, if possible, to find the value of the definite integral without obtaining the indefinite integral. The properties of definite integrals in the examples which follow are means to this end.

Example 1. $\displaystyle\int_a^b f(x)\,dx = -\int_b^a f(x)\,dx.$ (19.1)

For, if $\displaystyle\int f(x)\,dx = \phi(x) + C$, then

$$\int_b^a f(x)\,dx = \phi(a) - \phi(b)$$

and

$$\int_a^b f(x)\,dx = \phi(b) - \phi(a).$$

Example 2. $\displaystyle\int_a^b f(x)\,dx = \int_a^c f(x)\,dx + \int_c^b f(x)\,dx.$ (19.2)

623

For
$$\int_a^b f(x)\,dx = \phi(b) - \phi(a)$$

and
$$\int_a^c f(x)\,dx + \int_c^b f(x)\,dx = \phi(c) - \phi(a) + \phi(b) - \phi(c) = \phi(b) - \phi(a).$$

Example 3. $\displaystyle \int_a^b f(x)\,dx = \int_a^b f(y)\,dy.$ (19.3)

For each integral is equal to $\phi(b) - \phi(a)$. This self-evident property of definite integrals is used frequently in the work which follows.

Example 4. $\displaystyle \int_0^a f(x)\,dx = -\int_0^{-a} f(-x)\,dx.$ (19.4)

For, if $\displaystyle I = \int_0^a f(x)\,dx$, then putting $x = -y$, so that $dx/dy = -1$, $x = 0$ when $y = 0$, and $x = a$ when $y = -a$,

$$I = \int_0^{-a} f(-y)(-dy) = -\int_0^{-a} f(-y)\,dy = -\int_0^{-a} f(-x)\,dx.$$

Example 5. $\displaystyle \int_{-a}^a f(x)\,dx = \int_0^a [f(x) + f(-x)]\,dx.$ (19.5)

For
$$\int_{-a}^a f(x)\,dx = \int_{-a}^0 f(x)\,dx + \int_0^a f(x)\,dx$$

$$= -\int_a^0 f(-x)\,dx + \int_0^a f(x)\,dx$$

$$= \int_0^a f(-x)\,dx + \int_0^a f(x)\,dx$$

$$= \int_0^a [f(x) + f(-x)]\,dx.$$

Corollaries of equation (19.5) are

$$\int_{-a}^a f(x)\,dx = 2\int_0^a f(x)\,dx \quad \text{when } f(x) \text{ is an even function of } x,$$

$$\int_{-a}^{a} f(x)\,dx = 0 \text{ when } f(x) \text{ is an odd function of } x.$$

Example 6. $\quad \displaystyle\int_{0}^{a} f(x)\,dx = \int_{0}^{a} f(a-x)\,dx.$ $\hfill$ (19.6)

For, if $I = \displaystyle\int_{0}^{a} f(x)\,dx$, then putting $x = a - y$ so that $\dfrac{dx}{dy} = -1$, $y = 0$ when $x = a$, and $y = a$ when $x = 0$,

$$I = \int_{a}^{0} f(a-y)(-dy) = -\int_{0}^{a} f(a-y)(-dy) = \int_{0}^{a} f(a-y)\,dy = \int_{0}^{a} f(a-x)\,dx.$$

The relation between definite integrals and area, as defined in Chapter 3, is frequently helpful in the consideration of problems concerning definite integrals. Equation (19.2), for example, becomes self-evident when examined in the light of this relationship. It must be remembered that area above the x-axis as defined by the definite integral is positive and area below the x-axis is negative.

Example 1. Prove that $\quad \displaystyle\int_{ra}^{(r+1)a} f(x)\,dx = \int_{0}^{a} f(x+ra)\,dx.$

If $f(x + a) = cf(x)$ where a and c are constants, prove that, if $n \in \mathbb{N}$ and $c \neq 1$, then

$$(1-c)\int_{0}^{na} f(x)\,dx = (1-c^{n})\int_{0}^{a} f(x)\,dx.$$

State the corresponding result when $c = 1$.

If $I = \displaystyle\int_{ra}^{(r+1)a} f(x)\,dx$, then putting $x = y + ra$, so that $dx/dy = 1$, $x = ra$ when $y = 0$, and $x = ra + a$ when $y = a$,

$$\Rightarrow I = \int_{0}^{a} f(y+ra)\,dy = \int_{0}^{a} f(x+ra)\,dx.$$

$$f(x + a) = cf(x)$$
$$\Rightarrow f(x+ra) = cf[x+(r-1)a] = c^2 f[x+(r-2)a]$$
$$= \ldots = c^r f(x).$$

(This result could be formally proved by induction.)

$$\int_{0}^{na} f(x)\,dx = \int_{0}^{a} f(x)\,dx + \int_{a}^{2a} f(x)\,dx + \ldots + \int_{ra}^{(r+1)a} f(x)\,dx + \ldots + \int_{(n-1)a}^{na} f(x)\,dx$$

$$= \int_0^a f(x)dx + \int_0^a f(x+a)dx + \ldots + \int_0^a f[x+ra]dx + \ldots$$

$$+ \int_0^a f[x+(n-1)a]dx$$

$$= \int_0^a f(x)dx + \int_0^a cf(x)dx + \ldots + \int_0^a c^r f(x)dx + \ldots + \int_0^a c^{n-1} f(x)dx$$

$$= (1+c+c^2+\ldots c^r + \ldots c^{n-1}) \int_0^a f(x)dx$$

$$\Rightarrow \int_0^{na} f(x)dx = \frac{1-c^n}{1-c} \int_0^a f(x)dx \quad \text{if } c \neq 1,$$

$$\int_0^{na} f(x)dx = n \int_0^a f(x)dx \quad \text{if } c = 1.$$

Example 2. $I = \int_{-\pi/2}^{\pi/2} (2+x)^2 \sin 2x \, dx$

[using equation (19.5)]

$$= \int_0^{\pi/2} \{(2+x)^2 \sin 2x - (2-x)^2 \sin 2x\} \, dx$$

$$= \int_0^{\pi/2} 8x \sin 2x \, dx.$$

This also follows by use of the corollaries of Example 5, p. 624. Then integration by parts

$$\Rightarrow I = 8 \left[\frac{-x \cos 2x}{2} + \int \frac{\cos 2x}{2} dx \right]_0^{\pi/2}$$

$$= 8 \left[\frac{-x \cos 2x}{2} + \frac{\sin 2x}{4} \right]_0^{\pi/2} = 2\pi.$$

Example 3. Without evaluating the integral, show that

$$\int_0^a (ax-x^2)^3 \{x^3 + (x-a)^3\} \, dx = 0.$$

Use of equation (19.6) gives

$$\int_0^a x^3(a-x)^3 \{x^3 + (x-a)^3\} \, dx = \int_0^a (a-x)^3 x^3 \{(a-x)^3 + (-x)^3\} \, dx$$

$$= -\int_0^a x^3(a-x)^3\{(x-a)^3+x^3\}\,dx$$

$$\Rightarrow \int_0^a x^3(a-x)^3\{x^3+(x-a)^3\}\,dx = 0.$$

Example 4. Evaluate $I = \int_0^1 \dfrac{1}{x+\sqrt{(1-x^2)}}\,dx.$

Putting $x = \sin\theta$ gives

$$I = \int_0^{\pi/2} \frac{\cos\theta\,d\theta}{\sin\theta+\cos\theta}. \tag{1}$$

Putting $x = \cos\theta$ in the original integral gives

$$I = \int_{\pi/2}^0 \frac{-\sin\theta\,d\theta}{\sin\theta+\cos\theta} = \int_0^{\pi/2} \frac{\sin\theta\,d\theta}{\sin\theta+\cos\theta}. \tag{2}$$

Equation (2) can also be obtained by using equation (19.6) with $a = \pi/2$. Addition of equations (1) and (2) gives

$$2I = \int_0^{\pi/2} \frac{\sin\theta+\cos\theta}{\sin\theta+\cos\theta}\,d\theta = \int_0^{\pi/2} 1\,d\theta = \pi/2$$

$$\Rightarrow I = \pi/4.$$

Exercise 19.1

1. Prove that $\displaystyle\int_a^b f(x)dx = \int_a^b f(a+b-x)dx.$

2. Prove that $\displaystyle\int_a^b f(tx)dx = \frac{1}{t}\int_{ta}^{tb} f(x)dx$ where t is a constant.

3. Prove that $\displaystyle\int_0^a x^m(a-x)^n dx = \int_0^a x^n(a-x)^m dx.$

4. Prove that $\displaystyle\int_{-a}^a x^2 \sinh^3 x\,dx = 0.$

5. Prove that, if n is an even integer,

$$\int_{-\pi}^{\pi} x\sin^n x\,dx = 0,$$

and that, if n is an odd integer,

$$\int_{-\pi}^{\pi} x \sin^n x \, dx = \pi \int_{0}^{\pi} \sin^n x \, dx.$$

Hence evaluate $\displaystyle\int_{-\pi}^{\pi} x \sin^5 x \, dx.$

6. Prove that $\displaystyle\int_{0}^{\pi/2} (a \cos^2 \theta + b \sin^2 \theta) \, d\theta = \int_{0}^{\pi/2} (a \sin^2 \theta + b \cos^2 \theta) \, d\theta$, where a and b are constants.

7. Deduce from question 6 that $\displaystyle\int_{0}^{\pi/2} (a \cos^2 \theta + b \sin^2 \theta) \, d\theta = \tfrac{1}{4}(a + b)\pi.$

8. Prove that $\displaystyle\int_{0}^{\pi} x f(\sin x) \, dx = \frac{\pi}{2} \int_{0}^{\pi} f(\sin x) \, dx$ and evaluate $\displaystyle\int_{0}^{\pi} \frac{x \sin^3 x}{1 + \cos^2 x} \, dx.$

9. Evaluate $\displaystyle\int_{0}^{\pi} (1 + 2 \cos x)^3 \, dx.$

10. Show that, if $f(x) \equiv f(a - x)$,

$$\int_{0}^{a} x f(x) \, dx = \tfrac{1}{2} a \int_{0}^{a} f(x) \, dx.$$

Evaluate $\displaystyle\int_{0}^{\pi} \frac{x \, dx}{1 + \cos \alpha \sin x} \qquad (0 < \alpha < \pi).$

11. Evaluate $\displaystyle\int_{0}^{\pi} \cos x \sin^2 2x \, dx$ without first obtaining the indefinite integral.   (N.)

12. Evaluate each of the following:

(i) $\displaystyle\int_{0}^{3} x^2 \sqrt{(3 - x)} \, dx,$ (ii) $\displaystyle\int_{-\pi/2}^{\pi/2} \frac{\sin^3 x}{1 + \cos^2 x} \, dx,$

(iii) $\displaystyle\int_{-2}^{2} x^2 (2 - x)^6 \, dx,$ (iv) $\displaystyle\int_{-\pi/4}^{\pi/4} \sin x \sec^6 x \, dx.$

13. Prove (i) if $f(x) \equiv f(2a - x)$, then $\displaystyle\int_{0}^{2a} f(x) \, dx = 2 \int_{0}^{a} f(x) \, dx$; (ii) if $f(x) \equiv -f(2a - x)$, then $\displaystyle\int_{0}^{2a} f(x) \, dx = 0.$

14. Use the result of question 13 to evaluate $\displaystyle\int_{0}^{\pi} \sin^8 x \cos^3 x \, dx.$

15. Prove (i) $\displaystyle\int_0^{\pi/2} \sin^m x\,dx = \int_0^{\pi/2} \cos^m x\,dx,$

(ii) $\displaystyle\int_0^{\pi/2} \sin^m x \cos^n x\,dx = \int_0^{\pi/2} \sin^n x \cos^m x\,dx.$

16. Find the value of $\displaystyle\int_{-\alpha}^{\alpha} \frac{\sin^5 x}{1+\cos^2 x}\,dx.$

17. If $I = \displaystyle\int_0^\pi \frac{x \sin x\,dx}{1+\cos^2 x},$ prove that $I = \displaystyle\int_0^\pi \frac{(\pi - x)\sin x}{1+\cos^2 x}\,dx$

and hence deduce that $I = \pi^2/4.$ (L.)

19:2 THE CONVERGENCE OF INTEGRALS

Thus far, in order to obtain $\displaystyle\int_a^b f(x)\,dx$ we have made the assumptions

(i) that a and b are both finite,

(ii) that the function $f(x)$ is finite for all values of x in the range $a \leqslant x \leqslant b$. We now define

$$\int_a^\infty f(x)\,dx \text{ as } \lim_{t \to \infty} \int_a^t f(x)\,dx.$$

The existence of the *improper* integral thus defined is dependent upon the existence of the limit. If this limit exists, the integral is said to *converge*.

Thus $\displaystyle\int_0^\infty \frac{1}{1+x}\,dx$ does not exist because $\lim_{t \to \infty} \ln (1+t)$ is not finite. But

$$\int_0^\infty \frac{1}{(1+x)^{3/2}}\,dx = \lim_{t \to \infty} \int_0^t \frac{1}{(1+x)^{3/2}}\,dx = \lim_{t \to \infty} \int_1^{t+1} \frac{1}{y^{3/2}}\,dy$$

$$= \lim_{t \to \infty} \left[-2y^{-1/2} \right]_1^{t+1} = \lim_{t \to \infty} \left[-2(t+1)^{-1/2} + 2 \right] = 2$$

$$\Rightarrow \int_0^\infty \frac{1}{(1+x)^{3/2}}\,dx \quad \text{converges to the value 2.}$$

Area. The graph of $f(x) = (1+x)^{-3/2}$ is shown in Fig. 19.1 (i) and the graph of $f(x) = (1+x)^{-1}$ is shown in Fig. 19.1 (ii). In each case the shaded area is equal to $\int_0^t f(x)\,dx$ for the function concerned.

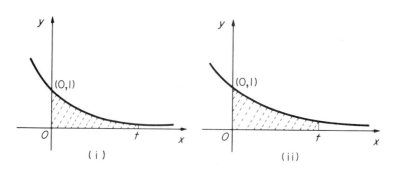

FIG. 19.1.

In the first case there is a limiting value to this area as $t \to \infty$ and $\int_0^\infty f(x)dx$ is *defined* as the area of the unbounded portion of the plane between the curve, the x-axis and the line $x = 0$.

In the second case $\int_0^t f(x)dx$ has no finite limit and the unbounded portion of the plane between the curve, the x-axis and the line $x = 0$ has no finite "area" as defined above.

Now consider $\int_a^b f(x)dx$ where $f(x) \to \infty$ as $x \to b$. If $\varepsilon > 0$ and $\lim_{\varepsilon \to 0+} \int_a^{b-\varepsilon} f(x)dx$ is finite and equal to L (say), then we *define* L as the value of $\int_a^b f(x)dx$. Similarly if $\phi(x) \to \infty$ as $x \to a$ and $\lim_{\varepsilon \to 0+} \int_{a+\varepsilon}^b \phi(x)dx = L'$, then L' is defined as the value of $\int_a^b \phi(x)dx$. If $f(x) \to \infty$ as $x \to c$ where $a < c < b$ and $f(x)$ is finite and continuous for all other values of x in the given range, then by definition

$$\int_a^b f(x)dx = \lim_{\varepsilon \to 0+} \int_a^{c-\varepsilon} f(x)dx + \lim_{\varepsilon' \to 0+} \int_{c+\varepsilon'}^b f(x)dx$$

if those limits exist, and where ε and ε' tend to zero independently.

These definitions are illustrated geometrically by the graphs of the functions $y = 1/\sqrt{(1-x^2)}$ [Fig. 19.2 (i)], $y = \tan x$ [Fig. 19.2 (ii)] and $y^3 = 1/x$ [Fig. 19.2 (iii)].

(i) The unbounded portion of the plane between the line $x = 1$, the x-axis and the curve $y = 1/\sqrt{(1-x^2)}$ has an "area" defined as

$$\lim_{\varepsilon \to 0+} \int_0^{1-\varepsilon} \frac{dx}{\sqrt{(1-x^2)}}.$$

This limit exists and is equal to

$$\lim_{\varepsilon \to 0+} \left[\sin^{-1}(1-\varepsilon) - \sin^{-1}0 \right] = \tfrac{1}{2}\pi.$$

(ii) The unbounded portion of the plane between the curve $y = \tan x$, the x-axis and the line $x = \tfrac{1}{2}\pi$ has no finite "area" as thus defined because $\int \tan x dx = \ln \sec x$ and $\lim_{\varepsilon \to 0+} \ln \sec(\tfrac{1}{2}\pi - \varepsilon)$ is not finite.

(iii) The unbounded portion of the plane between the curve $y^3 = 1/x$, the

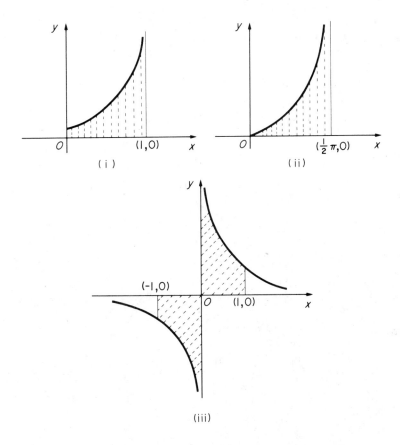

(i)

(ii)

(iii)

FIG. 19.2.

x-axis and the lines $x = -1$ and $x = +1$ has an "area" defined as

$$\lim_{\varepsilon \to 0+} \int_{-1}^{0-\varepsilon} x^{-1/3} \, dx + \lim_{\varepsilon' \to 0+} \int_{0+\varepsilon'}^{1} x^{-1/3} \, dx$$

$$= \lim_{\varepsilon \to 0+} \left[\frac{3}{2} x^{2/3} \right]_{-1}^{0-\varepsilon} + \lim_{\varepsilon' \to 0+} \left[\frac{3}{2} x^{2/3} \right]_{0+\varepsilon'}^{1} = -\frac{3}{2} + \frac{3}{2} = 0.$$

The area in the positive quadrant is 3/2.

If $I = \int_a^\infty f(x) dx$ and if the integral $\int_a^\infty |f(x)| dx$ is convergent, then I is said to be *absolutely convergent*. It can be proved that an absolutely convergent integral is convergent. If I is convergent but not absolutely convergent it is said to be *conditionally convergent*.

Example 1. $\displaystyle\int_0^\infty \cosh x \, dx$ does not exist because $\lim_{t \to \infty} [\sinh t]$ is not finite.

Example 2. $\displaystyle\int_0^\infty \frac{1}{1+x^2}\,dx = \lim_{t\to\infty}\left[\tan^{-1}t - \tan^{-1}0\right] = \tfrac{1}{2}\pi.$

Example 3. $\displaystyle\int_0^a \frac{1}{(x-a)^2}\,dx$ does not exist because $\displaystyle\lim_{\varepsilon\to 0+}\frac{1}{\varepsilon}$ is not finite.

Example 4. $\displaystyle\int_0^2 \frac{1}{(4x-x^2)^{1/2}}\,dx = \lim_{\varepsilon\to 0+}\left[\sin^{-1}\left(\frac{x-2}{2}\right)\right]_{0+\varepsilon}^2 = 0 - \left(-\tfrac{1}{2}\pi\right) = \tfrac{1}{2}\pi.$

Example 5. Evaluate $\displaystyle I = \int_0^1 \sqrt{\left(\frac{1+x}{1-x}\right)}\,dx.$

$$\int \sqrt{\left(\frac{1+x}{1-x}\right)}\,dx = \int \frac{1+x}{\sqrt{(1-x^2)}}\,dx$$

$$= -\frac{1}{2}\int \frac{-2x\,dx}{\sqrt{(1-x^2)}} + \int \frac{1}{\sqrt{(1-x^2)}}\,dx = -\sqrt{(1-x^2)} + \sin^{-1}x$$

$$\Rightarrow I = \lim_{\varepsilon\to 0+}\left[\sin^{-1}x - \sqrt{(1-x^2)}\right]_0^{1-\varepsilon}$$

$$= \tfrac{1}{2}\pi - (-1) = \tfrac{1}{2}\pi + 1.$$

Example 6. Evaluate $\displaystyle I = \int_0^\infty \frac{1}{(1+x^2)^2}\,dx.$

We can use the method of substitution with such a definite integral provided that the substitution is a valid one.

Let
$$I_1 = \int_0^t \frac{1}{(1+x^2)^2}\,dx.$$

Put $x = \tan\theta$. Then $dx/d\theta = \sec^2\theta$ and θ increases continuously from 0 to $\tan^{-1}t$ as x increases from 0 to t.

Then
$$I_1 = \int_0^{\tan^{-1}t} \cos^2\theta\,d\theta = \left[\tfrac{1}{2}\theta + \tfrac{1}{4}\sin 2\theta\right]_0^{\tan^{-1}t}$$

$$\Rightarrow I = \lim_{\theta\to\pi/2-}\left[\tfrac{1}{2}\theta + \tfrac{1}{4}\sin 2\theta\right] = \frac{\pi}{4}.$$

Example 7. Find the area of the region bounded by the curve $y = x/(1+x^2)$, the coordinate axes and the line $x = t$. Find the volume of rotation of this region about the x-axis. Show that, as $t \to \infty$, the function of t which represents the volume has a finite limit, but the function of t which represents the area has not a finite limit.

$$\text{The area} = \int_0^t \frac{x\,dx}{1+x^2} = \left[\tfrac{1}{2}\ln(1+x^2)\right]_0^t = \tfrac{1}{2}\ln(1+t^2),$$

and this has no finite limit as $t \to \infty$.

$$\text{Volume} = \pi \int_0^t \frac{x^2\,dx}{(1+x^2)^2} = \pi \int_0^{\tan^{-1}t} \frac{\tan^2\theta\sec^2\theta\,d\theta}{\sec^4\theta}$$

$$= \pi \int_0^{\tan^{-1}t} \sin^2\theta\,d\theta = \pi \left[\frac{\theta}{2} - \frac{\sin 2\theta}{4}\right]_0^{\tan^{-1}t} = \tfrac{1}{2}\pi\left(\tan^{-1}t - \frac{t}{1+t^2}\right).$$

Since $\lim\limits_{t\to\infty} \tfrac{1}{2}\pi\left[\tan^{-1}t - \dfrac{t}{1+t^2}\right] = \tfrac{1}{4}\pi^2$, as $t \to \infty$ the volume has the finite limit $\tfrac{1}{4}\pi^2$.

Exercise 19.2

Discuss the existence of each of the integrals in nos. 1–10 and, where possible, evaluate the integral.

1. $\displaystyle\int_0^1 \frac{1}{x}\,dx.$ 2. $\displaystyle\int_1^\infty \frac{1}{x}\,dx.$ 3. $\displaystyle\int_1^\infty \frac{1}{x^2}\,dx.$ 4. $\displaystyle\int_0^1 \frac{1}{x^2}\,dx.$ 5. $\displaystyle\int_0^\infty \frac{1}{4+9x^2}\,dx.$

6. $\displaystyle\int_0^4 \sqrt{\left(\frac{x}{4-x}\right)}\,dx.$ 7. $\displaystyle\int_1^4 \frac{1}{\sqrt{\{(x-1)(4-x)\}}}\,dx.$

8. $\displaystyle\int_0^1 x^2\ln x\,dx.$ (You may assume the result $\lim\limits_{x\to 0+} x^n\ln x = 0$ if $n > 0$.)

9. $\displaystyle\int_0^\infty xe^{-x}\,dx.$ 10. $\displaystyle\int_0^\infty \frac{x\,dx}{(x^2+4)(x^2+1)}.$

11. Evaluate $\displaystyle\int_{-2}^1 \sqrt{\left(\frac{2+x}{1-x}\right)}\,dx.$ 12. Evaluate $\displaystyle\int_1^2 \frac{1}{x\sqrt{(x-1)}}\,dx.$

13. Evaluate $\displaystyle\int_a^\infty \frac{1}{x^2\sqrt{(a^2+x^2)}}\,dx.$ 14. Evaluate $\displaystyle\int_0^2 \frac{(x^2+2)\,dx}{\sqrt{(4-x^2)}}.$

15. Evaluate $\displaystyle\int_0^\infty \operatorname{sech} x\,dx.$

16. Show that $\displaystyle\int_2^\infty \frac{(x^2+3)\,dx}{x^2(x^2+4)} = (\pi + 12)/32.$

17. Evaluate $\displaystyle\int_0^\infty e^{-ax}\sin x\,dx$ $(a > 0)$.   (L.)

18. Prove that $\displaystyle\int_{1}^{\infty} \frac{(x^2+2)dx}{x^2(x^2+1)} = 2 - \tfrac{1}{4}\pi.$ (L.)

19. Show that $\displaystyle\int_{0}^{1} \frac{1}{(1+x)\sqrt{(1-x^2)}}\,dx = 1.$ (L.)

20. Evaluate $\displaystyle\int_{1}^{2} \left(\frac{x-1}{2-x}\right)^{1/2} dx.$ (L.)

21. Sketch the curve $y^2 = \dfrac{x}{4-x}$ and find the area of the region between the curve and its asymptote. If V_b is the volume formed by the revolution about the x-axis of the region bounded by this curve and the line $x = b$, $(0 < b < 4)$, show that $\lim\limits_{b\to 4-} V_b$ is not finite.

22. Find the area of the unbounded region of the coordinate plane between the curves $y = \cosh x$ and $y = \sinh x$ and in the first quadrant.

23. Find the volume of the solid generated by rotation about the x-axis of the region between the curve $y = \dfrac{1}{1+x^2}$ and the x-axis.

24. Sketch the curve $y = 1/\{x\sqrt{(1+x)}\}$. Find the area of the unbounded region between the curve for values of $x \geqslant 1$, the line $x = 1$ and the x-axis and show that the volume formed by the rotation of this region about the x-axis is $\pi(1 - \ln 2)$.

25. Sketch the curve $y^2 = \dfrac{1}{(x-2)(4-x)}.$

Calculate the area of unbounded region between the curve and the lines $x = 2$, $x = 4$.

19:3 REDUCTION FORMULAE

If an integrand involves a constant $n \in \mathbb{Q}$, it may be possible to express the integral in terms of a similar integral which involves $(n-1)$ or some other number which is less than n. The formula which expresses the integral in this way is called a *reduction formula*, and by a repeated application of the reduction formula the original integral may be reduced to a form in which it can be obtained directly.

Example 1. Obtain a reduction formula for $\int x^n e^x \, dx$. Hence evaluate $\displaystyle\int_{0}^{1} x^5 e^x \, dx.$

Let
$$u_n = \int x^n e^x dx.$$

Then
$$u_n = x^n e^x - n \int x^{n-1} e^x dx$$

$$\Rightarrow u_n = x^n e^x - n u_{n-1},$$

which is the required reduction formula.

$$u_5 = \int x^5 e^x dx = x^5 e^x - 5u_4$$
$$= x^5 e^x - 5x^4 e^x + 20u_3$$
$$= x^5 e^x - 5x^4 e^x + 20x^3 e^x - 60u_2$$

$$= x^5 e^x - 5x^4 e^x + 20x^3 e^x - 60x^2 e^x + 120u_1$$

$$= x^5 e^x - 5x^4 e^x + 20x^3 e^x - 60x^2 e^x + 120xe^x - 120 \int e^x dx$$

$$= x^5 e^x - 5x^4 e^x + 20x^3 e^x - 60x^2 e^x + 120xe^x - 120e^x + C$$

$$\Rightarrow \int_0^1 x^5 e^x dx = (1 - 5 + 20 - 60 + 120 - 120)e - (-120)$$

$$= 120 - 44e.$$

Example 2. Obtain a reduction formula for $\int x^n \sinh x \, dx$ and hence obtain $\int x^5 \sinh x \, dx$.

Let $u_n = \int x^n \sinh x \, dx$.

Then $u_n = x^n \cosh x - n \int x^{n-1} \cosh x \, dx$

$$= x^n \cosh x - nx^{n-1} \sinh x + n(n-1) \int x^{n-2} \sinh x \, dx$$

$$\Rightarrow u_n = x^n \cosh x - nx^{n-1} \sinh x + n(n-1)u_{n-2}$$

$$\Rightarrow \int x^5 \sinh x \, dx = x^5 \cosh x - 5x^4 \sinh x + 20x^3 \cosh x - 60x^2 \sinh x$$

$$+ 120x \cosh x - 120 \sinh x + C.$$

Example 3. Evaluate $\displaystyle\int_0^{\pi/2} \cos^8 x \, dx$.

Let

$$u_n = \int_0^{\pi/2} \cos^n x \, dx \quad \text{where } n > 1.$$

Then

$$u_n = \int_{x=0}^{x=\pi/2} \cos^{n-1} x \cos x \, dx$$

$$= \left[\cos^{n-1} x \sin x \right]_0^{\pi/2} + (n-1) \int_0^{\pi/2} \cos^{n-2} x \sin^2 x \, dx$$

$$= 0 + (n-1) \int_0^{\pi/2} \cos^{n-2} x (1 - \cos^2 x) \, dx$$

$$= (n-1)u_{n-2} - (n-1)u_n$$

$$\Leftrightarrow nu_n = (n-1)u_{n-2}$$

$$\Rightarrow \int_0^{\pi/2} \cos^8 x \, dx = \frac{7}{8}\frac{5}{6}\frac{3}{4}\frac{1}{2} \int_0^{\pi/2} dx = \frac{35\pi}{256}.$$

Example 4. $\displaystyle\int_0^{\pi/2} \sin^m x \cos^n x \, dx$, where $m > 1, n > 1$.

Note that $\displaystyle\int_0^{\pi/2} \sin^m x \cos^n x \, dx = \int_0^{\pi/2} \sin^n x \cos^m x \, dx$ (see Ex. 19.1, Q. 15).

This integral can be obtained directly by substitution if either of m or n is odd as shown on p. 524. Thus

$$\int_0^{\pi/2} \sin^4 x \cos^5 x \, dx = \int_0^1 s^4 (1 - s^2)^2 \, ds \quad [\text{by the substitution } \sin x = s]$$

$$= \int_0^1 (s^4 - 2s^6 + s^8) \, ds = (\tfrac{1}{5} - \tfrac{2}{7} + \tfrac{1}{9}) = \tfrac{8}{315}.$$

If

$$u_{m,n} = \int_0^{\pi/2} \sin^m x \cos^n x \, dx,$$

then

$$u_{m,n} = \int_{x=0}^{x=\pi/2} \sin^m x \cos^{n-1} x \cos x \, dx$$

$$= \left[\sin^{m+1} x \cos^{n-1} x \right]_0^{\pi/2} - \int_0^{\pi/2} \{ m \sin^{m-1} x \cos^n x - (n-1) \sin^{m+1} x \cos^{n-2} x \} \sin x \, dx$$

$$= 0 - m \, u_{m,n} + (n-1) \int_0^{\pi/2} \sin^{m+2} x \cos^{n-2} x \, dx$$

$$= - m \, u_{m,n} + (n-1) \int_0^{\pi/2} (\sin^m x \cos^{n-2} x - \sin^m x \cos^n x) \, dx$$

$$= - m u_{m,n} + (n-1) u_{m,n-2} - (n-1) u_{m,n}$$

$$\Rightarrow u_{m,n} = \frac{n-1}{m+n} u_{m,n-2}.$$

This reduction formula reduces the integral $u_{m,n}$ to a multiple of either

$$(1) \int_0^{\pi/2} \sin^m x \, dx \quad \text{or} \quad (2) \int_0^{\pi/2} \sin^m x \cos x \, dx.$$

Integral (1) can be evaluated as in Example 3 above and

$$\text{integral } (2) = \left[\frac{\sin^{m+1} x}{m+1} \right]_0^{\pi/2} = \frac{1}{m+1}.$$

Thus

$$\int_0^{\pi/2} \sin^6 x \cos^4 x \, dx = \frac{3}{10} \cdot \frac{1}{8} \int_0^{\pi/2} \sin^6 x \, dx = \frac{3}{10} \cdot \frac{1}{8} \cdot \frac{5}{6} \cdot \frac{3}{4} \cdot \frac{1}{2} \cdot \frac{\pi}{2} = \frac{3\pi}{512}.$$

Wallis' formulae. The definite integrals

$$\int_0^{\pi/2} \sin^n \theta \, d\theta, \quad \int_0^{\pi/2} \cos^n \theta \, d\theta \quad \text{and} \quad \int_0^{\pi/2} \sin^m \theta \cos^n \theta \, d\theta$$

are of frequent occurrence and it is useful to remember the reduction formulae associated with them and to quote them where necessary.

$$\text{If } u_n = \int_0^{\pi/2} \sin^n \theta \, d\theta, \; n > 1, \quad \text{then} \quad u_n = \frac{n-1}{n} u_{n-2}. \tag{19.7}$$

$$\text{If } u_n = \int_0^{\pi/2} \cos^n \theta \, d\theta, \; n > 1, \quad \text{then} \quad u_n = \frac{n-1}{n} u_{n-2}. \tag{19.8}$$

$$\text{If } u_{m,n} = \int_0^{\pi/2} \sin^m \theta \cos^n \theta \, d\theta, \; m > 1, \; n > 1, \quad \text{then} \quad u_{m,n} = \frac{m-1}{m+n} u_{m-2,n}. \tag{19.9}$$

Example 1. $\displaystyle \int_0^{\pi/2} \sin^6 x \cos^4 x \, dx = \frac{3 \times 1}{10 \times 8} \int_0^{\pi/2} \sin^6 x \, dx$

$$= \frac{3 \times 1 \times 5 \times 3 \times 1}{10 \times 8 \times 6 \times 4 \times 2} \frac{\pi}{2} = \frac{3\pi}{512}.$$

Example 2. $\displaystyle \int_0^{\pi/2} \sin^3 x \cos^7 x \, dx = \frac{2}{10} \int_0^{\pi/2} \sin x \cos^7 x \, dx = \frac{2}{10} \left[\frac{-\cos^8 x}{8} \right]_0^{\pi/2}$

$$= \frac{2 \times 1}{10 \times 8} = \frac{1}{40}.$$

Example 3. $\displaystyle \int_0^a x^m (a^2 - x^2)^{n/2} \, dx.$ Writing $x = a \sin \theta$ gives

$$a^{m+n+1} \int_0^{\pi/2} \sin^m \theta \cos^{n+1} \theta \, d\theta \text{ which is of the above type.}$$

Example 4. Similarly the substitution $x = a \sin^2 \theta$ transforms

$$\int_0^a x^m (a - x)^n \, dx \quad \text{into} \quad 2a^{m+n+1} \int_0^{\pi/2} \sin^{2m+1} \theta \cos^{2n+1} \theta \, d\theta.$$

We now give illustrative examples of other types of reduction formulae.

Example 1. If n is a constant, prove that

$$\frac{d}{dx} \frac{x}{(1+x^2)^n} = \frac{2n}{(1+x^2)^{n+1}} - \frac{2n-1}{(1+x^2)^n}.$$

Hence obtain a reduction formula for

$$I_n = \int_0^1 \frac{1}{(1+x^2)^n}\,dx.$$

Evaluate I_3 and $I_{7/2}$.

$$\frac{d}{dx}\frac{x}{(1+x^2)^n} = \frac{(1+x^2)^n - 2nx^2(1+x^2)^{n-1}}{(1+x^2)^{2n}}$$

$$= \frac{1}{(1+x^2)^n} - \frac{2nx^2}{(1+x^2)^{n+1}}$$

$$= \frac{1}{(1+x^2)^n} - \frac{2n(1+x^2)}{(1+x^2)^{n+1}} + \frac{2n}{(1+x^2)^{n+1}}$$

$$= \frac{2n}{(1+x^2)^{n+1}} - \frac{2n-1}{(1+x^2)^n}$$

as required. Integrating this relation with respect to x between 0 and 1, and rearranging gives

$$2nI_{n+1} - (2n-1)I_n = \left[\frac{x}{(1+x^2)^n}\right]_0^1 = \frac{1}{2^n}, \tag{1}$$

which is a reduction formula for I_{n+1}. Changing n into $n-1$ gives the reduction formula for I_n:

$$2(n-1)I_n - (2n-3)I_{n-1} = \frac{1}{2^{n-1}}.$$

Putting $n = 2$ in (1) gives $4I_3 = 3I_2 + \frac{1}{4}$.
Putting $n = 1$ in (1) gives $2I_2 = I_1 + \frac{1}{2}$

$$\Rightarrow 4I_3 = \frac{1}{4} + \frac{3}{2}\left(\frac{1}{2} + I_1\right) = 1 + \frac{3}{2}\int_0^1 \frac{1}{1+x^2}\,dx = 1 + \frac{3\pi}{8}$$

$$\Rightarrow I_3 = (3\pi + 8)/32.$$

The recurrence formula is valid for non-integral values of n.

Putting $n = \dfrac{5}{2}$ in (1) gives $5I_{7/2} = 4I_{5/2} + \dfrac{1}{4\sqrt{2}}$.

Putting $n = \dfrac{3}{2}$ in (1) gives $3I_{5/2} = 2I_{3/2} + \dfrac{1}{2\sqrt{2}}$

$$\Rightarrow 5I_{7/2} = \frac{4}{3}\left(2I_{3/2} + \frac{1}{2\sqrt{2}}\right) + \frac{1}{4\sqrt{2}} = \frac{8}{3}I_{3/2} + \frac{11}{12\sqrt{2}}.$$

Putting $n = \frac{1}{2}$ in (1) gives $I_{3/2} = 1/\sqrt{2}$

$$\Rightarrow 5I_{7/2} = 43/12\sqrt{2} \Leftrightarrow I_{7/2} = 43/60\sqrt{2}.$$

Example 2. Prove that $\dfrac{d^2}{dx^2}\sin^n x + n^2 \sin^n x = n(n-1)\sin^{n-2} x$, where n is any constant.

If $I_n = \displaystyle\int_0^\pi e^{-x}\sin^n x\,dx$, show that $I_n = \dfrac{n(n-1)}{n^2+1}I_{n-2}$ where $n \in \mathbb{N}$, $n \neq 1$.

Show that $I_6 = \dfrac{144}{629}(1 - e^{-\pi})$.

$$\frac{d^2}{dx^2}\sin^n x = \frac{d}{dx}(n\sin^{n-1} x \cos x)$$

$$= n(n-1)\sin^{n-2} x \cos^2 x - n\sin^n x$$
$$= n(n-1)\sin^{n-2} x (1 - \sin^2 x) - n\sin^n x$$

$$\Rightarrow \frac{d^2}{dx^2}\sin^n x + n^2 \sin^n x = n(n-1)\sin^{n-2} x. \qquad (1)$$

$$\int e^{-x}\sin^n x\,dx = -e^{-x}\sin^n x + \int e^{-x}\frac{d}{dx}(\sin^n x)\,dx$$

$$= -e^{-x}\sin^n x - e^{-x}\frac{d}{dx}(\sin^n x) + \int e^{-x}\frac{d^2}{dx^2}(\sin^n x)\,dx.$$

Therefore, using equation (1),

$$\int e^{-x}\sin^n x\,dx = -e^{-x}(\sin^n x + n\sin^{n-1} x \cos x) + \int e^{-x}[n(n-1)\sin^{n-2} x - n^2 \sin^n x]\,dx$$

$$\Rightarrow (n^2+1)\int e^{-x}\sin^n x\,dx = -e^{-x}(\sin^n x + n\sin^{n-1} x \cos x) + n(n-1)\int e^{-x}\sin^{n-2} x\,dx$$

$$\Rightarrow (n^2+1)I_n = n(n-1)I_{n-2}, \quad \text{if} \quad n \geqslant 2,$$

$$\Rightarrow I_6 = \frac{6\times 5}{(6^2+1)}I_4, \quad I_4 = \frac{4\times 3}{(4^2+1)}I_2,$$

$$I_2 = \frac{2\times 1}{(2^2+1)}I_0,$$

$$I_0 = \int_0^\pi e^{-x}\,dx = \left[-e^{-x}\right]_0^\pi = 1 - e^{-\pi}$$

$$\Rightarrow I_6 = \frac{6!(1-e^{-\pi})}{(6^2+1)(4^2+1)(2^2+1)} = \frac{144}{629}(1-e^{-\pi}).$$

Exercise 19.3

In each of the questions 1–6 obtain a reduction formula and hence obtain the indefinite integral in the special case quoted.

1. $\int x^n e^{ax}\,dx; \qquad \int x^4 e^{3x}\,dx.$ 2. $\int x^n \sin x\,dx; \quad \int x^3 \sin x\,dx.$

3. $\int \tan^n x\,dx; \qquad \int \tan^6 x\,dx.$ 4. $\int x(\ln x)^n\,dx; \quad \int x(\ln x)^3\,dx.$

5. $\int \cosh^n x\,dx; \quad \int \cosh^6 x\,dx.$ 6. $\int \sec^n x\,dx; \quad \int \sec^5 x\,dx.$

Evaluate each of the integrals in questions 7–12.

7. $\int_0^{\pi/2} \sin^8 x\,dx.$ 8. $\int_0^{\pi/2} \cos^7 x\,dx.$ 9. $\int_0^{\pi/2} \sin^4 x \cos^3 x\,dx.$

10. $\int_0^\infty x^4 e^{-2x}\,dx.$ 11. $\int_0^a x^2(a^2 - x^2)^{3/2}\,dx.$ 12. $\int_{-\infty}^\infty \frac{x^6}{(a^2+x^2)^4}\,dx.$

13. Prove that $\int_0^{\pi/2} (2\sin^5 x - 3\sin^7 x)\cos^2 x\,dx = 0.$ (O.C.)

14. If $I_n = \int_0^1 x^{n+1/2}(1-x)^{1/2}\,dx,$

prove that

$$2(n+2)I_n = (2n+1)I_{n-1}, \quad (n \geqslant 1).$$

Hence or otherwise show that

$$\int_0^1 x^2 \sqrt{(x - x^2)} \, dx = 5\pi/128.$$

(L.)

15. If $I_n = \int_0^x \dfrac{t^n \, dt}{\sqrt{(1 + t^2)}}$, prove that, for $n > 1$,

$$nI_n + (n - 1)I_{n-2} = x^{n-1} \sqrt{(1 + x^2)}.$$

(L.)

16. If $I_n = \int_0^{\pi/3} \sin^n \theta \, d\theta$, prove that

$$I_n = \frac{n-1}{n} I_{n-2} - \frac{1}{2n} \left(\frac{\sqrt{3}}{2} \right)^{n-1}.$$

17. If $I_n = \int_0^\theta \dfrac{\sin(2n - 1)\theta}{\sin \theta} \, d\theta$, find $I_n - I_{n-1}$. Hence show that, if $n \in \mathbb{N}$,

$$\int_0^{\pi/2} \frac{\sin(2n - 1)\theta}{\sin \theta} \, d\theta = \frac{\pi}{2} \quad \text{and} \quad \int_0^{\pi/2} \frac{\sin^2 n\theta}{\sin^2 \theta} \, d\theta = \frac{n\pi}{2}.$$

18. (i) Prove that, if

$$I_n = \int_0^{\pi/2} \sin^n \theta \cos^2 \theta \, d\theta$$

and $n \geqslant 2$,

$$I_n = \frac{n-1}{n+2} I_{n-2}.$$

(ii) Prove that

$$\int_0^{\pi/2} \sin^4 \theta \cos^2 \theta \, d\theta = \frac{\pi}{32},$$

$$\int_0^{\pi/2} \sin^7 \theta \cos^2 \theta \, d\theta = \frac{16}{315}.$$

(iii) With the help of (ii), or otherwise, evaluate the integral

$$\int_0^1 x^4 (1 + x)^{7/2} (1 - x)^{1/2} \, dx.$$

(O.C.)

19. If $u_n = \int (\sec x + \tan x)^n \, dx$, prove that, if $n \neq 1$,

$$(n - 1)(u_n + u_{n-2}) = 2(\sec x + \tan x)^{n-1} + C.$$

Find

$$\int_{-\pi/6}^{\pi/6} (\sec x + \tan x)^{-1} \, dx,$$

and hence find

$$\int_{-\pi/6}^{\pi/6} (\sec x + \tan x)^{-3} \, dx.$$

(O.C.)

19:4 MEAN VALUES AND ROOT MEAN SQUARE

If $y = f(x)$ is a function of x which is continuous over the domain $a \leqslant x \leqslant b$ (Fig. 19.3), the *mean value of y with respect to x* over this domain is defined as

$$\frac{1}{b-a} \int_a^b f(x) \, dx. \tag{19.10}$$

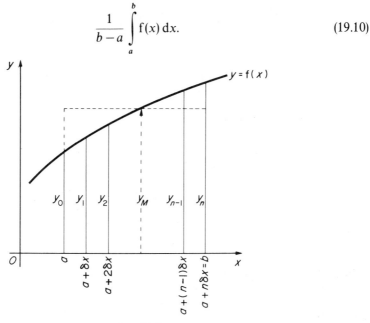

FIG. 19.3.

[*Note* that the *mean value of a function* is defined differently to that of a frequency distribution or a probability density function.]

If $y = y_0$ when $x = a$, and $y = y_r$ when $x = a + r\delta x$ where $n\delta x = b - a$, the arithmetical average value of the ordinates $y_0, y_1, y_2, \ldots, y_{n-1}$ is

$$y_M = \frac{1}{n} \sum_0^{n-1} y_r = \sum_{r=0}^{r=n-1} \frac{y_r \, \delta x}{n \, \delta x} = \frac{1}{b-a} \sum_{r=0}^{r=n-1} y_r \, \delta x$$

and the limiting value of this arithmetical average as $\delta x \to 0$, i.e., as $n \to \infty$, is

$$\frac{1}{b-a} \int_a^b f(x) \, dx.$$

The mean value as defined above is therefore identified with the limit of the arithmetical average of the ordinates as the interval between the ordinates tends to zero.

Also, since $\int_a^b f(x) \, dx$ represents the area under the curve between y_0 and y_n, this area is equal to $(b-a) \, y_M$, so that y_M is the height of the rectangle on base $(b-a)$ having an area equal to that under the curve.

If y is expressed in terms of a different variable u so that $y = \phi(u)$, $u = \alpha$ when $x = a$, $u = \beta$ when $x = b$ and $\phi(u)$ is a continuous function of u over the domain $\alpha \leqslant u \leqslant \beta$, the mean value of y with respect to u is

$$\frac{1}{\beta - \alpha} \int_\alpha^\beta \phi(u)\, du$$

and this mean value is not necessarily equal to the mean value of y with respect to x as defined above.

Example 1. With the usual notation for a particle moving in a straight line the equation of motion of a particle starting from the origin is

$$v = \omega \sin (pt + \alpha).$$

Calculate (a) the mean value of the speed with respect to the time for the first $\pi/2p$ s of the motion, (b) the mean value of the speed with respect to the distance for the first $\pi/2p$ s of the motion.

(a) The mean value of v with respect to t is

$$\frac{2p}{\pi} \int_0^{\pi/2p} v\, dt = \frac{2p}{\pi} \int_0^{\pi/2p} \omega \sin (pt + \alpha)\, dt = \frac{2p}{\pi} \left[-\frac{\omega}{p} \cos (pt + \alpha) \right]_0^{\pi/2p}$$

$$= \frac{2\omega}{\pi} (\cos \alpha + \sin \alpha).$$

(b) $s = \displaystyle\int_0^t v\, dt = \frac{\omega}{p} \{\cos \alpha - \cos (pt + \alpha)\}$; when $t = 0$, $s = 0$,

and when $t = \pi/2p$, $s = \dfrac{\omega}{p} \{\cos \alpha - \cos (\tfrac{1}{2}\pi + \alpha)\} = \dfrac{\omega}{p} (\cos \alpha + \sin \alpha)$

$$= s_1 \text{ (say)}.$$

Hence the mean value of v with respect to s over the interval $t = 0$ to $t = \pi/2p$ is

$$\frac{1}{s_1} \int_{t=0}^{t=\pi/2p} v\, ds = \frac{1}{s_1} \int_0^{\pi/2p} v \frac{ds}{dt}\, dt = \frac{1}{s_1} \int_0^{\pi/2p} v^2\, dt$$

$$= \frac{1}{s_1} \int_0^{\pi/2p} \omega^2 \sin^2 (pt + \alpha)\, dt$$

$$= \frac{\omega^2}{2s_1} \int_0^{\pi/2p} \{1 - \cos 2(pt + \alpha)\}\, dt$$

$$= \frac{\omega^2}{2s_1} \left[t - \frac{\sin 2(pt + \alpha)}{2p} \right]_0^{\pi/2p}$$

$$= \frac{\omega^2}{2s_1} \left(\frac{\pi}{2p} - \frac{\sin (\pi + 2\alpha)}{2p} + \frac{\sin 2\alpha}{2p} \right)$$

$$= \frac{\omega^2}{2s_1} \left(\frac{\pi}{2p} + \frac{\sin 2\alpha}{p} \right)$$

$$= \frac{\omega (\pi + 2 \sin 2\alpha)}{4 (\cos \alpha + \sin \alpha)}.$$

Example 2. Calculate the mean distance from a point P on the circumference of a circle of radius r of all the other points on the circumference.

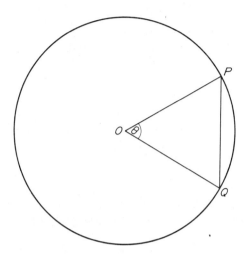

FIG. 19.4.

If Q (Fig. 19.4) is a point on the circumference such that $Q\hat{O}P = \theta$, $QP = 2r \sin\frac{1}{2}\theta$. Hence the mean distance of points on the circumference from P is

$$\frac{1}{2\pi} \int_0^{2\pi} 2r \sin\tfrac{1}{2}\theta \, d\theta = \frac{4r}{2\pi} \left[-\cos\tfrac{1}{2}\theta \right]_0^{2\pi} = \frac{4r}{\pi} \approx 1{\cdot}27 \, r.$$

The *root mean square* (R.M.S.) value of a function over a given interval is the square root of the mean value of its square over that interval. Thus the R.M.S. of $f(x)$ with respect to x for $a \leqslant x \leqslant b$ is

$$\sqrt{\left(\frac{1}{b-a} \int_a^b \{f(x)\}^2 \, dx \right)}. \qquad (19.11)$$

The practical importance of the root mean square lies in the necessity to measure the mean value of the square of such variables as alternating quantities in electricity where, for example, the heating effect of an electric current in a given time is proportional to the square of the current.

Example. Calculate the root mean square value over a period of an electric current given by the formula $I = I_0 \sin(\omega t + \alpha)$.

The *period* of the alternation is $2\pi/\omega$ and it is over this interval of time that the R.M.S. value of a periodic function is expressed.

$$\text{R.M.S. value} = \sqrt{\left\{ \frac{\omega}{2\pi} \int_0^{2\pi/\omega} I_0^2 \sin^2(\omega t + \alpha)\, dt \right\}}$$

$$= \sqrt{\left\{ \frac{\omega}{2\pi} I_0^2 \left[\frac{t}{2} - \frac{\sin(2\omega t + 2\alpha)}{4\omega} \right]_0^{2\pi/\omega} \right\}}$$

$$= \sqrt{\left\{ \frac{\omega}{2\pi} I_0^2 \left(\frac{\pi}{\omega} - \frac{\sin 2\alpha}{4\omega} + \frac{\sin 2\alpha}{4\omega} \right) \right\}}$$

$$= I_0/\sqrt{2}.$$

This result may be generalised as follows:

If $I = I_0 + I_1 \sin(\omega t + \varepsilon_1) + I_2 \sin(2\omega t + \varepsilon_2) + \ldots + I_n \sin(n\omega t + \varepsilon_n)$, the R.M.S. of I is

$$\sqrt{[I_0^2 + \tfrac{1}{2}(I_1^2 + I_2^2 + \ldots + I_n^2)]}.$$

Exercise 19.4

In questions **1–4** calculate the mean value of the function with respect to x over the stated range. Where necessary use Simpson's rule and give mean values to 2 significant figures.

1. $y = \tan x$; $x = 0$ to $x = \pi/4$.

2. $y = \ln \sec x$; $x = 0$ to $x = \pi/3$.

3. $y = x \sinh x$; $x = 0$ to $x = 1$.

4. $y = \sqrt{(a^2 - x^2)}$; $x = 0$ to $x = a$.

5. In the usual notation the equation of motion of a particle moving in a straight line is $x = 4 \sin t - 3 \cos t$. Calculate (i) the mean value of the speed with respect to the time from $t = 0$ to $t = \tfrac{1}{2}\pi$, (ii) the mean value of the speed with respect to the distance over this interval of time.

6. The density of a thin rod of length a at a distance x from one end is $\varrho\, x^2$. Calculate the mean density of the rod.

7. A piece of wire of length $2l$ is bent to form a rectangle. Find the mean value of the areas of all the rectangles which can be formed in this way.

8. Calculate the R.M.S. value over a period (π) of an electric current i which is given by

$$i = 3 \sin(2t + \tfrac{1}{3}\pi) + 4 \sin(4t + \tfrac{1}{4}\pi).$$

9. If $f(x) = a + bx + cx^2 + dx^3$, where a, b, c, d are constants, prove that

$$\int_{-h}^{h} f(x)\, dx = \frac{1}{3}h\, [f(-h) + 4f(0) + f(h)],$$

and deduce Simpson's rule for approximate integration.

Find the mean value of $(2 + \sin \theta)^{-1/2}$ in the range $0 \le \theta \le \pi/2$, using Simpson's rule with six strips, giving your answer to three significant figures.

Check that your answer is reasonable from a rough sketch of the graph of the function in the range.

(L.)

19:5 CENTRE OF MASS

The position $(\bar{x}, \bar{y})$ of the centre of mass of a system of coplanar particles of masses $m_1, m_2, m_3, \ldots, m_n$, whose positions in relation to rectangular axes Ox, Oy in the plane are $(x_1, y_1), (x_2, y_2), (x_3, y_3), \ldots, (x_n, y_n)$ respectively, is defined by

$$\bar{x} = \frac{\sum\limits_1^n m_r x_r}{\sum\limits_1^n m_r}, \quad \bar{y} = \frac{\sum\limits_1^n m_r y_r}{\sum\limits_1^n m_r}.$$

The position of the centre of mass of a continuous body, regarded as an aggregate of particles, can be found by using the methods of integration. The limits of the sums involved in the above formulae are replaced by definite integrals.

The *centroid of a region* coincides with the centre of mass of a uniform lamina in the shape of the region and the *centroid of a volume* of revolution similarly coincides with the centre of mass of a uniform solid in the shape of that volume.

Example 1. Find the position of the centre of mass of a uniform solid hemisphere of radius R.

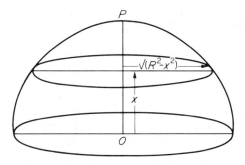

FIG. 19.5.

(Fig. 19.5) The centre of mass of the hemisphere will be on the axis of symmetry OP.

Let ϱ be the density of the hemisphere and consider a plane section of the hemisphere parallel to its base and distance x from it. The mass of a thin disc, of thickness δx, and with this section as base, is $\pi(R^2 - x^2)\delta x\,\varrho$ and therefore, if $\bar{x}$ is the distance of the centre of mass of the hemisphere from O,

$$\bar{x} = \frac{\lim\limits_{\delta x \to 0} \sum\limits_{x=0}^{x=R} \pi\varrho\,(R^2 - x^2)\delta x.x}{\lim\limits_{\delta x \to 0} \sum\limits_{x=0}^{x=R} \pi\varrho\,(R^2 - x^2)\,\delta x}$$

$$= \frac{\int_0^R \pi\varrho\,(R^2 - x^2)x\,dx}{\int_0^R \pi\varrho\,(R^2 - x^2)\,dx}$$

$$= \left[\frac{R^2 x^2}{2} - \frac{x^2}{4}\right]_0^R \Big/ \left[R^2 x - \frac{x^3}{3}\right]_0^R = 3R/8.$$

Example 2. Find the position of the centroid of the region bounded by the coordinate axes and the curve $y = \cos x$ from $x = 0$ to $x = \frac{1}{2}\pi$.

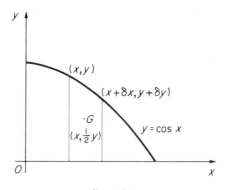

FIG. 19.6.

Consider a thin strip of the region parallel to the y-axis and bounded by the ordinates at (x, y) and $(x + \delta x, y + \delta y)$ on the curve (Fig. 19.6). The area of the strip is approximately $y\delta x$ and the coordinates of the centroid, G, of the strip are approximately $(x, \frac{1}{2}y)$. Hence if $(\bar{x}, \bar{y})$ are the coordinates of the centroid of the region,

$$\bar{x} = \int_0^{\pi/2} xy \, dx \bigg/ \int_0^{\pi/2} y \, dx = \int_0^{\pi/2} x \cos x \, dx \bigg/ \int_0^{\pi/2} \cos x \, dx$$

$$= \Big[x \sin x + \cos x \Big]_0^{\pi/2} \bigg/ \Big[\sin x \Big]_0^{\pi/2} = (\tfrac{1}{2}\pi - 1)/1 = \tfrac{1}{2}\pi - 1,$$

$$\bar{y} = \int_0^{\pi/2} \tfrac{1}{2} yy \, dx \bigg/ \int_0^{\pi/2} y \, dx = \tfrac{1}{2} \int_0^{\pi/2} \cos^2 x \, dx$$

$$= \tfrac{1}{2} \Big[\frac{x}{2} + \frac{\sin 2x}{4} \Big]_0^{\pi/2} = \tfrac{1}{8}\pi.$$

Thus the centroid is at $\big[(\tfrac{1}{2}\pi - 1), \tfrac{1}{8}\pi \big]$.

Example 3. Find the position of the centroid of the volume of revolution formed when the x-positive half of the ellipse $(x^2/a^2) + (y^2/b^2) = 1$ makes a complete revolution about the x-axis.

The centroid will be on the x-axis. Consider the thin disc formed by the revolution about the axis of the strip bounded by the ordinates at (x, y) and $(x + \delta x, y + \delta y)$, the curve and the x-axis. The volume of the disc is approximately $\pi y^2 \, \delta x$

$$\Rightarrow \bar{x} = \int_0^a \pi x y^2 \, dx \bigg/ \int_0^a \pi y^2 \, dx$$

$$= \int_0^a b^2 \Big(x - \frac{x^3}{a^2} \Big) dx \bigg/ \int_0^a b^2 \Big(1 - \frac{x^2}{a^2} \Big) dx$$

$$= \Big(\frac{a^2}{2} - \frac{a^2}{4} \Big) \bigg/ \Big(a - \frac{a}{3} \Big) = \frac{3a}{8}.$$

Thus the centroid of the volume is at $(3a/8, 0)$.

[Note that for a body whose dimensions are small compared with the radius of the earth, the centre of mass is the same point as the centre of gravity.]

Exercise 19.5

Find the position of the centre of mass of each of the solids in questions 1–4.

1. A uniform solid cone of height h.

2. A thin rod of length $2l$ such that the density at a point of the rod varies as the square of the distance from one end.

3. A wire of uniform density in the shape of a semicircle of radius r.

4. A square pyramid of height h.

In each of the questions 5–8 find the coordinates of the centroid of the region whose boundaries are as described.

5. The parabola $y^2 = 4ax$ and its latus rectum.

6. The curve $y = \sec x$ and the line $y = 2$.

7. The curve $x = a \cos^3 \theta$, $y = a \sin^3 \theta$ and the coordinate axes in the first quadrant.

8. The curves $y^2 = 4x$, $x^2 = 4y$ in the first quadrant.

In each of the questions 9–12 find the x-coordinate of the centroid of the volume of revolution about the x-axis of the region whose boundaries are as described.

9. The curve $y^2 = x^3$ and the double ordinate $x = 1$.

10. The curve $y = e^x$, the coordinate axes and the line $x = 2$.

11. The curve $y = \ln x$, for $x \geqslant 1$, the x-axis and the line $x = e$.

12. The curve $y^2 = x^2 (a^2 - x^2)$ for $x \geqslant 0$.

13. O is a fixed point on the circumference of a circular lamina of radius a. If N is a point on the diameter through O such that $ON = x$, the density at all points of the lamina on the chord through N perpendicular to ON is ρx, where ρ is a constant. Prove that the mass of the lamina is $\pi \rho a^3$, and that the distance from O of the centre of mass of the lamina is $\frac{5}{4}a$.   (C.)

19:6 THE THEOREM OF PAPPUS CONCERNING VOLUMES

If a plane closed curve makes a complete revolution about a line in its own plane which does not cut the curve, the volume of the solid generated is equal to the area of the region enclosed by the curve multiplied by the length of the path traced out by the centroid of that region.

Suppose that the only tangents to the curve which are parallel to the y-axis are $x = a$ and $x = b$ where $b > a$ (Fig. 19.7). Consider a strip $AA'B'B$ of the region enclosed by the curve of width δx and parallel to the y-axis, where A is the point (x, y_1) and B is (x, y_2). Then the volume of revolution of this strip about the x-axis is $\pi(y_2^2 - y_1^2)\delta x$. Hence the total volume of revolution about the axis of the area S enclosed by the curve is given by

$$V = \int_a^b \pi (y_2 - y_1)(y_2 + y_1)\, dx.$$

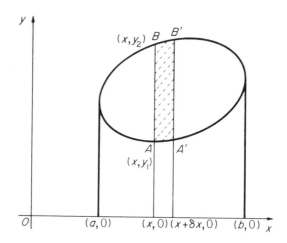

FIG. 19.7.

But $S = \displaystyle\int_a^b (y_2 - y_1)dx$ and, if $\bar{y}$ is the y-coordinate of the centroid of the region,

$$\bar{y} = \frac{\displaystyle\int_a^b (y_2 - y_1)\tfrac{1}{2}(y_2 + y_1)\,dx}{\displaystyle\int_a^b (y_2 - y_1)\,dx}$$

$$\Rightarrow V = 2\pi\bar{y}S. \tag{19.12}$$

The proof can be extended to any closed region which can be divided into portions each of which has only two tangents perpendicular to the axis of revolution, and it is clearly independent of the position in the plane of the axis of revolution provided that this axis does not cut the region (though it may be part of the boundary of the region as in Example 2 following).

The theorem may be useful *either* in finding V when y and S are known *or* in finding y when V and S are known.

Example 1. Calculate the volume of the solid generated by the complete revolution of the triangle ABC, where A is $(2, 3)$, B is $(-2, 4)$ and C is $(3, 5)$, about the line $4x - 3y = 12$.

Using the result of Volume 1, p. 5, the area of $\triangle ABC$ is

$$\tfrac{1}{2}|-10 - 12 + 9 - 10 + 8 + 6| = 4\tfrac{1}{2}.$$

The coordinates of the centroid of the triangle are $[\tfrac{1}{3}(2 - 2 + 3), \tfrac{1}{3}(3 + 4 + 5)]$, i.e. $(1, 4)$. The perpendicular distance of $(1, 4)$ from $4x - 3y - 12 = 0$ is

$$\left|\frac{4 - 12 - 12}{5}\right| = 4$$

and the triangle ABC lies completely on the same side of the given line as the origin. Hence the volume of revolution is $2\pi \times 4 \times 4\frac{1}{2} = 36\pi$.

Example 2. The region R bounded by the curve $y = \cos^2 x$ and the x-axis from $x = 0$ to $x = \frac{1}{2}\pi$ makes a complete revolution about the y-axis. Find the volume of revolution and deduce the value of $\int_0^1 (\cos^{-1} \sqrt{x})^2 \, dx$.

Let $(\bar{x}, \bar{y})$ be the centroid of the region R. The area of R is

$$\int_0^{\pi/2} \cos^2 x \, dx = \left[\tfrac{1}{2}x + \tfrac{1}{4} \sin 2x \right]_0^{\pi/2} = \tfrac{1}{4}\pi$$

$$\Rightarrow \bar{x} = \frac{4}{\pi} \int_0^{\pi/2} x \cos^2 x \, dx$$

$$= \frac{4}{\pi} \int_0^{\pi/2} x \left(\frac{1}{2} + \frac{\cos 2x}{2} \right) dx$$

$$= \frac{4}{\pi} \left[\frac{x^2}{4} + \frac{x \sin 2x}{4} - \int \frac{\sin 2x}{4} \, dx \right]_0^{\pi/2}$$

$$= \frac{4}{\pi} \left[\frac{x^2}{4} + \frac{x \sin 2x}{4} + \frac{\cos 2x}{8} \right]_0^{\pi/2}$$

$$= \frac{4}{\pi} \left(\frac{\pi^2 - 4}{16} \right) = \frac{\pi^2 - 4}{4\pi}.$$

Hence the volume of revolution about the y-axis is

$$2\pi \left(\frac{\pi^2 - 4}{4\pi} \right) \frac{\pi}{4} = \pi (\pi^2 - 4)/8$$

$$\Rightarrow \int_0^1 \pi x^2 \, dy = \int_0^1 \pi (\cos^{-1} \sqrt{y})^2 \, dy = \pi (\pi^2 - 4)/8$$

$$\Rightarrow \int_0^1 (\cos^{-1} \sqrt{x})^2 \, dx = (\pi^2 - 4)/8.$$

Example 3. Use the theorem of Pappus to find the position of the centroid of a semicircular lamina of radius r and centre O.

The centroid G clearly lies on the axis of symmetry of the lamina. Let $OG = y$.

When the semicircle makes a complete revolution about its bounding diameter, it generates a sphere of volume $4\pi r^3/3$.

By Pappus' theorem

$$4\pi r^3/3 = 2\pi \bar{y}.\tfrac{1}{2}\pi r^2$$

$$\Leftrightarrow \bar{y} = 4r/(3\pi).$$

Exercise 19.6

1. Use the theorem of Pappus to find the position of the centre of mass of a lamina in the shape of a quadrant of a circle of radius r.

2. Calculate the volume of the solid formed when the region bounded by the parabola $y^2 = 4ax$ and the latus rectum makes a complete revolution about the directrix.

3. A square $ABCD$ of side $2a$ makes a complete revolution about a line AP in its plane, where $A\hat{B}P = \theta$ and AP does not cut the square again. Find the volume of the solid formed and find its greatest and least values as θ varies.

4. Calculate the volume of revolution of the ellipse

$$4x^2 + 9y^2 - 8x - 36y + 4 = 0$$

about the line $4x + 3y + 5 = 0$.

5. Calculate the volumes of revolution of the triangle bounded by the lines $x - y = 0$, $4x - y - 6 = 0$, $5x - 2y - 3 = 0$ about each coordinate axis.

6. Calculate the area and the coordinates of the centroid of the region bounded by $y = \sinh x$, the x-axis and the line $x = 1$. Hence, calculate the volume of revolution of this region about the y-axis.

7. Use the theorem of Pappus to calculate the y-coordinate of the centroid of the region in the first quadrant bounded by the hyperbola $xy = c^2$ and the chord joining the points $(2c, \tfrac{1}{2}c)$, $(\tfrac{1}{2}c, 2c)$.

8. Use Simpson's rule and the theorem of Pappus to find an approximate value of the y-coordinate of the centroid of the region bounded by the x-axis, the extreme ordinates, and a curve joining the points

x	0	1	2	3	4
y	0	1·18	3·63	10·02	27·29

19:7 MOMENTS OF INERTIA

The moment of inertia, sometimes called the *second moment*, of a system of particles of masses $m_1, m_2, m_3, \ldots, m_n$ about a given straight line (the axis) is defined as

$$\sum_1^n m_r x_r^2,$$

where x_r is the distance of the particle of mass m_r from the axis. If

$$k^2 \sum_1^n m_r = \sum_1^n m_r x_r^2,$$

then k is called the *radius of gyration* of the system of particles about the axis.

The concept of moment of inertia is important in theoretical mechanics because the kinetic energy of a body rotating about a fixed axis is equal to $\frac{1}{2} I \omega^2$, where I is the moment of inertia about the axis and ω is the angular speed of the body.

In the case of a continuous distribution of mass the summation becomes an integral. Examples 1 to 4, which follow, involve the moments of inertia of bodies of standard shapes and it is useful to remember the results.

Example 1. *The moment of inertia of a thin uniform rod of length $2l$ and mass M about an axis through its centre perpendicular to its length.*

Let the line density of the rod be ϱ (Fig. 19.8). Then the M. of I. of a small increment δx of the rod, distance

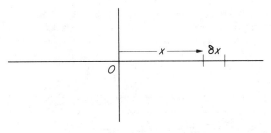

FIG. 19.8.

x from the axis, is approximately $\varrho\delta x.x^2$. Hence the total M. of I. of the rod about the axis is

$$I = \int_{-l}^{l} \varrho x^2 \, \mathrm{d}x = \frac{2\varrho l^3}{3} = \frac{Ml^2}{3}.$$

Example 2. *The moment of inertia of a thin uniform ring of mass M and radius r about an axis through its centre perpendicular to its plane.*
   Since every particle of the ring is at a distance r from the axis, $I = Mr^2$.

Example 3. *The moment of inertia of a uniform disc of radius r and mass M about an axis through its centre and perpendicular to its plane.*
   Let the surface density of the disc be ϱ. Consider a thin ring of the disc, concentric with the disc, of radius x and width δx (Fig. 19.9).

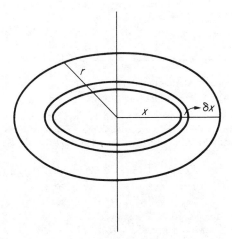

FIG. 19.9.

Then the M. of I. of this thin ring about the axis is approximately $2\pi\varrho x\delta x.x^2$

$$\Rightarrow I = \int_{0}^{r} 2\pi\varrho x.x^2 \, \mathrm{d}x = \frac{\pi\varrho r^4}{2} = \frac{Mr^2}{2}.$$

Example 4. *The moment of inertia of a uniform sphere of radius R and mass M about a diameter.*

Let the density of the sphere be ϱ and its mass M (Fig. 19.10). Consider a thin disc (of the sphere) whose plane faces are perpendicular to the axis at a distance x from the centre of the sphere. Let the thickness of the

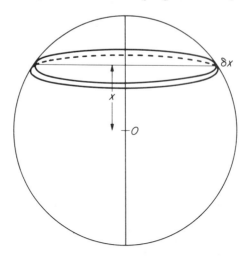

FIG. 19.10.

disc be δx. Then the mass of the disc is approximately $\pi (R^2 - x^2)\delta x \varrho$ and hence the M. of I. of the disc about the axis is approximately $\pi (R^2 - x^2)\delta x \varrho \frac{1}{2}(R^2 - x^2)$. Hence the M. of I. of the sphere about the axis is

$$I = \int_{-R}^{R} \tfrac{1}{2}\pi \varrho (R^2 - x^2)^2 \, dx = \tfrac{1}{2}\pi \varrho \left[R^4 x - \frac{2R^2 x^3}{3} + \frac{x^5}{5} \right]_{-R}^{R} = \frac{8\pi \varrho R^5}{15} = \frac{2\,MR^2}{5}.$$

Example 5. Find the moment of inertia about the axis of revolution of the solid (an ellipsoid) formed by the revolution of the ellipse $x^2/a^2 + y^2/b^2 = 1$ about the x-axis.

Let the density of the solid be ϱ and its mass be M. Then the M. of the I. of the disc formed when the strip of the area of the ellipse bounded by the curve, the x-axis and the ordinates at (x, y), $(x + \delta x, y + \delta y)$ is revolved about the axis is approximately $\pi y^2 \, \delta x \, \varrho.\tfrac{1}{2}y^2$. Hence the M. of I. of the ellipsoid about the axis is

$$\int_{-a}^{a} \frac{\pi \varrho}{2} \left(b^2 - \frac{b^2 x^2}{a^2} \right)^2 dx = \frac{\pi \varrho b^4}{2} \left[x - \frac{2x^3}{3a^2} + \frac{x^5}{5a^4} \right]_{-a}^{a} = \frac{8\pi \varrho ab^4}{15}.$$

Also

$$M = \int_{-a}^{a} \pi \varrho \left(b^2 - \frac{b^2 x^2}{a^2} \right) dx = \frac{4}{3}\pi \varrho ab^2.$$

Hence the M. of I. about the axis is $\dfrac{2Mb^2}{5}$.

Two theorems

(a) *The theorem of perpendicular axes for a lamina.* If Ox and Oy are any two rectangular axes in the plane of a lamina and Oz is an axis at right angles to the plane,

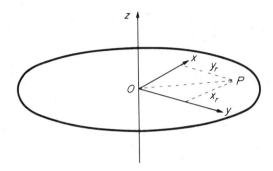

FIG. 19.11.

then the moment of inertia of the lamina about Oz is equal to the sum of the moments of inertia of the lamina about Ox and Oy. (Fig. 19.11.)

For, if m_r is the mass of a particle P of the lamina whose coordinates referred to Ox, Oy are (x_r, y_r), the M. of I. of the particle about Ox is $m_r y_r^2$ and the M. of I. of the particle about Oy is $m_r x_r^2$. But $m_r y_r^2 + m_r x_r^2 = m_r(y_r^2 + x_r^2) = m_r OP^2$ and this is equal to the M. of I. of P about Oz. Hence for the whole lamina, the sum of the moments of inertia about Ox and Oy is equal to the moment of inertia about Oz.

(b) *The theorem of parallel axes.* The moment of inertia of a body about any axis is equal to the moment of inertia about a parallel axis through the centre of mass together with the mass of the body multiplied by the square of the distance between the axes.

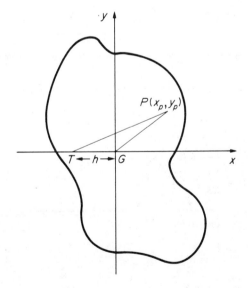

FIG. 19.12.

Fig. 19.12 shows a *section* of the body through the centre of mass G. This section is at right angles to the axis and meets the axis at T where $GT = h$. Take G as the origin and TG produced as the x-axis. A linear element of the solid of mass m_p which is parallel to the axis meets the section at $P(x_p, y_p)$. Then

$$m_p PT^2 = m_p[y_p^2 + (h + x_p)^2].$$

Hence the M. of I. of the body about the axis through T is

$$\Sigma m_p PT^2 = \Sigma m_p y_p^2 + \Sigma m_p h^2 + 2\Sigma m_p h x_p + \Sigma m_p x_p^2$$
$$= \Sigma m_p (x_p^2 + y_p^2) + \Sigma m_p h^2 + 2\Sigma m_p h x_p.$$

But $\Sigma m_p (x_p^2 + y_p^2) = $ M. of I. of the body about a parallel axis through G,

$\Sigma m_p h^2 = M h^2$ where M is the mass of the body,

$\Sigma m_p h x_p = h \Sigma m_p x_p = 0$ because the x-coordinate of the centre of mass of the system is given by $\bar{x} = \Sigma m_p x_p / \Sigma m_p$ and in this case $\bar{x} = 0$ since G is the origin of coordinates.

Hence the M. of I. of the body about the axis through T is equal to (the M. of I. of the body about the parallel axis through G) $+ M.TG^2$.

Example 1. Find the moment of inertia of a uniform disc of mass M and radius r about a tangent in the plane of the disc.

By the perpendicular axes theorem the sum of the moments of inertia of the disc about each of two diameters at right angles is equal to the moment of inertia of the disc about an axis through its centre perpendicular to its plane. Hence twice the M. of I. for the disc about a diameter is $Mr^2/2$

$\Rightarrow$ the M. of I. for the disc about a diameter is $Mr^2/4$.

Hence, by the parallel axes theorem the M. of I. for the disc about the tangent parallel to the chosen diameter is $Mr^2/4 + Mr^2 = 5Mr^2/4$

$\Rightarrow$ the moment of inertia of a uniform disc about a tangent is $5Mr^2/4$.

Example 2. Find the M. of I. of a uniform cylinder, of mass M, radius r and length l, about a diameter of one end.

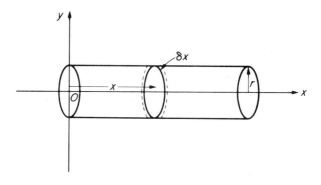

FIG. 19.13.

Consider a thin disc of the cylinder, thickness δx, perpendicular to the axis of the cylinder and distant x from Oy (Fig. 19.13). The M. of I. of this disc about the given axis is $\varrho \pi r^2 \, \delta x . \frac{1}{4} r^2 + \varrho \pi r^2 \, \delta x . x^2$. Hence the total M. of I. of the cylinder about a diameter of one end is

$$\int_0^l \varrho\pi\left(\frac{1}{4}r^4 + x^2r^2\right)dx = \varrho\pi\left[\frac{r^4x}{4} + \frac{x^3r^2}{3}\right]_0^{l}$$

$$= \varrho\pi\left(\frac{r^4l}{4} + \frac{l^3r^2}{3}\right) = M\left(\frac{r^2}{4} + \frac{l^2}{3}\right).$$

Example 3. Find the radius of gyration of an equilateral triangular lamina of side $2a$ (i) about an axis through one vertex parallel to the opposite side, (ii) about an axis through the centroid perpendicular to the lamina.   (L.)

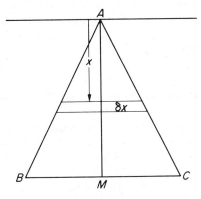

FIG. 19.14.

(i) Consider a thin strip of the lamina ABC parallel to BC, distance x from A and width δx (Fig. 19.14). Let the length of the side of the lamina be $2a$, its surface density be ϱ and its mass be m. Then the length of the strip is $2x/\sqrt{3}$ and the M. of I. of the strip about an axis through A and parallel to BC is approximately

$$\frac{2\varrho x\delta x}{\sqrt{3}}.x^2.$$

Hence the M. of I. of the lamina about this axis is

$$\int_0^{a\sqrt{3}} \frac{2\varrho x}{\sqrt{3}}x^2\,dx = \frac{2\varrho}{\sqrt{3}}\left[\frac{x^4}{4}\right]_0^{a\sqrt{3}} = \frac{3\sqrt{3}\varrho a^4}{2} = \frac{3\,ma^2}{2},$$

and the required radius of gyration is $a\sqrt{(3/2)}$.

(ii) By the parallel axes theorem the M. of I. of the lamina about an axis through the centroid parallel to BC is

$$\frac{3ma^2}{2} - m\left(\frac{2\sqrt{3}a}{3}\right)^2 = \frac{ma^2}{6}.$$

The M. of I. of the strip about the median AM is approximately

$$\frac{2\varrho x\delta x}{\sqrt{3}}\cdot\frac{x^2}{9}$$

and the M. of I. of the lamina about AM is

$$\frac{2\varrho}{9\sqrt{3}}\int_0^{a\sqrt{3}} x^3\,dx = \frac{2\varrho}{9\sqrt{3}}\frac{9a^4}{4} = \frac{ma^2}{6}.$$

Hence the M. of I. of the lamina about an axis through the centroid perpendicular to the lamina is

$$\frac{ma^2}{6} + \frac{ma^2}{6} = \frac{ma^2}{3},$$

giving the required radius of gyration as $a/\sqrt{3}$.

Exercise 19.7

1. Calculate the M. of I. about its axis of the uniform solid formed when the region bounded by $x^2 = 4ay$ and its latus rectum is revolved completely about the y-axis.

2. Calculate the M. of I. about its axis of the uniform solid formed when the region bounded by the curve $ay^2 = x^3$ and the line $x = a$ revolves completely about the x-axis.

3. Calculate the M. of I. about the x-axis of a uniform lamina in the shape of the region bounded by the curve $y = \sin x$ and the x-axis between $x = 0$ and $x = \pi$.

4. Calculate the M. of I. about each coordinate axis of a lamina in the shape of the region bounded by $y = 1/x$, $x = 1$, $x = 2$ and the x-axis.

5. Calculate the M. of I. of a uniform solid right circular cone of height h and base-radius r (i) about its axis, (ii) about a diameter of its base.

6. Calculate the radius of gyration about a diameter of a hollow sphere formed by removing a concentric sphere of radius b from a sphere of radius a.

7. Prove that the M. of I. of a thin uniform rod, of length $2l$ and mass M, about an axis through its centre at an angle θ with the rod is $\frac{1}{3}Ml^2 \sin^2 \theta$.

Calculate the M. of I. of a framework $ABCDEF$ consisting of a regular hexagon of rods each of length $2l$ and mass m,

(i) about AD,

(ii) about an axis through its centre perpendicular to its plane.

8. Calculate the M. of I. about an axis through one end and perpendicular to its length of a thin rod, of length l and of density ϱx at a point distance x from the axis.

9. Calculate the radius of gyration about its axis of revolution of the solid formed by the complete revolution of a circle of radius R about a tangent.

10. Calculate the limiting value of the radius of gyration of the solid formed by the rotation about the x-axis of the region bounded by the curve $y = 1/(1 + x^2)$, the x-axis and the lines $x = -a$, $x = +a$ as a increases indefinitely.

11. A uniform equilateral triangular lamina has side $2a$ and mass M. Find its moment of inertia about (i) a side; (ii) an axis through a vertex perpendicular to its plane; (iii) an axis parallel to the last one through its centre of mass. (O.C.)

12. A uniform solid right circular cone is of mass M and density ρ. Its base radius is r and its height kr. Prove that the moment of inertia of the cone about a line through its vertex parallel to its base is $\frac{1}{20}\pi\rho kr^5 (1 + 4k^2)$.

Express this result in terms of M, ρ and k, and deduce that if k can vary, the minimum value of this moment of inertia is $(9M/20)(3M/\pi\rho)^{2/3}$. (C.)

13. Sketch the curve with equation $y^2 = xe^{-x}$.

A uniform solid of revolution is generated by revolving the region bounded by the curve and the line $x = 1$ about the x-axis. Prove that, if the density of the solid is ρ, its moment of inertia about the x-axis is $\frac{1}{8}\pi\rho (1 - 5e^{-2})$.

19:8 MISCELLANEOUS EXAMPLES

Example 1. If $F(x) = \int_a^x f(t)\,dt$, where $f(t)$ is continuous, then $F'(x) = f(x)$.

For

$$F(x + \delta x) - F(x) = \int_a^{x+\delta x} f(t)\,dt - \int_a^x f(t)\,dt$$

$$= \int_x^{x+\delta x} f(t)\,dt = f(x)\,\delta x + O[(\delta x)^2].$$

Dividing by δx and letting $\delta x \to 0$ gives the stated result.
 This, of course, is the fundamental theorem of calculus.

---

*Example 2. Prove that, for $x > 0$.

$$\int_0^x [t]\,dt = (x - \tfrac{1}{2})\,[x] - \tfrac{1}{2}[x]^2,$$

where $[t]$ is the greatest integer $\leqslant t$.

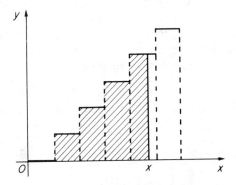

Fig. 19.15.

The graph of $y = [x]$ is sketched in Fig. 19.15.

Clearly $\int_0^x [t]\,dt$ is the sum of the areas of $[x]$ rectangles shown shaded in that figure. The first $[x - 1]$ of these rectangles are of areas $1, 2, \ldots, [x - 1]$ and the last rectangle is of area $(x - [x])\,[x]$. Hence

$$\int_0^x [t]\,dt = (1 + 2 + \ldots + [x - 1]) + (x - [x])\,[x]$$

$$= \tfrac{1}{2}[x - 1]\,[x] + (x - [x])\,[x]$$

and, using the relation $[x - 1] = [x] - 1$, we have the required result.

Example 3. In books on analysis the logarithm function is defined by

$$\ln x = \int_1^x \frac{dt}{t}, \qquad x > 0.$$

From this definition we can prove the rules given in §11:7. For example,

$$\ln (xy) = \int_1^{xy} \frac{dt}{t} = \int_1^x \frac{dt}{t} + \int_x^{xy} \frac{dt}{t}.$$

Writing $t = xu$ transforms the integral $\displaystyle\int_x^{xy} \frac{dt}{t}$ into $\displaystyle\int_1^y \frac{du}{u}$. Therefore

$$\ln (xy) = \int_1^x \frac{dt}{t} + \int_1^y \frac{du}{u} = \ln x + \ln y.$$

As an exercise the reader should prove from this definition the other properties of the logarithm function.

19:9 INEQUALITIES INVOLVING INTEGRALS

Although many indefinite integrals cannot be evaluated in closed form, numerical values of definite integrals involving the same integrands may be obtained by means of the techniques of §3:5.

However, some additional confidence in such results and rough approximations to other definite integrals can be obtained by considering inequalities. Also questions of convergence of integrals may sometimes be resolved by the use of inequalities.

$$\text{If } f(x) \geq 0 \quad \text{for} \quad a \leq x \leq b, \, a < b, \quad \text{then} \quad \int_a^b f(x)\, dx \geq 0.$$

This result follows at once by considering the definite integral as the limit of a sum.

Note that if $f(x) > 0$ in the domain $a \leq x \leq b$, $a < b$, then $\displaystyle\int_a^b f(x)\, dx > 0$.

Corollaries

(i) Since $f(x) \leq |f(x)|$, it follows that

$$\int_a^b \{|f(x)| - f(x)\}\, dx \geq 0$$

$$\Rightarrow \int_a^b |f(x)|\, dx \geq \int_a^b f(x)\, dx. \qquad\qquad (19.13)$$

Note that equation (19.13) implies the inequality

$$\int_a^b |f(x)| \, dx \geqslant \left| \int_a^b f(x) \, dx \right|. \tag{19.14}$$

(ii) If $\phi(x) \geqslant \psi(x)$ for $a \leqslant x \leqslant b$, then by considering $f(x) = \phi(x) - \psi(x)$ we have

$$\int_a^b \phi(x) \, dx \geqslant \int_a^b \psi(x) \, dx.$$

Also, if $\phi(x) > \psi(x)$ for $a \leqslant x \leqslant b$, $a < b$, *then*

$$\int_a^b \phi(x) \, dx > \int_a^b \psi(x) \, dx.$$

These results can be illustrated by considering the graphs of the various functions concerned.

Example 1. Without evaluating the integral, show that

$$1 > \int_0^1 \frac{1}{\sqrt{(1+3x^3)}} \, dx > \frac{1}{2}.$$

When
$$0 < x < 1,$$
$$1 < 1 + 3x^3 < 4$$
$$\Rightarrow 1 < \sqrt{(1+3x^3)} < 2 \quad \text{(taking positive square roots)}$$
$$\Rightarrow 1 > \frac{1}{\sqrt{(1+3x^3)}} > \frac{1}{2}$$

(taking reciprocals and reversing the inequality signs).
Integration between $x = 0$ and $x = 1$ now gives the required result.

Example 2. By considering

$$\int_a^b \left[f(x) + \lambda g(x) \right]^2 dx$$

show that

$$\left[\int_a^b fg \, dx \right]^2 \leqslant \int_a^b f^2 \, dx \times \int_a^b g^2 \, dx.$$

Since $\{f(x) + \lambda g(x)\}^2 \geqslant 0$ for real functions and real λ, it follows that

$$\int_a^b \{f(x) + \lambda g(x)\}^2 \, dx \geqslant 0$$

$$\Rightarrow \int_a^b [f(x)]^2 \, dx + 2\lambda \int_a^b f(x) \, g(x) \, dx + \lambda^2 \int_a^b [g(x)]^2 \, dx \geq 0. \tag{1}$$

Regarding the left-hand side of inequality (1) as a quadratic form in λ we see that it cannot have real distinct factors. The required result follows.

Example 3. Without evaluating the integrals, prove that, for $n > 0$,

$$\int_0^{\pi/2} \sin^n x \, dx > \int_0^{\pi/2} \sin^{n+1} x \, dx.$$

In the range $0 \leqslant x \leqslant \frac{1}{2}\pi$,

$$\sin^{n+1} x = \sin^n x . \sin x \leqslant \sin^n x$$

because $\sin x \leqslant 1$. If $n > 0$, $\sin^n (0) = \sin^{n+1} (0) = 0$ and $\sin^n (\frac{1}{2}\pi) = \sin^{n+1} (\frac{1}{2}\pi) = 1$. The graphs of $\sin^n x$ and $\sin^{n+1} x$ are therefore as shown in Fig. 19.16 and the area of the region enclosed by the curve $y = \sin^n x$, the x-axis and the ordinate $x = \frac{1}{2}\pi$ is greater than the area of the region enclosed by the curve $y = \sin^{n+1} x$, the x-axis and the ordinate $x = \frac{1}{2}\pi$

$$\Rightarrow \int_0^{\pi/2} \sin^n x \, dx > \int_0^{\pi/2} \sin^{n+1} x \, dx \qquad \forall n > 0.$$

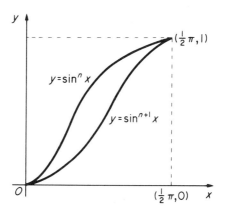

FIG. 19.16.

We can use this result in conjunction with Wallis' formulae, p. 636, to obtain numerical approximations for π. For example,

$$\int_0^{\pi/2} \sin^9 x \, dx > \int_0^{\pi/2} \sin^{10} x \, dx > \int_0^{\pi/2} \sin^{11} x \, dx$$

$$\Rightarrow \frac{8.6.4.2}{9.7.5.3} > \frac{9.7.5.3.1}{10.8.6.4.2} \frac{\pi}{2} > \frac{10.8.6.4.2}{11.9.7.5.3}$$

$$\Rightarrow 3 \cdot 30 > \pi > 3 \cdot 00.$$

Exercise 19.9

1. Show that, for $0 < x < \pi/2$,

$$x > \sin x > 2x/\pi.$$

Deduce that

$$1 - e^{-\pi/2} < \int_0^{\pi/2} e^{-\sin x}\, dx < (e-1)\pi/2e.$$

2. Prove that, if $0 \leqslant x < 1$,

$$\left(1 + \frac{x^2}{2}\right)^2 \leqslant \frac{1}{1-x^2}$$

and deduce that

$$x + \frac{x^3}{6} \leqslant \sin^{-1} x,$$

if $0 \leqslant x < 1$.

3. Prove that $\displaystyle\int_0^1 \frac{x^4 (1-x)^4\, dx}{1+x^2} = \frac{22}{7} - \pi.$

Show that $\displaystyle\frac{1}{2}\int_0^1 x^4 (1-x)^4\, dx < \int_0^1 \frac{x^4 (1-x)^4\, dx}{1+x^2} < \int_0^1 x^4 (1-x)^4\, dx,$

and hence deduce that $\displaystyle\frac{22}{7} - \frac{1}{1260} > \pi > \frac{22}{7} - \frac{1}{630}.$

Miscellaneous Exercise 19

1. (a) By means of the substitution $y = -x$, or otherwise, evaluate $\displaystyle\int_{-1}^1 x^2 \tan x\, dx.$

   (b) Determine the values of the constants A, B for which

$$\frac{x^4}{x^2+1} \equiv A(x^2 - 1) + \frac{B}{x^2+1}.$$

Show that this result may be used in the integration of the function $x^3 \tan^{-1} x$, and complete this integration. Deduce the area of the region bounded by the curve $y = x^3 \tan^{-1} x$ and the lines $x = 1$, $y = 0$.

2. Show that

$$\int_0^t \sin \omega x \cos \omega(t-x)\, dx = \tfrac{1}{2} t \sin \omega t.$$

3. (a) Evaluate (i) $\displaystyle\int_0^2 \frac{1}{\sqrt{(4-x^2)}}\, dx,$ (ii) $\displaystyle\int_{\pi/4}^{\pi/2} x \sin 2x\, dx.$

   (b) By means of the substitution $x = \tfrac{1}{2}\pi - y$, prove that

$$\int_0^{\pi/2} \frac{\cos x}{\cos x + \sin x}\, dx = \int_0^{\pi/2} \frac{\sin x}{\cos x + \sin x}\, dx.$$

By considering the sum of these integrals, or otherwise, determine their common value.   (N.)

4. Show that the area under the curve $y = 1/x$, from $x = n-1$ to $x = n+1$, is $\ln\{(n+1)/(n-1)\}$, provided that $n > 1$.

By applying Simpson's rule to this area, deduce that, approximately,

$$\ln\left(\frac{n+1}{n-1}\right) = \frac{1}{3}\left(\frac{1}{n-1} + \frac{4}{n} + \frac{1}{n+1}\right),$$

and show that the error in this approximation is $4/(15n^5)$, when higher powers of $1/n$ are neglected. (N.)

5. A particle describes simple harmonic motion in which the displacement x is given in terms of the time t by the equation

$$x = a \sin t.$$

Find, for the interval $t = 0$ to $t = \frac{1}{2}\pi$,
  (i) the mean value of the speed with respect to the time,
  (ii) the mean value of the speed with respect to the distance.

6. Prove that

$$\int_a^b f(x)\,dx = \int_a^b f(a+b-x)\,dx.$$

Hence evaluate

$$\int_{\pi/6}^{\pi/3} \frac{\cos^2 x}{x(\pi - 2x)}\,dx. \tag{C.}$$

7. Show that, if m and $n \in \mathbb{N}$,

$$\int_0^\pi \cos mx \cos nx \,dx = 0$$

when $m \neq n$, and find the value of the integral when $m = n$.
  Prove that

$$2 \sin x \,(\cos x + \cos 3x + \cos 5x + \cos 7x) = \sin 8x.$$

Hence, or otherwise, show that

$$\int_0^\pi \frac{\sin^2 8x}{\sin^2 x}\,dx = 8\pi.$$

8. Prove that

$$\int_{-a}^a f(x)\,dx = \int_0^a [f(x) + f(-x)]\,dx.$$

Hence show that

$$\int_{-\pi/2}^{\pi/2} \frac{e^x}{1+e^x} \sin^4 x \,dx = \int_0^{\pi/2} \sin^4 x \,dx$$

and evaluate the latter integral.   (N.)

9. If $0 < \alpha < \pi$, show that

$$\int_{\alpha}^{\pi-\alpha} \frac{x \, dx}{\sin x} = \int_{\alpha}^{\pi-\alpha} \frac{\pi - x}{\sin x} \, dx.$$

Prove that

$$\int_{\pi/3}^{2\pi/3} \frac{x \, dx}{\sin x} = \tfrac{1}{2}\pi \ln 3. \tag{N.}$$

10. Prove that $\displaystyle\int_{-a}^{a} f(x) \, dx = \int_{0}^{a} \left[f(x) + f(-x) \right] dx.$

Hence show that $\displaystyle\int_{-\pi/4}^{\pi/4} \frac{1}{1 + \sin x} \, dx = 2 \int_{0}^{\pi/4} \sec^2 x \, dx$, and evaluate this integral.

Use the first result to evaluate $\displaystyle\int_{-1}^{1} \frac{1}{1 + e^{-x}} \, dx. \tag{N.}$

11. Evaluate (i) $\displaystyle\int_{0}^{\pi/2} (\cos\theta - \sin\theta)^3 \, d\theta,$ (ii) $\displaystyle\int_{0}^{\pi/2} \theta \, (\cos\theta - \sin\theta)^2 \, d\theta.$ $\qquad$ (N.)

12. If I_n denotes $\displaystyle\int_{0}^{a} (a^2 - x^2)^n \, dx$, prove that, if $n > 0$,

$$I_n = \frac{2na^2}{2n+1} I_{n-1}. \tag{O.C.}$$

13. Prove that

$$\int_{0}^{\pi/2} f(\sin 2x) \sin x \, dx = \sqrt{2} \int_{0}^{\pi/4} f(\cos 2x) \cos x \, dx. \tag{O.C.}$$

14. Show that

$$\int_{0}^{a} f(x) \, dx = \int_{0}^{a} f(a - x) \, dx.$$

Deduce that

$$\int_{0}^{\pi/4} \left(\frac{1 - \sin 2x}{1 + \sin 2x} \right) dx = \int_{0}^{\pi/4} \tan^2 x \, dx,$$

and evaluate the integral. $\qquad$ (O.C.)

15. By means of the substitution $x = 1/t$ and considering the intervals -1 to 0 and 0 to 1 separately, prove that

$$\int_{-1}^{1} \frac{1}{1 - 2x \cos\theta + x^2} \, dx = \tfrac{1}{2} \int_{-\infty}^{\infty} \frac{1}{1 - 2x \cos\theta + x^2} \, dx$$

provided that θ is *not* an integral multiple of π.

Evaluate $\displaystyle\int_{-1}^{1} \frac{1}{1 - 2x\cos\theta + x^2}\,dx$ when θ lies between 0 and π.

What is the value of the integral (i) when $\theta = -\pi/4$, (ii) when $\theta = 5\pi/4$?   (N.)

16. Given that $\displaystyle C = \int_{0}^{\pi/2} \frac{\cos^2 x}{a^2\sin^2 x + b^2\cos^2 x}\,dx$

and

$$S = \int_{0}^{\pi/2} \frac{\sin^2 x}{a^2\sin^2 x + b^2\cos^2 x}\,dx,$$

prove that $C + S = \pi/2ab$ when $ab > 0$. Obtain another such relation between C and S and hence determine them when $a \neq b$.

Show that $\displaystyle\int_{0}^{\pi} \frac{\cos^2 x}{a^2\cos^2 x + b^2\sin^2 x}\,dx = \frac{\pi(b - a)}{ab(a + b)}.$ (N.)

17. Evaluate $\displaystyle\int_{0}^{\infty} x^2 e^{-x/2}\,dx$. It may be assumed that $x^n e^{-x/2} \to 0$ as $x \to \infty$.   (L.)

18. If $\displaystyle u_n = \int_{0}^{\pi/2} x^n \sin x\,dx$, and $n > 1$, show that

$$u_n = n(\pi/2)^{n-1} - n(n-1)u_{n-2},$$

and evaluate u_4.   (O.C.)

19. (i) Prove that, if

$$I_n = \int \sec^n x\,dx,$$

then

$$(n-1)I_n = \tan x\,\sec^{n-2} x + (n-2)I_{n-2}.$$

Use this formula to evaluate

$$\int_{0}^{\pi/4} \sec^5 x\,dx.$$

(ii) Find the positive value of x for which the definite integral

$$\int_{0}^{x} \frac{1-t}{\sqrt{(1+t)}}\,dt.$$

is greatest.

Evaluate the integral for this value of x.   (O.C.)

20. If $\displaystyle u_n = \int_{0}^{\pi/2} \frac{\sin(2n+1)\theta}{\sin\theta}$, evaluate $u_n - u_{n-1}$, and hence show that $u_n = \frac{1}{2}\pi$ for all positive integers n.

Using a similar method, prove that

$$\int_0^{\pi/2} \frac{\sin^2 n\theta}{\sin^2 \theta} \, d\theta = \tfrac{1}{2} n\pi,$$

and find the value of

$$\int_0^{\pi/2} \frac{(2n+1)\sin\theta - \sin(2n+1)\theta}{\sin^3 \theta} \, d\theta,$$

where n is a positive integer in each case.   (N.)

21. Prove that $\int \sec^n \theta \, d\theta = \dfrac{\sec^{n-2}\theta \tan\theta}{n-1} + \dfrac{n-2}{n-1} \int \sec^{n-2}\theta \, d\theta.$

By using the parameter θ given by $y = \tan^3 \theta$, or otherwise, find the area of the region between the curve

$$x^{2/3} = y^{2/3} + 1$$

and the ordinate $x = 2\sqrt{2}$.

22. If

$$u_n = \int_0^x \frac{dt}{(1+t^2)^n},$$

where $n \in \mathbb{N}$, prove that

$$2n u_{n+1} = (2n-1)u_n + \frac{x}{(1+x^2)^n}.$$

Evaluate

$$\int_0^\infty \operatorname{sech}^{2n-1}\theta \, d\theta,$$

where $n \in \mathbb{N}$.

23. The horizontal speed v of a point P is given by

$$v = v_0 \{\sin \omega t + |\sin \omega t|\}$$

where v_0 and ω are positive constants. Sketch the graphs of v and dv/dt from $t = 0$ to $t = 6\pi/\omega$.

If P is a point on one of the feet of a pedestrian whose average speed is V, find the ratio of v_0 to V.
                                                                             (N.)

24. If A denotes the region bounded by the coordinate axes, the ordinate $x = 1$ and the arc of the curve $y(1+x^2) = 1$ from $x = 0$ to $x = 1$, calculate the volumes obtained by rotating A about (i) the x-axis, (ii) the y-axis. Hence, by using Pappus' theorem, find the coordinates of the centroid of the region A.

25. (a) A number a can take any value between 0 and n, all values being equally likely. Prove that the mean value of $a(n-a)$ is $n^2/6$.

(b) In the quadratic equation $x^2 - px + q = 0$ both p and q are positive and p is fixed. If all positive values of q are equally likely, subject to the condition that the roots of the equation are real, prove that the mean value of the larger root is $5p/6$.
                                                                             (N.)

26. A particle moves along the x-axis, its speed dx/dt at time t being given by

$$dx/dt = \sqrt{(9t - t^3)}, \quad 0 \leqslant t \leqslant 3.$$

Use Simpson's rule to calculate approximately the distance travelled by the particle during the interval from $t = 0$ to $t = 3$. (Use steps of $0\cdot5$ in t.)

Show that the average speed is approximately three-quarters of the greatest speed of the particle in this interval.
                                                                             (N.)

27. Draw a rough sketch of the curve

$$y^2 = x^4 - 14ax^3 + 45a^2x^2.$$

The region enclosed by the loop of the curve is rotated about the x-axis. Find the distance from the origin of the centre of mass of the solid generated.   (O.C.)

28. The portion of the curve $y^2 = 4ax$ from $(a, 2a)$ to $(4a, 4a)$ revolves round the tangent at the origin. Prove that the volume bounded by the curved surface so formed and plane ends perpendicular to the axis of revolution is $\frac{62}{5}\pi a^3$ and find the square of the radius of gyration of this volume about its axis of revolution.
                                                                                                (O.C.)

29. Sketch the curve

$$y^2 (a + x) = x^2 (3a - x).$$

By means of the substitution $a + x = 4a \sin^2 \theta$, or otherwise, find the area of the region in the second and third quadrants between the curve and its asymptote.   (O.C.)

30. By the methods of the integral calculus find
(i) the area and also the coordinates of the centre of mass of a uniform plane lamina bounded by the axes of coordinates and that arc of the ellipse

$$\frac{x^2}{a^2} + \frac{y^2}{b^2} = 1$$

which lies in the positive quadrant;
(ii) the volume and also the position of the centre of mass of a uniform hemisphere whose base is a circle of radius a.   (O.C.)

31. The region enclosed between the curve $y = 3x - x^2$ and the straight line $y = x$ is rotated through four right angles about the axis of x. Prove that the volume of the solid generated is $56\pi/15$.
Calculate the distance from the origin of the centre of mass of this solid.   (C.)

32. Find the volume of the solid generated when the ellipse $\dfrac{x^2}{a^2} + \dfrac{y^2}{b^2} = 1$ is rotated completely about the axis of x.

The solid is cut into two portions by the plane formed by rotation of the line $x = a/2$ about the axis of x. Show that the centre of mass of the smaller of the two portions is distant $27a/40$ from the origin.

33. A sphere of radius r is cut into two portions by a plane which is distant c from the centre of the sphere. Show that the volume of the smaller of the two portions is

$$\tfrac{1}{3}\pi (r - c)^2 (2r + c).$$

Show also that the distance of the centre of mass of this portion from the centre of the sphere is $3(r + c)^2/4(2r + c)$.
Deduce the position of the centre of mass of a uniform solid hemisphere.

34. Show that $(\ln x)/\sqrt{x}$ has a maximum value when $x = e^2$.
The curve $y = (\ln x)/\sqrt{x}$ meets the x-axis at A, and B is the foot of the maximum ordinate PB. Show that the distance from the x-axis of the centroid of the region bounded by AB, BP and the arc AP of the curve is $1/3$.

35. The boundary of a uniform lamina consists of a straight part OA of length π and a curved part OPA, the equation of which, referred to rectangular axes OA and OB, is $y = \sin x$. Show that the square of the radius of gyration of the lamina about OB is $\tfrac{1}{2}\pi^2 - 2$.   (N.)

36. By means of Simpson's rule and taking unit intervals of x from $x = 8$ to $x = 12$, find approximately the area of the region enclosed by the curve $y = \ln x$, the lines $x = 8$ and $x = 12$, and the x-axis. Deduce the average value of $\ln x$ between $x = 8$ and $x = 12$.

37. The distance s fallen by a particle dropped from rest is given in terms of t by the relation $s = 16t^2$. Find the speed v both in terms of t and in terms of s. If the final speed on reaching the ground is V, find in terms of V, the mean value of v both with respect to time and with respect to distance, and show that the former is three-quarters of the latter.

38. The region bounded by the curve $y^2 = 4a(b - x)$ and the y-axis is rotated through two right angles about the x-axis. Show that the volume of the solid of revolution is $2\pi ab^2$, and find the radius of gyration of this solid about the y-axis.

39. Show that the centre of mass of a uniform semicircular disc of mass M and radius a is at a distance $4a/3\pi$ from its bounding diameter, AB, and find, by integration, the moment of inertia of the disc about AB.
Hence determine the moment of intertia of the disc about the tangent which is parallel to AB.   (L.)

40. A lamina in the x, y plane is bounded by the arc of the curve $y = a \cos(x/a)$, between the points $(0, a)$, $(\frac{1}{2}\pi a, 0)$, and the parts of the coordinate axes between the origin and these points. Show that the radii of gyration of the lamina about the axes of x and y are $\frac{1}{3}a\sqrt{2}$ and $\frac{1}{2}a\sqrt{(\pi^2 - 8)}$.

41. The equation of a curve is

$$y^2 = \frac{x-1}{x-2}.$$

Show that no part of the curve lies between $x = 1$ and $x = 2$, and sketch the curve for the remaining values of x.
The portion of the curve between $x = 0$ and $x = 1$ is rotated through $360°$ about the axis of x. Find the volume generated and the distance of its centroid from the origin.   (L.)

42. A solid of revolution is generated by rotation about the axis of x of the region bounded by the curve $y = \sec x$ and the lines $x = 0$, $x = \pi/4$ and $y = 0$. Find the volume of the solid, the position of its centroid and its radius of gyration about its axis.

43. Find the values of (i) $\displaystyle\int_0^\pi \cos pt \cos qt \, dt$, (ii) $\displaystyle\int_0^\pi \cos^2 pt \, dt$ where p, q are integers, $p \neq q$.

If an alternating current i is given by the formula $i = a_1 \cos pt + a_2 \sin pt$ where a_1, a_2 are constants, show that the mean value of i^2 taken over the half-period from $t = 0$ to $t = \pi/p$ is $\frac{1}{2}(a_1^2 + a_2^2)$.   (L.)

44. Evaluate

(i) $\displaystyle\int_0^\infty \frac{x \tan^{-1} x}{(1 + x^2)^2} \, dx$, (ii) $\displaystyle\int_1^\infty \frac{dx}{x^2(a^2 + x^2)^{1/2}}$, (iii) $\displaystyle\int_0^\infty \frac{dx}{\cosh^3 x}$,

(iv) $\displaystyle\int_0^\infty \frac{dx}{1 + e^{2x}}$, (v) $\displaystyle\int_0^\infty e^{-ax} \sin x \, dx$ $(a > 0)$.

45. Show that

(i) $\displaystyle\int_0^\infty \frac{x \, dx}{(1 + x^2)(4 + x^2)} = \frac{1}{3} \ln 2$, (ii) $\displaystyle\int_1^\infty \frac{(x^2 + 2) \, dx}{x^2(x^2 + 1)} = 2 - \frac{1}{4}\pi$,

(iii) $\displaystyle\int_0^1 \frac{1}{(1 + x)\sqrt{(1 - x^2)}} \, dx = 1$.

46. Show that the moment of inertia of a uniform disc of radius r and mass m about a diameter is $\frac{1}{4}mr^2$.
A solid of revolution is formed by rotating the interior of the circle $x^2 + y^2 = 2ax$ about the y-axis. Prove that the moment of inertia of the solid about the x-axis is $\frac{9}{8}Ma^2$ where M is the mass.   (O.C.)

47. A uniform cube of side a is cut by two planes parallel to a face and at distances x and $x + \delta x$ from the face, where δx is small. Show that the moment of inertia about a diagonal of that face of the cube of the thin section so formed is

$$\rho a^2 \, \delta x \, (x^2 + \tfrac{1}{12}a^2), \text{ approximately,}$$

where ρ is the density of the material of the cube.

Hence prove that the moment of inertia of a cube of mass M and side a about a diagonal of a face is $\frac{5}{12} Ma^2$.

(O.C.)

48. A solid right circular cone, of mass M, height h and whose base is of radius a, is made of uniform material of density ρ. Show by integration that $M = \frac{1}{3}\pi\rho a^2 h$ and calculate, in terms of M and a, the moment of inertia of the cone about its axis.

Show also that the moment of inertia of the cone about a line through its vertex and perpendicular to its axis is

$$\frac{3M(a^2 + 4h^2)}{20}.$$

Deduce, in terms of M, a and h, the moment of inertia of the cone about a diameter of its base.   (O.C.)

49. (i) $f(x)$ is a function of x which is positive but diminishes steadily as x increases. $I(n)$ denotes the integral $\int_1^n f(x)\,dx$. Prove from graphical considerations that the sum of the series

$$f(1) + f(2) + \ldots + f(n)$$

lies between $I(n+1)$ and $I(n) + f(1)$.

(ii) Without attempting to evaluate them, determine whether the following integrals are positive, negative, or zero:

(a) $\int_0^1 x^3(1-x)^3\,dx$;   (b) $\int_0^\pi \sin^2 x \cos^3 x\,dx$;   (c) $\int_0^\pi e^{-x}\sin x\,dx$.

20 Some Properties of Curves

20:1 THE LENGTH OF A CURVE

Definitions. If $P_1, P_2, P_3, \ldots, P_n$ are vertices of a polygon inscribed in a curve $P_1 P_n$ (Fig. 20.1), the length of the curve $P_1 P_n$ is defined as

$$\lim_{n \to \infty} \sum_{1}^{n-1} P_r P_{r+1}$$

if that limit exists and if the length of *every one* of $P_r P_{r+1}$ tends to zero as $n \to \infty$.

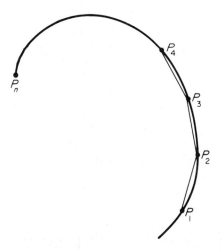

FIG. 20.1.

If the length of the arc measured from a chosen fixed point on the curve is denoted by s, the direction in which s increases may also be chosen arbitrarily. It is conventional to choose this direction so that s increases with x. When the equation of the curve is given in terms of a parameter t in the form $x = f(t), y = g(t)$, it is conventional to choose as the direction in which s increases the direction in which t increases.

The positive direction of the tangent at a point P on the curve is defined as the direction along the tangent which coincides with the direction of s increasing, and one of the angles which this direction makes with the positive direction of the x-axis is

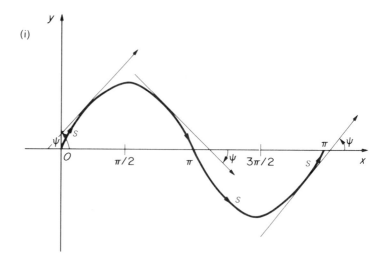

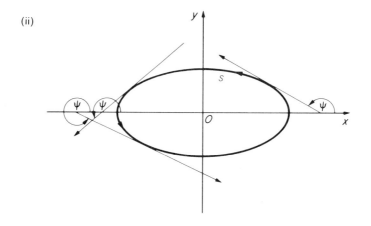

FIG. 20.2.

denoted by ψ. If possible, that value of ψ which ensures that ψ changes continuously with s is chosen. Thus, for example, Fig. 20.2(i) shows the direction of s increasing from the origin O, the positive direction of the tangent, and the angle ψ in three positions on the curve $y = \sin x$.

[Note that ψ is negative for the tangent at the point for which $\pi/2 < x < \pi$.]

Figure 20.2(ii) shows these three variables for the curve $x = a \cos t$, $y = b \sin t$.

Calculation of the length of a curve. (Fig. 20.3) If s is the length of an arc of the curve $y = f(x)$ measured from a fixed point $A[a, f(a)]$, P is the point (x, y) on the curve and Q the point $(x + \delta x, y + \delta y)$, then s is a function of x and the length of the arc PQ is δs, the increase in s which corresponds to the increase δx in x. Then

$$\sin Q\hat{P}M = \frac{\delta y}{\text{chord } PQ} = \frac{\delta y}{\delta s} \cdot \frac{\delta s}{PQ}.$$

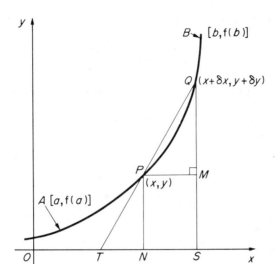

FIG. 20.3.

Similarly

$$\cos Q\hat{P}M = \frac{\delta x}{\delta s} \cdot \frac{\delta s}{PQ}.$$

If now we assume that $\lim\limits_{Q \to P} \dfrac{\delta s}{PQ} = 1$, which is an extension of the assumption we

make in arriving at the result $\lim\limits_{\theta \to 0} \sin\theta/\theta = 1$ in the case of a circle, then

$$\lim_{Q \to P} \sin Q\hat{P}M = \frac{dy}{ds},$$

$$\lim_{Q \to P} \cos Q\hat{P}M = \frac{dx}{ds}$$

$$\Rightarrow \sin\psi = \frac{dy}{ds}, \tag{20.1}$$

$$\cos\psi = \frac{dx}{ds}, \tag{20.2}$$

where ψ is the angle which the tangent at P makes with the positive Ox direction

$$\Rightarrow \left(\frac{ds}{dx}\right)^2 = \sec^2\psi = 1 + \tan^2\psi = 1 + \left(\frac{dy}{dx}\right)^2.$$

Similarly $$\left(\frac{ds}{dy}\right)^2 = 1 + \left(\frac{dx}{dy}\right)^2.$$

Since s has been defined so that it increases with x,

$$\frac{ds}{dx} = +\sqrt{\left\{1+\left(\frac{dy}{dx}\right)^2\right\}}.$$

Hence the length of the arc AB, where B is $[b, f(b)]$, is given by

$$s = \int_a^b \sqrt{\left\{1+\left(\frac{dy}{dx}\right)^2\right\}}\,dx. \tag{20.3}$$

Similarly

$$s = \int_{f(a)}^{f(b)} \sqrt{\left\{1+\left(\frac{dx}{dy}\right)^2\right\}}\,dy, \tag{20.4}$$

if s is specified so that it increases with y.

If the equation of the curve is given by $x = f(t)$, $y = g(t)$, where t increases steadily from t_1 at A to t_2 at B, then

$$s = \int_{t_1}^{t_2} \sqrt{\left\{1+\left(\frac{dy}{dx}\right)^2\right\}}\frac{dx}{dt}\,dt$$

(dx/dt is positive if x and t both increase with s)

$$\Rightarrow s = \int_{t_1}^{t_2} \sqrt{\left\{\left(\frac{dx}{dt}\right)^2+\left(\frac{dy}{dt}\right)^2\right\}}\,dt. \tag{20.5}$$

Care must be taken to ensure that (20.3) is not used over an interval which includes a point at which $\psi = \frac{1}{2}\pi$ since dy/dx would not exist there; (20.4) must not be used over an interval which includes a point $\psi = 0$.

Example 1. The curve into which a uniform heavy thin rope hangs when freely suspended from its ends is called a *catenary* and its equation, referred to horizontal and vertical axes of x and y chosen so that the coordinates of its lowest point are $(0, c)$, is $y = c \cosh(x/c)$. Show that the length of an arc of this catenary measured from the lowest point to the point (x, y) is given by

$$s = c \sinh(x/c)$$

and that, in the usual notation, $s = c \tan\psi$ and $y^2 = c^2 + s^2$.

$$y = c \cosh(x/c)$$

$$\Rightarrow \frac{dy}{dx} = \sinh(x/c)$$

$$\Rightarrow s = \int_0^x \sqrt{\{1+\sinh^2(x/c)\}}\,dx = \int_0^x \cosh(x/c)\,dx$$

$$\Rightarrow s = c \sinh(x/c).$$

$$\tan\psi = \frac{dy}{dx} = \sinh\left(\frac{x}{c}\right)$$

$$\Rightarrow s = c \tan\psi.$$

An equation in the form $s = f(\psi)$, where s and ψ have the meanings defined in this section, is known as *the intrinsic equation of the curve*.

$$y^2 - s^2 = c^2 \{\cosh^2 (x/c) - \sinh^2 (x/c)\} = c^2$$

$$\Rightarrow y^2 = c^2 + s^2.$$

Example 2. Find the length of the arc of the curve $x = a \sin^3 \theta$, $y = a \cos^3 \theta$, $a > 0$, in the first quadrant.

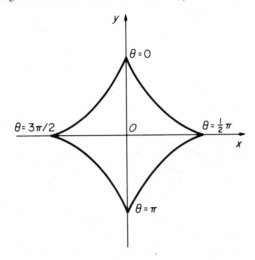

FIG. 20.4.

This curve, Fig. 20.4, is the *astroid*. The condition that s increases with θ is satisfied if the direction of s increasing is chosen as from $\theta = 0$ to $\theta = \frac{1}{2}\pi$. Then

$$\frac{dx}{d\theta} = 3a \sin^2 \theta \cos \theta, \quad \frac{dy}{d\theta} = -3a \cos^2 \theta \sin \theta$$

$$\Rightarrow s = \int_0^{\pi/2} 3a \sin \theta \cos \theta \sqrt{(\sin^2 \theta + \cos^2 \theta)} \, d\theta$$

$$= \int_0^{\pi/2} \frac{3a}{2} \sin 2\theta \, d\theta = \frac{3a}{2} \left[-\frac{\cos 2\theta}{2} \right]_0^{\pi/2} = \frac{3a}{2}$$

so that the length of the arc of the curve in the first quadrant is $3a/2$.

Example 3. Find the circumference of a circle of radius r.

The parametric equations of the circle are $x = r \cos \theta$, $y = r \sin \theta$, $0 \le \theta \le 2\pi$. The circumference is

$$\int_0^{2\pi} \sqrt{(r^2 \sin^2 \theta + r^2 \cos^2 \theta)} \, d\theta = 2\pi r.$$

20:2 THE CYCLOID

If P is a point on the circumference of a circle which rolls without slipping on a straight line, the path of P is called a *cycloid*. In Fig. 20.5 the initial position of P in contact with the line is taken as the origin and the line itself as the x-axis. The radius of the circle is a and the figure shows one of the positions of the circle along the line. N is the point of contact of the circle with the line, C is the centre of the circle, $N\hat{C}P = \theta$ and PT is perpendicular to NC. Then, because there is no slipping, $ON = $ arc NP and, if the coordinates of P are (x, y),

$$x = ON - PT = a\theta - a\sin\theta,$$
$$y = CN + CT = a - a\cos\theta.$$

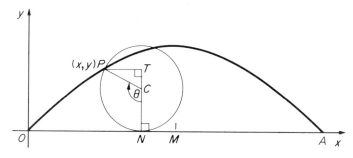

FIG. 20.5.

Hence the parametric equations of the cycloid referred to these axes are

$$x = a(\theta - \sin\theta), \quad y = a(1 - \cos\theta).$$

The length of the arc of the cycloid from $O(\theta = 0)$ to $A(\theta = 2\pi)$, i.e. the length of one *arch* of the cycloid, is

$$\int_0^{2\pi} a\sqrt{\{(1 - \cos\theta)^2 + \sin^2\theta\}}\, d\theta = a\int_0^{2\pi} \sqrt{(2 - 2\cos\theta)}\, d\theta$$

$$= a\int_0^{2\pi} 2\sin\left(\frac{\theta}{2}\right) d\theta = \left[-4a\cos\left(\frac{\theta}{2}\right)\right]_0^{2\pi} = 8a.$$

In general, $s = 4a(1 - \cos\frac{1}{2}\theta)$ if s is measured from O. If however s is measured from $M(\theta = \pi)$ on the cycloid, and in the reverse (opposite) sense, $s = 4a\cos\frac{1}{2}\theta$. But

$$\tan\psi = \frac{dy}{d\theta}\bigg/\frac{dx}{d\theta} = \frac{\sin\theta}{1 - \cos\theta} = \frac{2\sin\frac{1}{2}\theta\cos\frac{1}{2}\theta}{2\sin^2\frac{1}{2}\theta} = \cot\frac{1}{2}\theta$$

$$\Rightarrow \psi = \tfrac{1}{2}\pi - \tfrac{1}{2}\theta \quad \text{and} \quad \cos\tfrac{1}{2}\theta = \sin\psi.$$

Hence the intrinsic equation of the cycloid with this origin is

$$s = 4a\sin\psi.$$

Exercise 20.2

Calculate the lengths of the arcs of the curves specified in Nos. 1–5.

1. $y = x^2/2$ from $x = 0$ to $x = 1$.

2. $ay^2 = x^3$ from $x = 0$ to $x = a$ in the first quadrant.

3. $x = at^2$, $y = 2at$ from $t = 0$ to $t = 1$.

4. $y = \ln \sec x$ from $x = 0$ to $x = \dfrac{1}{4}\pi$.

5. $x = \tanh t$, $y = \operatorname{sech} t$ from $t = 0$ to $t = 1$.

6. Sketch the curve $y^2 = \dfrac{1}{6} x(2 - x)^2$ and find the length of the whole loop.

7. Prove that the length of the circumference of the ellipse

$$x = a \cos \theta, \quad y = b \sin \theta$$

is equal to

$$a \int_0^{2\pi} (1 - e^2 \cos^2 \theta)^{1/2} \, d\theta,$$

where $a^2 e^2 = a^2 - b^2$.   (O.C.)

20:3 AREAS OF SURFACES OF REVOLUTION

Definition. If the curve $y = f(x)$ [Fig. 20.6] from P_1 to P_n describes a complete revolution about the axis of x, each of the chords $P_r P_{r+1}$ will generate the surface of a frustum of a right circular cone. This surface is a developable one and its area S is given by the formula

$$S = 2\pi l \left(\frac{r_1 + r_2}{2} \right),$$

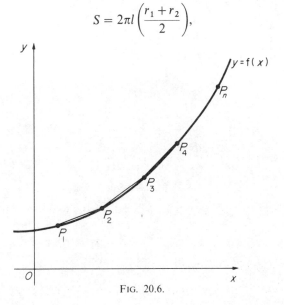

FIG. 20.6.

where r_1 and r_2 are the radii of the ends of the frustum and l is the slant height. The limit of the sum of the surface areas of these frusta as $n \to \infty$ is defined as the *surface area* generated by this part of the curve in its revolution about the x-axis.

In Fig. 20.7, P is (x, y) and Q is $(x + \delta x, y + \delta y)$ on the curve $y = f(x)$. A is the point $[a, f(a)]$ and B is the point $[b, f(b)]$. If arc $AP = s$, then s is a function of x and $PQ = \delta s$ where δs is the increment of s resulting from an increase δx in x. Then the surface generated by the arc PQ is approximately

$$2\pi \left(\frac{y + y + \delta y}{2} \right) \delta s$$

and hence the total surface area generated by AB is

$$S = \int_{x=a}^{x=b} 2\pi y \, ds \tag{20.6}$$

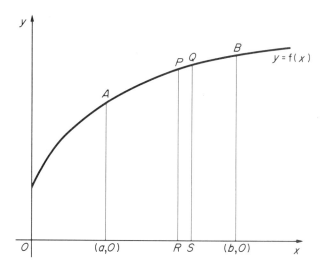

FIG. 20.7.

and since

$$\frac{ds}{dx} = \sqrt{\left\{ 1 + \left(\frac{dy}{dx} \right)^2 \right\}},$$

$$S = 2\pi \int_a^b y \sqrt{\left\{ 1 + \left(\frac{dy}{dx} \right)^2 \right\}} \, dx \tag{20.7}$$

$$= 2\pi \int_{f(a)}^{f(b)} y \sqrt{\left\{ 1 + \left(\frac{dx}{dy} \right)^2 \right\}} \, dy. \tag{20.8}$$

If $x = f(t)$, $y = g(t)$, $t = t_1$ when $x = a$ and $t = t_2$ when $x = b$, the total surface generated is

$$S = 2\pi \int_{t_1}^{t_2} g(t) \sqrt{\left\{ \left(\frac{dx}{dt} \right)^2 + \left(\frac{dy}{dt} \right)^2 \right\}} \, dt. \qquad (20.9)$$

In all cases we have chosen the sign of S so that it increases with the variable of integration.

Example 1. Find the area S of the surface of a sphere of radius R.

Consider the sphere generated by the rotation of $x = R \cos \theta$, $y = R \sin \theta$ about the x-axis. Then

$$\frac{dx}{d\theta} = -R \sin \theta, \quad \frac{dy}{d\theta} = R \cos \theta$$

$$\Rightarrow S = 2\pi \int_0^\pi R \sin \theta \sqrt{(R^2 \sin^2 \theta + R^2 \cos^2 \theta)} \, d\theta$$

$$= 2\pi \int_0^\pi R^2 \sin \theta \, d\theta = 2\pi R^2 \left[-\cos \theta \right]_0^\pi = 4\pi R^2.$$

Example 2. Find the surface area generated when the region enclosed by the curve

$$y^2 = \tfrac{1}{16} x^2 (2 - x^2)$$

is rotated completely about the x-axis.

$$\frac{dy}{dx} = \pm \frac{1 - x^2}{2\sqrt{(2 - x^2)}}.$$

The curve is symmetrical about both axes and $|x| \leqslant \sqrt{2}$. There are stationary points at $(\pm 1, \pm \tfrac{1}{4})$. Also $dx/dy = 0$ at $(\pm \sqrt{2}, 0)$, $dy/dx = \pm \sqrt{2}/4$ at $(0, 0)$. The curve is as shown in Fig. 20.8.

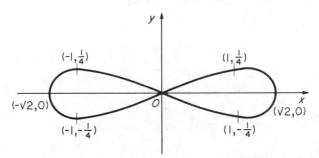

FIG. 20.8.

Consider the surface generated by the rotation of the portion of the curve $y = +\tfrac{1}{4} x \sqrt{(2 - x^2)}$ from $x = 0$ to $x = +\sqrt{2}$.

$$1 + \left(\frac{dy}{dx} \right)^2 = 1 + \frac{(1 - x^2)^2}{4(2 - x^2)} = \frac{x^4 - 6x^2 + 9}{4(2 - x^2)}.$$

Then the area of the surface generated is

$$2\int_0^{\sqrt{2}} 2\pi \frac{x\sqrt{(2-x^2)}}{4} \sqrt{\left\{\frac{x^4-6x^2+9}{4(2-x^2)}\right\}} dx = \frac{\pi}{2}\int_0^{\sqrt{2}} x(3-x^2)dx.$$

[Note that $\sqrt{(x^4-6x^2+9)}$ here is $(3-x^2)$ not (x^2-3), since, throughout the domain of x, $3-x^2 > 0$.]
Hence the surface area generated is

$$\frac{\pi}{2}\left[\frac{3x^2}{2}-\frac{x^4}{4}\right]_0^{\sqrt{2}} = \pi.$$

Exercise 20.3

Calculate the areas of the surfaces generated when the regions enclosed by the curves in questions 1–8, within the stated limits, are rotated completely about the x-axis.

1. $y = \cosh x$; $x = 0$ to $x = 1$.

2. One arch of the cycloid $x = a(\theta - \sin\theta)$, $y = a(1-\cos\theta)$.

3. $x = a\sin^3 t$, $y = a\cos^3 t$; $x = 0$ to $x = a$.

4. $y^2 = 4x$; $x = 0$ to $x = 1$.

5. $x = t^3$, $y = 3t^2$; $t = 0$ to $t = 2$.

6. $y = e^{-x}$; $x = 0$ to $x = 1$.

7. $x = \operatorname{sech} t$, $y = \tanh t$; $t = 0$ to $t = 1$.

8. $x^2 - y^2 = 1$; $x = 1$ to $x = \sqrt{5}$.

9. The region bounded by the ellipse with parametric equations $x = 3\cos\theta$, $y = 2\sqrt{2}\sin\theta$ and the positive x- and y-axes is rotated completely about the y-axis. Prove that the curved surface area of the solid thus formed is given by

$$6\pi\int_0^{\pi/2} \cos\theta\sqrt{(8+\sin^2\theta)}\,d\theta$$

and has the value $3\pi(4\ln 2 + 3)$.

10. Sketch the curve with parametric equations $x = e^t\cos t$, $y = e^t\sin t$ for values of t from 0 to 2π.
An inverted bowl is formed by rotating the portion of this curve between $t = 0$ and $t = \pi/2$ completely about the y-axis. Prove that the inner surface area of the bowl is $\frac{2}{5}\pi(e^\pi - 2)\sqrt{2}$.

20:4 THE THEOREM OF PAPPUS CONCERNING SURFACES OF REVOLUTION

This is a second theorem of Pappus corresponding to that of §19:6 and can be stated as follows.

The area of the surface generated by the rotation about an axis of an arc of a curve which does not cut that axis is equal to the length of the arc multiplied by the length of the path described by the centroid of the arc in the rotation.

The y-coordinate of the centroid of the arc AB of Fig. 20.7 is given by

$$\bar{y} = \frac{1}{s_1}\int_0^{s_1} y\,ds,$$

where s_1 is the length of the arc AB. The area of the surface of revolution about the x-axis is given by

$$S_1 = \int_0^{s_1} 2\pi y \, ds, \tag{20.6}$$

so that the area of the surface of revolution $S_1 = 2\pi \bar{y} s_1$, and in general

$$S = 2\pi \bar{y} s, \tag{20.10}$$

which proves the required result.

Example 1. Find the area of the surface generated by the complete revolution of a circle of radius r about a tangent.

The centroid of the circumference of the circle is at its centre. Hence the surface S of the solid of revolution is the length of the path traced by the centroid of the arc × the circumference of the circle

$$\Rightarrow S = 2\pi r \cdot 2\pi r = 4\pi^2 r^2.$$

Example 2. A uniform heavy chain is suspended from a point A, 8 m above the ground, and part of it hangs in the form of a catenary and part lies on the horizontal ground. The chain is horizontal at B, the point of contact with the ground. The length of the chain from A to B is 12 m. Calculate the height of the centre of mass of the suspended part of the chain above the ground and hence write down the area of the surface that would be generated by the complete revolution of this part of the chain about its projection on the ground.

From Example 1, p. 672, for the catenary $y^2 = c^2 + s^2$. Hence at A

$$(8+c)^2 = c^2 + 12^2$$
$$\Leftrightarrow 64 + 16c + c^2 = c^2 + 144$$
$$\Rightarrow c = 5$$

$$\Rightarrow \bar{y} = \frac{1}{12} \int_0^{12} y \, ds = \frac{1}{12} \int_0^{12} \sqrt{(25 + s^2)} \, ds.$$

Putting $s = 5 \sinh t$ gives

$$\bar{y} = \frac{25}{12} \int_{s=0}^{s=12} \cosh^2 t \, dt = \frac{25}{12} \left[\frac{t}{2} + \frac{\sinh 2t}{4} \right]_{s=0}^{s=12}.$$

$$s = 12 \Rightarrow \sinh t = \frac{12}{5}$$

$$\Rightarrow \cosh t = \frac{13}{5}, \quad \text{and} \quad \sinh 2t = \frac{2 \cdot 12}{5} \cdot \frac{13}{5} = \frac{312}{25}.$$

Also

$$t = \ln \left[\frac{12}{5} + \sqrt{\left\{ 1 + \left(\frac{12}{5} \right)^2 \right\}} \right] = \ln 5$$

$$\Rightarrow \bar{y} = \frac{25}{12} \left(\frac{1}{2} \ln 5 + \frac{78}{25} \right) = \frac{25}{24} \ln 5 + \frac{13}{2},$$

i.e. the height of the centre of mass of the suspended part of the chain above the ground is

$$\bar{y} - c = \left(\tfrac{25}{24} \ln 5 + \tfrac{3}{2} \right) \text{m},$$

and the required area of the surface of revolution is

$$\pi(25 \ln 5 + 36) \text{m}^2.$$

Exercise 20.4

1. Calculate the surface of the anchor ring generated by the revolution of a circle of radius r about a line in its plane and distance c from its centre. $(r < c.)$

2. Calculate the area of the surface of revolution of the circle $x^2 + y^2 - 4x + 2y - 4 = 0$ about the line $5x + 12y + 54 = 0$.

3. Use the answer obtained for Example 2, p. 673, to calculate the area of the surface of revolution of the astroid $x^{2/3} + y^{2/3} = a^{2/3}$ about the line $y = 2a$.

4. Calculate the area of the surface of revolution of an equilateral triangular framework of thin uniform wires each of length $2a$ about a line in its plane through one vertex at right angles to one of the sides which intersect there.

5. A groove is cut completely round a cylinder of radius a so that the plane of symmetry of the groove is at right angles to the axis of the cylinder and the section of the groove by a plane through the axis of the cylinder is a semicircle of radius $a/8$. Calculate the surface area of the groove. (The centroid of a semicircular arc of radius r is distant $2r/\pi$ from the centre of the circle.)

6. Prove that if the area of the surface of revolution of the ellipse $x = a \cos \theta$, $y = b \sin \theta$ about the tangent at $\theta = \pi/4$ is one third of the area of the surface of revolution of the ellipse about the line $x = 2a$, the eccentricity of the ellipse is $\sqrt{(5/7)}$.

7. Semicircles are described on each of the two shorter sides of a right-angled triangle in which the hypotenuse is of length a and one of the acute angles is θ. If the semicircular arcs are rotated completely about the hypotenuse, find an expression in terms of a and θ for the surface area generated and verify that it has a maximum value as θ varies when $\theta = 45°$. Calculate this maximum value.

20:5 AREAS IN POLAR COORDINATES

In Fig. 20.9, P_1 is the point (r, θ) and P_2 is the point $(r + \delta r, \theta + \delta \theta)$ on the arc AB of the curve $f(r, \theta) = 0$. It will be assumed that θ increases steadily from α at A to β at B. The circle with centre O and radius OP_1 cuts OP_2 at S, and the circle with centre O and radius OP_2 cuts OP_1 produced at R.

Then, in the case shown in Fig. 20.9,

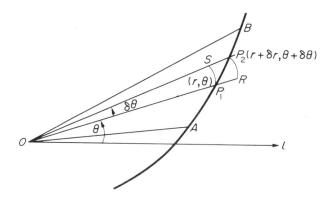

Fig. 20.9.

area of circular sector $OP_2R >$ area of sector $OP_2P_1 >$ area of circular sector OP_1S.

The area of the sector OAB is equal to the common value of

$$\lim_{\delta\theta \to 0} \sum_{\theta=\alpha}^{\theta=\beta} \tfrac{1}{2}r^2\delta\theta \quad \text{and} \quad \lim_{\delta\theta \to 0} \sum_{\theta=\alpha}^{\theta=\beta} \tfrac{1}{2}(r+\delta r)^2\delta\theta$$

if these limits exist

$$\Leftrightarrow \text{Area of sector} = \int_\alpha^\beta \tfrac{1}{2}r^2 \, d\theta. \tag{20.11}$$

This result is obtained for a part of the curve for which r increases steadily as θ increases. A similar result holds for a part of the curve for which r decreases steadily as θ increases. Then, by addition, the result (20.11) holds in general.

In the following examples and in Exercise 20.5, we adopt the convention that negative values of r are defined as values of r measured in a direction opposite to the direction of r − positive.

Example 1. Sketch the curve $r = a \cos 3\theta$, where $a > 0$, and find the area of the region enclosed by one loop of the curve.

Since $-1 \leqslant \cos 3\theta \leqslant 1$, this curve lies entirely inside the circle $r = a$ and these curves touch at points for which $3\theta = -2\pi, 0, 2\pi$. (For the values $\theta = -\pi/3, +\pi/3, r = -a$; this gives points coinciding with those already obtained.)

For values $\theta = -5\pi/6, -\pi/2, -\pi/6, \pi/6, \pi/2, 5\pi/6, r$ vanishes showing that there are three directions for the tangents at O, viz.

$$(\pi/6, -5\pi/6), (\pi/2, -\pi/2), (5\pi/6, -\pi/6).$$

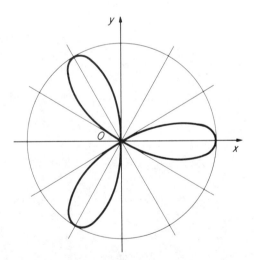

FIG. 20.10.

As θ runs from $-\pi$ to $+\pi$ the whole curve is traced twice, once with $r > 0$, viz.

$$-5\pi/6 < \theta < -\pi/2, \quad -\pi/6 < \theta < \pi/6, \quad \pi/2 < \theta < 5\pi/6,$$

and once with $r < 0$, viz.

$$-\pi < \theta < -5\pi/6, \quad -\pi/2 < \theta < -\pi/6, \quad \pi/6 < \theta < \pi/2, \quad 5\pi/6 < \theta < \pi.$$

The curve is shown in Fig. 20.10.

One loop is completed over the domain $-\pi/6 \leqslant \theta \leqslant \pi/6$. Hence the area of the region enclosed by one loop is

$$\int_{-\pi/6}^{\pi/6} \tfrac{1}{2}r^2\, d\theta = \int_{-\pi/6}^{\pi/6} \tfrac{1}{2}a^2 \cos^2 3\theta\, d\theta$$

$$= \tfrac{1}{2}a^2 \int_{-\pi/6}^{\pi/6} \tfrac{1}{2}(1 + \cos 6\theta)d\theta = \tfrac{1}{2}a^2 \left[\frac{\theta}{2} + \frac{\sin 6\theta}{12}\right]_{-\pi/6}^{\pi/6}$$

$$= \pi a^2/12.$$

Example 2. Calculate the area of the region enclosed between the loops of the limaçon $r = a(1 + 2 \cos \theta)$, where $a > 0$.

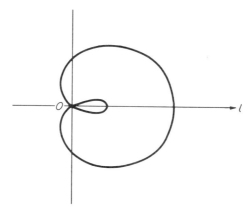

FIG. 20.11.

The inner loop is traced by the radius vector for $2\pi/3 \leqslant \theta \leqslant 4\pi/3$, (Fig. 20.11). Hence the area of the region enclosed by the inner loop is

$$\int_{2\pi/3}^{4\pi/3} \tfrac{1}{2}a^2(1 + 2 \cos \theta)^2\, d\theta$$

$$= \tfrac{1}{2}a^2 \int_{2\pi/3}^{4\pi/3} (3 + 2 \cos 2\theta + 4 \cos \theta)d\theta = \tfrac{1}{2}a^2 \left[3\theta + \sin 2\theta + 4 \sin \theta\right]_{2\pi/3}^{4\pi/3}$$

$$= \tfrac{1}{2}a^2(2\pi - 3\sqrt{3}).$$

The outer loop is traced in two parts $0 \leqslant \theta \leqslant 2\pi/3$ and $4\pi/3 \leqslant \theta \leqslant 2\pi$ and the areas of the two parts are equal. Hence the area of the region enclosed by the outer loop is

$$2 \int_0^{2\pi/3} \tfrac{1}{2}r^2\, d\theta = a^2 \left[3 + \sin 2\theta + 4 \sin \theta\right]_0^{2\pi/3}$$

$$= a^2(2\pi + \tfrac{3}{2}\sqrt{3}).$$

Thus the area of the region enclosed between the loops is $\tfrac{1}{2}a^2(2\pi + 6\sqrt{3})$.

Exercise 20.5

1. Calculate the area of the region enclosed by the cardioid $r = a(1 + \cos\theta)$, $a > 0$.

2. Calculate the area of the region enclosed by one loop of the lemniscate $r^2 = a^2 \cos 2\theta$, $a > 0$.

3. Calculate the area of the region enclosed by one loop of $r = a \cos 4\theta$, $a > 0$.

4. Calculate the total area of the region enclosed by the curve $r = a \sin n\theta$ where $n \in \mathbb{N}$, $a > 0$.

5. The region enclosed by the first revolution of the spiral $r = a\theta$, $a > 0$, and the initial line from $r = 0$ to $r = 2a\pi$ is divided into four parts by the lines $\theta = 0$, $\theta = \pi/2$, $\theta = \pi$ and $\theta = 3\pi/2$. Calculate the ratios of the areas of these four regions.

6. For the reciprocal spiral $r = a/\theta$, $a > 0$, prove that the area of the region enclosed by the lines $\theta = \beta$, $\theta = \alpha$ and the arc of the curve for $\beta \leqslant \theta \leqslant \alpha$ where $0 < \beta < \alpha < 2\pi$ varies as the increase of r from $\theta = \beta$ to $\theta = \alpha$.

7. Transform the equation of the parabola $y^2 = 4ax$, $a > 0$, to a polar equation with the focus at the pole and the axis of x as initial line. Hence find the area of the region enclosed between the curve and a focal chord which makes an angle $\pi/3$ with the positive x-direction.

8. Calculate the area of the region enclosed by one of the inner loops bounded by the two curves $r = a(1 + \cos\theta)$, $r = a(1 - \cos\theta)$, $a > 0$.

9. Calculate the area of the region bounded by the curve $r = a \tan\frac{1}{2}\theta$, where $a > 0$, and the lines $\theta = \pi/2$, $\theta = 3\pi/2$.

10. Sketch the curve with polar equation $r = 2a(\sec\theta - 1)$, where $a > 0$, showing its asymptote $r \cos\theta = 2a$.
Prove that the area of the region enclosed between the curve and the line $r \cos\theta = a$ is given by

$$a^2 \left[\sqrt{3} - 4 \int_0^{\pi/3} (\sec\theta - 1)^2 \, d\theta \right],$$

and hence is approximately equal to $1 \cdot 15 a^2$.

20:6 THE LENGTH OF AN ARC IN POLAR COORDINATES

If $P(r, \theta)$ is a point on a polar curve and if the cartesian coordinates of P referred to the pole as origin and the initial line as the positive x-axis are (x, y) then

$$x = r \cos\theta, \quad y = r \sin\theta$$

$$\Rightarrow \frac{dx}{d\theta} = -r \sin\theta + \frac{dr}{d\theta} \cos\theta,$$

$$\frac{dy}{d\theta} = r \cos\theta + \frac{dr}{d\theta} \sin\theta$$

$$\Rightarrow \left(\frac{ds}{d\theta}\right)^2 = \left(\frac{dx}{d\theta}\right)^2 + \left(\frac{dy}{d\theta}\right)^2 = r^2 + \left(\frac{dr}{d\theta}\right)^2. \qquad (20.12)$$

This result can be obtained by reference to Fig. 20.13, and reference to this figure probably affords the best way of remembering the result. In the triangle PQM, $MQ \approx \delta r$, $PM \approx r\delta\theta$ and $PQ \approx \delta s$, whence, using

$$PQ^2 = MQ^2 + MP^2,$$

$$\delta s^2 \approx \delta r^2 + r^2 \delta\theta^2$$

$$\Leftrightarrow \left(\frac{\delta s}{\delta \theta}\right)^2 \approx r^2 + \left(\frac{\delta r}{\delta \theta}\right)^2,$$

whence (20.12) follows.

Similarly,
$$\frac{dx}{dr} = \cos\theta - r\sin\theta\frac{d\theta}{dr},$$

$$\frac{dy}{dr} = \sin\theta + r\cos\theta\frac{d\theta}{dr}$$

$$\Rightarrow \left(\frac{dx}{dr}\right)^2 + \left(\frac{dy}{dr}\right)^2 = 1 + \left(r\frac{d\theta}{dr}\right)^2. \qquad (20.13)$$

From equations (20.5) and (20.12) the length of the arc of $f(r, \theta) = 0$ from $\theta = \theta_1$ to $\theta = \theta_2$, if s is measured so as to increase with θ, is

$$\int_{\theta_1}^{\theta_2} \sqrt{\left\{r^2 + \left(\frac{dr}{d\theta}\right)^2\right\}}\, d\theta. \qquad (20.14)$$

Similarly, if s is measured so as to increase with r, the length of the arc from $r = r_1$ to $r = r_2$ is

$$\int_{r_1}^{r_2} \sqrt{\left\{1 + \left(r\frac{d\theta}{dr}\right)^2\right\}}\, dr. \qquad (20.14a)$$

Example. Calculate the length of the arc of the spiral $r = a\theta$ from $\theta = 0$ to $\theta = 2\pi$.

$$\text{Length of arc} = \int_0^{2\pi} \sqrt{(a^2\theta^2 + a^2)}\, d\theta$$

$$= a\int_0^{\sinh^{-1} 2\pi} \sqrt{(1 + \sinh^2 t)}\cosh t\, dt = a\int_0^{\sinh^{-1} 2\pi} \cosh^2 t\, dt$$

$$= a\left[\frac{t}{2} + \frac{\sinh 2t}{4}\right]_0^{\sinh^{-1} 2\pi} = \tfrac{1}{2}a\left[\ln\{2\pi + \sqrt{(1 + 4\pi^2)}\} + 2\pi\sqrt{(1 + 4\pi^2)}\right].$$

20:7 VOLUMES OF REVOLUTION AND AREAS OF SURFACES OF REVOLUTION IN POLAR COORDINATES

These problems are best considered from first principles using the theorems of Pappus.

If the region bounded by the curve $r = f(\theta)$ and the radius vectors OA, OB ($\theta = \alpha$ and $\theta = \beta$, respectively, from Fig. 20.9) rotates through $360°$ about the initial line (Ox), then, by the first theorem of Pappus, the volume δV generated by the elementary sector $OP_1 P_2$ is given by

$$\delta V = \text{area } OP_1 P_2 \times 2\pi p,$$

where p is the distance of the centroid of the sector from Ox. But the area $OP_1 P_2 \approx \frac{1}{2} r^2 \delta\theta$ and since $OP_1 P_2$ is approximately a triangle with centroid at the intersection of the medians, $p \approx \frac{2}{3} r \sin\theta$

$$\Rightarrow \delta V \approx \tfrac{1}{2} r^2 \delta\theta \times 2\pi \tfrac{2}{3} r \sin\theta.$$

Hence, by addition, V, the total volume generated, is given by

$$V = \frac{2\pi}{3} \int_{\alpha}^{\beta} r^3 \sin\theta \, d\theta, \qquad (20.15)$$

provided that the region is not cut by Ox.

The area of the surface generated by revolution of the arc AB about the x-axis is given by

$$S = \int_{s_1}^{s_2} 2\pi y \, ds$$

and, if the conditions for a change of variable are satisfied, this becomes

$$S = \int_{\theta=\theta_1}^{\theta=\theta_2} 2\pi r \sin\theta \, \frac{ds}{d\theta} \, d\theta = \int_{\theta=\theta_1}^{\theta=\theta_2} 2\pi r \sin\theta \sqrt{\left\{ r^2 + \left(\frac{dr}{d\theta}\right)^2 \right\}} \, d\theta. \qquad (20.16)$$

Centroids and second moments of area and of volume are best found from first principles. The theorems of Pappus are used where it is expedient to do so.

Example. For the region enclosed by that part of the cardioid $r = a(1 + \cos\theta)$, $a > 0$, lying in the first quadrant and the coordinate axes, calculate
  (i) the volume of revolution about the initial line,
  (ii) the surface area of the solid obtained by the rotation of the curve,
  (iii) the distances of the centroid of the region from Ox and Oy.

The curve is symmetrical about Ox and is sketched in Fig. 20.12.
  (i) The volume of revolution

$$V = \frac{2\pi}{3} \int_0^{\pi/2} r^3 \sin\theta \, d\theta = \frac{2\pi a^3}{3} \int_0^{\pi/2} (1 + \cos\theta)^3 \sin\theta \, d\theta$$

$$= \frac{2\pi a^3}{3} \left[-\tfrac{1}{4}(1 + \cos\theta)^4 \right]_0^{\pi/2} = \frac{5\pi a^3}{2}.$$

  (ii) The surface area of the shell

$$S = \int_{\theta=0}^{\theta=\pi/2} 2\pi y \, ds = 2\pi \int_{\theta=0}^{\theta=\pi/2} r \sin\theta \sqrt{\left\{ r^2 + \left(\frac{dr}{d\theta}\right)^2 \right\}} \, d\theta.$$

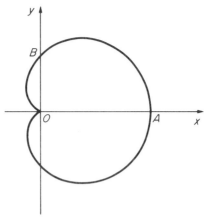

FIG. 20.12.

Here s is measured from $P(2a, 0)$ to $Q(a, \frac{1}{2}\pi)$, i.e. in the direction increasing with θ

$$\Rightarrow S = 2\pi a \int_0^{\pi/2} (1 + \cos\theta) \sin\theta \sqrt{\{a^2 + 2a^2 \cos\theta + a^2 \cos^2\theta + a^2 \sin^2\theta\}} \, d\theta$$

$$= 2\pi a^2 \int_0^{\pi/2} (1 + \cos\theta) \sin\theta \sqrt{(2 + 2\cos\theta)} \, d\theta$$

$$= 16\pi a^2 \int_0^{\pi/2} \cos^4 \tfrac{1}{2}\theta \sin \tfrac{1}{2}\theta \, d\theta = 32\pi a^2 \left[\frac{-\cos^5 \tfrac{1}{2}\theta}{5} \right]_0^{\pi/2}$$

$$= \frac{32\pi a^2}{5} \left(1 - \frac{\sqrt{2}}{8} \right).$$

(iii) The area A of the region bounded by the arc of the cardioid in the first quadrant and the coordinate axes is given by

$$A = \int_0^{\pi/2} \tfrac{1}{2} r^2 \, d\theta = \tfrac{1}{2} a^2 \int_0^{\pi/2} (1 + \cos\theta)^2 \, d\theta$$

$$= \tfrac{1}{2} a^2 \int_0^{\pi/2} (1 + 2\cos\theta + \cos^2\theta) \, d\theta$$

$$= \tfrac{1}{2} a^2 \left(\frac{\pi}{2} + 2 + \frac{\pi}{4} \right) = \frac{(3\pi + 8)a^2}{8}.$$

Hence, if the cartesian coordinates of the centroid, G, of the given region are $\bar{x}, \bar{y}$, then, by the first theorem of Pappus,

$$V = A \times 2\pi\bar{y}$$
$$\Rightarrow \bar{y} = 10a/(3\pi + 8).$$

To find $\bar{x}$ we take first moments about Oy. The first moment δM of the elementary sector OP_1P_2 of Fig. 20.9 about Oy is given by.

$$\delta M \approx \tfrac{1}{2} r^2 \, \delta\theta \times \frac{2}{3} r \cos\theta.$$

By addition the first moment of the whole region is

$$M = \frac{1}{3} \int_0^{\pi/2} r^3 \cos\theta \, d\theta = \frac{a^3}{3} \int_0^{\pi/2} (1+\cos\theta)^3 \cos\theta \, d\theta$$

$$= \frac{a^3}{3} \int_0^{\pi/2} (\cos\theta + 3\cos^2\theta + 3\cos^3\theta + \cos^4\theta) \, d\theta$$

$$= \frac{a^3}{3} \left[1 + \frac{3\pi}{4} + 3 \cdot \frac{2}{3} + \frac{3}{4} \cdot \frac{1}{2} \cdot \frac{\pi}{2} \right] = \frac{(16+5\pi)a^3}{16}.$$

But $M = A\bar{x}$

$$\Rightarrow \bar{x} = \frac{(16+5\pi)a}{(16+6\pi)}.$$

Exercise 20.7

1. Find the length of the arc of the spiral $r = ae^{k\theta}$, $a > 0$, $k > 0$, from $\theta = 0$ to $\theta = 2\pi$.

2. Find the total length of the cardioid $r = a(1+\cos\theta)$, $a > 0$.

3. Find the length of the arc of the curve $r = a\sin^3(\theta/3)$, $a > 0$, from $\theta = 0$ to $\theta = 3\pi$.

4. Find the length of the arc of the curve $r = a/\theta$, $a > 0$, from $r = 3a/4$ to $r = 12a/5$.

5. Calculate the distance of the centroid of one loop of the lemniscate $r^2 = a^2 \cos 2\theta$ from the pole.

6. Calculate the volume of rotation of the region bounded by the curve $r^2 = a^2 \cos\theta$ about the initial line.

7. Calculate the area of the surface of revolution of the curve $r = e^\theta$ from $\theta = 0$ to $\theta = \pi$ about the initial line.

20:8 THE ANGLE BETWEEN THE TANGENT AND THE RADIUS VECTOR

In Fig. 20.13, P is the point (r, θ) and Q the point $(r + \delta r, \theta + \delta\theta)$. The angle between OP and the tangent to the curve at P is ϕ and PM is perpendicular to OQ. In $\triangle OPM$,

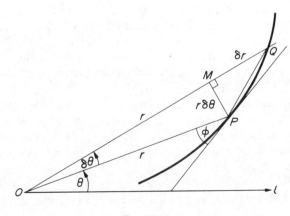

Fɪɢ. 20.13.

$MP = r \sin \delta\theta = r\delta\theta + O[(\delta\theta)^2]$ and $OM = r \cos \delta\theta = r + O[(\delta\theta)^2]$. Hence from $\triangle PMQ$,

$$\tan \phi = \frac{MP}{MQ} = \frac{r \, \delta\theta}{\delta r} + O(\delta\theta).$$

In the limit $\delta\theta \to 0$,

$$\tan \phi = r \frac{d\theta}{dr} = r \left/ \frac{dr}{d\theta} \right. . \tag{20.17}$$

Example 1. For the cardioid $r = a(1 - \cos\theta)$, discuss the shape of the curve at $\theta = 0$ and at $\theta = \pi$.

$$r\frac{d\theta}{dr} = \frac{a(1 - \cos\theta)}{a \sin\theta} = \frac{a \, 2\sin^2 \tfrac{1}{2}\theta}{a \, 2\sin \tfrac{1}{2}\theta \cos \tfrac{1}{2}\theta} = \tan \tfrac{1}{2}\theta.$$

Hence at $\theta = 0$ the curve is parallel to the initial line and at $\theta = \pi$ the curve is at right angles to the initial line. Note that near the pole, $r \approx 0$, $\Rightarrow 1 - \cos\theta \approx 0 \Rightarrow \theta \approx 0$ confirming the behaviour of the curve near the pole.

Example 2. Show that in the *equiangular spiral* $r = ae^{\theta \cot\alpha}$, $a > 0$, the angle between the tangent and the radius vector is constant and equal to α.

$$r = ae^{\theta \cot\alpha} \Rightarrow \frac{dr}{d\theta} = a \cot\alpha \, e^{\theta \cot\alpha}$$

$$\Rightarrow \tan\phi = r \frac{d\theta}{dr} = ae^{\theta \cot\alpha}/(a \cot\alpha \, e^{\theta \cot\alpha}) = \tan\alpha$$

$$\Rightarrow \phi = \alpha.$$

20:9 POINTS OF INFLEXION

In chapter I we considered points of inflexion on a curve $y = f(x)$ at which the tangents were parallel to the x-axis and also at which $f'(x)$ had either maximum or minimum values. We now *define* a point of inflexion on a curve as a point at which *the gradient has a maximum or minimum value*. Thus, $y = f(x)$ has a point of inflexion at the point (x_1, y_1) if $f''(x_1) = 0$ and if $f''(x_1)$ increases or decreases continuously as x passes through the value x_1. This is equivalent to the statement that, if $f''(x)$ is continuous at $x = x_1$, necessary and sufficient conditions for an inflexion at $x = x_1$ are $f''(x_1) = 0$ and $f''(x)$ *changes sign* as x passes through the value $x = x_1$. Sufficient conditions for an inflexion at $x = x_1$ are therefore $f''(x_1) = 0$, $f'''(x_1) \neq 0$. But, if $f''(x_1) = 0$ and $f'''(x_1) = 0$, it is necessary to investigate further the behaviour of the function $f''(x)$ as x passes through the value $x = x_1$.

In Fig. 20.14 (i) the tangent at the point $P(x_1, y_1)$ lies *above* the curve $y = f(x)$ for points on the curve near P. [The tangent at P is *above* the curve if $f(x_1 + \varepsilon) < f(x_1) + \varepsilon f'(x_1)$ however small $|\varepsilon|$ may be.] In such a case the curve is said to be concave downwards. In this case $f'(x)$ decreases continuously as x passes through the value $x = x_1$ and $f''(x_1)$ is therefore negative.

In Fig. 20.14 (ii) the tangent at P lies below the curve $y = f(x)$ for points of the curve near P. The curve is said to be *concave upwards* at P, $f'(x)$ increases continuously as x passes through the value x_1, and $f''(x_1)$ is positive.

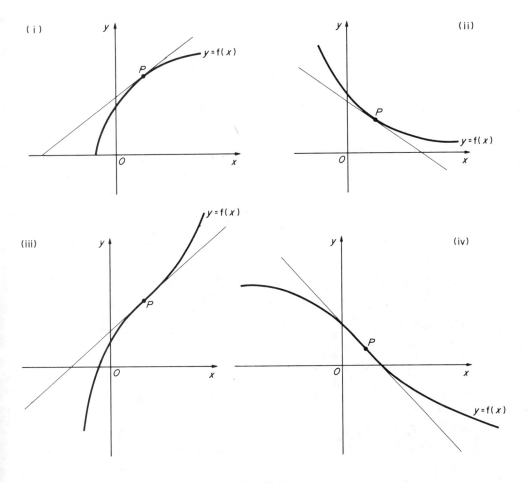

FIG. 20.14.

In Fig. 20.14 (iii) the tangent is above the curve for values of x less than x_1 and however near to x_1, the tangent is below the curve for values of x greater than x_1 and however near to x_1, i.e. $f''(x_1) < 0$ for $x = x_1 - \varepsilon$ and $f''(x_1) > 0$ for $x = x_1 + \varepsilon$ where $\varepsilon > 0$. Hence $f'(x)$ has a minimum at P, *the curve crosses the tangent at P and P is a point* of inflexion.

In Fig. 20.14 (iv) $f'(x)$ has a minimum at P which is a point of inflexion.

Example. Investigate the stationary points and the point(s) of inflexion on the curve $y = x^5 - 5x$ and sketch the curve.

Let $f(x) = x^5 - 5x$. Then

$$f'(x) = 5x^4 - 5.$$

Hence the stationary points occur where $x = \pm 1$. Also

$$f''(x) = 20x^3$$

so that $f(x)$ has a minimum at $(1, -4)$ and a maximum at $(-1, 4)$.

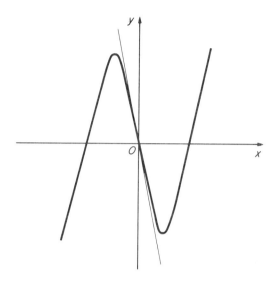

FIG. 20.15.

But $f''(x) = 0$ at $(0, 0)$ and $f'''(x) = 0$ at $(0, 0)$. However $f''(0 - \varepsilon) < 0$ and $f''(0 + \varepsilon) > 0$. Hence there is an inflexion at the origin.

At $(0, 0)$, $f'(x) = -5$ and the equation of the tangent at $(0, 0)$ is $y = -5x$. The curve is as shown in Fig. 20.15.

Exercise 20.9

1. Show that each of the following curves has an inflexion at the origin and in each case find the equation of the tangent at the origin:

(i) $y = x^7 + 7x$,   (ii) $y = \tan x$,   (iii) $y = \sinh x$,

(iv) $y = \sinh^{-1} x$,   (v) $y = \dfrac{x}{1 + x^2}$.

2. Find the coordinates of the stationary points, stating the nature of each, and the points of inflexion where they exist, for each of the following curves and, in each case, sketch the curve and its tangents at the points of inflexion:

(i) $y = x^3 - 12x^2 + 1$,   (ii) $y = x^4 - 2x^3$,   (iii) $y = x/(1 - x^2)$,

(iv) $y = ax^2/(a^2 + x^2)$,   (v) $y = \sin x + \cos x$.

3. Show that the curve $x^3 - y^3 = 1$ cuts the y-axis at a point of inflexion and calculate the angle at which it cuts the axis there.

4. Find the points of inflexion on the curve $y = e^{-x^2}$. Sketch the curve and its tangents at the points of inflexion.

5. Find the stationary points and the point of inflexion on the curve $y = xe^{-x}$. Find the equation of the tangent at the point of inflexion and sketch the curve. (Consider $\lim_{x \to 0} xe^{-x}$ as $\lim_{x \to 0} x/e^x$ and $\lim_{x \to \infty} xe^{-x}$ as $\lim_{x \to \infty} x/e^x$ in each case expanding the denominator and dividing by the numerator.)

6. For the curve $x = a(1 - \cos \theta)$, $y = a(\theta + \sin \theta)$, find dy/dx and d^2y/dx^2 each in terms of θ and hence find the coordinates of the stationary points and show that the curve has no points of inflexion.

7. Find the points of inflexion on the curve $y = x - \cos x$ and sketch the curve for $-2\pi \leqslant x \leqslant 2\pi$.

8. The equations of the curves in the shapes of beams supported and loaded in different ways are given as follows:

(i) resting on two supports at its ends with a load W suspended from its middle point,

$$y = \frac{W}{48\,EI}\,x\,(3l^2 - 4x^2),$$

(ii) fixed at one end and uniformly loaded,

$$y = \frac{W}{24\,EI}\,x^2\,(6l^2 - 4lx + x^2),$$

(iii) resting on supports at its extremities and uniformly loaded,

$$y = \frac{W}{384\,EI}\,(16x^4 - 24x^2 l^2 + 5l^4),$$

(iv) clamped horizontally at both ends and uniformly loaded,

$$y = \frac{W}{384\,EI}\,(4x^2 - l^2)^2,$$

where E and I are constants for the beam.

In each case investigate the positions of the inflexions (if any) in the curves of the loaded beams.

9. Find the stationary points and the points of inflexion on the curve $y = \cos 2x - 4 \sin x$. Sketch the curve.

10. Show that for the curve $y = x \cosh x$ there is a point of inflexion at $(a, a \cosh a)$, where

$$a = \tfrac{1}{2}\ln\left(\frac{2-a}{2+a}\right).$$

11. Determine how the number of stationary points on the curve $y = e^{-x^2}(x^2 + c)$ depends on the value of the real constant c.

Find the coordinates of the points of inflexion on the curve for which $c = 2$.

Sketch the curves for which (a) $c = 2$, (b) $c = 0$. (O.C.)

12. Find the coordinates of the points of inflexion of the curve $y = 3x/(1 + x^2)$. Show that the curve $y = x^3/(1 + x^2)$ has exactly the same points of inflexion and draw a rough sketch of the two curves. (O.C.)

13. Sketch the curve with polar equation $r = a(1 + 2\cos\theta)$.

Prove that the areas enclosed by the two loops of the curve are in the ratio $(4\pi + 3\sqrt{3}):(2\pi - 3\sqrt{3})$. (C.)

14. Sketch the curve with polar equation $r = a \sin 2\theta$.

A square enclosing the curve is drawn with two sides parallel to the line $\theta = 0$. Show that the area of the square is at least $\frac{64}{27}a^2$. (C.)

15. Sketch the curve whose polar equation is $r = \theta \sin\theta$ for $0 \leqslant \theta \leqslant 2\pi$.

Prove that the area of the region enclosed between the two loops of the curve is $\frac{1}{2}\pi^3$. (C.)

16. The equation of a curve in polar coordinates is $r = a \tan\theta$. Show that as θ increases from 0 to $\frac{1}{2}\pi$, $r \cos\theta$ increases to the value a and $r \sin\theta$ increases without limit. Sketch the curve for values of θ from 0 to π. If $0 < \phi < \frac{1}{2}\pi$ prove that the area enclosed by the lines $\theta = 0$, $\theta = \phi$, the curve and the line $r \cos\theta = a$ is $\frac{1}{2}a^2\phi$. (C.)

20:10 CURVATURE

The curvature (κ) at a point on the curve $y = f(x)$ is defined as the rate of change of ψ with respect to s, i.e. as $d\psi/ds$, for that point.

$$\kappa = \frac{d\psi}{ds}. \tag{20.18}$$

With the conventions we have adopted, s increases with x and therefore $d\psi/ds$ will be positive if ψ increases with x and negative if ψ decreases with x. If, however, the equations of the curve are given in such a way that these conventions do not apply the best course is to consider each case from first principles. For most purposes it is sufficient to obtain $|\kappa|$.

The curvature at any point on a circle of radius r

In Fig. 20.16, s is measured from the point A on the circle, CA is parallel to the y-axis, P is any point on the circle, the arc $AP = s$ and $A\hat{C}P = \psi$

$$\Rightarrow s = r\psi$$

$$\Rightarrow \frac{d\psi}{ds} = \frac{1}{r}.$$

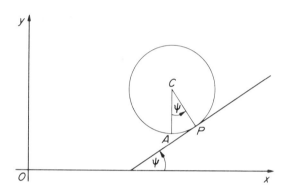

FIG. 20.16.

The curvature of a circle is inversely proportional to its radius

This result illustrates the definition of curvature given above. Curvature is the measure of the rate at which the tangent to the curve rotates compared with the distance moved along the curve, or *the rate at which the curve curves*. The reciprocal of the modulus of the curvature, $|\kappa|$, at a point P on a curve is the radius of a circle which has the same curvature as the curve at P. It is called the *radius of curvature* of the curve at P and is denoted by ϱ.

$$\varrho = \frac{1}{|\kappa|} = \left|\frac{ds}{d\psi}\right|. \tag{20.19}$$

Figure 20.17 (i) shows the point $P(x, y)$ on the curve $y = f(x)$ for which ψ increases with s at P. If the normal GP is produced to C so that $PC = \varrho$, C is called the *centre of curvature* for the curve at P and the circle with centre C and radius CP is called the *circle of curvature*. The centre of curvature is at a distance ϱ from P along the normal in the inward direction.

The positive direction of the normal is defined as that direction which makes an anticlockwise angle (i.e. a positive angle) of $\frac{1}{2}\pi$ with the positive direction of the tangent.

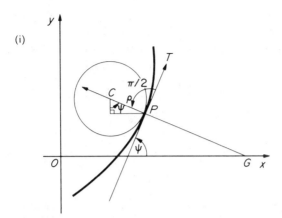

(i)

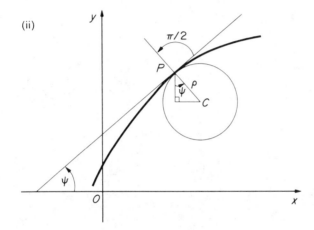

(ii)

Fig. 20.17.

When $\kappa > 0$, the inward direction along the normal is the positive direction. When $\kappa < 0$, the inward direction is opposite to the positive direction.

Figure 20.17 (ii) shows the centre and circle of curvature in a case in which the curvature is negative at the point.

If P is (x, y), the coordinates of the centre of curvature are (ξ, η) where $\xi = x \mp \varrho \sin \psi, \eta = y \pm \varrho \cos \psi$, the upper signs being taken when $\kappa > 0$, and the lower signs when $\kappa < 0$.

The locus of the centre of curvature as P moves on the curve is called the *evolute*.

Curvature in cartesian coordinates

$$\tan \psi = \frac{dy}{dx}$$

$$\Rightarrow \sec^2 \psi \, \frac{d\psi}{ds} = \frac{d^2 y}{dx^2} \frac{dx}{ds}.$$

But, from §20:1,
$$\frac{dx}{ds} = \cos\psi$$

$$\Rightarrow \kappa = \frac{d\psi}{ds} = \frac{\dfrac{d^2 y}{dx^2}}{\sec^3\psi} = \frac{\dfrac{d^2 y}{dx^2}}{(1 + \tan^2\psi)^{3/2}}$$

$$\Rightarrow \kappa = \frac{d^2 y}{dx^2} \bigg/ \left\{1 + \left(\frac{dy}{dx}\right)^2\right\}^{3/2} \tag{20.20}$$

$$\Rightarrow \varrho = \left|\left\{1 + \left(\frac{dy}{dx}\right)^2\right\}^{3/2} \bigg/ \frac{d^2 y}{dx^2}\right|. \tag{20.21}$$

The curvature is positive or negative according as $d^2 y/dx^2$ is positive or negative [the positive sign of the radical being taken in the formulae (20.20) and (20.21)], i.e. according to whether the curve is concave upwards, Fig. 20.14 (ii), or concave downwards, Fig. 20.14 (i). A necessary condition for an inflexion is $d^2 y/dx^2 = 0$; therefore the curvature at a point of inflexion is zero.

Curvature in parametric coordinates

If $x = f(t)$, $y = g(t)$ and if dx/dt is denoted by $\dot{x}$ and $d^2 x/dt^2$ by $\ddot{x}$, etc., then

$$\frac{dy}{dx} = \frac{\dot{y}}{\dot{x}}$$

$$\Rightarrow \frac{d^2 y}{dx^2} = \frac{d\left(\dfrac{dy}{dx}\right)}{dt} \bigg/ \frac{dx}{dt} = \frac{d}{dt}\left(\frac{\dot{y}}{\dot{x}}\right) \bigg/ \frac{dx}{dt}$$

$$= \frac{\dot{x}\ddot{y} - \dot{y}\ddot{x}}{\dot{x}^2} \frac{1}{\dot{x}} = \frac{\dot{x}\ddot{y} - \dot{y}\ddot{x}}{\dot{x}^3}$$

$$\Leftrightarrow \kappa = \frac{\dot{x}\ddot{y} - \dot{y}\ddot{x}}{\dot{x}^3} \frac{1}{\left\{1 + \dfrac{\dot{y}^2}{\dot{x}^2}\right\}^{3/2}}$$

$$\Leftrightarrow \kappa = \frac{\dot{x}\ddot{y} - \dot{y}\ddot{x}}{(\dot{x}^2 + \dot{y}^2)^{3/2}} \tag{20.22}$$

$$\Rightarrow \varrho = \left|\frac{(\dot{x}^2 + \dot{y}^2)^{3/2}}{\dot{x}\ddot{y} - \dot{y}\ddot{x}}\right|. \tag{20.23}$$

Example 1. The normal to the curve $4y = x^2$ at the point $P(2\sqrt{2}, 2)$ meets the curve again at Q. Show that PQ is the radius of curvature at P.   (N.)

$$4y = x^2$$

$$\Rightarrow \frac{dy}{dx} = \tfrac{1}{2}x, \qquad \frac{d^2 y}{dx^2} = \tfrac{1}{2}$$

$$\Rightarrow \varrho = 2(1 + \tfrac{1}{4}x^2)^{3/2}.$$

Hence at P,

$$\varrho = 6\sqrt{3}.$$

The normal at P is

$$y - 2 = -\frac{1}{\sqrt{2}}(x - 2\sqrt{2})$$

$$\Rightarrow x + \sqrt{2}y - 4\sqrt{2} = 0.$$

This meets the curve where

$$\sqrt{2}x^2 + 4x - 16\sqrt{2} = 0,$$

i.e. where $(x - 2\sqrt{2})(\sqrt{2}x + 8) = 0.$

Hence the normal meets the curve again at $(-4\sqrt{2}, 8)$. Thus the length of PQ is $\sqrt{\{(6\sqrt{2})^2 + 6^2\}} = 6\sqrt{3}$ so that PQ is the radius of curvature at P. Since PQ is the positive direction of the normal, Q is the centre of curvature corresponding to the point P.

Example 2. Find an expression for κ in terms of θ for the astroid $x = a\sin^3\theta$, $y = a\cos^3\theta$, $a > 0$.

$$\frac{dx}{d\theta} = 3a\sin^2\theta\cos\theta, \qquad \frac{dy}{d\theta} = -3a\cos^2\theta\sin\theta$$

$$\Rightarrow \frac{dy}{dx} = -\cot\theta.$$

$$\frac{d^2y}{dx^2} = \frac{d\left(\dfrac{dy}{dx}\right)}{d\theta} \cdot \frac{d\theta}{dx}$$

$$= \frac{\csc^2\theta}{3a\sin^2\theta\cos\theta} = \frac{1}{3a\sin^4\theta\cos\theta}$$

$$\Rightarrow \kappa = \frac{1}{3a\,(1+\cot^2\theta)^{3/2}\sin^4\theta\cos\theta} = \frac{1}{3a\sin\theta\cos\theta}.$$

Example 3. Find the radius of curvature for the curve

$$x^2 + y^2 - 5xy + 3y - 1 = 0$$

at the point $(1, 2)$.

Differentiation with respect to x gives

$$2x + 2y\frac{dy}{dx} - 5x\frac{dy}{dx} - 5y + 3\frac{dy}{dx} = 0. \qquad (1)$$

Hence at $(1, 2)$, $dy/dx = 4.$

Differentiating equation (1) with respect to x gives

$$2 + 2\left(\frac{dy}{dx}\right)^2 + 2y\frac{d^2y}{dx^2} - 5\frac{dy}{dx} - 5x\frac{d^2y}{dx^2} - 5\frac{dy}{dx} + 3\frac{d^2y}{dx^2} = 0.$$

Hence at $(1, 2)$, using $dy/dx = 4$ there,

$$2 + 32 + 4\frac{d^2y}{dx^2} - 20 - 5\frac{d^2y}{dx^2} - 20 + 3\frac{d^2y}{dx^2} = 0.$$

Thus at $(1, 2)$, $d^2y/dx^2 = 3$. Hence ϱ at $(1, 2)$ is $\dfrac{(1+16)^{3/2}}{3} = \dfrac{1}{3}17\sqrt{17}.$

20:11 NEWTON'S FORMULA FOR RADIUS OF CURVATURE AT THE ORIGIN

If the curve $y = f(x)$ touches the x-axis at the origin, $f(0) = 0$ and $f'(0) = 0$ and therefore $\varrho = |1/f''(0)|$. From the Maclaurin series

$$y = f(0) + xf'(0) + \frac{x^2}{2}f''(0) + \dots$$

and in our case

$$y = \frac{x^2}{2}f''(0) + \text{terms involving higher powers of } x$$

$$\Rightarrow \frac{2y}{x^2} = f''(0) + \text{terms containing } x \text{ as a factor}$$

$$\Rightarrow \kappa \text{ at the origin} = f''(0) = \lim_{x \to 0} \frac{2y}{x^2} \tag{20.24}$$

$$\Rightarrow \varrho \text{ at the origin} = \left| \lim_{x \to 0} \frac{x^2}{2y} \right|. \tag{20.25}$$

If the curve touches the y-axis at the origin,

$$\kappa \text{ at the origin} = \lim_{x \to 0} \frac{2x}{y^2} \tag{20.26}$$

$$\Rightarrow \varrho \text{ at the origin} = \left| \lim_{x \to 0} \frac{y^2}{2x} \right|. \tag{20.27}$$

Newton's formula has a limited application because it applies only to curves which *touch* one axis at the origin. It is sometimes convenient to move the origin in order to make use of the formula.

Example 1. Calculate the radius of curvature at the origin of the curve

$$ay^2 = x(a - x)^2.$$

The substitution $x = 0$ gives an equation in y with a repeated root. The curve, therefore, touches the y-axis at the origin. Hence ϱ at the origin is

$$\left| \lim_{x \to 0} \frac{y^2}{2x} \right| = \left| \lim_{x \to 0} \frac{x(a - x)^2}{2xa} \right| = \frac{a}{2}.$$

Example 2. Calculate the curvature of $y = \sin^2(x - \pi/6)$ at the point $(\pi/6, 0)$.

If we move the origin to $(\pi/6, 0)$, the equation becomes $Y = \sin^2 X$ and this curve touchès the X-axis at the new origin. Hence the curvature at the new origin is

$$\lim_{X \to 0} \frac{2Y}{X^2} = \lim_{X \to 0} 2\left(\frac{\sin X}{X}\right)^2 = 2$$

$$\Rightarrow \text{the curvature of } y = \sin^2(x - \pi/6) \text{ at } (\pi/6, 0) \text{ is } 2.$$

)

Exercise 20.11

Calculate the curvature for each of the curves in questions 1–10 at the points specified. Leave answers, where necessary, in surd form.

1. $y^2 = 4ax$; $(a, 2a)$.

2. $x = a \cos \theta$, $y = b \sin \theta$; $(\theta = \frac{1}{4}\pi)$.

3. $x = a \cosh t$, $y = b \sinh t$; $(t = 1)$.

4. $y = \ln x$; $(x = 1)$.

5. $s = a \sec^3 \psi$; $(\psi = \frac{1}{4}\pi)$.

6. $ay^2 = x^3$; $(4a, 8a)$.

7. $y = c \cosh (x/c)$; $(0, c)$.

8. $y = \ln \sin x$; $(x = \frac{1}{2}\pi)$.

9. $x^3 + y^3 + 3xy + 3x = 0$; $(1, -1)$.

10. $x^2 + 3xy + 3y^2 + 6x + 3y - 6 = 0$; $(0, 1)$.

Use Newton's formula to calculate the radius of curvature at the point specified in each of Nos. 11–15.

11. $y^2 = a^2 x/(a - x)$; $(0, 0)$.

12. $a^2 y^2 = x(a - x)^3$; $(0, 0)$.

13. $y = x^2/(1 + x^2)$; $(0, 0)$.

14. $y = c \cosh^2 (x/c)$; $(0, c)$.

15. $x^2/a^2 + y^2/b^2 = 1$; $(a, 0)$.

16. Show that the radius of curvature at any point of the cycloid

$$x = a(\theta - \sin \theta), \quad y = a(1 - \cos \theta)$$

is of magnitude $4a \sin (\theta/2)$ and that the corresponding centre of curvature is the point $\{a(\theta + \sin \theta), a(\cos \theta - 1)\}$.   (L.)

17. Show that the curves $x^2 - y^2 = 3a^2$ and $xy = 2a^2$ intersect at right angles, and that at the point of intersection their radii of curvature are in the ratio 4:3.   (L.)

18. Prove that the circle of curvature at the point $\theta = \pi/6$ on the astroid $x = a \cos^3 \theta$, $y = a \sin^3 \theta$ touches the y-axis.   (L.)

19. A curve is given parametrically by the equation

$$x = 4t^3 + 12t, \qquad y = 3t^4 + 6t^2.$$

Prove that the radius of curvature at the point with parameter t is $12(1 + t^2)^{5/2}$. If C is the centre of curvature at the point P prove that C and P lie on opposite sides of the y-axis.   (O.C.)

20. A curve is given parametrically by the equations

$$x = f'(t) \cos t + f(t) \sin t, \qquad y = f'(t) \sin t - f(t) \cos t, $$

where $f''(t) + f(t) > 0$ for all t. Prove that the slope of the curve at the point $t = u$ is $\tan u$. Prove also that the radius of curvature at this point is given by $\rho = f''(u) + f(u)$.

Given that $f(t) = a - b \cos 2t$, show that the centre of curvature at the point $t = u$ is the point $(4b \sin^3 u, 4b \cos^3 u)$.   (O.C.)

21. A curve is given by the parametric equations $x = e^{at} \cos t$, $y = e^{at} \sin t$. Prove that

$$\frac{dy}{dx} = \tan (t + \alpha)$$

where $a = R \cos \alpha$, $1 = R \sin \alpha$. Hence prove that the radius of curvature at the typical point is Re^{at}.

Find in terms of a the coordinates of the centre of curvature at the typical point. Show that, if $a = e^{-3\pi a/2}$, the locus of the centres of curvature is the original curve itself.   (O.C.)

22. Find the equation of the circle of curvature at the origin on the parabola $y = x^2$.
Show that the locus of the centre of curvature for a variable point on the parabola is

$$27x^2 = 2\,(2y - 1)^3.$$ (O.C.)

20:12 THE TANGENTIAL POLAR EQUATION—CURVATURE

In Fig. 20.18 ON is the perpendicular from the pole on to the tangent at $P(r, \theta)$ to the curve $f(r, \theta) = 0$. If $ON = p$,

$$p = r \sin \phi$$

$$\Rightarrow \frac{1}{p^2} = \frac{1}{r^2}\,\mathrm{cosec}^2\,\phi = \frac{1}{r^2}\,(1 + \cot^2 \phi)$$

$$\Leftrightarrow \frac{1}{p^2} = \frac{1}{r^2} + \frac{1}{r^4}\left(\frac{dr}{d\theta}\right)^2.$$ (20.28)

If θ is eliminated between this equation and the polar equation, the tangential-polar, pedal, or $p - r$ equation is obtained.

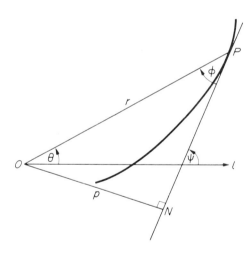

Fig. 20.18.

Example. Find the $p - r$ equation for the curve $r^2 = a^2 \sec 2\theta$.

$$r^2 = a^2 \sec 2\theta \qquad \Rightarrow 2r\,\frac{dr}{d\theta} = 2a^2 \sec 2\theta \tan 2\theta$$

$$\Rightarrow \left(\frac{dr}{d\theta}\right)^2 = \frac{a^4 \sec^2 2\theta \tan^2 2\theta}{r^2} = \frac{r^4\left(\dfrac{r^4}{a^4} - 1\right)}{r^2}.$$

$$\Leftrightarrow \frac{1}{r^4}\left(\frac{dr}{d\theta}\right)^2 = \frac{r^2}{a^4} - \frac{1}{r^2}$$

$$\Leftrightarrow \frac{1}{p^2} = \frac{1}{r^2} + \frac{r^2}{a^4} - \frac{1}{r^2} = \frac{r^2}{a^4}.$$

Hence the $p-r$ equation is $pr = a^2$.

Curvature. Calculation of curvature directly from a polar equation is an involved process and a better method is to obtain the $p-r$ equation and then to calculate the curvature from the formula given below. From Fig. 20.18,

$$\kappa = \frac{d\psi}{ds} = \frac{d(\theta+\phi)}{ds}.$$

From $\triangle PMQ$ of Fig. 20.13 and noting that $PQ = \delta s + O[(\delta\theta)^2]$,

$$\frac{d\theta}{ds} = \frac{1}{r}\sin\phi \quad \text{and} \quad \frac{dr}{ds} = \cos\phi$$

$$\Rightarrow \kappa = \frac{1}{r}\sin\phi + \cos\phi\,\frac{d\phi}{dr}$$

$$= \frac{1}{r}\left(\sin\phi + r\cos\phi\,\frac{d\phi}{dr}\right) = \frac{1}{r}\frac{d}{dr}(r\sin\phi)$$

$$\Rightarrow \kappa = \frac{1}{r}\frac{dp}{dr} \qquad\qquad (20.29)$$

and

$$\varrho = \left|r\,\frac{dr}{dp}\right|. \qquad\qquad (20.30)$$

Example. Calculate the radius of curvature of the lemniscate $r^2 = a^2\cos2\theta$, $a > 0$, at the point $\theta = 0$.

$$r^2 = a^2\cos2\theta \qquad \Rightarrow 2r\frac{dr}{d\theta} = -2a^2\sin2\theta$$

$$\Leftrightarrow \frac{dr}{d\theta} = \frac{-a^2\sin2\theta}{r}$$

$$\Rightarrow \left(\frac{dr}{d\theta}\right)^2 = \frac{a^4\sin^2 2\theta}{r^2} = \frac{a^4-r^4}{r^2}$$

$$\Leftrightarrow \frac{1}{p^2} = \frac{1}{r^2} + \frac{a^4-r^4}{r^6} = \frac{a^4}{r^6}.$$

Hence the $p-r$ equation is $r^3 = a^2 p$

$$\Rightarrow \kappa = \frac{1}{r}\frac{dp}{dr} = \frac{3r}{a^2}.$$

At $\theta = 0$, $r = a$; hence the radius of curvature there is $a/3$.

Exercise 20.12

Throughout these exercises ϕ is used to denote the angle between the tangent and the radius vector, ψ to denote the angle between the initial line and the tangent, and $a > 0$.

1. Calculate ϕ for each of the following curves at the point named:
(i) $r = a/(1 + \cos\theta)$; $(2a/3, \pi/3)$,
(ii) $r = a\sin 3\theta$; $(a/\sqrt 2, \pi/12)$,
(iii) $r = a(\sec\theta - \cos\theta)$; $(3a/2, \pi/3)$,
(iv) $r = a\theta$; $(\theta = \alpha)$,
(v) $r^2 - 12r\cos(\theta - \pi/4) + 5 = 0$; $(\theta = 7\pi/12)$.

2. Prove that, for the spiral $r = a\theta$, $\cos\phi = a/\sqrt{(a^2 + r^2)}$.

3. Calculate the angle between the two branches of $r^2 = a^2\cos 2\theta$ at the pole.

4. Show that tangents to the equiangular spiral $r = e^{\theta\cot\alpha}$ are parallel to the initial line at points where $\theta = n\pi - \alpha$.

5. For the curve $r = a\tan\frac 12\theta$ show that the branches corresponding to the ranges $0 \leqslant \theta \leqslant \pi$ and $\pi \leqslant \theta \leqslant 2\pi$ each touch the initial line at the pole and calculate the angle between the branches at their other point of intersection. Calculate also the θ-coordinates of the points at which the tangent to the curve is parallel to the line $\theta = \pi/2$. Sketch the curve.

6. Calculate the angle of intersection of the cardioids $r = a(1 + \cos\theta)$ and $r = a(1 - \cos\theta)$.

7. Prove that the curves $r = a(1 + \cos\theta)$ and $r = a/(1 + \cos\theta)$ intersect the initial line and also the line $\theta = \pi/2$ at equal angles.

8. Calculate ψ for the curve $r = \sin 2\theta$ at the point $\theta = \pi/4$ and find the coordinates of the points on the curve at which the tangent is parallel to the initial line.

9. Prove that the tangents to the cardioid $r = a(1 + \cos\theta)$ at the points $\theta = \alpha$, $\theta = \alpha + 2\pi/3$, $\theta = \alpha + 4\pi/3$ are parallel.

10. Sketch the polar curve $r^2 = a^2\sec 2\theta$ and find an expression for ϕ in terms of θ. Discuss the significance of the values of ϕ and of ψ when $\theta = \pi/4$ and $\theta = 3\pi/4$.

11. Calculate the angle between the tangents at the pole to $r = 1 + 2\cos\theta$.

12. Find the intrinsic equation of the cardioid $r = a(1 + \cos\theta)$. [The intrinsic equation is the $s - \psi$ equation.]

13. Find the $p - r$ equation for the spiral $r = a\theta$ and hence find the curvature at $\theta = 1$.

14. Find the $p - r$ equation for the reciprocal spiral $r = a/\theta$ and hence find the curvature at $\theta = 1$.

15. Show that, for the curve $r^2\cos 2\theta = a^2$, the radius of curvature at any point is proportional to r^3.

16. Find the $p - r$ equation of the curve $a = r\cosh\theta$ and hence find the curvature at $\theta = 0$.

17. For the parabola $r = a/(1 + \cos\theta)$ prove that $\varrho = a\operatorname{cosec}^3\phi$.

18. Find the values of θ at the points on the curve $r^2 = a^2\sin 2\theta$ at which $\phi = \pi/4$.

*20:13 ENVELOPES

A *family* of curves in two dimensions is represented by a set of equations of the form $f(x, y, t) = 0$, where t is a *parameter* for the family, so that each curve of the family is represented by this equation with its corresponding value of t. For example, $y = mx$ with m as the parameter represents the family of straight lines through the origin. Also $x^2 + y^2 = a^2$ with a as the parameter represents the family of circles centre the origin.

Consider two curves of the family $f(x, y, t) = 0$ corresponding to values $t, t + \delta t$ of the parameter. Any point of intersection of the two curves will have coordinates (x, y) which satisfy the equations

$$f(x, y, t) = 0, \qquad f(x, y, t + \delta t) = 0,$$

and therefore satisfy the equation

$$f(x, y, t + \delta t) - f(x, y, t) = 0,$$

and therefore satisfy the equation ·

$$\frac{f(x, y, t + \delta t) - f(x, y, t)}{\delta t} = 0, \qquad (\delta t \neq 0). \qquad (20.31)$$

If the left-hand side of equation (20.31) tends to a limit as $\delta t \to 0$, this limit is the partial derivative of $f(x, y, t)$ with respect to t, i.e. $\partial f/\partial t$.

If we can eliminate t between the equations

$$f(x, y, t) = 0, \qquad \frac{\partial}{\partial t} f(x, y, t) = 0,$$

or if we can solve these equations for x and y in terms of t, we obtain either the cartesian equation or the parametric equations of a curve which is called the *envelope* of the family, and which will be touched by every member of the family.

A practical illustration of the envelope of a family is given by the parabola of safety. [See *Applied Mathematics*, Volume 2, Part 2, Chapter 9.]

Example 1. Find the envelope of the family

$$f(x, y, t) \equiv y - \frac{x}{t} - at = 0.$$

$$\frac{\partial f}{\partial t} \equiv \frac{x}{t^2} - a = 0$$

$$\Rightarrow x = at^2, \ y = 2at \Rightarrow y^2 = 4ax.$$

Note: For this family, $dy/dx = 1/t$; thus the differential equation of which the family are solutions is

$$x \left(\frac{dy}{dx} \right)^2 - a = 0.$$

$$y^2 = 4ax \Rightarrow y\frac{dy}{dx} = 2a \Rightarrow \left(\frac{dy}{dx} \right)^2 = \frac{a}{x} \Rightarrow x \left(\frac{dy}{dx} \right)^2 = a.$$

Thus $y^2 = 4ax$ is a solution to the differential equation of the family—known as a "singular solution".

Example 2. Find the envelope of normals to the parabola $y^2 = 4ax$.

$$f(x, y, t) \equiv y + tx - 2at - at^3 = 0$$

$$\Rightarrow \frac{\partial f}{\partial t} \equiv x - 2a - 3at^2 = 0$$

$$\Rightarrow x = a(2 + 3t^2), \quad y = -2at^3 \Rightarrow 27ay^2 = 4(x - 2a)^3.$$

Note: It can be shown that the envelope of normals to a curve is also the locus of its centre of curvature—the evolute.

Exercise 20.13

In questions 1–4 find the envelopes of the given families of curves.

1. $\frac{x}{a} \cos \theta + \frac{y}{b} \sin \theta = 1.$ 2. $x + t^2 y = 2ct.$

3. $x \sec \theta + y \tan \theta = a$.   4. $y = t^2(x - 2a) - 2at^3$.

5. Find the evolute of the curve $x = at^2, \quad y = at^3$.

6. Find the evolute of the ellipse $\dfrac{x^2}{a^2} + \dfrac{y^2}{b^2} = 1$.

7. Show that the evolute of the rectangular hyperbola $x = ct, y = c/t$ is the curve

$$(x + y)^{2/3} - (x - y)^{2/3} = (4c)^{2/3}.$$

Miscellaneous Exercise 20

1. Determine
  (i) the area of the region enclosed by the axes of coordinates, the ordinate $x = \ln 2$, and the curve $y = \cosh x$;
  (ii) the x-coordinate of the centroid of this region;
  (iii) the length of the arc of the curve which bounds the region.

2. The parametric coordinates of a point on a curve are given by $x = a(\tan t - t), \quad y = a \ln \sec t$. Prove that the arc length s of the curve measured from a certain point O is $\sec t - a$ and give the coordinates of O.
   The arc of length a measured from O is rotated about the x-axis through four right angles. Find the surface area of the curved surface so formed.   (O.C.)

3. Determine the length of the arc of the curve given parametrically by $x = \frac{1}{3}t^3, \quad y = \frac{1}{2}t^2$ from the origin to the point $(\frac{1}{3}, \frac{1}{2})$.
   Find also the area of the curved surface obtained by rotating this arc through four right angles about the axis of x.   (C.)

4. The parametric equations of a curve are $x = ae^{-t} \cos t, y = ae^{-t} \sin t$. On this curve A and P are the points corresponding to $t = 0$ and $t = p$ respectively. Prove that the arc length AP of the curve is equal to $(\sqrt{2})(a - OP)$, where O is the origin of coordinates.
   The portion of the curve from $t = 0$ to $t = \pi$ is rotated through 2π radians about the x-axis. Prove that the area of the curved surface generated is $\frac{2}{5}(\sqrt{2})\pi a^2(1 + e^{-2\pi})$.   (C.)

5. The region enclosed by the curve with equation $y = x^3$, the y-axis and the line $y = 1$ is rotated through 2π radians about the y-axis. Find the centroid of the solid thus generated.
   Prove that the area of the curved surface of the solid is approximately 5·9 square units.   (C.)

6. A curve is defined in terms of a positive parameter t by

$$x = at \cos t, \quad y = at \sin t.$$

  (a) Give a rough sketch of the curve for $0 \leqslant t \leqslant 4\pi$.
  (b) Find the area of the region cut off between the x-axis and that part of the curve for which $0 \leqslant t \leqslant \pi$.
  (c) Find the arc-length along the curve measured from the origin to the point with parameter T.

7. Prove that the length of the arc of the parabola $y^2 = 4ax$, between $y = 0$ and $y = 2a$, is

$$a\{\sqrt{2} + \ln(1 + \sqrt{2})\}.$$

   This arc is rotated through 2π radians about the x-axis. Find the area of the surface of revolution generated. Hence, or otherwise, find the distance of the centroid of the arc from the line $y = 0$.   (L.)

8. Prove that the length s of the arc of the catenary $y = c \cosh(x/c)$ from the point where the curve cuts the y-axis to the point where $x = a$ is given by $s = c \sinh(a/c)$.
   From any point P on the catenary the ordinate PQ is drawn, meeting the x-axis at Q. From Q a perpendicular QR is drawn to the tangent at P, meeting the latter at R. Prove that the length of QR is constant and equal to c, and that the length of PR is equal to s.   (L.)

9. Sketch the curve given by $x = 4t, y = t^2 - 2 \ln t$ where t is a variable parameter, for values of t from 1 to 3.
   If A and B are the points corresponding to $t = 1$ and $t = 3$ respectively, calculate
  (i) the length of the arc AB,
  (ii) the area of the region bounded by the arc AB, the x-axis and the ordinates through A and B. (L.)

10. Sketch the curve $3ay^2 = x(x-a)^2$, where $a > 0$. Show that

$$\frac{dy}{dx} = \pm \frac{(3x-a)}{2\sqrt{(3ax)}}$$

and, hence or otherwise, show that the perimeter of the loop of the curve is $4a/\sqrt{3}$.   (L.)

11. Sketch the curve with parametric equations

$$3x = t^3 - 3t, \quad 3y = t^3.$$

The arc of this curve between the points where $t = -1$ and $t = 0$ is rotated completely about the y-axis. Prove that the area of the curved surface thus formed is

$$(5\sqrt{2\pi/24})\{\ln(1+\sqrt{2}) + \sqrt{2}\}.$$ (L.)

12. (i) A curve has polar equation $r = 2\cos(2\theta/3)$. Sketch the part of this curve for which $-3\pi/4 \leqslant \theta \leqslant 3\pi/4$. Show that the area of the finite region enclosed by this part of the curve is $3\pi/2$.

(ii) Show that the length of the arc of the curve $y = x^2 - \frac{1}{8}\ln x$ between the points where $x = 1$ and $x = 2$ may be expressed as

$$\int_1^2 \left(2x + \frac{1}{8x}\right)dx$$

and so find this arc length.   (L.)

13. Show that the perimeter of the ellipse given by $x = a\cos\theta$, $y = b\sin\theta$ is

$$2a \int_0^\pi (1 - e^2\cos^2\theta)^{1/2}\, d\theta,$$

where e is the eccentricity.

Determine the approximate perimeter when $e = \frac{1}{2}$ by Simpson's rule, using seven ordinates from $\theta = 0$ to $\theta = \pi$.   (L.)

14. Two curves C_1, C_2 have equations in polar coordinates given by $r = 3\cos\theta$ and $r = 1 + \cos\theta$ respectively. Sketch these curves in the same diagram.

Find the area of the region R which lies inside C_1 and outside C_2 and prove that the arc-length of the boundary of R is $(4 + 2\pi)$.

The region R is rotated about the line $\theta = 0$ through two right angles to produce a solid of revolution; find the surface area of this solid.   (O.C.)

15. Find the polar equation of the tangent to the curve

$$r^2 = a^2 \cos 2\theta$$

at the point $P(a/\sqrt{2}, \pi/6)$ on the curve, simplifying the result as far as possible.

Determine (a) the smallest area, (b) the largest area of the region enclosed by this tangent and the curve.

16. (i) Sketch the curves $r = a(1 + \cos\theta)$, $r = b(1 - \cos\theta)$, $a > 0$, $b > 0$, and show that they intersect orthogonally.

(ii) Find the area of the sector enclosed by the curve $r = a\sec^2\frac{1}{2}\theta$ and the radii $\theta = 0$, $\theta = \alpha$.

If the arc of this sector is revolved completely about the initial line, show that the area traced out is

$$\tfrac{8}{3}\pi a^2 (\sec^3\tfrac{1}{2}\alpha - 1).$$

17. Calculate the length of the arc of the curve whose parametric equation is

$$\begin{cases} x = a(3\cos t - \cos 3t) \\ y = a(3\sin t - \sin 3t) \end{cases}$$

between the points corresponding to $t = 0$ and $t = \pi$.

Find the area of the region between this arc and the x-axis.

What is the area of the surface of revolution formed when the arc is rotated through an angle of 2π about the x-axis?

18. Find the length of the curve $2y = c(e^{x/c} + e^{-x/c})$ between the points $(0, c)$ and $\left(c, \dfrac{ce^2 + c}{2e} \right)$.

Find also the curved surface and volume of the solid obtained by revolving the curve between these two points about the axis of x.

19. A curve is given in the form

$$x = \cosh t - t,$$
$$y = \cosh t + t.$$

Express t in terms of the length s of the arc of the curve measured from the point $(1, 1)$.

The coordinates of any point of the curve are expressed in terms of s and are then expanded in series of ascending powers of s. Prove that the first few terms of these expansions are

$$x = 1 - \frac{s}{\sqrt{2}} + \frac{s^2}{4} + \dots,$$

$$y = 1 + \frac{s}{\sqrt{2}} + \frac{s^2}{4} + \dots. \tag{O.C.}$$

20. Trace the curve $x = at$, $y = a(1 - \cos t)$ for values of t between -2π and 2π.

Determine the inflexions on the curve, and prove that if the tangents at two real points P and Q are at right angles, then both P and Q are inflexions. (O.C.)

21. An arc of a circle of radius c subtending an angle 2α at the centre is rotated about its chord. Prove that the area of the surface of revolution so formed is $4\pi c^2 (\sin \alpha - \alpha \cos \alpha)$, and that its volume is

$$\tfrac{2}{3} \pi c^3 (2 \sin \alpha + \sin \alpha \cos^2 \alpha - 3\alpha \cos \alpha). \tag{O.C.}$$

22. A cycloid, given by the equations

$$x = a(1 - \cos \theta), \quad y = a(\theta + \sin \theta),$$

revolves about its base $x = 2a$; find the area of the surface, and the volume of the solid of revolution so formed. (O.C.)

23. Show that the arc of the curve given by

$$x = a \cos t, \quad y = \tfrac{1}{4}a \cos 2t$$

between points for which $t = 0$, $t = \tfrac{1}{2}\pi$, is equal to

$$\int\limits_0^1 a(1 + z^2)^{1/2} \, dz. \tag{O.C.}$$

24. Sketch the curve $y^2 = (2 - x)/(x + 6)$.

Show that there are points of inflexion at $(0, \pm 3^{-1/2})$, and find the radius of curvature at the point $(-2, 1)$. (O.C.)

25. Show that the radius of curvature of the cycloid $x = a(\theta - \sin \theta)$, $y = a(1 - \cos \theta)$ at the point θ is $4a \sin (\theta/2)$. (L.)

26. If $f(x) = (2 - x^2) \sin x - 2x \cos x$, show that

  (i) the expansion of $f(x)$ in ascending powers of x is of the form
    $f(x) = kx^3 + \text{higher powers of } x,$
    and obtain the value of k;

  (ii) the graph of $f(x)$ has a stationary point at $x = 0$ and another at $x = \tfrac{1}{2}\pi$, and determine in each case whether the point is a maximum or minimum point or a point of inflexion;

  (iii) the equation $f(x) = 0$ has a root between $x = \tfrac{1}{2}\pi$ and $x = \pi$. Sketch the graph of $f(x)$ from $x = 0$ to $x = \pi$. (N.)

27. Determine the equations of the tangents at the points of inflexion on the curve $y = x^2 (x - 4)(x - 6)$ and the coordinates of the points where these tangents meet the curve again. (N.)

28. Sketch the arc of the curve $x = a \cos^3 t$, $y = a \sin^3 t$ for $0 \leqslant t \leqslant \tfrac{1}{2}\pi$.

Find the radius of curvature of the curve at the point t.

The normal at the point P whose parameter is t meets the x-axis at G, and C is the centre of curvature at P. Find the non-zero value of t for which P is the mid-point of CG. (N.)

29. Find the equation of the tangent to the curve $x = 3t^2$, $y = 2t^3$ at the point P whose parameter t is $\sqrt{2}$.

Determine the parameter of the point Q where this tangent meets the curve again and prove that PQ is the normal at Q.

Find the radius of curvature at Q. (N.)

30. Find the value of dy/dx in terms of t at a point on the curve whose parametric equations are

$$x = a(t - \tanh t), \quad y = a \operatorname{sech} t.$$

Prove that

$$a\frac{d^2 y}{dx^2} = \frac{\cosh^3 t}{\sinh^4 t}.$$

Hence find the length of the radius of curvature at the point whose parameter is t. (N.)

31. Given that $y = (x+1)e^{-x}$ find dy/dx and $d^2 y/dx^2$. Calculate the coordinates of the turning point and of the point of inflexion on the graph of y, and sketch the graph. (N.)

32. A curve is given by the parametric equations

$$x = a(t \sin t + \cos t - 1), \quad y = a(\sin t - t \cos t).$$

Find dy/dx and $d^2 y/dx^2$ in terms of t and prove that, at the point P whose parameter is t, the length of the radius of curvature is $|at|$.

Obtain the equation of the normal at P. Show that all the normals touch the circle $x^2 + y^2 + 2ax = 0$. (N.)

33. Obtain the equations of the tangents at the two points of inflexion on the curve $y = 3x^2 - 1/x^2$ and determine the coordinates of the points where these tangents meet the curve again. (N.)

34. A fixed point P of a circular disc of radius $2a$ is at a distance a from its centre. Show that the curve traced out by P when the disc rolls without slipping on the outside of a fixed circle of radius $6a$ may be expressed by the parametric equations

$$x = 8a \cos \theta - a \cos 4\theta, \quad y = 8a \sin \theta - a \sin 4\theta.$$

Show that the path of P has zero curvature at six points, situated at the vertices of two equilateral triangles. (N.)

35. Find the tangents at the origin and at the point $(-a, 0)$ to each of the following curves:
(i) $y^2(a-x) = x^2(a+x)$,
(ii) $a^3 y^2(a-x) = x^2(a+x)^4$.

Sketch the graphs of the above two curves. Find (by Newton's method or otherwise) the radius of curvature of the curve (i) at the point $(-a, 0)$. (N.)

36. The line $x = 4a$ cuts the parabola $y^2 = 4ax$ at the points A and B. Find
(i) the gradient of the curve at A and at B;
(ii) the area of the region enclosed between the chord AB and the arc AB of the parabola;
(iii) the length of the arc AB. (O.C.)

37. The coordinates of a point of a curve are given in terms of a parameter t by the equations $x = te^t$, $y = t^2 e^t$. Find dy/dx in terms of t, and prove that

$$\frac{d^2 y}{dx^2} = \frac{t^2 + 2t + 2}{(t+1)^3} e^{-t}.$$

Prove also that the radii of the circles of curvature at the two points at which the curve is parallel to the x-axis are in the ratio $e^2 : 1$. (O.C.)

38. Prove that the curve

$$y = x^2 + \frac{32}{x+3}$$

has only one turning point, which is at $x = 1$, and state whether this point determines a maximum or a minimum. Prove also that the curve has one and only one point of inflexion.

Draw a rough sketch of the curve.   (N.)

39. Find the equation connecting s and ψ, where s is the length of arc OP and $\tan \psi$ is the slope of the tangent at P, if P is any point on the cycloid

$$x = a(\theta + \sin \theta), \quad y = a(1 - \cos \theta).$$

Hence or otherwise show that the radius of curvature at P is twice the distance PN where N is the point of intersection of the line $y = 2a$ with the normal at P.   (O.C.)

40. Sketch the curve $y = \sin^2 2x$ between $x = -\frac{1}{2}\pi$ and $x = \frac{1}{2}\pi$. Use Newton's formula to find the radius of curvature at the point $(0, 0)$.   (N.)

41. A small bead P is threaded on a thin rod. The rod rotates in a plane with uniform angular speed 2 radians/sec about one end A, and P is made to move along the rod and away from A so that its speed relative to the rod at any instant is proportional to the distance AP. Initially P is at unit distance from A and has a speed relative to the rod equal to 2 units. Taking A as the pole and the initial position of the rod as the initial line,

   (i) show that the bead is moving parallel to the initial line when $\theta = 3\pi/4$;

   (ii) find the values of θ in the range $0 \leqslant \theta \leqslant 2\pi$ for which the bead is moving in a direction perpendicular to the initial line;

   (iii) find the equation of the path of the bead and sketch the path for values of θ from 0 to 2π.   (N.)

42. A loop of the curve $r^2 = a^2 \cos 2\theta$ rotates round the line $\theta = \frac{1}{2}\pi$. Prove that the area of the surface generated is $2\sqrt{2\pi a^2}$, and find the volume contained by it.   (O.C.)

43. Find the area of the region between the parabola $r = 2a/(1 + \cos \theta)$ and the radii $\theta = \pi/4$ and $\theta = 3\pi/4$.

Find the distance of the centroid of the same region from the axis.

44. Prove that the curve with polar equation $r = a \cos \theta \sin 2\theta$ is symmetrical about the line $\theta = \pm \pi/2$. Find the values of r at points on the curve (other than the pole) where $dr/d\theta = 0$, and explain the nature of these points. Sketch the curve.

Prove that the total area enclosed by the curve is $\frac{1}{8}\pi a^2$.

45. Show that $r = 2(\cos \theta - \sin \theta)$ is the polar equation of a circle.

Sketch in separate diagrams the curves with polar equations $r = 2(\cos \theta - \sin \theta)$, $r = 2(\cos 2\theta - \sin 2\theta)$. Prove that the area enclosed by the second curve is twice that enclosed by the circle.

46. Prove that the curve with polar equation

$$r = a (\cos 2\theta + \sin \theta)$$

is symmetrical about the line $\theta = \pi/2$, and that $r = 0$ when $\theta = \pi/2$, $\theta = 7\pi/6$ and $\theta = 11\pi/6$. Sketch the curve.

Prove that the area enclosed by the loop of this curve which is symmetrical about the line $\theta = \pi/2$ is $a^2 (16\pi + 27 \sqrt{3})/48$.

47. Sketch that part of the graph of

$$y^2 = \frac{a^2 x^2}{x^2 + a^2}$$

which lies in the first quadrant.

Prove that the polar equation of this curve referred to the origin as pole and the axis of x as initial line is

$$r^2 = 4a^2 \cot 2\theta \ \mathrm{cosec} \ 2\theta,$$

and show that the area enclosed by the curve and the line $\theta = \frac{1}{8}\pi$ is $a^2 (\sqrt{2} - 1)$.   (C.)

48. Sketch separately the curves with polar equations

   (a) $r = a \sin 3\theta$;   (b) $r = a \sin 2\theta$.

Prove that the total area enclosed by the curve (b) is twice the total area enclosed by the curve (a).

49. The equation of a curve in polar coordinates is $r = a (2 \sec \theta - 1)$. Show that as θ increases from 0 to $\frac{1}{2}\pi$,

$r \cos \theta$ increases to the value $2a$ and $r \sin \theta$ increases without limit. Sketch the curve for values of θ from $-\pi$ to $+\pi$, showing that it consists of two distinct branches. Prove that if a straight line through the pole meets the branches in P and Q then PQ is of constant length.

Prove that the area of the sector contained by the lines $\theta = 0, \theta = \frac{1}{4}\pi$, and the branch of the curve given by $0 \leqslant \theta \leqslant \frac{1}{4}\pi$ is about $0.63a^2$.

21 Geometry of Two Dimensions

21:1 STRAIGHT LINES THROUGH THE ORIGIN

Consider the equation

$$\prod_{r=1}^{n} (y - m_r x) = 0 \quad (m_r \in \mathbb{R}) \tag{21.1}$$

in which the symbol $\prod_{r=1}^{n}$ represents the product of n factors, the rth being $(y - m_r x)$. Corresponding to each of these real linear factors there is a straight line through the origin, and thus the locus represented by (21.1) consists of n straight lines through the origin.

Clearly if $\quad f(x, y) \equiv \prod_{r=1}^{n} (y - m_r x), \quad$ then $\quad f(tx, ty) \equiv t^n f(x, y).$

A polynomial in x and y in which every term is of the form $kx^r y^s$, where $k \in \mathbb{R}$ and $r + s = n$, is said to be *homogeneous* in x and y and of the nth degree.

Example. $x^6 + 3x^4 y^2 + 7xy^5$ is homogeneous and of the sixth degree in x and y.

A necessary and sufficient condition for $g(x, y)$ to be homogeneous of degree n in x and y is

$$g(tx, ty) \equiv t^n g(x, y).$$

From the above definition, a homogeneous polynomial $g(x, y)$ will have *at most* n real linear factors of the form $ax + by$; other real factors must be of the form $p^2 x^2 + q^2 y^2$ which cannot be zero for any x, y except $x = y = 0$. Hence the locus $g(x, y) = 0$ consists of l straight lines through the origin where $l \leqslant n$.

Example 1. The equation $y^3 + 3xy^2 - x^2 y - 3x^3 = 0 \equiv (y - x)(y + x)(y + 3x) = 0$ and therefore represents the three straight lines $y - x = 0$, $y + x = 0$, $y + 3x = 0$.

Example 2. The equation $x^4 + x^3 y - 6x^2 y^2 = 0 \equiv x^2 (x - 2y)(x + 3y) = 0$ and therefore represents four straight lines: $x - 2y = 0$, $x + 3y = 0$ and two coincident lines $x = 0$.

Example 3. The equation $x^4 - 81y^4 = 0 \equiv (x - 3y)(x + 3y)(x^2 + 9y^2) = 0$ and therefore represents two straight lines $x - 3y = 0$, $x + 3y = 0$.

THE LINE-PAIR

The equation

$$ax^2 + 2hxy + by^2 = 0 \qquad (21.2)$$

is a homogeneous equation of the second degree.

After division by x^2, equation (21.2) can be regarded as a quadratic equation in (y/x), and if $h^2 \geqslant ab$ so that the equation has real roots, say $y/x = m_1$ and $y/x = m_2$, it represents the two lines $y = m_1 x$ and $y = m_2 x$ through the origin. If it has no real roots, it is satisfied only by the point $(0, 0)$.

The roots of the equation in y/x may be irrational and, if only for this reason, it is usually more convenient to consider the equation of the *line-pair* than to consider the separate equations of the lines.

The angle between the lines $ax^2 + 2hxy + by^2 = 0$, $h^2 \geqslant ab$.
If

$$ax^2 + 2hxy + by^2 = b(y - x\tan\theta_1)(y - x\tan\theta_2),$$

then

$$\tan\theta_1 + \tan\theta_2 = -2h/b, \quad \tan\theta_1 \tan\theta_2 = a/b.$$

Therefore, if α is the angle between the lines,

$$\tan\alpha = \frac{\tan\theta_1 - \tan\theta_2}{1 + \tan\theta_1 \tan\theta_2} = \sqrt{\left\{\frac{4h^2}{b^2} - \frac{4a}{b}\right\}} \bigg/ \left(1 + \frac{a}{b}\right) = \frac{2\sqrt{(h^2 - ab)}}{a + b}.$$

The lines are (i) at right angles if $a = -b$, (ii) coincident if $h^2 = ab$.

The equation $ax^2 + 2hxy + by^2 + 2gx + 2fy + c = 0$, with a, b, c, f, g and $h \in \mathbb{R}$, represents two straight lines if it has a rational solution for y in terms of x. The equation can be expressed as

$$by^2 + (2hx + 2f)y + (ax^2 + 2gx + c) = 0$$

and it represents the two loci

$$y = [-hx - f \pm \sqrt{\{(hx + f)^2 - b(ax^2 + 2gx + c)\}}]/b.$$

These equations will represent two straight lines if $b \neq 0$ and

$$(hx + f)^2 - b(ax^2 + 2gx + c)$$

is a perfect square.

Example 1. Find the acute angle between the lines

$$3x^2 + 2xy - 2y^2 = 0.$$

If the lines are $y = m_1 x$ and $y = m_2 x$, then $m_1 m_2 = -3/2$, $m_1 + m_2 = 1$

$$\Rightarrow m_1 - m_2 = \sqrt{[(m_1 + m_2)^2 - 4m_1 m_2]} = \sqrt{7}.$$

Therefore the acute angle between the lines is

$$\tan^{-1}\left|\frac{m_1 - m_2}{1 + m_1 m_2}\right| = \tan^{-1} 2\sqrt{7} \approx 79\cdot 3°.$$

Example 2. Find k, if the equation

$$2x^2 - xy - ky^2 + 2y + x - 1 = 0$$

represents a line-pair.

The equation of the locus is

$$ky^2 + (x - 2)y - (2x^2 + x - 1) = 0$$

and this represents a line-pair if

$(x - 2)^2 + 4k(2x^2 + x - 1)$ is a perfect square and $k \neq 0$

$\Leftrightarrow x^2(1 + 8k) + x(4k - 4) + (4 - 4k)$ is a perfect square and $k \neq 0$

$\Leftrightarrow 16(k - 1)^2 = 4(4 - 4k)(1 + 8k)$ and $k \neq 0$

$\Leftrightarrow k(k - 1) = 0$ and $k \neq 0$

$\Leftrightarrow k = 1$.

The examples which follow illustrate an important application of a principle used in Chapter 2 when the intersection of circles and of lines and circles was discussed.

Example 1. Prove that the acute angle between the lines joining the origin to the intersection of $x + y = 3$ with $x^2 - 3xy + y^2 + x + y - 2 = 0$ is $\tan^{-1}(3/4)$.

The equation

$$x^2 - 3xy + y^2 + (x + y)\left(\frac{x + y}{3}\right) - 2\left(\frac{x + y}{3}\right)^2 = 0$$

is satisfied by the coordinates of the points of intersection (implied real) of the line and the curve. It is homogeneous of the second degree, and therefore it represents two straight lines through the origin. It is, therefore, the equation of the line-pair which joins the origin to the points of intersection of the straight line with the curve.

The equation is

$$10x^2 - 25xy + 10y^2 = 0$$
$$\Leftrightarrow 2x^2 - 5xy + 2y^2 = 0,$$

representing the lines $2x - y = 0$ and $x - 2y = 0$.

The acute angle between these lines is $\tan^{-1}\left(\dfrac{2 - \frac{1}{2}}{1 + 1}\right) = \tan^{-1}\dfrac{3}{4}$.

Example 2. Find the equation of the pair of lines joining the origin to the points of intersection of the line $x + ky = 1$ and the circle

$$x^2 + y^2 + 4x - 4y - 1 = 0.$$

Determine the value of k for which the equation represents perpendicular lines. Find also the values of k for which the equation represents coincident lines and deduce the equation of the tangents to the circle from the point $(1, 0)$.   (N.)

The equation of the line-pair is

$$x^2 + y^2 + (4x - 4y)(x + ky) - (x + ky)^2 = 0$$
$$\Leftrightarrow 4x^2 + (2k - 4)xy + (1 - 4k - k^2)y^2 = 0.$$

These lines are perpendicular if

$$1 - 4k - k^2 = -4$$
$$\Leftrightarrow k^2 + 4k - 5 = 0$$
$$\Leftrightarrow k = -5 \quad \text{or} \quad k = 1.$$

The lines are coincident if

$$(2k-4)^2 = 16(1-4k-k^2)$$
$$\Leftrightarrow 5k^2 + 12k = 0$$
$$\Leftrightarrow k = 0 \quad \text{or} \quad -12/5.$$

The line $x + ky = 1$ goes through $(1, 0)$ for all values of k. When the lines joining the origin to the intersection of $x + ky = 1$ with the circle are coincident, $x + ky = 1$ is a tangent to the circle. Therefore $x = 1$ and $5x - 12y = 5$ are the equations of the tangents from $(1, 0)$ to the circle.

___

Exercise 21.1

1. Find the angles between each of the following line-pairs.
   (i) $3x^2 - 5xy + y^2 = 0$;   (ii) $x^2 + xy - 5y^2 = 0$;
   (iii) $2x^2 + xy - 4y^2 = 0$;   (iv) $x^2 - 3xy - y^2 = 0$.

2. State which of the following equations represent line-pairs, and, in the cases in which they do, find the separate equations of the lines.
   (i) $2x^2 - xy - y^2 - 3x + 6y - 5 = 0$;
   (ii) $3x^2 + xy - 2y^2 + 5x - 15y - 28 = 0$;
   (iii) $x^2 - 3xy + 2y^2 + 10x - 6y - 8 = 0$;   (iv) $2x^2 + xy - y^2 + 3x - 3y - 2 = 0$.

3. Show that if $h^2 > ab$, $ax^2 + 2hxy + by^2 = 0$ represents a pair of straight lines through the origin. Find the condition that the lines should be perpendicular.

   If either bisector of the angles between the lines makes an angle θ with the axis of x, prove that $(a-b)\tan 2\theta = 2h$. Prove also that the equation of the two bisectors is

   $$h(x^2 - y^2) = (a-b)xy.$$

4. Find the equation of the pair of lines joining the origin to the points in which the pair of lines

   $$4x^2 - 15xy - 4y^2 + 39x + 65y - 169 = 0$$

   are met by the line $x + 2y - 5 = 0$.
   Show that the quadrilateral having the first pair, and also the second pair, as adjacent sides is cyclic.

5. Find the equation of the pair of straight lines joining the origin of coordinates to the intersection of the circle $x^2 + y^2 + 2gx + 2fy + c = 0$ and the straight line

   $$lx + my = 1.$$

Hence, or otherwise, find the coordinates of the circumcentre of the triangle formed by the lines

$$ax^2 + 2hxy + by^2 = 0, \quad lx + my = 1.$$

If the lines $ax^2 + 2hxy + by^2 = 0$ vary, but are equally inclined to the axes, show that the circumcentre moves on a line through the origin.

(L.)

21:2 FURTHER CURVE SKETCHING

Algebraic curves

The procedures suggested in Chapter 6 for sketching a curve can still be used when we consider curves with more complicated equations, including cases when y is not a function of x; but in two respects some amplification is required. We now consider in more detail asymptotes and form near the origin.

Asymptotes

We have shown how to find the equations of asymptotes parallel to either axis of coordinates. Now we need a more general procedure, and start with a general statement based on the following.

At great distances from the origin the terms of highest degree in the equation of an algebraic curve must balance one another and so a first approximation to the behaviour of a curve can be obtained by equating the terms of highest degree to zero.

A curve with equation of degree n in x and y has *at most* n asymptotes; if $ax + by = 0$ is the line through the origin parallel to an asymptote, then $(ax + by)$ will be a factor of the terms of highest degree in the equation.

The converse of this statement, that a linear factor of the highest degree terms gives the direction of an asymptote, is generally true, though there are exceptional cases when the factor is repeated [see Example 1 (iii) below]; however, we can investigate such factors for possible asymptotes.

Example 1. Find the asymptotes of the curves:

$$\text{(i) } (y+x)(2y-3x) = x+4, \quad \text{(ii) } y^2(y+4x) = x+y, \quad \text{(iii) } y^2(y-2x) = x+y.$$

(i) The terms of highest degree equated to zero give

$$(y+x)(2y-3x) = 0.$$

The asymptotes are therefore *parallel to* the lines

$$y + x = 0, \quad 2y - 3x = 0.$$

To examine the behaviour a long way from the origin we use successive approximations in the equation of the curve (using each approximation to obtain the next).

To find the equation of the asymptote parallel to the line $y + x = 0$, we write the equation of the curve as

$$y + x = \frac{x+4}{2y-3x}$$

$$\Leftrightarrow y = -x + \frac{x+4}{2y-3x} \tag{1}$$

$$\Rightarrow y \approx -x + \frac{x+4}{-2x-3x}.$$

[Note that the first approximation of $y \approx -x$ is used to obtain the second approximation by substitution in the second term on the right-hand side of (1).]

Since
$$\lim_{x \to \infty} \frac{x+4}{-5x} = -\frac{1}{5},$$

$$y = -x - \tfrac{1}{5} \equiv 5x + 5y + 1 = 0$$

is the equation of one asymptote. Similarly, for the other asymptote,

$$y = \frac{3x}{2} + \frac{x+4}{2(y+x)} \approx \frac{3x}{2} + \frac{x+4}{2\left(\dfrac{3x}{2}+x\right)} = \frac{3x}{2} + \frac{x+4}{5x}.$$

Hence the other asymptote is

$$y = 3x/2 + \tfrac{1}{5} \equiv 15x - 10y + 2 = 0.$$

(ii) In this case the approximations are:

$$y = -4x + \frac{x+y}{y^2} \approx -4x + \frac{-3x}{16x^2} \to -4x \text{ as } x \to \infty,$$

$$y = 0 \pm \sqrt{\left(\frac{x+y}{y+4x}\right)} \approx \pm \sqrt{\left(\frac{1}{4}\right)} = \pm \frac{1}{2}.$$

Hence the asymptotes are $y + 4x = 0$, $2y + 1 = 0$, $2y - 1 = 0$.

(iii) The approximations are:

$$y = 2x + \frac{x+y}{y^2} \approx 2x + \frac{3x}{4x^2} \to 2x \text{ as } x \to \infty,$$

$$y = 0 \pm \sqrt{\left(\frac{x+y}{y-2x}\right)} \approx 0 \pm \sqrt{\left(\frac{x}{-2x}\right)}, \text{ giving no real values for } y \text{ as } |x| \to \infty.$$

Hence the only asymptote is $y = 2x$.

Example 2. Determine the form of the curve near its asymptotes in Example 1(i).

Near the asymptote $y = -x - \frac{1}{5}$,

$$y = -x - \frac{1}{5} + \left(\frac{x+4}{2y-3x} + \frac{1}{5}\right) = -x - \frac{1}{5} + \frac{2x+2y+20}{5(2y-3x)}$$

$$\approx -x - \frac{1}{5} + \frac{2x + 2(-x - \frac{1}{5}) + 20}{5(-2x - \frac{2}{5} - 3x)}$$

$$= -x - \frac{1}{5} - \frac{98}{5(25x+2)}$$

$$= -x - \frac{1}{5} - \frac{98}{125x} + O\left(\frac{1}{x^2}\right).$$

Hence,   as $x \to +\infty$, $y < -x - \frac{1}{5}$, so that the curve is below the asymptote;

       as $x \to -\infty$, $y > -x - \frac{1}{5}$, so that the curve is above the asymptote.

Near the asymptote $y = \frac{3}{2}x + \frac{1}{5}$,

$$y = \frac{3x}{2} + \frac{1}{5} + \left(\frac{x+4}{2(y+x)} - \frac{1}{5}\right) = \frac{3x}{2} + \frac{1}{5} + \frac{3x-2y+20}{10(y+x)}$$

$$\approx \frac{3x}{2} + \frac{1}{5} + \frac{3x - 3x - \frac{2}{5} + 20}{15x + 2 + 10x} = \frac{3x}{2} + \frac{1}{5} + \frac{98}{5(25x+2)}$$

$$= \frac{3x}{2} + \frac{1}{5} + \frac{98}{125x} + O\left(\frac{1}{x^2}\right)$$

$\Rightarrow$ as $x \to +\infty$ the curve is above the asymptote;

    as $x \to -\infty$ the curve is below the asymptote.

The student should carry out this procedure for parts (ii) and (iii) of Example 1.

Form of a curve near the origin

If a curve passes through the origin, tangents to the various branches are obtained by equating to zero any linear factors of the terms of *lowest* degree in the equation of the curve. Further approximations can be made as necessary.

If it is required, similar information about the behaviour of the curve at a point (x_0, y_0) other than the origin can be obtained by choosing new coordinate axes having this point as origin.

Singular points. A singular point on a curve is a point at which dy/dx is indeterminate (taking the form $0/0$).

Classification of singular points at the origin. When the origin lies on the curve, the equation of the curve must be of the form

$$(a_1x + a_2y) + (b_1x^2 + b_2xy + b_3y^2) + (c_1x^3 + c_2x^2y + c_3xy^2 + c_4y^3) + \ldots = 0,$$

where the a's, b's, etc., are constants. If at least one of a_1, a_2 is non-zero, then the origin O is an ordinary point on the curve, the tangent there having equation $a_1x + a_2y = 0$. If $a_1 = a_2 = 0$ then the origin is a singular point of the curve. Further, if not all of b_1, b_2, b_3 are zero this singular point is a *double point* of the curve. Similarly if $a_1 = a_2 = b_1 = b_2 = b_3 = 0$ and not all of c_1, c_2, c_3, c_4 are zero the curve has a *triple point at O*, etc. When the curve has a double point at O, it behaves near O like the pair of straight lines

$$b_1x^2 + b_2xy + b_3y^2 = 0.$$

1. If these lines are real and distinct the curve has two distinct branches through O and the double point is a *node*.

2. If these lines are real and coincident,

i.e. if $$b_1x^2 + b_2xy + b_3y^2 \equiv k(\beta_1x + \beta_2y)^2,$$

the curve has in general a single tangent at the origin and the curve has a *cusp* there if it does not continue through the origin, and has a *tacnode* if the curve does continue through the origin.

3. If the lines are imaginary, then the origin is an *isolated point*. An isolated point is a point P whose coordinates satisfy the equation of the curve but which is such that there are no other points of the curve in the vicinity of P. For example, the curve $y^2 = x^2(x - a)$, where $a > 0$, has an isolated point at the origin. The various cases are illustrated in Fig. 21.1.

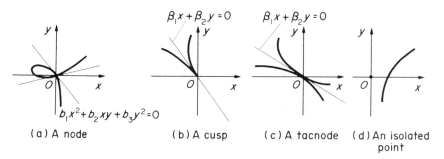

(a) A node   (b) A cusp   (c) A tacnode   (d) An isolated point

FIG. 21.1.

Singular points can be located by obtaining dy/dx in the form $f(x)/g(x)$, without cancellation of common factors, and finding the common roots of $f(x) = 0, g(x) = 0$. If the point (h, k) is a singular point, it can be investigated most easily by choosing new coordinate axes with this point as origin. (See § 21:3.)

Example 1. The curve $a^2y^2 = x^2(a^2 - x^2)$, where $a > 0$, has a node at O. Note that the curve is symmetrical about each of the coordinate axes. Also that the curve is contained within the rectangle

$$|x| = a, \quad |y| = a/2.$$

The curve is shown in Fig. 21.2.

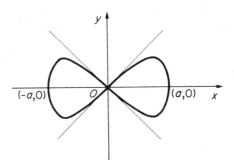

FIG. 21.2.

Example 2. The curve $ay^2 = x^3$, $a > 0$, has a cusp at O.

Example 3. The curve $a^2(x+y)^2 = x^2y^2$, $a > 0$, has a tacnode at O. [See Fig. 21.3(a).] Its equation can be written $xy = \pm a(x+y)$. The $+$ sign corresponds to a hyperbola with asymptotes $x = a$, $y = a$; the $-$ sign to a similar hyperbola with asymptotes $x = -a$, $y = -a$. The tacnode at the origin has $x+y = 0$ as tangent.

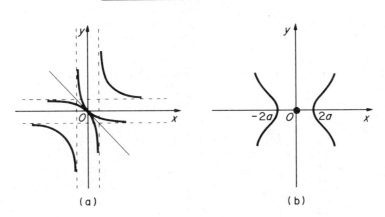

(a)   (b)

FIG. 21.3.

Example 4. The curve $a^2y^2 = x^2(x^2 - 4a^2)$, $a > 0$, has an isolated point at O. [See Fig. 21.3(b).]

Example 5. Sketch the form near the origin of the curves

$$\text{(i) } x^2 - y^2 + 2x + 3y = 0, \qquad \text{(ii) } x^3 + y^3 + x - y = 0.$$

(i) A first approximation to the equation of the curve near O is $2x + 3y = 0$, which is the tangent to the curve at O. A second approximation is obtained as follows:

$$y = -\frac{2x}{3} + \frac{1}{3}(y^2 - x^2)$$

$$\Rightarrow y \approx -\frac{2x}{3} + \frac{1}{3}\left[\left(\frac{-2x}{3}\right)^2 - x^2\right]$$

$$\Rightarrow y \approx -\frac{2x}{3} - \frac{5x^2}{27}.$$

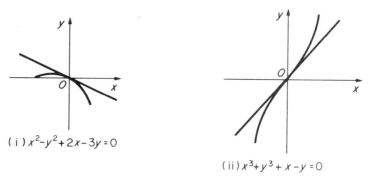

(i) $x^2-y^2+2x-3y=0$

(ii) $x^3+y^3+x-y=0$

FIG. 21.4.

The curve, therefore, lies below its tangent, since $y < -2x/3$ for both $x > 0$ and $x < 0$.

The curve and its tangent near O are sketched in Fig. 21.4(i).

(ii) A first approximation is $x - y = 0$ which is the tangent at O. A second approximation is obtained as follows:

$$y = x + (x^3 + y^3)$$
$$\Rightarrow y \approx x + 2x^3.$$

The curve lies, therefore, above the tangent when $x > 0$ and below the tangent for $x < 0$. The curve therefore crosses its tangent at O which is a point of inflexion as shown in Fig. 21.4(ii).

Example 6. Sketch the form near the origin of the curves (i) $y^2 = x^3 + y^3$; (ii) $x^3 + y^3 = 3(x^4 + y^4)$.

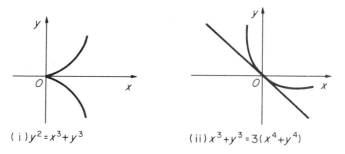

(i) $y^2 = x^3 + y^3$

(ii) $x^3 + y^3 = 3(x^4 + y^4)$

FIG. 21.5.

(i) A first approximation near O is $y^2 = 0$. There are two branches touching $y = 0$ at the origin; the next approximation near O is $y^2 \approx x^3$. Hence the shape near O is as shown in Fig. 21.5(i), $(x \not< 0)$.

(ii) A first approximation near O is $x^3 + y^3 = 0$, the only real tangent there being $x + y = 0$. Then, near O,

$$y = -x + \frac{3(x^4 + y^4)}{x^2 - xy + y^2} \approx -x + \frac{6x^4}{3x^2} = -x + 2x^2.$$

Hence, the next approximation near O is the parabola $y = x(2x - 1)$.

The shape near O is shown in Fig. 21.5(ii).

Example 7. Sketch the curve

$$(x^2 + y^2)^2 = 16axy^2, \quad a > 0. \tag{1}$$

In cases such as this it is most convenient to convert to polar coordinates, the polar equation of this curve being

$$r^3(r - 16a \cos \theta \sin^2 \theta) = 0.$$

Since the pole ($r = 0$) lies on the curve

$$r = 16a \cos \theta \sin^2 \theta = 16a \cos \theta(1 - \cos^2 \theta), \tag{1}$$

we need sketch this latter curve only. Note the following points.

(i) Since equation (1) implies $x \geqslant 0$ and the curve, C say, is symmetrical about Ox ($\theta = 0$), we need consider the range $0 \leqslant \theta \leqslant \pi/2$ only.

(ii) *Near* the pole, $\cos \theta \sin^2 \theta$ is *nearly* zero and as $\theta \to 0$, $\pm \pi/2$ or π, $r \to 0$; the tangents at the pole are
$$xy^2 = 0, \text{ i.e. } x = 0, y = 0 \text{ (twice)}.$$

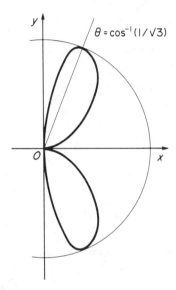

FIG. 21.6.

(iii) The curve has no real asymptotes.

(iv) Since $dr/d\theta = 16a \sin \theta (3 \cos^2 \theta - 1)$, r increases as θ increases from 0 to $\cos^{-1}(1/\sqrt{3})$, has a maximum $32a/(3\sqrt{3})$ when $\theta = \cos^{-1}(1/\sqrt{3})$, then decreases to a minimum $-32a/(3\sqrt{3})$ when $\theta = -\cos^{-1}(1/\sqrt{3})$ and then increases to zero as θ increases to π. In fact the curve C touches the circle $r = 32a/(3\sqrt{3})$ at the point where $\theta = \cos^{-1}(1/\sqrt{3})$. The curve is shown in Fig. 21.6.

Example 8. Find the coordinates, in terms of t, of the point (other than the origin) in which the line $y = tx$ meets the curve $x^3 + y^3 = 3axy$, $a > 0$. Hence sketch the curve.

The coordinates are easily found to be

$$x = \frac{3at}{1 + t^3}, \quad y = \frac{3at^2}{1 + t^3}, \tag{1}$$

and that part of the curve in the first quadrant is given parametrically by equations (1) as t increases from 0 to ∞. [Note that the only real asymptote of the curve is parallel to the line $x + y = 0$ and is, in fact, the line $x + y = -a$. The curve does not go off to infinity in the first quadrant as is clear from equation (1) since $t > 0$ for points in the first quadrant.]

Since
$$\frac{dy}{dx} = \frac{t(2 - t^3)}{1 - 2t^3},$$

(i) the x-axis, touches the curve at the origin $t = 0$;

(ii) when $t = 2^{-1/3}$, dy/dx is infinite and the tangent is vertical, i.e. the tangent is parallel to Oy at $(2^{2/3}a, 2^{1/3}a)$;

(iii) y has a maximum (considered as a function of x) when $t = 2^{1/3}$, i.e. at the point $(2^{1/3}a, 2^{2/3}a)$;

(iv) $dy/dx \to \infty$ as $t \to \infty$ so that the y-axis is the tangent as the curve approaches the origin again.

[Results (i) and (iv) are obvious from the fact that near the origin the curve behaves like $xy = 0$.]

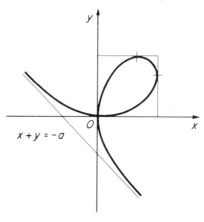

$x + y = -a$

Fig. 21.7.

The curve is clearly symmetrical about the line $y = x$ (because its equation is symmetrical in x and y). The curve is shown in Fig. 21.7.

Exercise 21.2(a)

In questions **1–8** sketch the form of the curve near the origin.

1. $x^3 = x - y$.   **2.** $y^2 = x - x^4$.   **3.** $2xy = x^3 - y^3$.   **4.** $y^2 = x^2 + 3x^5$.

5. $(x + y)^3 = (x - y)^2$.   **6.** $y^3 = 2(x - y)^2$.   **7.** $y(1 + x)^2 = x^3$.

8. $(y - x)^2 (y + x) = 4x^2$.

In questions **9–16** find the equations of the asymptotes and sketch the curves.

9. $(y - x)(y - 2x) = 4$.   **10.** $y(y + x) = 2y - x$.   **11.** $x^2 y = x - y$.

12. $x^2(y - x) = 2y$.   **13.** $y^2 - 4x^2 = x + 1$.   **14.** $xy(x^2 + y^2) = x^3 - y^3$.

15. $y(y^2 - x^2) = 3x + 4y$.   **16.** $x^4 - y^4 = 4xy^2$.

17. By putting $y = tx$ find parametric equations for the curve $x^3 + y^2 = xy$. Hence show that the curve has a loop. Prove that the area of the loop is $1/60$.

18. If $(x^2 - 1)(y^2 - 1) = 2xy$, prove that (for finite values of x and y) dy/dx is always negative. Find the equations of the asymptotes of the curve, and sketch the curve.   (O.C.)

19. Sketch the curve

$$6y^2 = x(x - y)(x - 3y).$$

20. Sketch the curve

$$x^5 + y^5 = 5ax^2 y^2 \qquad (a > 0).$$

By writing $y = tx$ prove that the area of the loop is $5a^2/2$.

21. Sketch the curve

$$r(1 - 2\cos\theta) = 3a\cos 2\theta, \qquad (a > 0),$$

and find the equations of its asymptotes.

* SOME OTHER CURVES

Many of the techniques discussed for algebraic curves can be applied when the equation of a curve involves transcendental functions (such as e^x). We illustrate some other techniques in the following examples.

Example 1. Show that the curve

$$y = f(x) = ae^{-kx} \sin(nx + \beta),$$

where a, k, n, β are positive constants, is contained between the curves $y = \pm ae^{-kx}$. Show also that

$$f'(x) = pae^{-kx} \cos(nx + \beta + \theta),$$

where $p = \sqrt{(n^2 + k^2)}$, $\theta = \tan^{-1}(k/n)$. Deduce that all the maxima of $f(x)$ lie on the curve $y = (an/p)e^{-kx}$ and the minima of $f(x)$ lie on the curve $y = -(an/p)e^{-kx}$.

Since $\sin(nx + \beta) \leqslant 1 \ \forall x$, it follows that

$$ae^{-kx} \sin(nx + \beta) \leqslant ae^{-kx}$$

$\forall x$, equality holding when $nx + \beta = 2r\pi + \pi/2$, where $r \in \mathbb{Z}$. It follows that the curves $y = f(x)$, $y = ae^{-kx}$ have common points at $x = (2r\pi + \frac{1}{2}\pi - \beta)/n$, $y = a \exp[-k\{(2r\pi + \frac{1}{2}\pi - \beta)/n\}]$ but do not cross. Hence the curves touch at these points. [*Note* that these points are *not* maxima for the curve $y = f(x)$ since the slopes at these points must be the same as the corresponding slopes of $y = ae^{-kx}$, i.e. negative.] Similarly, the curve $y = f(x)$ touches the curve $y = -ae^{-kx}$ where $\sin(nx + \beta) = -1$, i.e. where $x = [2s\pi - \frac{1}{2}\pi - \beta]/n$, $s \in \mathbb{Z}$.

$$f'(x) = ae^{-kx}\{n \cos(nx + \beta) - k \sin(nx + \beta)\}$$
$$= a\sqrt{(n^2 + k^2)}e^{-kx}\{\cos\theta \cos(nx + \beta) - \sin\theta \sin(nx + \beta)\},$$

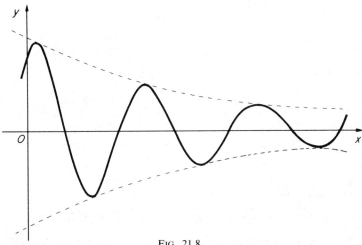

Fig. 21.8.

where θ is the acute angle $\tan^{-1}(k/n)$,

$$\Rightarrow f'(x) = pae^{-kx} \cos(nx + \beta + \theta)$$

as stated. Hence the maxima of $f(x)$ occur where

$$nx + \beta + \theta = (2r + \frac{1}{2})\pi, \quad r \in \mathbb{Z}.$$

At these points $\sin(nx + \beta) = \cos\theta = n/p$ and $y = (an/p)e^{-kx}$, i.e. the maxima of $f(x)$ lie on the stated curve. Similarly, the minima of $f(x)$ lie on the curve $y = -(an/p)e^{-kx}$.

The curve is shown on Fig. 21.8.

Example 2. Sketch the curve

$$y = \tfrac{1}{2}a\{\sin(m+n)x + \sin(m-n)x\}, \tag{1}$$

where $m/n \gg 1$.

The equation of this curve can be written

$$y = a \sin mx \cos nx \tag{2}$$

and clearly (using arguments similar to those of Example 1 above) touches the curves $y = a \cos nx$, $y = a \sin mx$, but not, in general, at stationary points of any of these curves. However, since $m/n \gg 1$, we can consider the given curve from a different point of view by regarding equation (2) as representing a rapidly varying harmonic function $\sin mx$ with a slowly varying amplitude factor $a \cos nx$. The period $2\pi/m$ can be regarded as the period of the oscillations, the period $2\pi/n$ as the (much slower) period of amplitude variation.

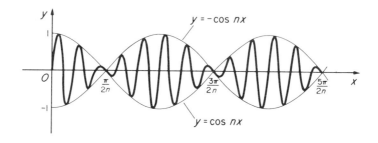

FIG. 21.9.

Note that (1) is the result of combining two harmonic vibrations of equal amplitudes but slightly differing frequencies and illustrates the mathematical background to the physical phenomenon of beats or amplitude modulation in electronics. The curve is sketched in Fig. 21.9.

Example 3. A circle, of radius a and centre C, rolls on the outside of a fixed circle of radius $4a$ and centre O. A marked point P on the rolling circle is initially in contact with a marked point A on the fixed circle. Taking rectangular cartesian axes Oxy with the x-axis along OA, show that the locus of P is the *epicycloid* given by the equations

$$x = a(5\cos\theta - \cos 5\theta), \quad y = a(5\sin\theta - \sin 5\theta),$$

where θ is the angle AOC. Sketch this epicycloid.

Prove that the length of the arc AP of the epicycloid is $s = 10a \sin^2\theta$ for $0 \leqslant \theta \leqslant \pi/2$ and calculate, in terms of θ, the angle ψ which the tangent at P to the epicycloid makes with Ox. By calculating $ds/d\psi$ or otherwise, show that the radius of curvature of the epicycloid at P is $|(10\,a\sin 2\theta)/3|$.

With reference to Fig. 21.10(a), the condition for rolling is

$$\text{arc } IP = \text{arc } IA \Rightarrow 4a\theta = a\phi,$$

so that $\phi = 4\theta$. The coordinates of P are

$$x = OI\cos\theta + CP\cos(\pi - \phi - \theta) = a(5\cos\theta - \cos 5\theta),$$
$$y = OI\sin\theta - CP\sin(\pi - \phi - \theta) = a(5\sin\theta - \sin 5\theta).$$

The locus of P is sketched in Fig. 21.10 (b).

The locus of a marked point on a circle which rolls on the *outside* of a fixed circle is called an *epicycloid*. In our case this locus is a closed curve; P retraces its path after one complete revolution. However, an epicycloid is only a closed curve if the ratio of the radii of the fixed and moving circles is a rational number. The locus of a point on a circle rolling *inside* a fixed circle is called a *hypocycloid*.

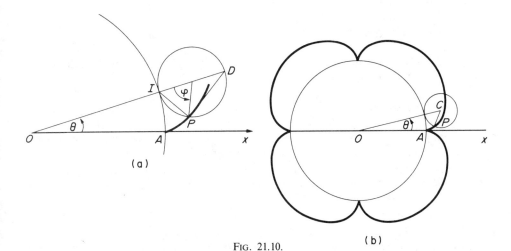

Fɪɢ. 21.10.

For $0 \leqslant \theta \leqslant \pi/2$,

$$\left(\frac{ds}{d\theta}\right)^2 = \left(\frac{dx}{d\theta}\right)^2 + \left(\frac{dy}{d\theta}\right)^2$$

$$= 25a^2\{(\sin\theta - \sin 5\theta)^2 + (\cos\theta - \cos 5\theta)^2\}$$

$$= 25a^2\{2 - 2(\cos 5\theta \cos\theta + \sin 5\theta \sin\theta)\}$$

$$= 50a^2(1 - \cos 4\theta) = 100a^2 \sin^2 2\theta$$

$$\Rightarrow \frac{ds}{d\theta} = 10a \sin 2\theta$$

$$\Rightarrow s = \int_0^\theta 10a \sin 2\theta \, d\theta$$

$$= 5a(1 - \cos 2\theta) = 10a \sin^2 \theta.$$

Also

$$\tan\psi = \frac{dy}{d\theta} \bigg/ \frac{dx}{d\theta} = \frac{\cos\theta - \cos 5\theta}{\sin 5\theta - \sin\theta} = \tan 3\theta$$

$$\Rightarrow \psi = 3\theta$$

$$\Rightarrow \varrho = \left|\frac{ds}{d\psi}\right| = \left|\frac{1}{3}\frac{ds}{d\theta}\right| = \left|\frac{(10a \sin 2\theta)}{3}\right|.$$

---

*Exercise 21.2(b)

1. Let
$$f(x) = \begin{cases} \dfrac{\sin x}{x}, & x \neq 0, \\ 1, & x = 0. \end{cases}$$

Show that $f(x)$ is an even function. Show also that the curve $y = f(x)$ touches the lines $y = x$ and $y = -x$ alternately. Sketch the curve $y = f(x)$.

2. Sketch the curves

(i) $y = x \sin x$,   (ii) $y = \dfrac{\cos x}{x}$,   (iii) $y = |x \sin x|$.

3. Let $f(x) = \begin{cases} x & , 0 \leqslant x \leqslant 1, \\ px + q, & 1 < x < 2, \\ 3/2, & x \geqslant 2. \end{cases}$

Show that it is possible to choose p, q such that $f(x)$ is continuous for all $x \geqslant 0$, but that for these values of p and q, $f(x)$ is not differentiable at $x = 1$ or $x = 2$.

Let $g(x) = \begin{cases} x, & 0 \leqslant x \leqslant 1, \\ ax^2 + bx + c, & 1 < x < 2, \\ 3/2, & x \geqslant 2. \end{cases}$

Show that it is possible to choose a, b, c such that $g(x)$ is differentiable for all $x > 0$.
For these values of a, b, c, sketch the graphs of $g(x)$ and $g'(x)$ for $0 < x < 3$.

4. A circle of radius a rolls inside a fixed circle of radius $3a$. Show that, referred to axes through the centre of the fixed circle, the parametric equations to the curve C described by a point P on the circumference of the rolling circle may be expressed in the form

$$x = 2a \cos \theta + a \cos 2\theta,$$
$$y = 2a \sin \theta - a \sin 2\theta.$$

Find the angle ψ which the tangent to C at P makes with the axis of x, and show that, if A is the point of contact of the two circles, AP is the normal to C at P.

By calculating $ds/d\psi$, or otherwise, show that the radius of curvature at P is numerically equal to $8a \sin (3\theta/2)$, and that the centre of curvature is the point K on PA produced such that $AK = 3PA$.

5. Sketch for *positive* values of x the graphs of each of the following functions

$$\text{(i) } y = e^{(x-1)/x^2}, \quad \text{(ii) } y = \frac{1}{x} - 1 + \ln x.$$

Indicate carefully in each case the behaviour of the function and its derivative both for very large and for very small positive values of x, and the positions of any stationary points there may be.

6. Sketch the curves $y = f(x)$, $y = f'(x)$, for a function $f(x)$ for which

$$f'(0) < 0, \quad f''(x) \geqslant 0;$$

$$\frac{f(x)}{x} \to 1 \quad \text{as} \quad x \to +\infty$$

in the range $x \geqslant 0$.

21:3 TRANSLATION AND ROTATION OF AXES

In many geometrical figures the axes of coordinates have no direct connection with the figure; the axes are introduced as a means of representing the geometrical points and properties in terms of variables, x, y, and equations. The properties are unaltered if different axes are used, although the equations and representations of these properties are altered. Nevertheless, the position of the origin and the directions of the axes are at our disposal and success in the solution of any particular problem in coordinate geometry frequently lies in a suitable choice of axes. To sum up: "Fit the axes to the problem and not the problem to the axes."

When we change the axes of coordinates we are transforming, under T say, the axes relative to the plane; this is equivalent to transforming points of the plane, relative to the axes, under T^{-1} and keeping the axes fixed. So to shift the origin to (α, β) without changing the directions of the axes we carry out the translation $\begin{pmatrix} -\alpha \\ -\beta \end{pmatrix}$.

$$\begin{pmatrix} x' \\ y' \end{pmatrix} = \begin{pmatrix} x \\ y \end{pmatrix} + \begin{pmatrix} -\alpha \\ -\beta \end{pmatrix} \Rightarrow \begin{pmatrix} x \\ y \end{pmatrix} = \begin{pmatrix} x' \\ y' \end{pmatrix} + \begin{pmatrix} \alpha \\ \beta \end{pmatrix}.$$

To rotate the axes through an angle θ (anticlockwise) about O requires a rotation of $-\theta$ about O applied to points (x, y)

$$\Rightarrow \begin{pmatrix} x' \\ y' \end{pmatrix} = \begin{pmatrix} \cos\theta & -\sin\theta \\ \sin\theta & \cos\theta \end{pmatrix}^{-1} \begin{pmatrix} x \\ y \end{pmatrix} \Rightarrow \begin{pmatrix} x \\ y \end{pmatrix} = \begin{pmatrix} \cos\theta & -\sin\theta \\ \sin\theta & \cos\theta \end{pmatrix} \begin{pmatrix} x' \\ y' \end{pmatrix}.$$

Hence to obtain the equation $f(x', y') = 0$ referred to new axes we substitute for x and y in $f(x, y) = 0$ as above.

If $O \to O'$, $f(x, y) \to f(x' + \alpha, y' + \beta)$.

If $O \to O$ but axes are rotated through θ,

$$f(x, y) \to f(x' \cos\theta - y' \sin\theta,\ x' \sin\theta + y' \cos\theta).$$

21:4 CROSS-RATIO

We define the *cross-ratio* of four ordered points A, B, C, D on a straight line as

$$\frac{AB.CD}{AD.CB} = (ABCD), \tag{21.3}$$

where the lengths are directed so that, for example, $AB = -BA$.

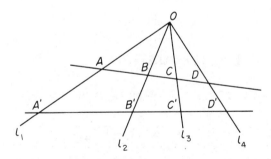

FIG. 21.11.

If l_1, l_2, l_3, l_4 are four straight lines concurrent at O, through A, B, C, D respectively, and another straight line meets l_1, l_2, l_3, l_4 at A', B', C', D' respectively, it can be shown that $(A'B'C'D') = (ABCD)$. Hence $(ABCD)$ can be called *the* cross-ratio of the ordered concurrent lines l_1, l_2, l_3, l_4 (Fig. 21.11).

In the particular case when $(ABCD) = -1$ we say that $(ABCD)$ is a *harmonic range*, or that A and C *harmonically separate* B and D. Since

$$(ABCD) = -1 \Leftrightarrow \frac{AB}{BC} = -\frac{AD}{DC},$$

B and D divide AC internally and externally in the same (numerical) ratio.

The name harmonic range derives from the fact that, if $\dfrac{1}{a} + \dfrac{1}{c} = \dfrac{2}{b}$, then b is called the

harmonic mean of a and c; and it can easily be shown that

$$\frac{1}{AB} + \frac{1}{AD} = \frac{2}{AC} \Leftrightarrow (ABCD) = -1.$$

21:5 THE GENERAL CONIC

A conic section, or *conic*, is a section of a right circular cone by a plane. Although we still use the term "conic", this definition is not a convenient starting point from which to study the properties of these important curves, and we use various alternative definitions which can of course be shown to be equivalent. So far (see Chapters 2, 5) we have studied particular types of conic, usually defined by their parametric equations; but all conics have certain common properties, and it is convenient to establish some of these. To do so, using coordinate methods, we require a form of equation which can represent all types of conic—i.e. circle, ellipse, parabola, hyperbola, and line-pair. It can be shown that an equation of the second degree in x and y always represents a conic, and that all conics have a second degree equation, regardless of the choice of coordinate axes. We use the general form

$$S \equiv ax^2 + 2hxy + by^2 + 2gx + 2fy + c = 0,$$

where a, b, c, f, g, $h \in \mathbb{R}$ and not all of a, b, h are zero.

21:6 JOACHIMSTHAL'S EQUATION

If $P_1 \equiv (x_1, y_1)$ and $P_2 \equiv (x_2, y_2)$, the point dividing $P_1 P_2$ in the ratio $k_2 : k_1$ has coordinates

$$\left(\frac{k_1 x_1 + k_2 x_2}{k_1 + k_2}, \frac{k_1 y_1 + k_2 y_2}{k_1 + k_2} \right).$$

This point lies on $S = 0$ if

$$a(k_1 x_1 + k_2 x_2)^2 + 2h(k_1 x_1 + k_2 x_2)(k_1 y_1 + k_2 y_2) + b(k_1 y_1 + k_2 y_2)^2$$
$$+ 2g(k_1 x_1 + k_2 x_2)(k_1 + k_2) + 2f(k_1 y_1 + k_2 y_2)(k_1 + k_2)$$
$$+ c(k_1 + k_2)^2 = 0$$
$$\Leftrightarrow k_1^2(ax_1^2 + 2hx_1 y_1 + by_1^2 + 2gx_1 + 2fy_1 + c) + 2k_1 k_2 [ax_1 x_2 + h(x_1 y_2 + x_2 y_1)$$
$$+ by_1 y_2 + g(x_1 + x_2) + f(y_1 + y_2) + c] + k_2^2(ax_2^2 + 2hx_2 y_2 + by_2^2 + 2gx_2 + 2fy_2 + c)$$
$$= 0. \tag{21.4}$$

This is known as *Joachimsthal's equation*.

We can abbreviate the coefficients of k_2^2, $2k_1 k_2$ and k_2^2 thus:

$$S_{11}k_1^2 + 2S_{12}k_1 k_2 + S_{22}k_2^2 = 0. \tag{21.4a}$$

If $k_1 \neq 0$, this equation can be regarded as a quadratic for $k_2:k_1$, showing that a straight line $(P_1 P_2)$ meets a conic $(S = 0)$ in two points, which may be real and distinct, coincident or imaginary.

Polar of a point

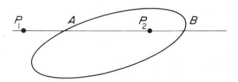

FIG. 21.12.

Suppose that $P_1 P_2$ meets the conic at the points A, B (Fig. 21.12). The condition for $(P_1 A P_2 B) = -1$ is that the sum of the ratios in which A and B divide $P_1 P_2$ shall be zero. We know that the sum of the roots of the quadratic $ax^2 + bx + c = 0$ is $-b/a$ whether or not the roots are real, so, whether or not A and B are real points,

$$(P_1 A P_2 B) = -1 \Leftrightarrow S_{12} = 0.$$

Now, if P_1 is a fixed point not on $S = 0$, and P_2 can vary subject to the above condition, the coordinates of P_2 must satisfy the equation obtained by writing x, y for x_2, y_2—that is by dropping the suffix 2 in $S_{12} = 0$, giving $S_1 = 0$

$$\equiv ax_1 x + h(y_1 x + x_1 y) + by_1 y + g(x + x_1) + f(y + y_1) + c = 0.$$

Since this equation is first degree, it represents a straight line—the *polar* of P_1 with respect to the conic. The point P_1 is the *pole* of $S_1 = 0$ with respect to $S = 0$. The symmetrical form of S_{12} shows that if P_2 lies on the polar of P_1, then P_1 lies on the polar of P_2.

Tangents through a point

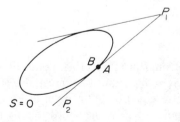

FIG. 21.13.

If P_2 lies on a tangent to $S = 0$ through P_1 (Fig. 21.13), A and B will coincide at the point of contact; this implies that the roots of (21.4a) will be equal, for which the condition is $S_{12}^2 = S_{11} . S_{22}$. Hence for all positions of P_2,

$$S_1^2 = S_{11} . S. \tag{21.5}$$

Since this is a second-degree equation, there must be two tangents, if any, through P_1, and the equation (21.5) will represent this line-pair. If P_1 is "inside" the conic, there will be no real values of x and y satisfying (21.5).

Tangent at a point

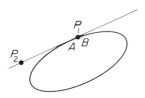

FIG. 21.14.

If P_1 lies on $S = 0$, $S_{11} = 0$, so that one root of (21.4a) is zero since A coincides with P_1 (Fig. 21.14). If the sum of the roots is zero, i.e. if $S_{12} = 0$, the second root must also be zero, meaning that B as well as A must coincide with P_1; this means that P_2 must lie on the tangent to $S = 0$ at P_1. Hence, if P_1 is on $S = 0$, the polar of P_1 with equation $S_1 = 0$ is the tangent at P_1.

Chord of contact

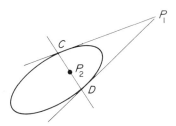

FIG. 21.15.

If there are real tangents from P_1 to $S = 0$, their points of contact C and D (Fig. 21.15) must satisfy the equation $S_1^2 = S_{11} . S$; but they must also satisfy $S = 0$, and therefore must satisfy $S_1 = 0$. This is the equation of the polar of P_1; so if there are real tangents to $S = 0$ through P_1, the polar of P_1 is the chord joining their points of contact.

Chord joining two given points

If P_1 and P_2 lie on $S = 0$, they clearly satisfy the equation $S_1 + S_2 = S_{12}$, since $S_{11} = S_{22} = 0$. Since this is a first-degree equation, it must represent the chord $P_1 P_2$.

Chord with given mid-point

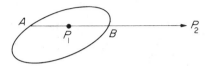

FIG. 21.16.

If P_1 is the mid-point of the chord AB of $S = 0$ (Fig. 21.16), let P_2 on AB be such that $(P_1AP_2B) = (AP_1BP_2) = -1$, i.e. let P_2 be the point where AB meets the polar of P_1. But since $AP_1 = P_1B$, this requires that $AP_2 = BP_2$, which is impossible. Hence, AB does not meet $S_1 = 0$; so, being parallel to $S_1 = 0$, AB must have an equation of the form $S_1 = k$, where k is a constant. But P_1 lies on AB

$$\Rightarrow S_{11} = k \Rightarrow AB \equiv S_1 = S_{11}. \tag{21.6}$$

Example 1. Show that the equation of the pair of tangents from the point $P(\alpha, \beta)$ to the circle $x^2 + y^2 = r^2$ can be expressed in the form

$$(\beta x - \alpha y)^2 = r^2[(x - \alpha)^2 + (y - \beta)^2].$$

These tangents cut the x-axis at the points A and B.
　(i) If the mid-point of AB is the fixed point $(k, 0)$, show that the locus of P is the parabola $ky^2 = r^2(k - x)$.
　(ii) If AB is of length $4r$, show that P lies on the quartic curve

$$x^2y^2 = (y^2 - r^2)(3y^2 - 4r^2).$$

Using equation (21.5), where $S \equiv x^2 + y^2 - r^2 = 0$, the pair of tangents from (α, β) have the equation

$$(x^2 + y^2 - r^2)(\alpha^2 + \beta^2 - r^2) = (\alpha x + \beta y - r^2)^2$$
$$\Leftrightarrow \beta^2 x^2 + \alpha^2 y^2 - 2\alpha\beta xy = r^2[x^2 + y^2 + \alpha^2 + \beta^2 - 2(\alpha x + \beta y)]$$
$$\Leftrightarrow (\beta x - \alpha y)^2 = r^2[(x - \alpha)^2 + (y - \beta)^2].$$

(i) The x-coordinates of A and B are the roots x_1 and x_2 of the quadratic $\beta^2 x^2 = r^2[(x - \alpha)^2 + \beta^2]$ obtained by putting $y = 0$ in the above. For the mid-point of AB,

$$\tfrac{1}{2}(x_1 + x_2) = -\frac{\alpha r^2}{\beta^2 - r^2} = k$$

$$\Rightarrow \text{Locus of } P \text{ is } \quad ky^2 = r^2(k - x).$$

(ii) $AB^2 = (x_1 - x_2)^2 = (x_1 + x_2)^2 - 4x_1x_2 = \dfrac{4\alpha^2 r^4}{(\beta^2 - r^2)^2} + \dfrac{4r^2(\alpha^2 + \beta^2)}{\beta^2 - r^2}$

$$\Rightarrow 4\alpha^2 r^4 + 4r^2(\alpha^2 + \beta^2)(\beta^2 - r^2) = 16r^2(\beta^2 - r^2)^2$$
$$\Rightarrow \text{Locus of } P \text{ is } \quad x^2r^2 + (x^2 + y^2)(y^2 - r^2) = 4(y^2 - r^2)^2$$
$$\Leftrightarrow x^2y^2 = (y^2 - r^2)(3y^2 - 4r^2).$$

Example 2. Obtain the equation of the pair of lines from the origin to the points of intersection of the circle

$$x^2 + y^2 + 2gx + 2fy + c = 0$$

and the straight line $lx + my + n = 0$.
　Prove that the locus of the middle points of chords of the circle

$$x^2 + y^2 - 2ax + 2b^2 = 0$$

which subtend a right angle at the origin, is the circle

$$x^2 + y^2 - ax + b^2 = 0. \tag{O.C.}$$

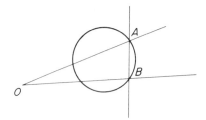

FIG. 21.17.

$$lx + my + n = 0 \Leftrightarrow -\frac{lx + my}{n} = 1.$$

Consider the equation

$$x^2 + y^2 - 2(gx + fy)(lx + my)/n + c(lx + my)^2/n^2 = 0.$$

It is homogeneous of the second degree in x and y, and therefore represents a line-pair through the origin; but the coordinates of the points A and B where $lx + my + n = 0$ meets the circle make

$$-\frac{lx + my}{n} = 1 \quad \text{and} \quad x^2 + y^2 + 2gx + 2fy + c = 0,$$

and therefore satisfy the equation that we have constructed.

Using the form $S_1 = S_{11}$, (21.6), the chord of the given circle with mid-point (x_1, y_1) has the equation

$$xx_1 + yy_1 - a(x + x_1) + 2b^2 = x_1{}^2 + y_1{}^2 - 2ax_1 + 2b^2$$
$$\equiv x(x_1 - a) + yy_1 = x_1{}^2 + y_1{}^2 - ax_1.$$

The line-pair joining O to the points of intersection A, B of this chord with the circle, Fig 21.17, is

$$x^2 + y^2 - \frac{2ax[x(x_1 - a) + yy_1]}{x_1{}^2 + y_1{}^2 - ax_1} + \frac{2b^2[x(x_1 - a) + yy_1]^2}{(x_1{}^2 + y_1{}^2 - ax_1)^2} = 0.$$

The chord subtends a right-angle at O if the sum of the coefficients of x^2 and y^2 is zero

$$\Rightarrow 2(x_1{}^2 + y_1{}^2 - ax_1)^2 - 2a(x_1 - a)(x_1{}^2 + y_1{}^2 - ax_1) + 2b^2[(x_1 - a)^2 + y_1{}^2] = 0.$$

Hence (x_1, y_1) lies on $(x^2 + y^2 - ax + b^2)(x^2 + y^2 - 2ax + a^2) = 0$.

But $(x - a)^2 + y^2 \neq 0$ except at the point $(a, 0)$.

Hence the locus of mid-points is the circle

$$x^2 + y^2 - ax + b^2 = 0.$$

Example 3. Find the locus of the point of intersection of perpendicular tangents to the ellipse $b^2x^2 + a^2y^2 - a^2b^2 = 0$.

Using the form $S_{11}.S = S_1{}^2$, the pair of tangents from (x_1, y_1) to the ellipse have the equation

$$(b^2x_1{}^2 + a^2y_1{}^2 - a^2b^2)(b^2x^2 + a^2y^2 - a^2b^2) = (b^2xx_1 + a^2yy_1 - a^2b^2)^2.$$

These are perpendicular if the sum of the coefficients of x^2 and y^2 is zero, i.e. if

$$(a^2 + b^2)(b^2x_1{}^2 + a^2y_1{}^2 - a^2b^2) - b^4x_1{}^2 - a^4y_1{}^2 = 0.$$

Hence the locus of (x_1, y_1) is

$$a^2b^2(x^2 + y^2) - a^2b^2(a^2 + b^2) = 0$$
$$\Leftrightarrow x^2 + y^2 = a^2 + b^2.$$

This is a circle concentric with the ellipse—the *director circle* of the ellipse.

Exercise 21.6

1. Write down the equation of the polar of each of the following points with respect to the conic whose equation is given:

(i) $(0, 0)$, $x^2 + y^2 - 6x + 2y - 4 = 0$,   (ii) $(5, 6)$, $5(x-1)^2 + 2(y-4)^2 = 10$.

2. Find the coordinates of the pole of each of the following lines with respect to the conic whose equation is given:

(i) $3x + 5y - 15 = 0$, $x^2 + y^2 = 9$,   (ii) $ax + y - 4a^2 = 0$, $y^2 = 4a(x + 2a)$,

(iii) $4x + y - 7 = 0$, $x^2 + y^2 - 2x + 4y = 0$,   (iv) $x + y - 1 = 0$, $(x + 2)(y - 1) = 16$,

(v) $13x + 24y - 3 = 0$, $3x^2 + 4xy - 2y^2 + x - 8y + 7 = 0$.

3. In each of the following cases write down and simplify the equations of the pair of tangents from the point to the conic:

(i) $(1, 1)$, $3x^2 + 4y^2 - 1 = 0$,   (ii) $(0, 0)$, $xy = 4$,

(iii) $(0, -1)$, $3x^2 + 4xy + y^2 = 10$,   (iv) $(1, -1)$, $4x^2 - y^2 = 10$,

(v) $(0, 4)$, $(y + 1)^2 = 4(x - 1)$.

4. Prove that the equation of the chord of contact of the tangents drawn from the origin of coordinates to the circle

$$x^2 + y^2 + 2gx + 2fy + c = 0$$

is

$$gx + fy + c = 0.$$

Prove that the area of the triangle formed by the two tangents from the origin of coordinates and the chord of contact is

$$\{c^{3/2} \sqrt{(g^2 + f^2 - c)}\}/(g^2 + f^2).$$

5. Find the locus of the poles of normal chords of the parabola

$$x = at^2, \quad y = 2at.$$

6. A point P moves so that the chord of contact of the tangents from P to the ellipse $b^2x^2 + a^2y^2 = a^2b^2$ touches the ellipse $4(b^2x^2 + a^2y^2) = a^2b^2$. Find the locus of P.   (L.)

7. Show that the line $lx + my + n = 0$ touches the parabola $y^2 = 4ax$, if $am^2 = nl$.

If the polar of a point P with respect to the ellipse $(x + a)^2/a^2 + y^2/b^2 = 1$ touches the parabola $y^2 = 4ax$, show that the locus of P is the hyperbola $x(x + a)/a^4 - y^2/b^4 = 0$.   (L.)

8. Write down the equation of the polar of the point (x_1, y_1) with respect to the parabola $y^2 = 4ax$.

Show that the polar of every point of the parabola $x^2 = 4by$ with respect to the parabola $y^2 = 4ax$ touches the same rectangular hyperbola of the form $xy = k$, and express k in terms of a and b.   (O.)

9. Show that if m_1, m_2 are the gradients of the two tangents from the point (α, β) to the hyperbola $b^2x^2 - a^2y^2 = a^2b^2$, then m_1, m_2 are the roots of the equation

$$m^2(\alpha^2 - a^2) - 2\alpha\beta m + b^2 + \beta^2 = 0.$$

Show that the locus of points from which the two tangents to the curve are equally inclined to a line of gradient $\tan\theta$ is the rectangular hyperbola

$$x^2 - y^2 - 2xy \cot 2\theta = a^2 + b^2. \qquad \text{(L.)}$$

10. Show that the equation of the chord joining the points $P(2t_1, 2/t_1)$ and $Q(2t_2, 2/t_2)$ of the hyperbola $xy = 4$, is

$$x + yt_1t_2 = 2(t_1 + t_2).$$

If the chord varies in such a way as to be always a tangent to the circle

$$x^2 + y^2 = 1,$$

prove that the point of intersection of the tangents at P and Q to the hyperbola lies always on a circle with centre at the origin and radius 8.   (L.)

21:7 IDENTIFICATION OF THE GENERAL CONIC

We have already seen (p. 487) how to refer the equation of the form $ax^2 + 2hxy + by^2 = k$ to principal axes so that it reduces to the form $a'x'^2 + b'y'^2 = k'$. We now consider how to reduce the general quadratic form to a homogeneous form. When the origin is changed to (x_1, y_1), keeping the directions of the axes unchanged, the equation

$$ax^2 + 2hxy + by^2 + 2gx + 2fy + c = 0$$

becomes

$$a(x' + x_1)^2 + 2h(x' + x_1)(y' + y_1) + b(y' + y_1)^2 + 2g(x' + x_1) + 2f(y' + y_1) + c = 0.$$

Choose (x_1, y_1), if possible, so that the coefficients of x' and y' are zero.

We require

$$ax_1 + hy_1 + g = 0,$$
$$hx_1 + by_1 + f = 0.$$

These equations are satisfied by

$$x_1 = \frac{hf - bg}{ab - h^2}, \quad y_1 = \frac{gh - af}{ab - h^2}$$

if $ab - h^2 \neq 0$; in which case the equation of the conic becomes

$$ax'^2 + 2hx'y' + by'^2 + S_{11} = 0.$$

Clearly the curve has point symmetry about (x_1, y_1); this point is called the *centre* of the conic. All chords of the conic through the centre are bisected there, and are called *diameters* of the conic.

If $S_{11} \neq 0$, three cases arise.

(i) If $h^2 > ab$, the real factors of $ax'^2 + 2hx'y' + by'^2$ give the directions of the asymptotes, and the conic is a hyperbola.

(ii) If $h^2 < ab$, there are no asymptotes, and the conic is in general an ellipse; in the special case $h = 0$, $a = b$ the conic is a circle.

(iii) If $h^2 = ab$, the conic has no centre, and is a parabola (cf. $y = x^2$, $y^2 = 4ax$), unless the condition obtained below for a line pair is also satisfied; in this case the conic is a pair of parallel straight lines.

$$S_{11} \equiv x_1(ax_1 + hy_1 + g) + y_1(hx_1 + by_1 + f) + gx_1 + fy_1 + c.$$

Hence, if (x_1, y_1) is the centre as defined above,

$$S_{11} = 0 \Leftrightarrow gx_1 + fy_1 + c = 0.$$

But for (x_1, y_1) to satisfy the three equations

$$\Delta \equiv \begin{vmatrix} a & h & g \\ h & b & f \\ g & f & c \end{vmatrix} = 0.$$

This is then the condition for $S = 0$ to reduce to the form

$$ax'^2 + 2hx'y' + by'^2 = 0.$$

If $h^2 > ab$, $ax'^2 + 2hx'y' + by'^2$ has real factors, giving two straight lines through (x_1, y_1).

If $h^2 < ab$, there will be no points of the locus.

The usual notation in which capital letters denote cofactors of Δ enables us to write the coordinates of the centre as $(G/C, F/C)$. Notice also that Δ is the determinant of the symmetric matrix

$$\begin{pmatrix} a & h & g \\ h & b & f \\ g & f & c \end{pmatrix},$$

in terms of which $S = 0$ becomes

$$(x \quad y \quad 1) \begin{pmatrix} a & h & g \\ h & b & f \\ g & f & c \end{pmatrix} \begin{pmatrix} x \\ y \\ 1 \end{pmatrix} = 0.$$

Example. Identify the following equations:

    (i) $xy - 3x - 2y + 2 = 0$;

   (ii) $6x^2 - xy - 2y^2 + x + 18y - 40 = 0$;

  (iii) $4x^2 - 4xy + y^2 + 2x - 6y + 9 = 0$;

  (iv) $3x^2 + 4xy + 3y^2 - 2x + 12y + 7 = 0$.

Each equation is second-degree in x and y, and therefore represents a conic.

(i) $xy - 3x - 2y + 2 = 0 \rightarrow (x' + 2)(y' + 3) - 3(x' + 2) - 2(y' + 3) + 2 = 0 \equiv x'y' = 4$.

This is a rectangular hyperbola, centre $(2, 3)$, with asymptotes parallel to the coordinate axes.

(ii)
$$\begin{vmatrix} 6 & -\frac{1}{2} & \frac{1}{2} \\ -\frac{1}{2} & -2 & 9 \\ \frac{1}{2} & 9 & -40 \end{vmatrix} = 0 \quad \text{and} \quad h^2 > ab.$$

This is a line-pair, as can be seen by factorising

$$(2x + y - 5)(3x - 2y + 8) = 0.$$

(iii) $h^2 = ab$ but $\Delta \neq 0$; therefore this is a parabola.

By comparison with $y^2 = 4ax$, we write the equation in the form

$$(2x - y + k)^2 = 2(-1 + 2k)x + 2(3 - k)y + k^2 - 9$$

and choose k so that $2x - y + k = 0$ is perpendicular to

$$2(-1 + 2k)x + 2(3 - k)y = 0;$$

hence
$$2 = \frac{3 - k}{-1 + 2k} \Rightarrow k = 1.$$

The equation can now be written

$$\left(\frac{2x - y + 1}{\sqrt{5}}\right)^2 = \frac{2}{\sqrt{5}}\left(\frac{x + 2y - 4}{\sqrt{5}}\right);$$

hence it represents a parabola with axis $y = 2x + 1$, and $2y + x = 4$ as tangent at the vertex.

(iv) The centre is $(3, -4)$. Changing the origin to $(3, -4)$ gives

$$3(x' + 3)^2 + 4(x' + 3)(y' - 4) + 3(y' - 4)^2 - 2(x' + 3) + 12(y' - 4) + 7 = 0$$
$$\equiv 3x'^2 + 4x'y' + 3y'^2 = 20.$$

The characteristic equation of the matrix $\begin{pmatrix} 3 & 2 \\ 2 & 3 \end{pmatrix}$ is

$$(3 - \lambda)^2 - 4 = 0 \Leftrightarrow \lambda = 1, 5.$$

By the method illustrated on p. 487 we can diagonalise the matrix so that the equation becomes

$$(x \quad y) \begin{pmatrix} 1 & 0 \\ 0 & 5 \end{pmatrix} \begin{pmatrix} x \\ y \end{pmatrix} = 20 \equiv x^2 + 5y^2 = 20.$$

This represents an ellipse with semi-axes $2\sqrt{5}, 2$.

Exercise 21.7

1. Identify the type of conic represented by each of the following equations and where appropriate find its centre:

(i) $x^2 - 6xy - 7y^2 - 2x + 4y - 1 = 0$.

(ii) $13x^2 - 18xy + 36y^2 + 2x + 4y - 2 = 0$.

(iii) $x^2 + 6xy + 9y^2 + 3x + 2y + 4 = 0$.

(iv) $x^2 - 4xy + 5y^2 + 2x - 8y - 6 = 0$.

2. Find the coordinates of the centre of the conic whose equation is

$$3x^2 + 8xy + 6y^2 - 14x - 20y - 26 = 0.$$

Also obtain the equations of the tangent and the normal to this conic at the point $(2, 3)$.

3. Show that the conic

$$x^2 - 6xy - 7y^2 - 2x + 4y - 1 = 0$$

is a hyperbola, and find its centre and the equation of each axis and each asymptote.

4. Obtain the equation of the conic

$$x^2 + 4xy + y^2 - 6(x + y) + 9 = 0$$

referred to its centre as origin. Also obtain the equation of its major axis.

5. A diameter of the conic

$$2x^2 + xy + 3y^2 = 6x + 7y + 44$$

is parallel to the line $y = 2x$. Find where this diameter, produced if necessary, meets the y-axis.

6. A given line $lx + my = 1$ meets the hyperbola

$$\frac{x^2}{a^2} - \frac{y^2}{b^2} = 1$$

at A, B and its asymptotes at C, D. Prove that AB and CD have the same mid-point M. Prove that M lies on the line

$$\frac{mx}{a^2} + \frac{ly}{b^2} = 0.$$

If the given line passes through the fixed point (α, β), prove that the equation of the locus of M is

$$\frac{x^2 - \alpha x}{a^2} - \frac{y^2 - \beta y}{b^2} = 0.$$ (O.C.)

7. The line $lx + my + n = 0$ meets the circle

$$x^2 + y^2 + 2gx + 2fy + c = 0$$

at the points A and B. Show that the equation

$$x^2 + y^2 - 2(gx + fy)\left(\frac{lx + my}{n}\right) + c\left(\frac{lx + my}{n}\right)^2 = 0$$

represents the pair of straight lines OA and OB, where O is the origin.
If the line $2x + y - 1 = 0$ meets the circle

$$x^2 + y^2 + 2x - 4y - 2 = 0$$

at the points L and M, prove that $\tan L\hat{O}M = -\frac{1}{4}\sqrt{34}$. (C.)

8. Prove that the latus rectum of the parabola

$$9x^2 + 6xy + y^2 + 2x + 3y + 4 = 0$$

is $7\sqrt{10}/100$.

21:8 CONICS THROUGH FOUR POINTS

If $S = 0$ and $S' = 0$ are the equations of two conics, $S + \lambda S' = 0$ $(\lambda \in \mathbb{R}, \lambda \neq 0)$ is a second-degree equation which is satisfied by the coordinates of any point lying both on $S = 0$ and $S' = 0$. Two second-degree equations have four solutions in complex algebra, so $S + \lambda S' = 0$ represents a conic through the four points of intersection (not necessarily all real) of $S = 0$ and $S' = 0$.

Given four points, it is often convenient to express the equation of a conic through them in the form $S + \lambda S' = 0$, where $S = 0$ and $S' = 0$ are two other conics through the points.

Example. If a circle meets a central conic (i.e. an ellipse or a hyperbola) in four real points A, B, C, D, prove that a pair of common chords AC, BD are equally inclined to the axes of the central conic.

Choose the axes of the central conic as coordinate axes, so that the curve has an equation of the form $ax^2 + by^2 + c = 0$. Let a pair of common chords be $lx + my + n = 0$, $l'x + m'y + n' = 0$. Then the equation of the circle must of the form

$$(lx + my + n)(l'x + m'y + n') + \lambda(ax^2 + by^2 + c) = 0.$$

For this equation to represent a circle, it is necessary for the coefficient of xy to be zero

$$\Rightarrow lm' + l'm = 0 \Rightarrow \frac{l}{m} = -\frac{l'}{m'}.$$

Hence the chords are equally inclined to the x- and y-axes which are the axes of the conic.

Figures 21.18 illustrate some special cases of conics through four points.
(i) $S + \lambda LM = 0$ is a conic through the four (distinct) points of intersection of $S = 0$ with the lines $L = 0$, $M = 0$.

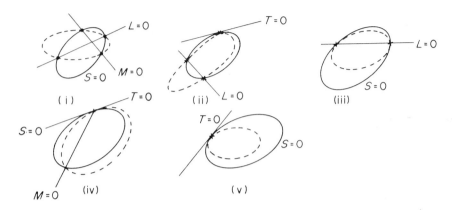

FIG. 21.18.

(ii) $S + \lambda LT = 0$ is a conic through the points of intersection of $S = 0$ and $L = 0$, and touching $T = 0$ where this line touches $S = 0$.

(iii) $S + \lambda L^2 = 0$ is a conic having "double contact" with $S = 0$ at its points of intersection with $L = 0$.

(iv) $S + \lambda MT = 0$ is a conic having "three-point contact" with $S = 0$ at the point where the chord $M = 0$ meets the tangent $T = 0$.

(v) $S + \lambda T^2 = 0$ is a conic having "four-point contact" with $S = 0$ at the point of contact of its tangent $T = 0$.

Exercise 21.8

1. Prove that the points of intersection of the conics

$$x^2 - 4xy + 4y^2 + 42x - 9y - 109 = 0,$$
$$17x^2 + 12xy + 8y^2 - 46x - 9y - 13 = 0,$$

lie on a circle.

2. Find the equation of the circle through the origin and through the points of intersection of $y + x^2 + 1 = 0$ and $y - x + 2 = 0$.

3. Prove that a circle cannot have four-point contact with the parabola $y^2 = 4ax$ except at the vertex.

4. The equation of a conic is $S = 0$, where

$$S \equiv ax^2 + 2hxy + by^2 + 2gx + 2fy + c.$$

If the equation of the tangent at a point P of the conic is $L = 0$, interpret (giving a brief justification) the equation

$$S + \lambda L^2 = 0,$$

where λ is a constant.

Prove that the centre of the rectangular hyperbola which has four-point contact with a parabola at a point P is the reflection of P in the directrix of the parabola.   (O.C.)

5. Write down the equation of the general conic which passes through the intersections of the conic $S = 0$ and the lines $L = 0$, $M = 0$.

The lines

$$x \cos \alpha + y \sin \alpha = p, \qquad x \cos \alpha - y \sin \alpha = q$$

meet the conic $ax^2 + by^2 = 1$ at points A, B, C, D. Prove that

(i) A, B, C, D are concyclic;

(ii) every conic through A, B, C, D has axes parallel to the axes of the given conic;

(iii) there is in general just one rectangular hyperbola through A, B, C, D, but that under certain conditions, which should be stated, every conic through A, B, C, D is a rectangular hyperbola.   (O.C.)

21:9 CIRCLE OF CURVATURE

The circle of curvature at a point P of a curve can be defined as the "circle of closest fit" to the curve at P. If we consider Q and R as (non-collinear) points on the the curve near to P, we can find the equation of the circle through P, Q, R. As Q and $R \to P$ the circle PQR becomes in the limit the circle having three-point contact with the curve at P, and this in general is the highest order of contact possible between a circle and a curve. To find the equation of the circle of curvature, it may be convenient to use (iv) above.

Example. Find the equation of the circle of curvature at the point $(at^2, 2at)$ of the parabola $y^2 = 4ax$.

Let the circle meet the parabola again at the point $(am^2, 2am)$. Then its equation is of the form

$$y^2 - 4ax + \lambda\left(y - \frac{x}{t} - at\right)[2x - y(t + m) + 2atm] = 0.$$

For this equation to represent a circle, the coefficients of x^2 and y^2 must be equal

$$\Rightarrow -\frac{2\lambda}{t} = t\lambda(t + m) \Rightarrow \lambda(t^2 + tm - 2) = t;$$

also the coefficient of xy must be zero

$$\Rightarrow 2 + \frac{t + m}{t} = 0 \Rightarrow m = -3t \Rightarrow \lambda = -\frac{t}{2(1 + t^2)}.$$

Equation of circle

$$\equiv 2(1 + t^2)(y^2 - 4ax) - (ty - x - at^2)(2x + 2ty - 6at^2) = 0$$
$$\equiv x^2 + y^2 - 2ax(2 + 3t^2) + 4at^3y - 3a^2t^4 = 0.$$

Exercise 21.9

1. If $S = 0$ is the equation of a conic, $T = 0$ is the equation of the tangent at a point P of the conic, and $L = 0$ is the equation of any line through P, interpret (giving a brief justification) the equation

$$S + \lambda TL = 0.$$

The circle of curvature at a variable point P of the parabola $y^2 = 4ax$ meets the parabola again at Q. Prove that PQ touches the parabola $y^2 + 12ax = 0$.   (O.C.)

2. The circles of curvature of $y^2 = 4ax$ at the points P, Q, R and S meet the parabola again at P', Q', R' and S' respectively. If P, Q, R and S lie on a circle centre $(-g, -f)$, prove that P', Q', R' and S' also lie on a circle and find its centre.

3. Find the equation of the circle of curvature of $xy = c^2$ at the point $(ct, c/t)$.

4. Find the centre and radius of curvature at the point $(b/2, b^2/4a)$ on the parabola $x^2 = bx - ay$.

21:10 COAXAL CIRCLES

Let $S \equiv x^2 + y^2 + 2gx + 2fy + c = 0$ and $S' \equiv x^2 + y^2 + 2g'x + 2f'y + c' = 0$ be the equations of two circles. Then $S + \lambda S' = 0$, $\forall \lambda \in \mathbb{R}$, is the equation of a set of circles through the two common points (if the circles intersect in real points) of $S = 0$ and $S' = 0$, each value of λ corresponding to one circle except $\lambda = -1$; in this case $S - S' = 0$ is a straight line, called the *radical axis* of the *coaxal system* of circles determined by $S = 0$ and $S' = 0$. Clearly if $S = 0$ and $S' = 0$ are intersecting circles, the radical axis is the common chord of every pair of circles in the system. If $L \equiv S - S'$, the coaxal system can be represented by the equation $S + \lambda(S - L) = 0$ or $S + \mu L = 0$. It can be shown that the lengths of the tangents from any point on the radical axis to the circles of the system are equal. The square of this length is the *power* of the point with respect to the circle.

The line joining any pair of centres of a coaxal system of circles is at right angles to the radical axis. All the centres of the circles of a coaxal system are therefore collinear. It is convenient, in order to discuss the properties of a coaxal system of circles, to choose the line of centres as the x-axis and the radical axis as the y-axis. In this case the equation of any circle of the system is expressible in the form

$$x^2 + y^2 + 2gx + c = 0$$

and since the power of the origin (a point on the radical axis) is the same with respect to every circle of the system, c is constant for the coaxal system.

Orthogonal coaxal systems and limiting points

If c is negative, the coaxal circles $x^2 + y^2 + 2gx + c = 0$ cut the y-axis at the points $(0, \pm \sqrt{(-c)})$ and the circles constitute an intersecting set of coaxal circles with the y-axis as common chord.

If c is positive, none of the circles intersects the y-axis. Therefore in this case the circles constitute a non-intersecting set of coaxal circles. The radius of $x^2 + y^2 + 2gx + c = 0$ is equal to $\sqrt{(g^2 - c)}$ and as $g \to \pm \sqrt{c}$ the radius of the circle of the coaxal system tends to zero. The points $(\pm \sqrt{c}, 0)$ are called the *point circles* or the *limiting points* of the system. If P is any point on the radical axis and L_1 and L_2 are the limiting points of the system, then $PL_1^2 = PL_2^2 =$ the power of P with respect to the system.

The equation of the intersecting coaxal system of circles through $(\pm \sqrt{c}, 0)$, where c is positive, is

$$x^2 + y^2 + 2fy - c = 0$$

where f varies. Any circle of this system is orthogonal with any circle of the system

$$x^2 + y^2 + 2gx + c = 0$$

because the square on the distance between the centres is $g^2 + f^2$ and the sum of the squares on the radii is $g^2 - c + f^2 + c = g^2 + f^2$. It follows that *a non-intersecting system of coaxal circles is orthogonal to the intersecting system which passes through its limiting points.*

Example 1. Find the coordinates of the common points of the coaxal system of circles given by the equation

$$x^2 + y^2 + 2gx - 16 = 0,$$

where g varies. Show that only one circle of the system touches the line $x + 2y = 8$, and find its equation. If S is any member of the system and S' is the circle

$$x^2 + y^2 + 6x + 8y + 24 = 0,$$

prove that the radical axis of S and S' passes through a fixed point on the axis of y. (N.)

The circles of the system $x^2 + y^2 + 2gx - 16 = 0$ cut the y-axis at $(0, \pm 4)$, and these are the common points of the system. The condition that $x + 2y = 8$ is a tangent to $x^2 + y^2 + 2gx - 16 = 0$ is that the distance of $(-g, 0)$ from $x + 2y = 8$ should be numerically equal to the radius of the circle

$$\Rightarrow \left| \frac{-8 - g}{\sqrt{5}} \right| = \sqrt{(g^2 + 16)}$$

$$\Rightarrow g^2 + 16g + 64 = 5g^2 + 80.$$

This equation has the repeated root $g = 2$, and therefore the only circle of the system which touches $x + 2y = 8$ is $x^2 + y^2 + 4x - 16 = 0$.
The radical axis of S and S' is

$$(x^2 + y^2 + 2gx - 16) - (x^2 + y^2 + 6x + 8y + 24) = 0$$

$$\Rightarrow (g - 3)x - 4y - 20 = 0,$$

and this passes through the fixed point $(0, -5)$ for all values of g.

Example 2. Find the equation of the circle passing through the point $(1, -1)$ and cutting orthogonally all circles of the coaxal system

$$x^2 + y^2 + 2gx + 2gy + 2 = 0.$$ (N.)

The radius of any circle of the system is $\sqrt{(2g^2 - 2)}$. Hence the limiting points of the system are $(1, 1)$ and $(-1, -1)$. Therefore any circle through $(1, 1), (-1, -1)$ cuts all the circles of the system orthogonally. The circle which passes through these points and the point $(1, -1)$ has centre at the origin, radius $\sqrt{2}$, and equation

$$x^2 + y^2 = 2.$$

Example 3. Find the coordinates of the limiting points of the system of circles which is coaxal with

$$S_1 \equiv x^2 + y^2 - 2x - 2y + 1 = 0,$$
$$S_2 \equiv x^2 + y^2 + 4x + 4y + 4 = 0$$

and find the equation of that circle of the system which passes through $(3, 1)$.

The radical axis of S_1 and S_2 is $2x + 2y + 1 = 0$ and so the coaxal system is

$$x^2 + y^2 - 2x - 2y + 1 + \lambda(2x + 2y + 1) = 0.$$

The radius of a circle of this system is

$$\sqrt{\{(1 - \lambda)^2 + (1 - \lambda)^2 - (1 + \lambda)\}} = \sqrt{(2\lambda^2 - 5\lambda + 1)}.$$

Therefore the limiting points are given by $\lambda = \alpha$ and $\lambda = \beta$, where α and β are the roots of $2\lambda^2 - 5\lambda + 1 = 0$; so the limiting points are $(1 - \alpha, 1 - \alpha)$ and $(1 - \beta, 1 - \beta)$, i.e. $[\frac{1}{4}(\sqrt{17} - 1), \frac{1}{4}(\sqrt{17} - 1)]$ and $[-\frac{1}{4}(\sqrt{17} + 1), -\frac{1}{4}(\sqrt{17} + 1)]$.

The circle $x^2 + y^2 - 2x - 2y + 1 + \lambda(2x + 2y + 1) = 0$ goes through $(3, 1)$ if $\lambda = -\frac{1}{3}$, and so the circle of the system which goes through $(3, 1)$ is

$$x^2 + y^2 - 2x - 2y + 1 - \frac{1}{3}(2x + 2y + 1) = 0$$
$$\Rightarrow 3x^2 + 3y^2 - 8x - 8y + 2 = 0.$$

Exercise 21.10

1. Write down the equation of the radical axis of each of the following pairs of circles. State, in each case, whether the circles are intersecting or non-intersecting circles and write down an equation for the coaxal system to which the circles belong.

(i) $x^2 + y^2 + 4x - 6y + 1 = 0$, $x^2 + y^2 - 5x + y - 4 = 0$,

(ii) $2x^2 + 2y^2 - 5x + 4y = 0$, $4x^2 + 4y^2 - x + 6y = 0$,

(iii) $x^2 + y^2 + 5x - 2y - 1 = 0$, $3x^2 + 3y^2 - x + y - 1 = 0$,

(iv) $x^2 + y^2 + 2ax + 2by + c = 0$, $x^2 + y^2 - 2ax - 2by + c = 0$.

2. Find the coordinates of the points, the powers of which, with respect to each of the circles

$$x^2 + y^2 + 10x + 8y + 16 = 0,$$
$$x^2 + y^2 - 4x + 2y - 4 = 0$$

are each $+40$.

3. Find the coordinates of the points, the powers of which, with respect to each of the circles

$$x^2 + y^2 - 2x + y - 24 = 0,$$
$$x^2 + y^2 - 4x - y - 18 = 0$$

are each -1.

4. Find the radical centre of the circles

$$x^2 + y^2 - 4x + 1 = 0,$$
$$x^2 + y^2 + 2x + 1 = 0,$$
$$x^2 + y^2 - 5x + y - 5 = 0$$

and hence find the equation of the circle which cuts the three circles orthogonally.

5. Prove that the origin is one of the limiting points of the coaxal system to which the circles

$$x^2 + y^2 - 2x + 6y + 4 = 0,$$
$$5x^2 + 5y^2 - 4x + 12y + 8 = 0$$

belong. Find the other limiting point and the equation of the orthogonal system of circles.

6. Find the coordinates of the limiting points of the system of circles to which the circles

$$x^2 + y^2 - 9x - 5y + 1 = 0,$$
$$x^2 + y^2 + 11x + 7y + 17 = 0$$

belong, and find the equation of the circle which cuts these two circles orthogonally and passes through the origin.

7. Find the coordinates of the limiting points of the coaxal system to which the circles

$$x^2 + y^2 - 3x + 4y + 4 = 0,$$
$$x^2 + y^2 - 6x + 8y + 24 = 0$$

belong. Give a geometrical explanation of the fact that in this case the limiting points coincide.

8. Find the equations of the circles touching the line $x + y + 10 = 0$ and belonging to the coaxal system with limiting points $(1, -2)$ and $(-3, 1)$.

9. The limiting points of a coaxal system are $(\pm k, 0)$, where k is a positive constant. Find the coordinates of the centres of the circles of the system which have a radius $k\sqrt{3}$ units.

10. A system of coaxal circles has limiting points $(0, -2)$ and $(1, 0)$. Find the equations of the two members of the system with radius $5/4$.

Find the equation of the member of the system which touches the line $x + 2y = 0$.　　　　(C.)

11. One limiting point of a system of coaxal circles is $(1, -1)$, and a circle of the system has the equation

$$x^2 + y^2 - 6x + 14y + 38 = 0.$$

Find the coordinates of the other limiting point.

Find the equation of the circle through the origin which cuts every circle of the system orthogonally.

(C.)

12. Prove that in general there are two circles of a non-intersecting coaxal system touching a given straight line.

Find the equations of the two circles belonging to the same coaxal system as

$$x^2 + y^2 + 8y + 8 = 0$$

and

$$x^2 + y^2 + 8x + 16y + 4 = 0$$

which touch the line $x = 2$.

(C.)

13. Show that, by suitable choice of coordinate axes, the equation of a system of non-intersecting coaxal circles may be expressed in the form

$$x^2 + y^2 - 2\lambda x + a^2 = 0,$$

where a is constant and λ is a variable parameter.

If any line through one of the limiting points L of the system meets any circle of the system in points P_1, P_2, with coordinates (x_1, y_1), (x_2, y_2), prove that

(i) $x_1 x_2 = a^2$,

(ii) if M is the other limiting point, ML bisects the angle $P_1 M P_2$.

Miscellaneous Exercise 21

1. Find the equations of the tangents at the origin to the curve

$$y^2 = \frac{3x^2(3 - x)}{x + 1},$$

and the coordinates of the points at which the tangents to the curve are parallel to the x-axis. Sketch the curve.

(L.)

2. The equation of a curve is

$$y^2 = \frac{x(2 - x)^2}{4 - x}.$$

Sketch the curve and show that the angle between the tangents to the curve at the point $(2, 0)$ is $\pi/2$.

Find, in terms of π, the volume generated when the area enclosed by the loop of the curve is rotated through π radians about the x-axis.

(L.)

3. By transformation to polar coordinates, sketch the curve

$$a^2(x^2 - y^2) = (x^2 + y^2)^2$$

where $a > 0$, and find the total area of the region enclosed by it.

(L.)

4. Show that the equation

$$kx^2 + 2\lambda xy - ky^2 = 0$$

represents two perpendicular straight lines.

Show that the pair of straight lines joining the origin O to the intersections A and B of the line $lx + my = 1$ with the conic $a^2x^2 + b^2y^2 = 1$ has the equation

$$(a^2 - l^2)x^2 - 2lmxy + (b^2 - m^2)y^2 = 0.$$

Deduce that if AOB is a right angle, then the line AB touches the circle

$$(a^2 + b^2)(x^2 + y^2) = 1.$$

(L.)

5. Prove that the circles

$$x^2 + y^2 - 10x + 16 = 0 \text{ and } x^2 + y^2 - 40x + 256 = 0$$

touch externally, and find the equations of their three common tangents.   (C.)

6. Sketch the curve represented by the equations $x = 3 \sin 2t$, $y = 2 \cos t$. Find the area of the region enclosed by one loop of the curve and find also the coordinates of the centroid of this region.   (L.)

7. Find a general expression for the roots of the equation $\sin(x^2) = 0$, and show that the distance between successive roots approaches 0 as $x \to \infty$. Find the slope of the curve $y = \sin(x^2)$ at the origin. Draw a rough sketch of the curve, indicating its principal features.   (L.)

8. Sketch the curves

(i) $xy^2 = 1$, (ii) $xy^2 = (x-4)^3$, (iii) $xy^2 = (x-1)^2$, (iv) $x(y^2 + 1) = y - 1$.

9. Sketch the curve whose equation is

$$x^3 - axy + y^3 = 0.$$

Prove that one branch of the curve at the origin touches the x-axis there and find its curvature at the origin.

Obtain the coordinates of the point (other than the origin) at which the tangent to this curve is parallel to the x-axis, and find the curvature at this point.   (O.C.)

10. Show that the curve $x^3 + y^3 = 3axy$ is symmetrical about the line $y = x$ and also that it has a loop. The solid of revolution formed by rotating the loop about the axis of symmetry has volume V. By rotating the coordinate axes through $\frac{1}{4}\pi$ to become OX, OY prove that

$$V = \frac{1}{3}\pi \int_0^{3c} \frac{(3c - X)X^2}{c + X} \, dX,$$

where $c = a/\sqrt{2}$. Hence find V in terms of a.   (O.C.)

11. Find the asymptotes of the curve

$$y^3 - 3x^2 y + 6x - 4 = 0$$

and the slope of the curve at the finite points where it crosses its asymptotes.

Prove that the curve has a double point at the point $(1, 1)$ and that any line $y = h \, (h \neq 1)$ meets the curve in two distinct points.

Draw a rough sketch of the curve.   (O.C.)

12. Sketch the curve $y = \ln(x^n)$, where n is a positive integer, and show that for all values of n the tangent to the curve at the point given by $x = e$ passes through the origin.

Show also that the x-coordinate of the centroid of the region defined by the inequalities.

$$1 \leqslant x \leqslant e, \quad 0 \leqslant y \leqslant \ln(x^n)$$

is independent of n, and calculate the y-coordinate of this centroid.   (L.)

13. Write down the 2×2 matrix $\mathbf{R}$ such that $\mathbf{R} \begin{pmatrix} x \\ y \end{pmatrix} = \begin{pmatrix} X \\ Y \end{pmatrix}$ represents a rotation of the coordinate axes

through $\pi/4$ in an anticlockwise direction.

S is the curve whose equation in terms of x and y is

$$x^2 + y^2 - 2xy - 2x - 2y = 0.$$

Determine the equation of S in terms of X and Y.

Sketch the two sets of axes and the curve S.   (N.)

14. The matrix $\mathbf{M}$, where

$$\mathbf{M} = \begin{pmatrix} a & b \\ c & d \end{pmatrix}, \ (a, b, c, d \text{ real and non-zero})$$

and its transpose $\mathbf{M}^T$ satisfy

$$\mathbf{M}^T \mathbf{M} = \begin{pmatrix} k^2 & 0 \\ 0 & k^2 \end{pmatrix}, \ (k \text{ real and non-zero}).$$

If $\begin{pmatrix} X \\ Y \end{pmatrix} = \mathbf{M} \begin{pmatrix} x \\ y \end{pmatrix}$, show that

$$X^2 + Y^2 = k^2(x^2 + y^2).$$

Show also that

(i) $a^2 - d^2 = b^2 - c^2$,

(ii) $\dfrac{a^2}{d^2} = \dfrac{c^2}{b^2}$,

and deduce that $a^2 = d^2$ and $b^2 = c^2$.

Write down the two matrices $\mathbf{M}_1$, $\mathbf{M}_2$ of the above form for which $a = 3, b = 4$. Describe the geometric transformation represented by either $\mathbf{M}_1$ or $\mathbf{M}_2$, interpreting the significance of the determinant of the matrix.

15. Show that the equation of the bisectors of the angles between the lines

$$ax^2 + 2hxy + by^2 = 0 \quad \text{is} \quad h(x^2 - y^2) = (a - b)xy.$$

Find the equation of the pair of lines which meet at right angles at $P(h, k)$ and are such that each contains the pole of the other with respect to the parabola $y^2 = 4ax$. The join of P (assumed not on the line $x = a$) to the fixed point (a, c) is a bisector of the angle between these lines. Show that P lies on the circle

$$(x - a)^2 + y^2 = c^2. \tag{C.}$$

16. Write down the equation of the polar of the point (x', y') with respect to the conic

$$\alpha x^2 + \beta y^2 = 1,$$

and deduce that the coordinates of the pole of the line

$$lx + my = 1$$

with respect to this conic are $(l/\alpha, m/\beta)$.

A system of conics is given by the equation

$$\frac{x^2}{a^2 + \lambda} + \frac{y^2}{b^2 + \lambda} = 1,$$

where λ is a parameter. Prove that the poles of a fixed line L with respect to the conics of this system lie on another fixed line M perpendicular to L.

If now the line L is varied in such a way that it always passes through a fixed point on one of the coordinate axes, prove that M also passes through a fixed point on the same axis. (O.C.)

17. A variable chord CD of the ellipse $b^2x^2 + a^2y^2 = a^2b^2$ passes through a focus $(ae, 0)$ of the ellipse. The tangents at C and D to the ellipse meet at T. Find the equation of the locus of T.

Prove that the locus of the mid-point of CD has the equation $b^2x^2 + a^2y^2 - ab^2ex = 0$. (C.)

18. Prove that the coordinates of the limiting points of the coaxal system determined by the circles with equations

$$x^2 + y^2 - 14x + 6y - 2 = 0 \quad \text{and} \quad x^2 + y^2 - 20x + 8y - 4 = 0$$

are $(1, -1)$ and $(-2, 0)$.

Find the equation of a circle which is orthogonal to the two given circles and which touches the x-axis. (C.)

19. Interpret the equation $S + kLM = 0$, where $S = 0$ is the equation of a conic, $L = 0, M = 0$ are the equations of two lines, and k is constant.

The line $px + qy = 1$ meets the conic

$$ax^2 + 2hxy + by^2 + 2gx + 2fy + c = 0$$

at A and B; and from the origin O lines OA, OB are drawn to meet the conic again at C and D. Find the equation of the pair of lines OA, OB, and hence, or otherwise, show that the equation of the line CD is

$$c(px + qy + 1) + 2(gx + fy) = 0. \tag{O.C.}$$

20. By changing the origin to the point C with coordinates $(-g/a, -f/b)$, show that, if $a \neq 0$ and $b \neq 0$, the equation

$$S \equiv ax^2 + by^2 + 2gx + 2fy + c = 0$$

represents a conic whose centre is at C.

The equation of a second conic is

$$S' \equiv \alpha x^2 + \beta y^2 - 1 = 0.$$

Write down the general equation of a conic through the intersections of the conics $S = 0$ and $S' = 0$; and show that these points of intersection lie on a circle when $\alpha\beta \neq b\alpha$.

Find the locus of the centre of a variable conic through the intersections of the conics $S = 0$ and $S' = 0$.

(O.C.)

21. Show that the equation of the circle with the line joining the points $(x_1, y_1), (x_2, y_2)$ as diameter can be expressed in the form

$$(x - x_1)(x - x_2) + (y - y_1)(y - y_2) = 0.$$

The sides AB, BC, CD, DA of the quadrilateral $ABCD$ have equations $x - y = 1$, $3x - y = 1, y = 2x, x = 2$ respectively. AB and DC when produced meet at E, BC and AD produced meet at F. Find the equations of the circles on AC, BD, EF as diameters. Prove that these three circles are coaxal, giving the equation of their radical axis.

(C.)

22. Three coplanar circles are such that their centres are not collinear. Prove that the radical axes of the three pairs of circles meet in a point – the *radical centre* of the three circles.

Prove that a system of circles with equations

$$x^2 + y^2 + 2\lambda x + 2\mu y + c^2 = 0,$$

where λ and μ vary but c has the same value for every circle of the system is such that all sets of three circles with non-collinear centres have the same radical centre, and determine the coordinates of this point.

(C.)

23. Find the coordinates of the limiting points of the coaxal system defined by the circles $x^2 + y^2 + 8x - 10y - 9 = 0$, $3(x^2 + y^2) + 12x - 14y - 5 = 0$. Find the equations of any circles of the system touching the line $y + x = 0$, and comment on the result.

(L.)

24. Chords are drawn through the focus of the parabola P_1 whose equation is $y^2 = 4ax$; show that their mid-points lie on a parabola P_2 coaxial with P_1, and find its focus and directrix.

Parabolas $P_3, P_4, \ldots, P_m, \ldots$ are derived in a similar way from $P_2, P_3, \ldots, P_{n-1}, \ldots$ respectively. Show that their foci tend to the point $(2a, 0)$ as limit. What is the limiting position of their directrices?

25. Find the equation of the normal at the point with parameter t on the rectangular hyperbola $x = ct$, $y = c/t$. Show that, if the normals at the points t_1, t_2, t_3, t_4 are concurrent, then $t_1 t_2 t_3 t_4 = -1$ and the line joining any two of the four points is perpendicular to the line joining the other two.

26. Show that the orthocentre of a triangle inscribed in the hyperbola $x = kt$, $y = k/t$ lies on the curve. P_1, P_2, P_3, P_4 are four points on the hyperbola. H_1 is the orthocentre of the triangle $P_2 P_3 P_4$ and H_2, H_3, H_4 are the orthocentres of the other three triangles defined by three of the points P_1, P_2, P_3, P_4. Show that if P_1, P_2, P_3, P_4 lie on the circle

$$x^2 + y^2 + 2gx + 2fy + c = 0,$$

then H_1, H_2, H_3, H_4 are also concyclic.

27. If P is the point $(cp, c/p)$ on the rectangular hyperbola $xy = c^2$, show that there are two circles, each of which touches the hyperbola at P and another point.

Show that each of the circles has radius equal to OP, where O is the origin.

28. Two parabolas have four real points of intersection P_i ($i = 1, 2, 3, 4$), and their axes of symmetry are perpendicular. If these are taken as coordinate axes, and the coordinates of P_i are (x_i, y_i), show that

(i) $x_1 + x_2 + x_3 + x_4 = 0$,

(ii) $x_2 x_3 x_4 + x_1 x_3 x_4 + x_1 x_2 x_4 + x_1 x_2 x_3$ is unchanged if either parabola is moved parallel to its axis of symmetry.

22 Algebraic Structure

22:1 ALGEBRAIC STRUCTURE

The structure of algebra consists of elements combined by means of procedures which can be classified under two main headings, namely *relations* and *operations*. There are usually conventional signs available for specific procedures; we shall use $\mathscr{R}$ and $\mathscr{C}$ to denote a relation and an operation respectively.

22:2 RELATIONS

In mathematics a meaningful proposition is one that can be shown or assumed to be either true (T) or false (F); we say that the "truth-value" of a proposition is T or F.

A (binary) relation $\mathscr{R}$ is defined on a set S if

$$\forall a, b \in S; \ \exists \mathscr{R}; \ a \, \mathscr{R} \, b \quad \text{or} \quad a \, \not\mathscr{R} \, b,$$

where $a \, \mathscr{R} \, b$ means that the proposition "a has the relation $\mathscr{R}$ to b" is true and $a \, \not\mathscr{R} \, b$ means that the proposition is false.

For example, "is greater than" is defined on $\mathbb{R}$ since $3 > -1, 2 \not> \sqrt{5}, 7 \cdot 2 \not> 7 \cdot 2$; but the relation is not defined on $\mathbb{C}$ since $(3 - 2i) > (2 + i)$ is meaningless.

There are certain general properties which a relation defined on a set may or may not possess.

(i) $\mathscr{R}$ on S is **reflexive** $\Leftrightarrow \forall a \in S, \ a \, \mathscr{R} \, a$.
For example "has the same names as" on {people}, "is equal to" on $\mathbb{R}$.
(ii) $\mathscr{R}$ on S is **symmetric** $\Leftrightarrow \forall \ a, b \in S; \ a \, \mathscr{R} \, b \Leftrightarrow b \, \mathscr{R} \, a$.
For example, "plays golf with" on {people}, "is supplementary to" on {angles}.
(iii) $\mathscr{R}$ on S is **transitive** $\Leftrightarrow \forall \ a, b, c \in S; \ (a \, \mathscr{R} \, b \text{ and } b \, \mathscr{R} \, c) \Rightarrow a \, \mathscr{R} \, c$.
For example "is less than" on $\mathbb{R}$, "is a subset of" on {sets}.

We can construct a "**combination table**" for a relation on a set, in which the first element is read from the master column $\downarrow$ and the second from the master row $\rightarrow$. For example,

$$\forall p, q \in \{1, 2, 3, 4, 5, 6\}; \quad p \, \mathscr{R} \, q \Leftrightarrow p + q \geq 7.$$

TABLE 22.1

	1	2	3	4	5	6
1	F	F	F	F	F	T
2	F	F	F	F	T	T
3	F	F	F	T	T	T
4	F	F	T	T	T	T
5	F	T	T	T	T	T
6	T	T	T	T	T	T

The relation is clearly symmetric, but not reflexive or transitive.

A relation which is reflexive, symmetric and transitive on a given set is called an **equivalence relation** on that set; for example "has the same number of elements as" on {sets}, "equals" on $\mathbb{Q}$. An equivalence relation on S **partitions** the elements of S into subsets (**equivalence classes**) such that each member of any equivalence class S_r has the relation to every member of S_r but to no other element of S,

i.e. $\forall a \in S \quad \exists S_r : a \in S_r.$

(The reflexive property ensures that each element is at least related to itself, so that it can be the single element in an equivalence class.)

For "equals" on $\mathbb{Q}$, the equivalence classes are such as

$$\left\{ \frac{1}{2}, \frac{2}{4}, \frac{3}{6}, \cdots \right\}, \left\{ 3, \frac{6}{2}, \frac{9}{3} \cdots \right\}.$$

Exercise 22.2

1. State, in each case, whether the given relation is reflexive, symmetric or transitive on the given set:
(a) "differs by more than 6 from" on $\mathbb{Z}$;
(b) "has the same father as" on {children};
(c) "is perpendicular to" on {straight lines in a plane};
(d) "leaves the same remainder when divided by 3 as" on $\mathbb{N}$;
(e) "is parallel to" on {straight lines};
(f) "is the square of" on $\mathbb{R}$.
Hence state which, if any, are equivalence relations.

2. Explain why "is not separated by sea from" on {towns} is an equivalence relation. Give the equivalence classes for the relation on {Birmingham, Chicago, Paris, Berlin, Glasgow, Tokyo, Rome, Cardiff, Madrid}.

3. Give one example in each case of an equivalence relation on S which partitions S into (a) three, (b) four, (c) $n \in \mathbb{N}$ equivalence classes.

4. Construct a relation which is:
 (i) symmetric, transitive, not reflexive;
 (ii) symmetric, not reflexive, not transitive;
 (iii) not symmetric, reflexive, transitive;
 (iv) not symmetric, not reflexive, transitive;
 (v) not symmetric, not reflexive, not transitive.

5. Determine which of the following are equivalence relations, and for each of these describe the equivalence classes:
(a) "is greater than the square of" on $\mathbb{Q}$;

(b) "is similar to" on {triangles};
(c) "is coplanar with" on {straight lines};
(d) "likes" on {people};
(e) "differs by 2 from" on $\mathbb{Z}$;
(f) "is equal in magnitude to" on {two-dimensional vectors};
(g) "is a subset of" on {sets}.

6. The relation ρ on the set of positive integers is such that $x\rho y \Leftrightarrow$ "a positive integer n can be found such that $x^n = y$".
Prove that
(i) $(x\rho y$ and $y\rho z) \Rightarrow x\rho z$,
(ii) $(x\rho y$ and $y\rho x) \Rightarrow (x = y)$.
Suppose that the relation ρ is defined on the set $\{1, 2, 3, 4, 5, 6\}$ and that multiplication is carried out modulo 7; i.e. x^n now denotes the remainder when the normal value of x^n is divided by 7. Find the values of x such that $x\rho 1$, and determine all pairs (x, y) with $x \neq y$ for which both $x\rho y$ and $y\rho x$.   (C.)

7. Determine whether the following are reflexive, symmetric, or transitive, and hence which, if any, are equivalence relations:
(a) "is prime to" on $\mathbb{N}$;
(b) "is a factor of" on $\mathbb{N}$;
(c) "is a prime factor of" on $\mathbb{N}$;
(d) "is a multiple of" on $\mathbb{N}$;
(e) "is less than half of" on $\mathbb{Q}$.

22:3 THEORY OF NUMBERS

We considered in Exercise 22.2 a few examples of relations, defined on $\mathbb{Z}$, occurring in the branch of mathematics called the theory of numbers; this subject can be a fascinating one, but we can only refer to it briefly. For convenience, we shall usually only consider the set $\mathbb{N}$.

22:4 FACTORS

We say that a is a factor of b, or a divides b, or b is divisible by a, if and only if $b = ka$ where $a, b, k \in \mathbb{N}$; unless otherwise stated we assume $a \neq 1$. This relation, written $a|b$, is reflexive and transitive but not symmetric. If $b \in \mathbb{N}$ has no factors other than b and 1, we say that b is a **prime number**; if b is not a prime number, it has at least two prime factors (not necessarily distinct), and it can be proved that the expression of a non-prime (composite) number in prime factors is unique, for example,

$$2352 = 2^4 \times 3 \times 7^2.$$

The **highest common factor** (HCF) of $a, b \in \mathbb{N}$ is the largest number $h \in \mathbb{N}$ which divides both a and b. There are obvious advantages, e.g. in reducing a fraction to its lowest terms, in being able to find a HCF; we shall illustrate the general procedure by means of an example.

Example. Find the HCF of 2760 and 2093.

$$
\begin{array}{llll}
3 & |2093 & 2760| & 1 \\
  & 2001 & 2093 & \\
4 & \overline{\;\;\;92} & \overline{667|} & 7 \\
  & \;\;\;92 & 644 & \\
  & \overline{} & \overline{} & \\
  & \;\; .\;. & 23 &
\end{array}
$$

Now $23|92 \Rightarrow 23|644 \Rightarrow 23|(644+23) \Rightarrow 23|2001 \Rightarrow 23|(2001+92) \Rightarrow 23|2093$ and $23|(2093+667) \Rightarrow 23|2760$. Conversely $k|2093$ and $k|2760 \Rightarrow k|23$.

Hence 23 is the greatest divisor—i.e. the HCF—of 2760 and 2093. If the final remainder in the above process of successive division is 1, the integers have no common factor—two such integers are prime to each other.

From the above working, $23 = 667 - 644 = 667 - 7 \times 92$

$$= 667 - 7(2093 - 2001) = 667 - 7 \times 2093 + 21 \times 667$$

$$= 22(2760 - 2093) - 7 \times 2093 = 22 \times 2760 - 29 \times 2093.$$

Thus the HCF of 2760 and 2093 can be expressed in the form $2760x + 2093y$ where $x, y \in \mathbb{Z}$.

The corresponding general result is that if the HCF of a and b is h, then $\exists x, y \in \mathbb{Z} : ax + by = h$. This result is known as **Euclid's algorithm**.

The **least common multiple** (LCM) of $a, b \in \mathbb{N}$ is the smallest number l such that $a|l$ and $b|l$. It can easily be shown that, if h is the HCF of a and b so that $a = a_0 h, b = b_0 h$ for some $a_0, b_0 \in \mathbb{N}$, then $l = a_0 b_0 h$. Thus in the above example, the LCM of 2760 and 2093 is $120 \times 91 \times 23$.

22:5 MODULAR ARITHMETIC

If $n \in \mathbb{N}$ is divided by $m \in \mathbb{N}$, where $m < n$, then $n = qm + r$, where $q \in \mathbb{N}$ and $r \in \{0, 1, 2, \ldots, (m-1)\}$. The set $\{0, 1, 2, \ldots, (m-1)\}$ of possible remainders is called the set of **residues modulo** m. If $a, b \in \mathbb{N}$ give the same remainder when divided by m, we say that a and b are **congruent modulo** m, written $a \equiv b \pmod{m}$. Clearly

$$a \equiv b \pmod{m} \Leftrightarrow a - b \equiv 0 \pmod{m}$$
$$\Leftrightarrow a - b = km \;(k \in \mathbb{Z}).$$

The reader should easily be able to establish the following useful properties of congruences, in which $a, b, \ldots, \in \mathbb{N}$; we shall therefore only prove the first.

(i) $a \equiv b \pmod{m} \Rightarrow a - b = km \Rightarrow p(a-b) = pkm \Rightarrow pa \equiv pb$.

It should not be necessary to include $\pmod{m}$ at all stages in working with congruences.

(ii) $a \equiv b \pmod{m}$ and $c \equiv d \pmod{m} \Rightarrow pa \pm qc \equiv pb \pm qd \pmod{m}$.
(iii) $a \equiv b$ and $c \equiv d \Rightarrow ac \equiv bd$.
(iv) $a \equiv b \Rightarrow a^n \equiv b^n \;(n \in \mathbb{N})$.
(v) $ab \equiv 0 \Rightarrow a \equiv 0$ or $b \equiv 0$ *if* m *is prime*.
(vi) $ax \equiv bx \Rightarrow a \equiv b$ *if* m *is prime to* x.

It can easily be verified that the relation of congruence mod m on $\mathbb{N}$ is reflexive, symmetric and transitive, and is therefore an equivalence relation. The residues mod m partition $\mathbb{N}$ into m equivalence classes. For example, if $m = 3$, these are

$$\{0, 3, 6, \ldots\}, \{1, 4, 7, \ldots\}, \{2, 5, 8, \ldots\}.$$

We give some examples to show how the above results can be used in connection with congruences; in particular, it can be shown that a **linear congruence** of the form $ax \equiv b \pmod{m}$ will, *if m is prime,* have a unique solution set, which will be an equivalence class for congruence mod m.

Example 1. Solve for x, $75x \equiv 2 \pmod{13}$

$$
\begin{aligned}
&\Leftrightarrow 75x \equiv 2 + 13 && \text{(ii)}\\
&\Leftrightarrow 5x \equiv 1 && \text{(vi)}\\
&\Leftrightarrow 5x \equiv 1 + 3 \cdot 13 && \text{(ii)}\\
&\Leftrightarrow x \equiv 8 && \text{(vi)}\\
&\Leftrightarrow x = \{8, 21, 34, \ldots\}.
\end{aligned}
$$

In the above example it is not difficult to find a first stage in reducing the coefficient of x; when this is not clear, we can always use the procedure for finding a HCF as illustrated in §22:4, from which successive remainders can be used as successive coefficients of x for as many steps as may be required.

Example 2. Solve for x, $43x \equiv 5 \pmod{61}$.
The procedure for finding the HCF of 43 and 61 is

$$
\begin{array}{ll}
2|43 & \qquad 61|1 \\
\underline{36} & \qquad \underline{43} \\
1|\ 7 & \qquad 18|2 \\
\underline{4} & \qquad \underline{14} \\
3 & \qquad 4 \\
 & \qquad \underline{3} \\
 & \qquad 1
\end{array}
$$

We use the first remainder, 18, as the first reduced coefficient of x.

$$
\begin{aligned}
18x &= 61x - 43x \\
\Rightarrow 18x &\equiv -43x \equiv -5 \\
\Rightarrow 36x &\equiv -10.
\end{aligned}
$$

By subtraction, $7x \equiv 15 \Rightarrow 42x \equiv 90.$

By addition, $-x \equiv 85 \Rightarrow x \equiv -85 \equiv 37.$

$$x \equiv 37 \pmod{61}.$$

Example 3. Prove that $5^{132} \equiv 4 \pmod{127}$.

$$
\begin{aligned}
5^3 &\equiv -2 \pmod{127} && \text{[we allow the negative}\\
 & && \text{remainder for convenience]}\\
\Rightarrow 5^{21} &\equiv -128 \equiv -1 && \text{(iv)}\\
\Rightarrow 5^{63} &\equiv -1 && \text{(iv)}\\
\Rightarrow 5^{66} &\equiv -125 \equiv 2 && \text{(iii)}\\
\Rightarrow 5^{132} &\equiv 4 \pmod{127}. && \text{(iv)}
\end{aligned}
$$

Example 4. (a) Prove that, if $11n \equiv 37 \pmod{111}$, then n is a multiple of 37.

Prove that, if the number $100a + 10b + c$ is divisible by 37, so are $100b + 10c + a$ and $100c + 10a + b$.
(b) Prove that the number of primes is infinite. (C.)

$$(a)\ 11n = 37 + 111\,k = 37(1 + 3k) \Rightarrow 37|11n$$
$$\Rightarrow n \text{ is a multiple of 37.}$$
$$100a + 10b + c + 11(100b + 10c + a) = 111a + 1110b + 111c$$
$$= 37(3a + 30b + 3c).$$

Hence $\qquad\qquad\qquad 37|(100a + 10b + c) \Rightarrow 37|(100b + 10c + a).$

Similarly, $\qquad 11(100a + 10b + c) + 100c + 10a + b = 37(30a + 3b + 3c)$
$$\Rightarrow 37|(100c + 10a + b).$$

(b) Suppose that there is a greatest prime, p.
Let $n = 2.3.5 \ldots p + 1$ where $2.3.5 \ldots p$ is the product of all primes. Then, if n is prime, $n > p$; but if n is not prime, it can only have prime factors greater than p. Hence in either case we are led to a contradiction of our original assumption, which must therefore be false.
[It is instructive to study the form of the argument in (b) above.
Let $r \equiv$ "there is a greatest prime p";
$\qquad s \equiv$ "$n > p$ is composite";
$\qquad t \equiv$ "the prime factors of n are less than p".
The argument is as follows:
$\quad r \Rightarrow (s \text{ and } t)$, that is "If r is true, it is *necessary* for both s and t to be true".
Either s and t are both false or s is true and t is false. Therefore r is false.]

Example 5. Solve for x and y the congruences
$$4x - 5y \equiv 1 \quad (\text{mod } 7), \tag{1}$$
$$x + 2y \equiv 3 \quad (\text{mod } 7). \tag{2}$$

From (2) $\qquad\qquad\qquad 4x + 8y \equiv 12 \equiv 5. \tag{3}$

Subtracting (1) from (3) $\qquad \Rightarrow 13y \equiv 4 \equiv 39$
$$\Rightarrow y \equiv 3 \ (\text{mod } 7), \quad 4x \equiv 16 \Rightarrow x \equiv 4 \ (\text{mod } 7).$$

Exercise 22.5

1. Determine (giving reasons) which of the following properties define equivalence relations, and state the corresponding equivalence classes:
    (i) parallelism among the set of straight lines;
    (ii) possession of a factor in common among the set of positive integers;
    (iii) equality of modulus among the set of complex numbers;
    (iv) equality of radius among the set of circles. (O.C.S.M.P.)

2. The relation R on the set of all points in the x-y plane is defined by $(x, y)\,R\,(u, v) \Leftrightarrow y^2 - v^2 = x - u$. Show that R is an equivalence relation.
Sketch the equivalence class determined by the point $(1, 2)$.

3. The ordered pairs (a, b) and (c, d), where a, b, c, d are positive integers (greater than zero), are said to be in the relationship $\mathscr{R}$ if
$$ad - bc = 0.$$
Prove that $\mathscr{R}$ is an equivalence relation in the set of all such pairs.
    With what type of number can each equivalence class be associated? (O.C.S.M.P.)

4. Solve the congruences
$$\text{(i) } 31x \equiv 1 \ (\text{mod } 71); \quad \text{(ii) } 98x \equiv 1 \ (\text{mod } 139).$$

5. Find integers A and B such that $60A + 13B = 1$.
Of two 12-hour clocks, one (F) gains 9 seconds and the other (S) loses 4 seconds, every day. At noon on one

day, F shows the correct time, S is 17 seconds fast. At noon n days later, the second hands on the clocks mark the same number of seconds. Find the smallest value of n and state, in hours, minutes and seconds, the times shown by the clocks after this number of days.

(L.)

6. Solve the congruence

$$x \equiv 3 \ (\text{mod } 7) \equiv 5 \ (\text{mod } 11).$$

7. Use the fact that $10 \equiv 1 \ (\text{mod } 9)$ to prove that an integer n is divisible by 9 if and only if the sum of the digits of the decimal form for n is divisible by 9.

Use a similar method to establish a test for divisibility by 11.

8. (a) The simultaneous congruences

$$x + 2y \equiv 2(\text{mod } 5), \qquad 4x + 3y \equiv 3(\text{mod } 5)$$

have 5 solutions, but

$$x + 2y \equiv 2(\text{mod } 5), \qquad 4x + y \equiv 4(\text{mod } 5)$$

have a unique solution.

Explain why this is so, and find the solutions.

(b) If the simultaneous congruences

$$3x + 2y \equiv a(\text{mod } 5), \qquad 4x + y \equiv 4(\text{mod } 5)$$

do not have any solution, find the possible values of a.

(L.)

9. Prove that $2^{23} - 1$ is divisible by 47.

10. Prove that $3^{2n+4} - 2^{2n}$ is divisible by 5.

11. In each of the following establish whether or not the relation $\mathcal{R}$ defined on the set $S = \{x, y, \ldots\}$ is reflexive, symmetric or transitive. Also, if it forms an equivalence relation, state the partitioning of S which it achieves, and, if it is an ordering relation, describe the ordering which it achieves.

(i) $S = \{0, 1, 2, 3\}$, $x \mathcal{R} y$ if $(x + y)(\text{mod } 4) = |x - y|$.

(ii) $S = \{0, 1, 2, 3\}$, $x \mathcal{R} y$ if $(x + y)(\text{mod } 2) = |x - y| \ (\text{mod } 2)$.

(iii) S is the set of integers from 2 to 20 inclusive, $x \mathcal{R} y$ if y is exactly divisible by x.

(iv) $S = \{2, 3, 4, 5, 7, 8, 9\}$, $x \mathcal{R} y$ if x and y have a common factor. [Regard the factors of a number as including the number itself, but excluding 1.]

(C.)

22:6 OPERATIONS

A (binary) operation, when defined on a certain set, combines two ordered elements of the set to form a unique third element, which may or may not be a member of the given set. For example, multiplication is defined on $\mathbb{R}$, with results such as $7 \times 2 \cdot 3 = 16 \cdot 1$; union is defined on the set of all sets, giving results such as $A \cup B = C$

$$\forall a, b \in S; \quad a \ \mathcal{O} \ b = c.$$

An operation defined on a set may or may not obey certain laws; such information is usually important in dealing with a structure containing the operation.

(i) **Closure.** A set is said to be **closed** under a certain operation if, when any two elements of the set are combined under the operation, the answer is itself an element of the set, i.e.

$$\forall a, b \in S; \quad \exists c \in S: a \ \mathcal{O} \ b = c.$$

For example, $\mathbb{Z}$ is closed under addition, since the sum of two integers is an integer, but is not closed under division, since the ratio of two integers is not **necessarily** an integer.

(ii) **Commutativity.** A set S is **commutative** under $\mathcal{O}$ if the order in which two elements of S are combined is immaterial.

$$\forall\, a, b \in S; \quad a\,\mathcal{O}\,b = b\,\mathcal{O}\,a.$$

For example, $\mathbb{R}$ is commutative under multiplication, the set of 2×2 matrices is not.

(iii) **Associativity.** As it stands, $a\,\mathcal{O}\,b\,\mathcal{O}\,c$ is meaningless; we can make it meaningful in two different ways by using brackets to indicate priority—either $(a\,\mathcal{O}\,b)\,\mathcal{O}\,c$ or $a\,\mathcal{O}\,(b\,\mathcal{O}\,c)$. If the final answer in each case is the same, we say that $\mathcal{O}$ is associative on S, i.e.

$$\forall\, a, b, c \in S; \quad (a\,\mathcal{O}\,b)\,\mathcal{O}\,c = a\,\mathcal{O}\,(b\,\mathcal{O}\,c).$$

For example, addition is associative on $\mathbb{R}$ since $(\sqrt{2}+3)+4{\cdot}9 = \sqrt{2}+(3+4{\cdot}9)$ but subtraction is not associative on $\mathbb{R}$ since $(\sqrt{2}-3)-4{\cdot}9 \neq \sqrt{2}-(3-4{\cdot}9)$.

(iv) **Distributivity.** If $\mathcal{O}_1$, $\mathcal{O}_2$ are two binary operations defined on S, it may be possible to distribute one operation over the other. $\mathcal{O}_1$ is **distributive** over $\mathcal{O}_2$ if

$$\forall\, a, b, c \in S; \quad a\,\mathcal{O}_1(b\,\mathcal{O}_2\,c) = (a\,\mathcal{O}_1\,b)\,\mathcal{O}_2\,(a\,\mathcal{O}_1\,c).$$

For example, if $S = \mathbb{R}$,

$$a \times (b+c) = (a \times b) + (a \times c)$$

but

$$a + (b \times c) \neq (a+b) \times (a+c).$$

Hence multiplication is distributive over addition, but addition is not distributive over multiplication.

(v) **Identity elements.** We say that e is an **identity** (or neutral) element of a set S under an operation $\mathcal{O}$ if, for any element $a \in S$, $a\,\mathcal{O}\,e = e\,\mathcal{O}\,a = a$. As familiar examples, 0 is an identity element of $\mathbb{Z}$ or $\mathbb{Q}$ under addition, as is 1 under multiplication; the "stay-put" transformation (I) is an identity element under "product" of transformations. Notice that 0 is *not* an identity element of $\mathbb{Z}$ under subtraction since $a - 0 = a$ but $0 - a \neq a$.

(vi) **Inverse elements.** If $a \in S$, there may or may not be an element $a^{-1} \in S$ such that $a\,\mathcal{O}\,a^{-1} = a^{-1}\,\mathcal{O}\,a = e$; if so, a^{-1} is an **inverse** of a. For example, there is no inverse of an element of $\mathbb{N}$ under addition, but in $\mathbb{Z}$ an element a has the inverse $-a$. Clearly if q is the inverse of p, then p is the inverse of q, a symmetric relation.

We can show the results of an operation on a finite set by means of a combination table, and hence may be able to establish various properties of the structure.

Example 1. Give the combination table for multiplication modulo 5 on the residue $\{0, 1, 2, 3, 4\}$.

TABLE 22.2

$x \pmod 5$	0	1	2	3	4
0	0	0	0	0	0
1	0	1	2	3	4
2	0	2	4	1	3
3	0	3	1	4	2
4	0	4	3	2	1

The set is closed, and symmetry about the leading diagonal shows commutativity. The identity element is 1, the element 4 is self-inverse, and 2 and 3 are inverses; 0 has no inverse.

Example 2. Give the combination table for multiplication on $\{1, i, -1, -i\}$, and state any properties of the structure.

TABLE 22.3

$\times$	1	i	-1	$-i$
1	1	i	-1	$-i$
i	i	-1	$-i$	1
-1	-1	$-i$	1	i
$-i$	$-i$	1	i	-1

The set is closed, and the symmetry of Table 22.3 shows commutativity; the associative law holds for multiplication on $\mathbb{C}$. 1 is the identity element, -1 is its own inverse, i and $-i$ are inverses.

Exercise 22.6

1. If aHb, aLb are, respectively, the HCF and LCM of $a, b \in \mathbb{N}$, (i) is H associative?; (ii) is L associative?; (iii) is H distributive over L?; (iv) is L distributive over H?

2. A law of combination for two vectors **a** and **b** is defined by $\mathbf{a} \circ \mathbf{b} = ab\sin\theta$, where a, b are the magnitudes of **a**, **b** and where θ is the angle between the vectors. Assuming that $\circ$ is distributive over addition show, by combining $\mathbf{a} + \mathbf{b}$ with itself, that $\circ$ is not commutative. Suggest a suitable convention for measuring θ.   (O.C.S.M.P.)

3. Compile operation tables for
   (i) $\{1, 2\}$ under multiplication modulo 3;
   (ii) $\{0, 1\}$ under addition modulo 2.
Write down all possible ordered pairs whose first element belongs to the set $\{1, 2\}$ and whose second element belongs to the set $\{0, 1\}$.
Compile an operation table for these ordered pairs under the operation defined by

$$(x_1, y_1) * (x_2, y_2) = (x_1 x_2, y_1 + y_2),$$

where the multiplication and addition are as defined in (i) and (ii).
State the identity element and give a general proof of the associativity of the operation $*$. (Multiplication and addition in (i) and (ii) may be assumed to be associative.)   (O.C.S.M.P.)

4 (a) Is the operation $a * b = a + b - ab$ (i) associative, (ii) distributive over addition, (iii) distributive over multiplication?
   (b) Is (i) addition, (ii) multiplication distributive over $*$?
   (c) Is $*$ distributive over itself?

5. The binary operation of "multiplication" $\otimes$ is defined on the set of ordered pairs (z, w) of complex numbers z and w by setting

$$(z_1, w_1) \otimes (z_2, w_2) = (z_1 z_2 - w_1{}^* w_2, z_1 w_2 + w_1 z_2{}^*),$$

where the star denotes the complex conjugate. Write out the multiplication table for the subset of elements

$$(1, 0), (i, 0), (0, 1) \text{ and } (0, i),$$

where $i^2 = -1$.
On the original set of all pairs (z, w),
   (i) prove that the operation $\otimes$ is not commutative;
   (ii) obtain the "identity element" for the operation $\otimes$.

6. S denotes the set of matrices of the form

$$\begin{pmatrix} x & y \\ y & x \end{pmatrix},$$

where x and y are real numbers. Investigate whether, for the operation of matrix multiplication,
   (i) S is closed;
   (ii) the identity belongs to S;
   (iii) every element of S has an inverse.

                          (O.C.S.M.P.)

7. A set $\{a, b\}$ of two distinct elements is subjected to an operation $*$ which has the properties both of closure and of associativity, and for which the products $a * b, b * a$ are different. Obtain all the allowable tables of products in the form

$*$	a	b
a	w	x
b	y	z

where w, x, y, z are elements of the set.
   [For full credit, a logical argument is required.]   (C.)

8. E, O represent an even and an odd integer respectively. Construct the combination tables for (a) addition, (b) subtraction, (c) multiplication on $\{E, O\}$.
   Is addition distributive over multiplication? Is multiplication distributive over addition?

9. The binary operation $*$ on the set $\mathbb{R}$ of all real numbers is defined by $a*b = a + b + 2$. Determine whether this operation is (i) commutative, (ii) associative. Investigate also the existence of a neutral element and of an inverse element to a.

Another binary operation $\circ$ is defined on $\mathbb{R}$ by
$a \circ b = \frac{1}{2}(a + b)$. Investigate whether
   (a) $\circ$ is distributive over $*$,

   (b) $*$ is distributive over $\circ$.

10. If $x \circ y = x + |y|$ on $\mathbb{Z}$, determine whether the structure has the properties of (a) closure, (b) commutativity, (c) associativity, (d) an identity element.

11. aGb means "the greater of a and b if $a \neq b$, a if $a = b$", defined on $\{-2, -1, 0, 1, 2\}$. Construct the combination table. Is the set closed under this operation? Does the associative law hold? Does the commutative law hold? Is there an identity element? Name any inverse elements.

12. Repeat question 11 for the following structures:

$a \circ b =$
     (i) $\frac{1}{2}(a + 3b)$ on $\{0, 2, 4, 6\}$,
     (ii) $\sqrt{(ab)}$ on $\{0, 4, 9\}$,
     (iii) $10 - (a + b)$ on $\{0, 5, -5\}$.

22:7 SOLUTION OF A LINEAR EQUATION

We already know that there is a unique solution in $\mathbb{R}$ to the equation $a \times x = b$ for a, $b \in \mathbb{R}$ ($a \neq 0$). We shall now consider the conditions under which we can solve for $x \in S$ the equation $a \,\mathcal{C}\, x = b$, where $\mathcal{C}$ is defined on S and $a, b, e \in S$, e being an identity element. At each stage in the solution we note the assumption necessary in order to reach that stage.

$$a \mathcal{O} x = b$$
$$\Rightarrow a^{-1} \mathcal{O} (a \mathcal{O} x) = a^{-1} \mathcal{O} b \quad [\exists a^{-1} \in S : a^{-1} \mathcal{O} a = e]$$
$$\Leftrightarrow (a^{-1} \mathcal{O} a) \mathcal{O} x = a^{-1} \mathcal{O} b \quad [\text{associativity}]$$
$$\Leftrightarrow e \mathcal{O} x = c \qquad\qquad [\text{closure}; \exists c \in S : a^{-1} \mathcal{O} b = c \in S]$$
$$\Leftrightarrow x = c. \qquad\qquad\quad [\text{identity element}].$$

It is not obvious that the first step is reversible, but we can check our solution by direct substitution: $a \mathcal{O} (a^{-1} \mathcal{O} b) = (a \mathcal{O} a^{-1}) \mathcal{O} b = b$.

To show that the solution is **unique** we need to prove that every element of S has a **unique** inverse in S. Suppose that $a_1, a_2 \in S$ are both inverses of a. Then

$$a_1 \mathcal{O} a = e \Rightarrow (a_1 \mathcal{O} a) \mathcal{O} a_2 = e \mathcal{O} a_2$$
$$\Rightarrow a_1 \mathcal{O} (a \mathcal{O} a_2) = a_2$$
$$\Rightarrow a_1 = a_2.$$

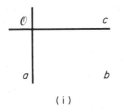   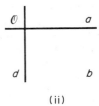

(i)   (ii)

FIG. 22.1.

Hence the equation $a \mathcal{O} x = b$ has a unique solution, c say, within the given structure, which is assumed to be closed, associative and to have identity and inverse elements, and the corresponding entries in the combination table must be as shown in Fig. 22.1 (i); it follows that every "b" (i.e. every element $\in S$) must appear once and only once in the row headed by every "a" (i.e. in every row). We leave it to the reader to show that by a similar argument and with the same assumptions as above, we obtain a unique solution, d say, within S to the equation $x \mathcal{O} a = b$. Since we do **not** assume commutativity this is of course an essentially different equation for which the corresponding entry will be as in Fig. 22.1 (ii), and it is quite possible that $c \neq d$. Thus every element $\in S$ must appear once and only once in each column of the table; since the row and column headed by an identity element must be the same as the master row and column forming the headings, it therefore follows that there can only be one identity element.

22:8 GROUP STRUCTURE

It is convenient to define a structure, which turns out to be an essential feature of many branches of mathematics and science, such that within the structure equations of the form $a \mathcal{O} x = b$ or $x \mathcal{O} a = b$ are uniquely soluble. Such a structure is called a **group**, and has the properties required for solving these equations; these may be formally stated as follows, assuming that an operation $\mathcal{O}$ is defined on a set S.

(i) **Closure:** $\qquad\qquad \forall a, b \in S; \; \exists c \in S : a \; \mathcal{O} \; b = c.$
(ii) **Associativity:** $\qquad \forall a, b, \, c \in S; \; a \; \mathcal{O} \; (b \; \mathcal{O} \; c) = (a \; \mathcal{O} \; b) \; \mathcal{O} \; c.$
(iii) **Identity element:** $\quad \exists e \in S : \forall a \in S; \; a \; \mathcal{O} \; e = e \; \mathcal{O} \; a = a.$
(iv) **Inverse elements:** $\quad \forall a \in S; \; \exists a^{-1} \in S : a \; \mathcal{O} \; a^{\pm 1} = a^{-1} \; \mathcal{O} \; a = e.$

[Strictly speaking, given that $a \, \mathcal{O} \, e = a$ and $a \; \mathcal{O} a^{-1} = e$, it can be proved, using (i) and (ii) that $e \; \mathcal{O} a = a$ and $a^{-1} \; \mathcal{O} a = e$. This is left as an exercise.]

If, in addition to these four properties, the commutative law holds,

$$\text{i.e.} \quad \forall a, b \in S; \; a \mathcal{O} b = b \mathcal{O} a,$$

the group is called **Abelian**.

It is important to remember that in the definitions of closure and associativity the elements need not be distinct, e.g. for closure $a \; \mathcal{O} a \in S$, and for associativity $a \; \mathcal{O} (b \; \mathcal{O} b) = (a \; \mathcal{O} b) \; \mathcal{O} b$.

Example. The symmetry transformations of a square.

Consider a square $ABCD$ cut from a sheet of cardboard. Since the square has rotational symmetry of order 4, and four axes of symmetry Ox, Oy, AC, BD (Fig. 22.2), the square can be replaced in the "hole" left in the sheet in eight different ways (including its original position). We can describe these ways in terms of the transformations to be carried out on the square before replacement.

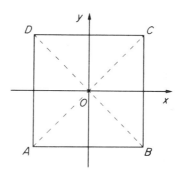

FIG. 22.2.

Thus, besides the identity transformation (I) corresponding to the original position, we have rotations through 90° (Q), 180° ($QQ = Q^2$), and 270° (Q^3); also reflections in Ox (R_x), Oy (R_y), $y + x = 0$ (R_1) and $y - x = 0$ (R_2). Now consider the binary operation "followed by", forming the "product" of two transformations, carried out on the set $\{I, Q, Q^2, Q^3, R_x, R_y, R_1, R_2\}$, where T_1 followed by T_2 is written $T_2 T_1$.

Table 22.4 shows, as the reader should verify, that the product of any two transformations from the set is equivalent to a single such transformation, e.g. $R_x Q = R_1$, $Q R_x = R_2$.

Note that, in reading the above table, we use our convention that the first element is taken from the master column and the second from the master row; but we also use the convention that, for example, $R_x Q$ means "Q followed by R_x".

TABLE 22.4

	I	Q	Q^2	Q^3	R_x	R_y	R_1	R_2
I	I	Q	Q^2	Q^3	R_x	R_y	R_1	R_2
Q	Q	Q^2	Q^3	I	R_1	R_2	R_x	R_y
Q^2	Q^2	Q^3	I	Q	R_y	R_x	R_2	R_1
Q^3	Q^3	I	Q	Q^2	R_2	R_1	R_y	R_x
R_x	R_x	R_2	R_y	R_1	I	Q^2	Q	Q^3
R_y	R_y	R_1	R_x	R_2	Q^2	I	Q^3	Q
R_1	R_1	R_x	R_2	R_y	Q	Q^3	I	Q^2
R_2	R_2	R_y	R_1	R_x	Q^3	Q	Q^2	I

Hence the set is closed. The associative law must always hold for the product of transformations, since $(RQ)P$ and $R(QP)$ both involve the same three transformations carried out in the same order: "P, then Q then R" or "P then Q, then R". Every element has an inverse since I appears in say the rth row and sth column whenever it appears in the sth row and the rth column. Thus the structure satisfies all the requirements for a group—clearly not Abelian.

22:9 IDENTIFICATION OF GROUPS

Obviously it may be a very tedious process to check that the associative law holds for every permutation of three elements from a set. In practice this is seldom necessary, since the commoner operations involved in group structures are known to be associative, for example, addition or multiplication on numbers or on matrices, or the product of transformations as in the preceding example. However, associativity is an essential group property, and to identify a structure as a group it must be verified either in general or in particular.

In a group the equations $a \oslash x = b$ and $x \oslash a = b$ clearly have unique solutions so it is **necessary** for each element of S to appear just once in each row and column of a combination table for $\oslash$ on S as pointed out in §22:7. If this requirement is not satisfied then $\oslash$ on S cannot be a group structure; unfortunately, the requirement is not also a sufficient one (see Example 1 below), so that we cannot regard its fulfilment as a guarantee of a group.

Example 1. Do the given combination tables represent groups?

TABLE 22.5

$\circ$	a	b	c		$*$	p	q	r
a	c	b	a		p	r	q	p
b	b	c	c		q	q	p	r
c	a	c	b		r	p	r	q
	(i)					(ii)		

(i) clearly does not represent a group, since the element c appears twice in a row and also twice in a column;

(ii) satisfies the above condition, each element appearing just once in each row and column; but $(p*q)*r = q*r = r$, and $p*(q*r) = p*r = p$. Hence the structure does not obey the associative law, and therefore cannot be a group. (Incidently, another requirement for a group is more obviously not satisfied; in this case there is no identity element.)

Example 2. An operation $*$ is defined on a set of numbers, S:

$$\forall a, b \in S; \quad a*b = a+b+ab.$$

Determine whether the structure is a group when (i) $S = \mathbb{Z}$, (ii) $S = \mathbb{Q}$.

Clearly $\mathbb{Z}$ and $\mathbb{Q}$ are both closed under $*$.

$$(a*b)*c = (a+b+ab)*c = a+b+ab+c+c(a+b+ab);$$
$$a*(b*c) = a+b+c+bc+a(b+c+bc).$$

Hence the associative law holds.

$$a*0 = 0*a = a, \text{ so that } 0 \text{ is the identity element.}$$

$$a+b+ab = 0 \Rightarrow b = -\frac{a}{1+a} = a^{-1} \text{ (the inverse of } a\text{)}.$$

Since $\dfrac{a}{1+a} \notin \mathbb{Z}$, (i) is not a group.

$$a^{-1} \in \mathbb{Q} \quad \text{provided that } a \neq -1.$$

Hence (ii) is only a group if -1 is excluded from S.

Example 3. Show that the set of three-dimensional vectors forms a group under vector addition.

The set is closed; since $(\mathbf{a}+\mathbf{b})+\mathbf{c} = \mathbf{a}+(\mathbf{b}+\mathbf{c})$ the associative law holds. The zero vector $\mathbf{0}$ is the identity element, and the inverse of $\mathbf{a}$ is $-\mathbf{a}$. Hence the structure forms a group.

Exercise 22.9

1. Determine whether the following operations, considered as $a \text{ } \mathcal{C} \text{ } b$, where $a, b \in \mathbb{N}$, are (i) closed, (ii) associative, (iii) commutative.

(a) $\frac{1}{2}(a+b)$; (b) $a+b+1$; (c) lesser of a and b;

(d) $\sqrt{(ab)}$; (e) $[\sqrt{(ab)}]$; (f) $a+3b$;

(g) $|a-b|$; (h) b^2-a^2; (j) 2^{a+b}.

2. Determine whether, for the following operations on the set of all sets

(a) $\cap$ is distributive over $\cup$,

(b) $\cup$ is distributive over $\sim$ where $A \sim B = \{x : x \in A, x \notin B\}$,

(c) $\sim$ is distributive over $\cap$.

3. Which of the following are groups? State which, if any, of the conditions for a group do not hold.
(a) addition on $\mathbb{Z}$; (b) multiplication on $\mathbb{C}$; (c) subtraction on $\mathbb{N}$; (d) scalar product on three-dimensional vectors; (e) multiplication mod n on $\{1, 2, 3, \ldots, (n-1)\}$ when (i) n is prime, (ii) n is composite; (f) "followed by" on rotations about a given point; (g) multiplication on 2×2 matrices; (h) multiplication on $\{1, -1, i, -i\}$; (j) multiplication mod 10 on $\{1, 3, 7, 9\}$; (k) multiplication mod 10 on $\{1, 3, 5, 7, 9\}$.

4. Show that the following are groups:
(a) "followed by" on symmetry transformations of (i) a rectangle, (ii) an equilateral triangle; (b) multiplication mod 6 on $\{0, 1, 2, 3, 4, 5\}$; (c) "followed by" on rotations about a fixed point $\{0°, 60°, 120°, 180°, 240°, 300°\}$.

5. Construct a combination table for multiplication mod 15 on $\{1, 4, 7, 13\}$, and hence show that the structure is a group.

6. A is the set of positive rational numbers that can be written as x/y, where neither x nor y is divisible by 7.
(a) Is A closed with respect to the operation of addition?
(b) Is A closed with respect to the operation of multiplication?
(c) Is A a group with respect to multiplication?
B is the set of positive rational numbers that can be written as z/t where z is divisible by 7 but t is not. Answer questions (a), (b) and (c) above for the set B.
Give reasons for all your answers.   (L.)

7. Explain why the set of all complex numbers is neither a group under the operation of multiplication nor a group under the operation of subtraction.   (O.C.S.M.P.)

8. Three distinct elements a, b, c are subject to a law of combination $*$ for which

$$a * c = c * a = a, \tag{1}$$
$$b * c = c * b = b, \tag{2}$$
$$a * a = c. \tag{3}$$

Prove that they do not form a group.
Keeping the relations (1) and (2), suggest an alternative value for $a * a$ under which the three elements do form a group.   (O.C.S.M.P.)

9. Construct a combination table for addition mod 5 on $\{0, 1, 2, 3, 4\}$, and prove that the structure is a group.

10. The set S consists of all transformations T of the real numbers given by $T(x) = ax + b$ ($a \neq 0, a, b \in \mathbb{R}$).
Show that if $T_1, T_2 \in S$ then $T_1 T_2 \in S$. Prove that the set S forms a group under composition of functions. Find the elements T of S of order 2 (i.e. $TT = I$ and $T \neq I$ where I is the identity transformation). Prove that these elements are all reflections. Show that there are no elements of order 3 in S.   (N.)

11. Let G be the set of all matrices of the form $\begin{pmatrix} 1 & x \\ 0 & 1 \end{pmatrix}$, where $x \in \mathbb{R}$:

(i) Show that G does not form a group under matrix addition;
(ii) Show that G does form a group under matrix multiplication.
[You may assume associativity of matrix addition and multiplication.]   (N.)

12. Prove that within a group

$$ax = bx \Rightarrow a = b.$$

22:10 ISOMORPHISM

Naturally the form of a structure does not depend on the "labels" which we attach to its elements. If two structures are basically the same, we say that they are **isomorphic**. To show that two finite group structures are isomorphic we must ensure that we can carry out a consistent "relabelling" process on the elements of one structure so that its combination table can be made identical with that of the other structure. Formally, this requires a one-to-one correspondence between the elements of the two sets, including in particular the identity elements:

$$a \Leftrightarrow a^*, \ b \Leftrightarrow b^*, \ldots, \ e \Leftrightarrow e^*, \ldots, \ p \Leftrightarrow p^*, \ q \Leftrightarrow q^*, \ r \Leftrightarrow r^*, \ldots$$

such that

$$p \circ q = r \Leftrightarrow p^* * q^* = r^* \ \forall p, q \in S; \quad p^*, q^* \in S^*.$$

The "pattern" in the two combination tables will then be identical provided that the elements in the master row and column of one table are arranged in the same order as their corresponding elements in the other master row and column.

Clearly the concept of isomorphism is of great importance, since it means that when we have studied the properties of one particular group structure, we can apply our knowledge to any other structure to which it is isomorphic, even though the set and the operation involved may be different.

If two structures are of the same form but if the correspondence between elements is many-one: $a_1, a_2, \ldots, \leftrightarrow a^*, b_1, b_2, \ldots, \leftrightarrow b^*, \ldots$, we say that the structures are **homomorphic**: for example, addition on $\mathbb{N}$ is homomorphic to addition mod 3 on $\mathbb{N}$, where

$$5 + 7 = 12, \ 17 + 4 = 21, \ldots, \ \leftrightarrow 2 + 1 = 0 \ (\text{mod } 3) \ldots.$$

Example 1. Show that the set of matrices $\begin{pmatrix} \cos\theta & -\sin\theta \\ \sin\theta & \cos\theta \end{pmatrix}$ under multiplication is isomorphic to the set of complex numbers $\cos\theta + \mathrm{i}\sin\theta$ under multiplication. *Name* a set of geometrical transformations, together with an operation, which is isomorphic to each of these sets.   (O.C.S.M.P.)

Since $\begin{pmatrix} \cos\phi & -\sin\phi \\ \sin\phi & \cos\phi \end{pmatrix} \begin{pmatrix} \cos\theta & -\sin\theta \\ \sin\theta & \cos\theta \end{pmatrix} = \begin{pmatrix} \cos(\theta+\phi) & -\sin(\theta+\phi) \\ \sin(\theta+\phi) & \cos(\theta+\phi) \end{pmatrix}$

and $(\cos\phi + \mathrm{i}\sin\phi)(\cos\theta + \mathrm{i}\sin\theta) = \cos(\theta+\phi) + \mathrm{i}\sin(\theta+\phi)$, the two sets are isomorphic under multiplication.

The matrices represent rotations about O, so that these transformations under "followed by" are isomorphic to each of the given groups.

Example 2. Three indistinguishable coins are placed in a line on a table and the following operations defined on them.

P: turn the left-hand one over and interchange the other two; Q: turn the right-hand one over and interchange the other two. If I is the identity operation show that $P^2 = I$ and $Q^2 = I$. Show also that, if PQ means "P and then Q", then $PQ \neq QP$ and prove that $PQP = QPQ$.

Prove that only six possible operations are generated by P and Q, and state one isomorphism between the group so formed and the group of symmetries of an equilateral triangle.   (O.C.S.M.P.)

Let the two faces of each coin be represented by H_1, T_1; H_2, T_2; H_3, T_3 respectively. If $I = H_1 H_2 H_3$, $P = T_1 H_3 H_2$, $Q = H_2 H_1 T_3$

$$\Rightarrow P^2 = H_1 H_2 H_3 = I, \quad Q^2 = H_1 H_2 H_3 = I.$$
$$PQ = H_3 T_1 T_2, \quad QP = T_2 T_3 H_1 \neq PQ.$$
$$PQP = T_3 T_2 T_1 = QPQ.$$
$$PQPQ = QPQ^2 = QP, \quad PQP^2 = PQ.$$

Hence the only possible operations generated by P and Q are $\{I, P, Q, PQ, QP, PQP\}$, forming a group under composition. The symmetries of an equilateral triangle are anticlockwise rotation through $2\pi/3$ (T) and through $4\pi/3$ (T^2), reflections in the three axes of symmetry (R_1, R_2, R_3) and the identity transformation (I). (See the example on p. 759.)

The combination table for the structure is:

TABLE 22.6

	I	P	Q	PQ	QP	PQP
I	I	P	Q	PQ	QP	PQP
P	P	I	PQ	Q	PQP	QP
Q	Q	QP	I	PQP	P	PQ
PQ	PQ	PQP	P	QP	I	Q
QP	QP	Q	PQP	I	PQ	P
PQP	PQP	PQ	QP	P	Q	I

22:11 SUBGROUPS: LAGRANGE'S THEOREM

If $G \equiv (\mathcal{O}, S)$, it may be that there is a subset $S_1 \subset S$ such that $(\mathcal{O}, S_1)$ is a group, G_1. We call G_1 a **subgroup** of G; clearly S_1 must include e the (unique) identity element of G, and $(\mathcal{O}, \{e\})$ must always be a subgroup of G, as must G itself. We refer to **proper subgroups** (cf. proper subsets) as excluding these two trivial cases.

An important theorem due to Lagrange, which we shall not prove, states that a group of order n can only have (but does not **necessarily** have) proper subgroups of orders which are factors of n. Thus if n is prime, the group can have no proper subgroup. Clearly Lagrange's theorem will be useful in any consideration of subgroups, but we postpone further discussion of these until later.

Example. Explain, with reasons, why the symmetries of an equilateral triangle form a group of order 6, with a suitable operation which should be clearly stated.

[If you define any of the elements of the group as a rotation (or reflection) you should explain whether you are regarding it as a rotation about a line fixed in the triangle or a line fixed in space.]

Give the group operation table and identify *all* the subgroups.

If G is the largest subgroup with less than six elements, find a subgroup of the complex numbers under multiplication which is isomorphic to G.   (C.)

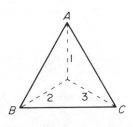

FIG. 22.3.

The symmetries of an equilateral triangle are as stated in Example 2, §22:10, where in Fig. 22.3 the axes 1, 2, 3 are to be considered as fixed in space; carrying out any two symmetry transformations in succession will be equivalent to a single transformation, for example T followed by R_1 ($R_1 T$) is equivalent to R_2. The combination table is shown in Table 22.7(i), and for comparison the table from Example 2, is shown rearranged in (ii).

TABLE 22.7

	I	T	T^2	R_1	R_2	R_3
I	I	T	T^2	R_1	R_2	R_3
T	T	T^2	I	R_2	R_3	R_1
T^2	T^2	I	T	R_3	R_1	R_2
R_1	R_1	R_3	R_2	I	T^2	T
R_2	R_2	R_1	R_3	T	I	T^2
R_3	R_3	R_2	R_1	T^2	T	I

(i)

	I	QP	PQ	P	Q	PQP
I	I	QP	PQ	P	Q	PQP
QP	QP	PQ	I	Q	PQP	P
PQ	PQ	I	QP	PQP	P	Q
P	P	PQP	Q	I	PQ	QP
Q	Q	P	PQP	QP	I	PQ
PQP	PQP	Q	P	PQ	QP	I

(ii)

The group properties can be verified from (i); T and T^2 are inverses, all reflections are self-inverse. Subgroups are the group itself, $\{I\}$, and proper subgroups formed from the subsets $\{R_1, I\}$, $\{R_2, I\}$, $\{R_3, I\}$ and $\{T, T^2, I\}$ (G).

If ω is a complex cube root of $z^3 = 1$, the set $\{\omega, \omega^2, 1\}$ forms a group under multiplication which is isomorphic to G.

Comparison of (i) and (ii) shows that the two groups are isomorphic.

Exercise 22.11

1. In the group G of symmetry transformations of the square let R denote a 90° rotation about O, I the identity transformation, X and Y reflections in the axes indicated in the figure.

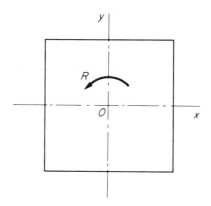

FIG. 22.4.

Show that $\{I, X, Y, R^2\}$ is a subgroup of G of order 4. Name another subgroup of order 4 which is not isomorphic to this subgroup, giving a simple reason for the lack of isomorphism.

Prove that if A and B are two subgroups of any group H, then $A \cap B$ is also a subgroup of H. Deduce a subgroup of G of order 2.

2. T is the set of triangular matrices of the form

$$\begin{pmatrix} 1 & a \\ 0 & 1 \end{pmatrix}$$

where $a \in \mathbb{Z}$. Prove that T is a group with respect to matrix multiplication and that this group is isomorphic with the additive group of integers. (The associativity of matrix multiplication may be assumed.)

Is the same true for the set of triangular matrices of the form

$$\begin{pmatrix} a & 1 \\ 0 & a \end{pmatrix}$$

with $a \in \mathbb{Z}$? Justify your answer.   (L.)

3. Show that the set $\{1, -1, i, -i\}$ under multiplication is a group and that the set of matrices $\{A, B, C, D\}$, where

$$A = \begin{pmatrix} 0 & 0 \\ 1 & 1 \end{pmatrix}, \qquad B = \begin{pmatrix} 0 & 0 \\ -1 & -1 \end{pmatrix},$$

$$C = \begin{pmatrix} 0 & 0 \\ i & i \end{pmatrix}, \qquad D = \begin{pmatrix} 0 & 0 \\ -i & -i \end{pmatrix},$$

under matrix multiplication is also a group. Are the two groups isomorphic? Give reasons to justify your answer.

[The associativity of multiplication may be assumed in both cases.]   (L.)

4. Show whether or not multiplication mod 6 on $\{1, 2, 3, 4, 5\}$ is a group. Find any subgroups.

5. A binary operation $\otimes$ is defined on the set of all ordered pairs (x, y) of real numbers such that $y \neq 0$, by setting

$$(x_1, y_1) \otimes (x_2, y_2) = (x_1 + y_1 x_2, y_1 y_2).$$

Prove that the set of pairs with this operation forms a group.

Is the group abelian?

Does the subset consisting of all pairs of the form $(0, y)$ with $y \neq 0$, form a subgroup?   (O.C.S.M.P.)

6. Let $A \Delta B = \{x : x \in A, x \notin B\} \cup \{x : x \in B, x \notin A\}$. Prove that the set of all subsets of $\{A, B, C\}$ forms a group under Δ of order 8. Find the proper subgroups.

7. Show that $\{1, 3, 7, 9\}$ is a group under multiplication mod 10. Find any subgroups.

8. Show that product of rotations on $\{\frac{1}{2}\pi, \pi, \frac{3}{2}\pi, 2\pi\}$ is isomorphic to matrix multiplication on

$$\left\{ \begin{pmatrix} 0 & -1 \\ 1 & 0 \end{pmatrix}, \begin{pmatrix} -1 & 0 \\ 0 & -1 \end{pmatrix}, \begin{pmatrix} 0 & 1 \\ -1 & 0 \end{pmatrix}, \begin{pmatrix} 1 & 0 \\ 0 & 1 \end{pmatrix} \right\}.$$

Show that each group is isomorphic to addition mod 4 on $\{0, 1, 2, 3\}$.

9. Show that the following are groups:

(i) the set G of complex numbers z such that $|z| = 1$, under the operation of multiplication;

(ii) the set H of real numbers x such that $0 \leqslant x < 2\pi$, under the operation $\circ$ defined by:

$$x \circ y = r,$$

where r is the number in H satisfying

$$x + y = 2k\pi + r$$

for some integer k.

Using the fact that each element of G can be uniquely expressed in the form $\cos x + i \sin x$, where $x \in H$, show that the groups G and H are isomorphic.   (C.)

10. A group G consists of six elements $h_1, h_2, h_3, u_1, u_2, u_3$, where h_1, h_2, h_3 form a sub-group H and the elements u_1, u_2, u_3 form a set U. Prove that the products within G denoted by

$$h_i u_j \quad \text{and} \quad u_i h_j \quad (i, j = 1, 2, 3)$$

are members of U and that the products

$$u_i u_j$$

are members of H.   (C.)

11. Prove carefully from the definition that, under the operation of addition modulo p, the p numbers $0, 1, 2, \ldots, p-1$ form a group. Call this group G_p.

Find a sub-group of G_{96} that is isomorphic to G_4, and prove that there is no sub-group of G_{96} isomorphic to G_9.

Determine whether there is a proper sub-group of G_{96} isomorphic to a subgroup of G_9.   (C.)

12. The operation of *multiplication modulo* 15 is defined on the set of integers by defining the product "mn" of two integers m and n to be the remainder after dividing the ordinary arithmetical product by 15. Prove that the set $\{3, 6, 9, 12\}$, together with this rule of multiplication modulo 15, is a group.

Is this group isomorphic with the group consisting of the four numbers $1, -1, i, -i$ subject to the operation of ordinary multiplication of complex numbers?   (O.C.S.M.P.)

13. Show that there is an infinite number of infinite subgroups in the group formed by multiplication on $\{x : x = \mathbb{Q}, x \neq 0\}$.

14. Copy and complete the following multiplication table for a group G with four elements e, a, b, c.

	e	a	b	c
e	e	a	b	c
a	a	b	.	.
b	b	.	.	.
c	c	.	.	.

Write down a set of four complex numbers which, with respect to ordinary multiplication of complex numbers, forms a group isomorphic with G.

Another group G' has four elements E, A, B, C with multiplication table

	E	A	B	C
E	E	A	B	C
A	A	E	C	B
B	B	C	E	A
C	C	B	A	E

Show that G' is not isomorphic with G.   (L.)

15. Three tumblers are placed side by side on a table; they are altered by turning any two over at once. In how many ways can this be done?

If these alterations together with the identity alteration, which leaves the tumblers alone, form the set A, and if $\circ$ is the operation "followed by", compile an operation table for $(A, \circ)$. Explain carefully why you recognize it as a group table.

Sketch geometrical figures whose complete group of symmetries
(i) is isomorphic to $(A, \circ)$;
(ii) has a proper subgroup isomorphic to $(A, \circ)$.   (O.C.S.M.P.)

16. a and b are elements of a group with identity element e. If $a^3 = b^2 = e$ and $ba = a^2 b$,
(i) prove that $(ba)^2 = e$;
(ii) find the simplest form of $(bab)^{-1}$.   (O.C.S.M.P.)

17. The set of ordered pairs (a, b), where a, b are real numbers and $a \neq 0$, forms a group under the

operation ∗ defined by

$$(a_1, b_1) * (a_2, b_2) = (a_1 a_2, a_1 b_2 + b_1).$$

Find: (i) the identity element; and (ii) the inverse of (a, b).   (O.C.S.M.P.)

18. Prove that the integers

$$\{\ldots, -3, -2, -1, 0, 1, 2, \ldots\}$$

form a group $\{G, +\}$ under the operation of ordinary addition.
    Give an example of a proper subgroup.
    Prove that the powers of 2

$$\{\ldots, \tfrac{1}{8}, \tfrac{1}{4}, \tfrac{1}{2}, 1, 2, 4, \ldots\}$$

form a group $\{H, \times\}$ under the operation of ordinary multiplication.
    Establish an isomorphism between $\{G, +\}$ and $\{H, \times\}$, and give the sub-group of H corresponding under this isomorphism to your previous subgroup of G.   (C.)

19. T is the set of all 2×2 real matrices

$$\begin{pmatrix} a & b \\ c & d \end{pmatrix} \text{ with } ad - bc = 1.$$

Prove that T is a group under matrix multiplication.
    U is the set of 2×2 matrices:

$$U = \left\{ \begin{pmatrix} p & -q \\ q & p \end{pmatrix} : \ p, q \text{ real but not both zero} \right\}.$$

Show that U is also a group under matrix multiplication. Establish whether either T or U is abelian.
    Four more sets of matrices are defined as follows:

$$V = \left\{ \begin{pmatrix} p & 0 \\ 0 & p \end{pmatrix} : \ p \neq 0, p \text{ real} \right\},$$

$$W = \left\{ \begin{pmatrix} 0 & -q \\ q & 0 \end{pmatrix} : \ q \neq 0, q \text{ real} \right\},$$

$$X = \left\{ \begin{pmatrix} r & 0 \\ 0 & 1/r \end{pmatrix} : \ r \neq 0, r \text{ real} \right\},$$

$$Y = \left\{ \begin{pmatrix} p & -q \\ q & p \end{pmatrix} : \ p^2 + q^2 = 1; p, q \text{ real} \right\}.$$

For each of V, W, X and Y, under matrix multiplication, state whether (a) it is a group, (b) it is a subgroup of T, (c) it is a subgroup of U.   (O.C.S.M.P.)

20. Points $A, B, C, \ldots$, in the euclidean plane, are represented by their position vectors $\mathbf{a}, \mathbf{b}, \mathbf{c}, \ldots$. H_a denotes the half-turn with centre A, and T_c the translation in which $O \to C$. Show that
    (i) $T_a T_b = T_{(a+b)}$;
    (ii) $H_a T_b = H_{(a-\frac{1}{2}b)}$;
    (iii) $H_b H_a = T_{2(b-a)}$.
    (Note. FG means that the transformation G acts first and is then followed by F.)
    Write down the inverses of H_a and T_c and deduce that the set of all translations and half-turns forms a group.
    Prove that $H_a H_b H_c$ is a half-turn. Hence, or otherwise, prove that

$$H_a H_b H_c = H_c H_b H_a.$$ (O.C.S.M.P.)

21. Find elements which with 5 form a group of order 8 under multiplication modulo 24, and give the combination table.

22. $p \rightarrow q$ is a conditional proposition (I); we define the converse (C) as $q \rightarrow p$, the inverse (V) as $\sim p \rightarrow \sim q$, and the contrapositive (P) as $\sim q \rightarrow \sim p$. Show that the four propositions form a group under "succession", where for instance $CV \equiv \sim q \rightarrow \sim p$. State the converse of the contrapositive of the inverse of I, and the inverse of the converse of the contrapositive of I.

22:12 CYCLIC GROUPS

Choose any element a, other than the identity element e, from the set S_n of a finite group $G_n \equiv (\mathcal{O}, S_n)$ of order n. Then if $a \mathcal{O} a = b$ we may write $b = a^2$. Again if $a \mathcal{O} a \mathcal{O} a = a \mathcal{O} b = c$ we may write $c = a^3$, and so on. This process must terminate; the powers of a cannot all be distinct as S_n is finite, and so there exist integers $t > s \geqslant 1$ such that $a^t = a^s$ and so $a^{t-s} = e$. Let r be the smallest positive integer such that $a^r = e$. Then two cases can arise; either $1 < r < n$ or $r = n$. Notice that the elements $a, a^2, \ldots, a^{r-1}, e$ must be distinct.

In either case, we have expressed the elements of a subset of S_n, including e, in terms of a single element. When $r = n$, this subset is S_n itself, and we say that G_n is a **cyclic group** with $\mathcal{O}$ defined on $\{a, a^2, a^3, \ldots, a^{n-1}, e\}$; when $r < n$ we have a subset S_r such that $G_r \equiv (\mathcal{O}, S_r)$ is clearly a subgroup of G_n, since, in $\{a, a^2, \ldots, a^{r-1}, e\}$, a^j is the inverse of a^{r-j}. G_r is a cyclic subgroup of G_n; we say that G_r is **generated** by a, and that a is a **generator** of order r. The process of choosing a generator a and finding the smallest r such that $a^r = e$ provides a practical method of looking for subgroups. Clearly for any $n \in \mathbb{N}$ we can construct a cyclic group of order n as shown in Table 22.8; such a group must be Abelian since $a^r a^s = a^s a^r$.

TABLE 22.8

	e	a	a^2	a^3	$\ldots$	a^{n-2}	a^{n-1}
e	e	a	a^2	a^3	$\ldots$	a^{n-2}	a^{n-1}
a	a	a^2	a^3	a^4	$\ldots$	a^{n-1}	e
a^2	a^2	a^3	a^4	a^5	$\ldots$	e	a
a^3	a^3						
$\vdots$	$\vdots$						
a^{n-2}	a^{n-2}						
a^{n-1}	a^{n-1}	e	a	a^2	$\ldots$	a^{n-3}	a^{n-2}

By Lagrange's theorem, if G_r is a subgroup of G_n, r is a factor of n. Hence if n is prime, the cyclic group structure must be the only possible structure for a group of order n, so that every group of this order must be isomorphic to the cyclic group.

Example 1. Determine the proper subgroups of the symmetry group for the square (see Table 22.4, p. 755).

For subgroups of order 2 we require generators of order 2 such that $a^2 = e$. In this case we have

$$(Q^2)^2 = R_x^2 = R_y^2 = R_1^2 = R_2^2 = I;$$

hence subgroups of order 2 are formed by

$$\{Q^2, I\}, \{R_x, I\}, \{R_y, I\}, \{R_1, I\}, \{R_2, I\}.$$

We know that there is a cyclic subgroup of order 4 which is clearly formed by $\{Q, Q^2, Q^3, I\}$; any other subgroup of order 4 must contain generators of order 2. Such subgroups are $\{Q^2, R_x, R_y, I\}$, $\{Q^2, R_1, R_2, I\}$. Notice that these are symmetry groups for the rectangle and the rhombus respectively which are clearly isomorphic.

Example 2. Form a cyclic group of order 3.
   If ω is one of the complex cube roots of 1, multiplication on $\{1, \omega, \omega^2\}$ is such a group.

Example 3. Form a cyclic rotational group of order n, when n is prime.
   If T represents a rotation about O through $2\pi/n$, $T^n = 2\pi$, and the product of rotations on $\{T, T^2, T^3, \ldots, T^{n-1}, 2\pi\}$ is a cyclic group of order n.

Example 4. If A, B, C, D are four collinear points and if the cross-ratio $(ABCD) = x$, it can be shown that the only other possible values of the cross-ratio, when the points are taken in a different order, are

$$1-x, \quad \frac{1}{x}, \quad 1-\frac{1}{x}, \quad \frac{1}{1-x}, \quad \frac{x}{x-1}.$$

If $\qquad\qquad f_1: x \to x, \; f_2: x \to 1-x, \ldots, f_6: x \to \dfrac{x}{x-1}$,

show that $\{f_1, f_2, f_3, f_4, f_5, f_6\}$ forms a group under composition of functions, and find any subgroups.

   The combination table is

TABLE 22.9

	f_1	f_2	f_3	f_4	f_5	f_6
f_1	f_1	f_2	f_3	f_4	f_5	f_6
f_2	f_2	f_1	f_4	f_3	f_6	f_5
f_3	f_3	f_5	f_1	f_6	f_2	f_4
f_4	f_4	f_6	f_2	f_5	f_1	f_3
f_5	f_5	f_3	f_6	f_1	f_4	f_2
f_6	f_6	f_4	f_5	f_2	f_3	f_1

   The table shows closure and the nature of the operation implies associativity; f_1 is the identity element, and each element has an inverse. Hence this structure is a group.
   Since $f_2^2 = f_3^2 = f_6^2 = f_1$, there are three subgroups of order 2 given by

$$\{f_2, f_1\}, \{f_3, f_1\}, \{f_6, f_1\}.$$

Since

$$f_4^2 = f_5 \text{ and } f_4^3 = f_1,$$

$\{f_4, f_5, f_1\}$ forms a subgroup of order three, isomorphic to the cyclic subgroup.
   There are in fact just two different structures forming groups of order 6, the other one being of course the cyclic group.

Exercise 22.12

1. Show that $\{1, 2, 4, 8\}$ under multiplication mod 15 forms a cyclic group; find a non-cyclic group of order 4 under the same operation.

2. Show that $\{1, 3, 5, 9, 11, 13\}$ under multiplication mod 14 forms a cyclic group.

3. The group G with operation $*$ is a cyclic group of order n with generator g. Given that $g^5 = g^{17}$, determine, with reasons, the set of possible values of n.

In the case $n = 6$, determine the order of each element of G and list all the generators of G.   (C.)

4. Let Z_6 denote the set of integers $\{0, 1, 2, 3, 4, 5\}$ and $+_6$ denote addition modulo 6 (so that $a +_6 b$ is the remainder when $a + b$ is divided by 6). Show that, with this operation, Z_6 is a cyclic group with exactly two generators.

Describe all the homomorphisms of Z_6 under $+_6$ into the group $Z_3 = \{0, 1, 2\}$ under $+_3$ (addition modulo 3).   (C.)

5. If

$$f : x \to \frac{x - 1}{x + 1},$$

show that $f^5 : x \to x$. Give the combination table and show that the structure is a cyclic group of order 5.

6. Prove that every subgroup of a cyclic group is cyclic.

7. Find a cyclic group of order 6 under multiplication mod 18 in which the identity element is 10.

8. Prove that the order of any element in a group is equal to the order of its inverse.

9. Show that 2 is a generator of order 12 for a cyclic group under multiplication modulo 13.

10. Show that the matrix

$$\mathbf{T} = \frac{1}{2} \begin{pmatrix} 1 & -\sqrt{3} \\ \sqrt{3} & 1 \end{pmatrix}$$

under matrix multiplication is a generator for a cyclic group of order 6. Find the proper subgroups of this group.

11. Find non-zero values for a, b, c, d such that $g : x \to \dfrac{ax + b}{cx + d}$ generates a cyclic group of order 2.

22:13 PERMUTATION GROUPS

Any arrangement of a finite set of elements is a **permutation** of these elements; it is convenient to use 1, 2, 3, . . . as a general representation of these elements, where we take the initial arrangement to be the numerical order. So there is an identity permutation of order n which may be written

$$\begin{bmatrix} 1 & 2 & 3 \ldots (n-1) & n \\ 1 & 2 & 3 \ldots (n-1) & n \end{bmatrix},$$

where the upper row remains in the original order and maps onto the lower row.

Consider the set $\{1, 2, 3, 4, 5, 6\}$; the permutation

$$\begin{bmatrix} 1 & 2 & 3 & 4 & 5 & 6 \\ 2 & 5 & 1 & 4 & 6 & 3 \end{bmatrix}$$

can be described "$1 \to 2, 2 \to 5, 5 \to 6, 6 \to 3, 3 \to 1$" (implying that $4 \to 4$) and can be written more concisely $(1 \quad 2 \quad 5 \quad 6 \quad 3)$ where this is understood to be a cyclic mapping, with 4

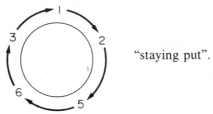

"staying put".

Another permutation is

$$\begin{bmatrix} 1 & 2 & 3 & 4 & 5 & 6 \\ 4 & 3 & 2 & 1 & 6 & 5 \end{bmatrix}$$

where $1 \leftrightarrow 4$, $2 \leftrightarrow 3$ and $5 \leftrightarrow 6$; this can be written (1 4) (2 3) (5 6).

We can form the "product" of these permutations by the mapping

$$1 \to 2 \to 3, \; 3 \to 1 \to 4, \; 4 \to 4 \to 1, \; 2 \to 5 \to 6, \; 6 \to 3 \to 2, \; 5 \to 6 \to 5,$$

i.e.   (1 4) (2 3) (5 6) . (1 2 5 6 3) = (1 3 4) (2 6).

Thus we have an operation defined on permutations. There are $n!$ permutations of n elements, including an identity permutation, and the set is clearly closed under the "product" operation. The associative law holds, and each permutation will have an inverse, in which each "cycle" of more than two elements will be reversed in sense, for example, (1 2 5 6 3)$^{-1}$ = (1 3 6 5 2). The permutations of n elements therefore form a group of order $n!$, which is called the **symmetric group of degree** n.

We might expect a permutation group to have subgroups; at any rate Lagrange's theorem ensures their possibility since $n!$ cannot be prime for $n > 2$. In fact, a theorem due to Cayley states that every finite group of order n is isomorphic to a subgroup of the symmetric group of degree n.

To prove the theorem, we must first set up a one-to-one correspondence between the n elements in $G_n \equiv (\circ, S)$ and n permutations of $1, 2, 3, \ldots, n$; to do so, choose an order $a_1, a_2, a_3, \ldots, a_n \; \forall \, a \in S$ and then make

$$a_j \leftrightarrow \begin{bmatrix} 1 & 2 & 3 & \ldots & n \\ a_1 \circ a_j & a_2 \circ a_j & a_3 \circ a_j & \ldots & a_n \circ a_j \end{bmatrix} = p_j$$

and in particular

$$e \leftrightarrow \begin{bmatrix} 1 & 2 & 3 & \ldots & n \\ a_1 & a_2 & a_3 & \ldots & a_n \end{bmatrix}$$

(knowing that in a group structure $a_1 \circ a_j, \ldots a_n \circ a_j$ are all distinct).

To show that the structures are isomorphic we must show that if $a_j \circ a_k = a_r$, then $p_k p_j = p_r$. But

$$a_r \leftrightarrow \begin{bmatrix} 1 & 2 & \ldots & n \\ a_1 \circ (a_j \circ a_k) & a_2 \circ (a_j \circ a_k) & & a_n \circ (a_j \circ a_k) \end{bmatrix} = p_r$$

$$\leftrightarrow \begin{bmatrix} 1 & \ldots & n \\ (a_1 \circ a_j) \circ a_k & \ldots & (a_n \circ a_j) \circ a_k \end{bmatrix} \; \text{(associativity in } G_n)$$

$$\leftrightarrow p_k p_j.$$

Hence the set of permutations thus chosen forms a group isomorphic to G_n.

Example. Find permutation groups isomorphic to each subgroup of order four in the symmetry group for the square. (Example, p. 754.)

Take the elements of the cyclic subgroup $\{Q, Q^2, Q^3, I\}$ in the given order; then according to the scheme explained above,

$$p_1 = \begin{bmatrix} 1 & 2 & 3 & 4 \\ Q.Q & Q^2.Q & Q^3.Q & I.Q \end{bmatrix} = \begin{bmatrix} 1 & 2 & 3 & 4 \\ 2 & 3 & 4 & 1 \end{bmatrix} = (1 \quad 2 \quad 3 \quad 4);$$

$$p_2 = \begin{bmatrix} 1 & 2 & 3 & 4 \\ Q.Q^2 & Q^2.Q^2 & Q^3.Q^2 & I.Q^2 \end{bmatrix} = \begin{bmatrix} 1 & 2 & 3 & 4 \\ 3 & 4 & 1 & 2 \end{bmatrix} = (1 \quad 3)(2 \quad 4);$$

$$p_3 = \begin{bmatrix} 1 & 2 & 3 & 4 \\ Q.Q^3 & Q^2.Q^3 & Q^3.Q^3 & I.Q^3 \end{bmatrix} = \begin{bmatrix} 1 & 2 & 3 & 4 \\ 4 & 1 & 2 & 3 \end{bmatrix} = (1 \quad 4 \quad 3 \quad 2);$$

$$p_4 = \begin{bmatrix} 1 & 2 & 3 & 4 \\ Q.I & \cdots & \cdots & \cdots \end{bmatrix} = \begin{bmatrix} 1 & 2 & 3 & 4 \\ 1 & 2 & 3 & 4 \end{bmatrix}.$$

Note that p_1 and p_3 are inverses under product of permutations, p_2 is self-inverse, and p_4 is the identity permutation. Comparing combination tables, the isomorphism of the two structures is clear.

TABLE 22.10

	Q	Q^2	Q^3	I			p_1	p_2	p_3	p_4
Q	Q^2	Q^3	I	Q		p_1	p_2	p_3	p_4	p_1
Q^2	Q^3	I	Q	Q^2		p_2	p_3	p_4	p_1	p_2
Q^3	I	Q	Q^2	Q^3		p_3	p_4	p_1	p_2	p_3
I	Q	Q^2	Q^3	I		p_4	p_1	p_2	p_3	p_4

(i)   (ii)

For the Klein symmetry subgroup, take the elements in the order $\{Q^2, R_x, R_y, I\}$; then

$$p_1 = \begin{bmatrix} 1 & 2 & 3 & 4 \\ Q^2.Q^2 & R_x.Q^2 & R_y.Q^2 & I.Q^2 \end{bmatrix} = \begin{bmatrix} 1 & 2 & 3 & 4 \\ 4 & 3 & 2 & 1 \end{bmatrix} = (1 \quad 4)(2 \quad 3)$$

$$p_2 = \begin{bmatrix} 1 & 2 & 3 & 4 \\ Q^2.R_x & R_x.R_x & R_y.R_x & I.R_x \end{bmatrix} = \begin{bmatrix} 1 & 2 & 3 & 4 \\ 3 & 4 & 1 & 2 \end{bmatrix} = (1 \quad 3)(2 \quad 4)$$

$$p_3 = \begin{bmatrix} 1 & 2 & 3 & 4 \\ Q^2.R_y & R_x.R_y & R_y.R_y & I.R_y \end{bmatrix} = \begin{bmatrix} 1 & 2 & 3 & 4 \\ 2 & 1 & 4 & 3 \end{bmatrix} = (1 \quad 2)(3 \quad 4)$$

$$p_4 = \begin{bmatrix} 1 & 2 & 3 & 4 \\ 1 & 2 & 3 & 4 \end{bmatrix}.$$

Here each permutation is self-inverse, and again the two structures are clearly isomorphic.

TABLE 22.11

	Q^2	R_x	R_y	I			p_1	p_2	p_3	p_4
Q^2	I	R_y	R_x	Q^2		p_1	p_4	p_3	p_2	p_1
R_x	R_y	I	Q^2	R_x		p_2	p_3	p_4	p_1	p_2
R_y	R_x	Q^2	I	R_y		p_3	p_2	p_1	p_4	p_3
I	Q^2	R_x	R_y	I		p_4	p_1	p_2	p_3	p_4

(i)　　　　　　　　　　　　　(ii)

Exercise 22.13

1. Form the cyclic subgroups for the group of permutations of $\{a, b, c, d\}$.

2. Construct the group table for the permutations of $\{a, b, c\}$.

3. Find permutations of A, B, C, D to correspond to the six different values of the cross-ratio of four collinear points. Verify that these form a group.

22:14 SPECIAL GROUPS

We can use our knowledge of cyclic groups and properties of a group structure to consider particular structures of low order:

(i) **Order 2.** If the elements are $\{a, e\}$ we know that $a^2 = e$, giving a unique group structure as shown in Table 22.12(i).

TABLE 22.12

	e	a			e	a	a^2
e	e	a		e	e	a	a^2
a	a	e		a	a	a^2	e
				a^2	a^2	e	a

(i)　　　　　　　　　　　　　(ii)

Thus all groups of order two are isomorphic.

(ii) **Order 3.** Table 22.12(ii) shows the cyclic group, and, since 3 is prime, any group of this order must be isomorphic to this one.

(iii) **Order 4:** We know that there is a cyclic group of this order with elements that can be labelled $\{e, a, a^2, a^3\}$, for which the group table will be as in Table 22.13(i).

Table 22.13

	e	a	a^2	a^3
e	e	a	a^2	a^3
a	a	a^2	a^3	e
a^2	a^2	a^3	e	a
a^3	a^3	e	a	a^2

(i)

	e	a	b	c
e	e	a	b	c
a	a	e		
b	b		e	
c	c			e

(ii)

	e	a	b	c
e	e	a	b	c
a	a	e	c	b
b	b	c	e	a
c	c	b	a	e

(iii)

The only other possible group structure will contain elements a, b, c of order 2 so that $a^2 = b^2 = c^2 = e$. We can thus begin to compile the table as in (ii), and it is easy to verify that the table can only be completed in one way, (iii), to ensure that each element occurs just once in each row and column. We still cannot say that this is a group structure until we have checked for associativity; this can easily be done, particularly as the table is symmetrical about its leading diagonal. This structure is known as the **Klein group**.

The Klein four group has already arisen in this chapter. The subgroups $\{Q^2, R_x, R_y, I\}$ and $\{Q^2, R_1, R_2, I\}$ of the symmetry groups of the square, Example 1, p. 764, are examples. Either of these groups has the given multiplication table.

(iv) **Order 6.** Besides the cyclic group, there is only one other, as found in Example 4 (p. 765).

22:15 A FIELD

Since in the algebra of numbers the two most frequently employed operations are addition and multiplication, and since multiplication is distributive over addition but not vice versa, it is convenient to define a kind of "double-group" structure embodying the group properties of both operations and the distributivity of one over the other. There is a difficulty here, however, since the identity element (0) required for a group under addition will not have an inverse under multiplication; we therefore exclude it from the set on which multiplication is defined, and to indicate this will write the set as S^-. For convenience, the two operations involved in our formal definition are referred to as "addition" and "multiplication", and the symbols "$+$" and "$.$" are used for them; but any operations having the required properties can replace them.

A **field** is defined as a structure with the following properties:
 (i) $(+, S)$ is an Abelian group;
 (ii) $(., S^-)$ is an Abelian group;
 (iii) "." is distributive over "+".
For convenience we may say "S is (or is not) a field", with the assumption that the operations are addition and multiplication.

Example: (i) $\mathbb{R}$ is a field;
             (ii) $\mathbb{Q}$ is a field [a "subfield" of (i)];
             (iii) $\mathbb{C}$ is a field.

The above can easily be verified.
             (iv) Addition and scalar multiplication on two-dimensional vectors is not a field, since the scalar product of vectors is not a vector and thus the set is not closed.
             (v) Addition and vector multiplication on two-dimensional vectors is not a field, since vector multiplication is not commutative.

22:16 A RING

Another structure is defined by somewhat relaxing the requirements for a field; this is a **ring**, which can loosely be termed "a group and a half", since properties (i) and (iii) of a field still hold, but instead of (ii) we only require "multiplication" on S to be closed and associative. If "multiplication" on S is also commutative we say that the ring is **commutative**.

Although $\mathbb{Z}$ is not a field, since integers have no inverses under multiplication, we can see that $\mathbb{Z}$ is a ring.

Example 1. Name a "subring" of $\mathbb{Z}$.
 The even integers form a ring, since the set is closed under addition and multiplication and the other requirements are clearly satisfied; but of course the odd integers are not closed under addition.

Example 2. Show that $\{n \times n \text{ matrices}\}$ form a ring for any $n \in \mathbb{N}$.
 The set forms an Abelian group under addition on matrices since (say)

$$\begin{pmatrix} a & b \\ c & d \end{pmatrix} \text{ has the inverse } \begin{pmatrix} -a & -b \\ -c & -d \end{pmatrix}.$$

Multiplication on matrices is closed and associative on these sets (but not of course commutative).

Exercise 22.16

 1. Prove that $\{0, 1, 2, 3, 4\}$ is a field under addition and multiplication modulo 5.

 2. Prove that $\{0, 1, 2, 3\}$ is a commutative ring under addition and multiplication modulo 4.

 3. Prove that the set of all diagonal 2×2 matrices is a field under matrix addition and multiplication.

 4. Determine the kind of structure formed by the set of all 2×2 matrices under matrix addition and multiplication.

 5. Identify the structures formed by:

             (a) $\mathbb{Z}$ under $+, \times$;   (b) $\{E, 0\}$ under $+, \times$.

6. Let $S = \{(a, b): a, b \in \mathbb{R}\}$. Define $(a, b) = (c, d)$ if and only if $a = c$, $b = d$; $(a, b) + (c, d) = (a + c, b + d)$; $(a, b) . (c, d) = (ac - bd, ad + bc)$. Show that S is a field under $+$ and $.$ defined as above.

7. Show that $\mathbb{C}$ is a field under addition and multiplication of complex numbers.

22:17 A VECTOR SPACE

In order to make satisfactory use of vectors we have so far defined (somewhat loosely) two operations on them; addition of vectors ($\oplus$) with substraction ($-$) as its inverse, and multiplication ($p\mathbf{a}$) by a real number (a scalar). We now consider such a structure from a more formal point of view, to see just what our definitions have been taken to imply; for example, let V_3 be the set of three-dimensional vectors, and check that this set forms a commutative group under vector addition. We have already seen that $\mathbb{R}$ is a field under addition and multiplication, so that our structure combines a group and a field; it remains to investigate the combination of $\mathbf{a}, \mathbf{b} \in V_2$ and $p, q \in \mathbb{R}$, which we can write $p \circ \mathbf{a}$. We have already assumed the following properties of this operation which we now formally state:

(i) $(p + q) \circ \mathbf{a} = p\mathbf{a} + q\mathbf{a}$;

(ii) $p \circ (\mathbf{a} \oplus \mathbf{b}) = p\mathbf{a} \oplus p\mathbf{b}$;

(iii) $p(q\mathbf{a}) = pq\mathbf{a}$;

(iv) $1 \circ \mathbf{a} = \mathbf{a}$.

Here (i) and (ii) form a kind of "double distributive law"—but using $\oplus$ for vector addition emphasises that this is not a case of mutual distributivity between addition and multiplication; (iii) is an associative law involving a product of two or more scalars, and (iv) requires that the identity element of the field (1 in our example) also acts as an identity element under $\circ$. We call this formalised structure a **vector space**, but having used our experience with vectors to suggest the form of the structure we do not restrict either the group to vector addition or the field to $\mathbb{R}$; a vector space is any structure in which an Abelian group and a field are combined under an operation with the above properties.

If (V, F) is a vector space, and if a subset $V_s \subset V$ forms a vector space with F, then (V_s, F) is a **vector subspace** of (V, F); for example (V_2, F) is a subspace of (V_3, F).

Given a vector $\mathbf{v} \in V_2$ and a subset $\{\mathbf{a}, \mathbf{b}, \mathbf{c}, \mathbf{d}\}$ of V_2, it may be possible to express $\mathbf{v}$ in the form

$$\mathbf{v} = p_1\mathbf{a} + p_2\mathbf{b} + p_3\mathbf{c} + p_4\mathbf{d},$$

where $p_1, p_2, p_3, p_4 \in \mathbb{R}$.

For example $\quad \begin{pmatrix} 2 \\ 5 \end{pmatrix} = 5\begin{pmatrix} 1 \\ 1 \end{pmatrix} - 3\begin{pmatrix} -1 \\ 2 \end{pmatrix} - 2\begin{pmatrix} 3 \\ 0 \end{pmatrix} - 6\begin{pmatrix} 0 \\ -1 \end{pmatrix}.$

It may also be possible to do this leaving out some of the vectors $\mathbf{a}, \mathbf{b}, \mathbf{c}, \mathbf{d}$, say

$$\mathbf{v} = q_1\mathbf{a} + q_2\mathbf{b} + q_3\mathbf{c},$$

where $q_1, q_2, q_3 \in \mathbb{R}$.

For example,
$$\begin{pmatrix} 2 \\ 5 \end{pmatrix} = \begin{pmatrix} 1 \\ 1 \end{pmatrix} + 2\begin{pmatrix} -1 \\ 2 \end{pmatrix} + \begin{pmatrix} 3 \\ 0 \end{pmatrix}$$

and
$$\begin{pmatrix} 2 \\ 5 \end{pmatrix} = 3\begin{pmatrix} 1 \\ 1 \end{pmatrix} + \begin{pmatrix} -1 \\ 2 \end{pmatrix}.$$

We know that in V_2, the subsets $\{a, b, c, d\}$ and $\{a, b, c\}$ are linearly dependent, i.e. $\exists\, \lambda_1, \lambda_2, \lambda_3, \lambda_4, \mu_1, \mu_2, \mu_3 \in \mathbb{R}$, not all zero:

$$\lambda_1 a + \lambda_2 b + \lambda_3 c + \lambda_4 d = 0 = \mu_1 a + \mu_2 b + \mu_3 c;$$

but $\{a, b\}$ may be linearly independent as they are in this example,

i.e. $\not\exists\, v_1, v_2$ not both zero: $v_1 a + v_2 b = 0.$

Any subset S of V such that every $v \in V$ can be expressed in terms of elements of S in this way is said to **span the vector space**, i.e.

$$\forall\, v \in V; \quad \exists\, r \in \mathbb{N}; \quad \exists\, s_1, s_2, \ldots, s_r \in S; \quad \exists\, p_1, p_2, \ldots, p_r \in F:$$
$$v = p_1 s_1 + p_2 s_2 + \cdots + p_r s_r.$$

A linearly independent subset which spans a vector space is called a **basis** for that space; a basis is not unique, but it can be proved that for a given vector space, every basis contains the same number of vectors, called the **dimension** of the space. If V has dimension n, then any n linearly independent vectors in V form a basis for V.

Notice that, although the infinite vector set K_n, in which the vectors have n components, is always of dimension n, a subset of K_n may well be of dimension r, where $r < n$; for example, the subset

$$\begin{pmatrix} 1 \\ 2 \\ 1 \end{pmatrix}, \begin{pmatrix} -2 \\ -7 \\ 1 \end{pmatrix}, \begin{pmatrix} 5 \\ 7 \\ 8 \end{pmatrix}$$

of V_0 has

$$\begin{pmatrix} 1 \\ 2 \\ 1 \end{pmatrix}, \begin{pmatrix} -2 \\ -7 \\ 1 \end{pmatrix}$$

as a basis since

$$\begin{pmatrix} 5 \\ 7 \\ 8 \end{pmatrix} = 7\begin{pmatrix} 1 \\ 2 \\ 1 \end{pmatrix} + \begin{pmatrix} -2 \\ -7 \\ 1 \end{pmatrix},$$

this being, of course, because they are coplanar.

Example 1. Show that polynomials of the form $ax^2 + bx + c$, where $a, b, c \in \mathbb{R}$, under addition and $\mathbb{R}$ are a vector space under multiplication. Find a basis and the dimension of the space.

The polynomials form a group P under addition, with an identity element given by $a = b = c = 0$ and $-(ax^2 + bx + c)$ as the inverse of $ax^2 + bx + c$. $\mathbb{R}$ is a field and when combined with P under multiplication satisfies the properties (i)–(iv).

A set such as $\{x^2, x, 1\}$ forms a basis, so that the dimension of the vector space is three. Clearly the set P_n of polynomials of degree n forms a vector space of dimension $n + 1$.

Example 2. Show that

$$\begin{pmatrix} 1 \\ 1 \\ 1 \end{pmatrix}, \begin{pmatrix} 0 \\ 1 \\ 1 \end{pmatrix}, \begin{pmatrix} -1 \\ 0 \\ 1 \end{pmatrix}$$

form a basis for a vector space.

$$l\begin{pmatrix} 1 \\ 0 \\ 1 \end{pmatrix} + m\begin{pmatrix} 0 \\ 1 \\ 1 \end{pmatrix} + n\begin{pmatrix} -1 \\ 0 \\ 1 \end{pmatrix} = \begin{pmatrix} 0 \\ 0 \\ 0 \end{pmatrix} \Leftrightarrow l - n = m = l + m + n = 0$$

$$\Rightarrow l = m = n = 0.$$

Hence the given vectors are linearly independent, and form a basis for a vector space.

Exercise 22.17

1. Prove that each of the following sets of vectors forms a basis for the vector space stated:

(a) $\begin{pmatrix} 2 \\ 3 \end{pmatrix}, \begin{pmatrix} -1 \\ 6 \end{pmatrix}, \begin{pmatrix} 4 \\ 5 \end{pmatrix}$ for V_2;   (b) $\begin{pmatrix} 1 \\ 2 \\ 3 \end{pmatrix}, \begin{pmatrix} -1 \\ 1 \\ 1 \end{pmatrix}, \begin{pmatrix} 1 \\ -1 \\ 1 \end{pmatrix}$, for V_3;

(c) $\begin{pmatrix} 1 \\ 1 \\ 1 \end{pmatrix}, \begin{pmatrix} 1 \\ 2 \\ 3 \end{pmatrix}, \begin{pmatrix} 3 \\ 4 \\ 5 \end{pmatrix}, \begin{pmatrix} 6 \\ 7 \\ 8 \end{pmatrix}$, for V_2.

2. Show that the following are vector spaces, stating in each case the operations involved:
(a) $\mathbb{R}$ over $\mathbb{R}$;   (b) $\mathbb{C}$ over $\mathbb{R}$;
(c) solutions of $\dfrac{d^2 y}{dx^2} + n^2 y = 0$ over $\mathbb{R}$.

Give a basis for each space, and state its dimension.

3. Prove that the set of matrices $\begin{pmatrix} a & b \\ c & d \end{pmatrix}$, $\forall a, b, c, d \in \mathbb{R}$, is a vector space over $\mathbb{R}$.

4. Determine whether the following sets of matrices form bases for the vector space of question **3**:

(a) $\begin{pmatrix} 1 & 0 \\ 0 & 1 \end{pmatrix}, \begin{pmatrix} 1 & 0 \\ 0 & 0 \end{pmatrix}, \begin{pmatrix} 0 & 0 \\ 0 & 1 \end{pmatrix}$;   (b) $\begin{pmatrix} 1 & 0 \\ 0 & 0 \end{pmatrix}, \begin{pmatrix} 0 & 1 \\ 0 & 0 \end{pmatrix}, \begin{pmatrix} 0 & 0 \\ 1 & 0 \end{pmatrix}, \begin{pmatrix} 0 & 0 \\ 0 & 1 \end{pmatrix}$;

(c) $\begin{pmatrix} -1 & 0 \\ 0 & 0 \end{pmatrix}, \begin{pmatrix} 0 & 1 \\ 0 & 0 \end{pmatrix}, \begin{pmatrix} 0 & 0 \\ -1 & 0 \end{pmatrix}, \begin{pmatrix} 0 & 0 \\ 0 & 1 \end{pmatrix}$.

5. Prove that each of the following forms a basis for the vector space stated:
(a) $x^2 - 5x + 2$, $2x^2 + 3x - 1$, $x^2 + 4$, for P_2;
(b) $3x - 1$, $2x + 7$, $4x - 9$, for P_1;
(c) $x^2 - 2x + 4$, $x^2 + 5x - 13$, $3x^2 + x - 5$, for P_2.

Miscellaneous Exercise 22

1. (i) A relation R on a set S has the following *two* properties:
   (1) R is reflexive, i.e. aRa for all $a \in S$,
   (2) for all a, b, $c \in S$, if aRb and bRc, then cRa.
   Prove that R is symmetric and transitive, so that (1) and (2) in fact define an equivalence relation.
   (ii) A relation R is defined on the set of points (x, y) in a plane by

   $$(x_1, y_1) R(x_2, y_2) \Leftrightarrow x_1^2 - x_2^2 = y_2^2 - y_1^2.$$

   Prove that R is an equivalence relation and describe geometrically the equivalence classes.   (L.)

2. M is the set of non-singular 2×2 real matrices.
   (a) An equivalence relation R_1 is defined on M by

   $$XR_1Y \Leftrightarrow \text{(the determinants of } X \text{ and } Y \text{ are equal)}.$$

   For each of the following subsets M_1 and M_2, state whether all its members are in the same equivalence class. If so, give the value of the determinant; if not, give a counter-example.
   $M_1 = \{$Matrices representing rotations of the plane about an origin$\}$,
   $M_2 = \{$Matrices which are equal to their inverses$\}$.
   (b) A relation R_2 is defined on M by

   $$XR_2Y \Leftrightarrow XY = YX.$$

   Show by means of a counter-example that R_2 is not an equivalence relation.   (L.)

3. Let Z denote the set of integers (positive, negative and zero) and N denote the set of positive integers. Show that, if a, $b \in N$ are given and h is the highest common factor of a, b, then

   $$\{xa + yb : x \in Z, \, y \in Z\} = \{zh : z \in Z\}.$$

   Determine the solution set $(x, y \in Z)$, in each case, of

   $$\text{(i) } 60x + 24y = 108, \qquad \text{(ii) } 60x + 24y = 20.   \text{(C.)}$$

4. Prove that the number of primes is infinite.
   By considering $2(2.3.5. \ldots .p) - 1$, where the numbers in the bracket are the complete set of primes up to p, prove that the number of primes of the form $4n + 3$ is infinite.
   Prove also that the number of primes of the form $6n + 5$ is infinite.   (C.)

5. Prove that
   (a) if $p \equiv q$ (mod m) and $r \equiv s$ (mod m), then $pr \equiv qs$ (mod m);
   (b) if $p \equiv q$ (mod m), then $p^n \equiv q^n$ (mod m) for any positive integer n;
   (c) the sum of the geometric series $\sum_{r=0}^{29} 2(3^r)$ is divisible by 7.   (L.)

6. *In this question all calculations are in modulo 3 arithmetic.*
   The set S is defined by

   $$S = \{(x, y) : x, \, y \in \{0, 1, 2\}\}$$

   and the relation R is defined on S such that

   $$(a, b) R (c, d) \Leftrightarrow a \equiv kc \text{ and } b \equiv kd, \text{ where } k = 1 \text{ or } 2.$$

   Show that R is an equivalence relation and exhibit the set P of equivalence classes into which S is partitioned by R.
   Assume that the operation $*$ on P may be defined by

   $$(a, b) * (c, d) = (ac, ad + bc),$$

   where these ordered pairs are representatives of the equivalence classes. Identify a group $(G, *)$, of order 3, where $G \subset P$, listing the members of G and giving the group table for the operation $*$.   (L.)

7. Solve the congruences:
   (a) $5x \equiv 6$ (mod 7); (b) $2^{41} \equiv x$ (mod 23); (c) $18^{96} \equiv x$ (mod 11).

8. The binary operation $*$ on the set S of all real numbers is defined by $a * b = |a - b|$. Investigate whether $*$ is (i) commutative, (ii) associative.

The relation R on S is defined by $aRb \Leftrightarrow a * b = 1$. State, with reasons, whether or not R is (i) reflexive, (ii) symmetric, (iii) transitive. (N.)

9. How many axes of symmetry has an equilateral-triangular lamina (i) in its own plane, and (ii) otherwise?
With each of these axes of symmetry can be associated a finite group of rotations of the triangle into itself. What is the order of each group? (O.C.S.M.P.)

10. Two elements u, v of a group $(G, *)$ are said to *commute* if $u * v = v * u$. Prove that the set consisting of all the elements of G that commute with a given element a is a sub-group H_a of G.
Prove also that

$$[(x \in H_a) \text{ and } (x \in H_b)] \Rightarrow (x \in H_{a*b}).$$ (C.)

11. a, b are elements of a group. Prove that

$$ab = ba \Rightarrow ba^2 = a^2 b.$$

Show that the converse is false by taking a to be an element of order 2 in the group of symmetries of an equilateral triangle. (O.C.S.M.P.)

12. The set S consists of all transformations T of the real numbers given by

$$T(x) = ax + b \qquad (a \neq 0, a, b \text{ real}).$$

Show that if $T_1, T_2 \in S$ then $T_1 T_2 \in S$. Prove that the set S forms a group under composition of functions. [The associative law may be assumed.]
Find the elements T of S of order 2 (i.e. $TT = I$ and $T \neq I$, where I is the identity transformation). Prove that these elements are all reflections.
Show also that there are no elements of order 3 in S. (N.)

13. The cartesian product of a set with itself, $G \times G$, is the set of all ordered pairs (x, y) with $x \in G$ and $y \in G$. Prove that if G is a multiplicative group, then $(G \times G, \circ)$ is a group, where the binary operation $\circ$ on $G \times G$ is defined by

$$(p, q) \circ (r, s) = (pr, qs).$$

In the case when G is the group with exactly two elements, determine whether or not $(G \times G, \circ)$ is isomorphic to the group with the four elements $\{1, -1, i, -i\}$, the operation being multiplication of complex numbers. (C.)

14. Let G be a non-empty set and $\circ$ be an associative binary operation on G. Suppose that
(a) there is an element u in G such that, for all x in G,

$$u \circ x = x,$$

(b) for each x in G, there is an element x' in G such that

$$x' \circ x = u.$$

Prove that
(i) if $x \in G$ and $x \circ x = x$, then $x = u$,
(ii) if $x, y \in G$ and $y \circ x = u$, then $x \circ y = u$,
(iii) for all $x \in G$, $x \circ u = x$,
(iv) $(G, \circ)$ is a group. (C.)

15. The matrices (with complex elements)

$$E = \begin{pmatrix} 1 & 0 \\ 0 & 0 \end{pmatrix}, \quad A = \begin{pmatrix} -1 & 0 \\ 0 & 0 \end{pmatrix}, \quad B = \begin{pmatrix} i & 0 \\ 0 & 0 \end{pmatrix}, \quad C = \begin{pmatrix} x & y \\ z & t \end{pmatrix}$$

are given to form a group under the operation of matrix multiplication. Find x, y, z and t, confirm that $A(BC)$ is then equal to $(AB)C$, and exhibit the group operation table.
[The associative law for matrix multiplication may be assumed except for the relation quoted.] (C.)

16. Let $\mathbb{Z}$ be the set of all integers and n a fixed positive integer. The equivalence relation $\sim$ is defined on $\mathbb{Z}$ by putting $a \sim b$, for $a, b \in \mathbb{Z}$, if and only if $(b - a)$ is divisible by n. Let $\hat{a} = \{x : x \in \mathbb{Z}, x \sim a\}$ be the equivalence class determined by a. Show that, if $\hat{a} = \hat{c}$ and $\hat{b} = \hat{d}$ $(a, b, c, d \in \mathbb{Z})$, then

$$\widehat{a+b} = \widehat{c+d} \text{ and } \widehat{a.b} = \widehat{c.d.}$$

On the set $\mathbb{Z}_n$ of all equivalence classes as given above, multiplication $\odot$ and addition $\oplus$ are defined by $\hat{a} \odot \hat{b} = \widehat{ab}$ and $\hat{a} \oplus \hat{b} = \widehat{a+b}$. Show that these definitions are independent of the particular elements a, b chosen to represent the classes $\hat{a}, \hat{b}$.

Using the expression for the positive highest common factor of two integers as a linear combination of them, show that, if n is a prime, $\hat{a} \in \mathbb{Z}_n$, and $\hat{a} \neq \hat{0}$, then there exists $\hat{b} \in \mathbb{Z}_n$ such that $\hat{a} \odot \hat{\beta} = \hat{1}$ in $\mathbb{Z}_n$. (L.)

17. A group G of order m consists of elements $\{x_1, x_2, \ldots, x_m\}$ with an operation $\circ$; the identity element is e, the inverse of x is x'. A group H of order n consists of elements $\{y_1, y_2, \ldots, y_n\}$ with an operation $*$; the identity element is f, the inverse of y is y'. Prove that the set of ordered pairs $\{(x, y) | x \in G, y \in H\}$ with the operation $\square$ defined by $(x_1, y_1) \square (x_2, y_2) = (x_1 \circ x_2, y_1 * y_2)$ forms a group K, stating its order, the identity element and the inverse of (x, y).

If the elements of G are $\{e, a\}$ and those of H are $\{f, p, q\}$, write down the group tables for G, H and hence that for K. Identify one sub-group of K of order 3 and one of order 2. By geometrical consideration of the number of subgroups decide whether K is isomorphic to the group of rotations of a regular hexagon or to the group of symmetries of an equilateral triangle. (L.)

18. L is the subgroup $\{e, p, s\}$ of the group G for which the combination table under the operation $*$ is as follows (where, for example, $t * q = s$):

$*$	e	p	q	r	s	t
e	e	p	q	r	s	t
p	p	s	r	t	e	q
q	q	t	e	s	r	p
r	r	q	p	e	t	s
s	s	e	t	q	p	r
t	t	r	s	p	q	e

If a, b denote general elements of G, then a relation R relating a to b is defined by $aRb \Leftrightarrow a * b^{-1} \in L$. In which of the following pairs of elements is the first related to the second: (i) t, q; (ii) q, r; (iii) s, s; (iv) t, p? To what elements is t related?

If G is any group of which L is any subgroup then for R (defined as above) to be an equivalence relation it is necessary that $a * b^{-1} \in L \Rightarrow b * a^{-1} \in L$, for all $a, b \in G$. Write down (in a similar way) two other necessary conditions. State with reasons which of these three conditions are true. (O.C.S.M.P.)

19. Find 2×2 matrices $\mathbf{I}, \mathbf{J}$ independent of a and b such that the matrix

$$\mathbf{M} = \begin{pmatrix} a & b \\ -b & a \end{pmatrix}$$

can be written in the form $a\mathbf{I} + b\mathbf{J}$ and show that $\mathbf{J}^2 = -\mathbf{I}$ (a and b are real numbers).

Prove also that $\det \mathbf{M} = 0$ if and only if $\mathbf{M}$ is the zero matrix. Hence prove that the set S of all matrices of the form $\mathbf{M}$, excluding the zero matrix, forms a group under matrix multiplication. (The associativity of matrix multiplication may be assumed.)

Deduce that if

$$f: \begin{pmatrix} a & b \\ -b & a \end{pmatrix} \rightarrow a + bi \text{ where } i^2 = -1$$

then f is an isomorphism of this set of matrices on to the set of non-zero complex numbers under multiplication.

Show that the set of matrices of the form

$$\begin{pmatrix} \cos\theta & \sin\theta \\ -\sin\theta & \cos\theta \end{pmatrix}$$

forms a subgroup of S. To what set of complex numbers is this subgroup isomorphic?

Write down a cyclic subgroup of S of order 3.   (N.)

20. (i) Prove that the set of non-zero real numbers forms a group under the operation $\circ$ defined by

$$x \circ y = 2xy.$$

State the identity element for this system, and give the inverse of x.

(ii) A set of four 2×2 matrices forms a group under matrix multiplication. Two members of the set are

$$\begin{pmatrix} -1 & 0 \\ 0 & 1 \end{pmatrix} \quad \text{and} \quad \begin{pmatrix} 1 & 0 \\ 0 & -1 \end{pmatrix}.$$

Find the other members and write out the group table.   (L.)

21. If n is a given natural number, show that the set of nth roots of unity forms a group under multiplication.

If S is the set of all the roots of all the equations

$$z^n = 1 \text{ for all } n \in \mathbb{N},$$

show that S forms a group under multiplication.   (L.)

22. Let G be a group with the operation of multiplication and let Z be the set of all elements x of G with the property $xy = yx$ for all y in G.

Show that Z is not empty and that multiplication in Z is commutative. Prove that Z is a subgroup of G. Under what circumstances does Z coincide with G?

[You may assume that the identity in G and inverses in G are two-sided.]   (N.)

23. The set $\{1, 2, 3, 4, 5, 6\}$ of residue classes modulo 7 forms a group G under multiplication. State the order of each element of the group, and name the elements which are generators. List the proper sub-groups of G.

A group S, isomorphic to G, is generated by the permutation

$$p = \begin{pmatrix} a\,b\,c\,d\,e\,f \\ b\,c\,d\,e\,f\,a \end{pmatrix},$$

under the operation of successive application. State another generator of S.   (N.)

24. A binary operation $\circ$ is defined on the set M of 3×3 matrices $\mathbf{A}, \mathbf{B}, \ldots$ by

$$\mathbf{A} \circ \mathbf{B} = \mathbf{AB} - \mathbf{BA},$$

where $\mathbf{AB}$ is the usual matrix product. Show that

$$\mathbf{A} \circ \mathbf{B} = -\mathbf{B} \circ \mathbf{A},$$
$$\mathbf{A} \circ (\mathbf{B} + \mathbf{C}) = \mathbf{A} \circ \mathbf{B} + \mathbf{A} \circ \mathbf{C},$$
$$(\mathbf{A} \circ \mathbf{B}) \circ \mathbf{C} + (\mathbf{B} \circ \mathbf{C}) \circ \mathbf{A} + (\mathbf{C} \circ \mathbf{A}) \circ \mathbf{B} = \mathbf{O}$$
$$\text{and} \quad (a\mathbf{A}) \circ \mathbf{B} = a(\mathbf{A} \circ \mathbf{B})$$

for any number a.

The set of vectors $\mathbf{u}, \mathbf{v}, \ldots$ is mapped by the one-to-one relation

$$\mathbf{u} = u_1\mathbf{i} + u_2\mathbf{j} + u_3\mathbf{k} \leftrightarrow \begin{pmatrix} 0 & -u_3 & u_2 \\ u_3 & 0 & -u_1 \\ -u_2 & u_1 & 0 \end{pmatrix} = \mathbf{U}$$

onto the set of matrices $\mathbf{U}, \mathbf{V}, \ldots$ which is a subset of M. Show that, in this mapping, if $\mathbf{u} \leftrightarrow \mathbf{U}$ and $\mathbf{v} \leftrightarrow \mathbf{V}$ then $\mathbf{u} \times \mathbf{v} \leftrightarrow \mathbf{U} \circ \mathbf{V}$.   (N.)

25. If R, T, I are distinct elements of a set on which is defined an associative binary operation for which I is the identity element, and if $RR = TT = I$, prove that RT must be a fourth distinct element of the set.

Let a be a complex number of unit modulus, with $a \neq 1$, and define the following four functions on the complex numbers

$$I : z \to z,$$
$$R : z \to z^*, \text{ the complex conjugate of } z,$$
$$S : z \to az,$$
$$T : z \to az^*.$$

Show that under the operation of composition of functions, which may be assumed associative, I is an identity element and $RR = TT = I$. Prove that a necessary and sufficient condition for the four functions I, R, S, T to form a group under composition is that $a = -1$. In this case, determine whether the group so formed is isomorphic to the group with elements $\{1, -1, i, -i\}$ and the operation of multiplication of complex numbers.   (N.)

26. In each case below, determine whether the given set with the given operations forms (a) a ring, (b) a field, (c) neither a ring nor a field. Give your reasons, *briefly*.

(i) The integers (positive, negative and zero) under the usual arithmetic operations (addition and multiplication).

(ii) The integers $\{0, 1, 2, 3, 4, 5\}$ under the usual arithmetic operations *modulo* 6.

(iii) The integers $\{0, 1, 2, 3, 4\}$ under the usual arithmetic operations *modulo* 5.

(iv) The set of vectors in three dimensions under vector addition and scalar (or dot) product.

(v) The set of invertible (non-singular) 2×2 matrices under the usual matrix operations of addition and multiplication.   (C.)

27. Let $(R, +, .)$ be a commutative ring in which R is a finite set. Suppose that $xy \neq 0$ for any pair x, y of non-zero elements of R. If a is a non-zero element of R, show, by considering the set

$$\{ax : x \in R \text{ and } x \neq 0\},$$

that there is an element e in R such that $ae = a$. Deduce that R is a ring with unity.

Show further that every non-zero element of R has a (multiplicative) inverse.   (C.)

28. Find a basis for the vector space spanned by the vectors

$$(1, 2, -1), (3, -1, 2), (2, -10, 8), (7, -7, 8).$$

What is the dimension of this space?

For what value (or values) of a does $(1, 4, a)$ belong to the space?   (C.)

29. Show that the set M of all 2×2 matrices with real elements forms a non-commutative ring with unity under the operations of matrix addition and multiplication.

[The associativity of matrix addition and multiplication may be assumed, together with the fact that matrix multiplication is distributive over matrix addition.]

Show that the subset S of M consisting of all matrices of the form

$$\begin{pmatrix} a & b \\ -b & a \end{pmatrix}$$

forms a field.

Show that, in S, the equation $\mathbf{X}^2 + \mathbf{I} = \mathbf{O}$ has exactly two solutions, where $\mathbf{I}$ is the unit matrix and $\mathbf{O}$ is the zero matrix.   (C.)

30. Assuming that the set M of 2×2 real matrices forms a vector space V under the operations of addition of matrices and multiplication of matrices by real numbers, give the dimension of, and a basis for, V.

One of the following subsets of M forms a vector space under the same operations; give its dimension and a basis. Give a counter-example to show that the other subset is not a vector space under the same operations.

(a) $\left\{ \begin{pmatrix} x & -y \\ y & x \end{pmatrix} : x \in \mathbb{R}, \ y \in \mathbb{R} \right\}$,

(b) singular 2×2 real matrices.

Show, by using the correspondence

$$\begin{pmatrix} x & -y \\ y & x \end{pmatrix} \leftrightarrow (x + yi),$$

that the set (a) and the set of complex numbers, each with addition and multiplication defined in the usual way, are isomorphic. Name the structure so established for the set (a) and deduce that one property, not generally true of matrix multiplications, must apply within this set.   (L.)

Answers to the Exercises

Note. Constants of integration have been omitted from answers to indefinite integrals.

Exercise 13.1 (p. 407)

1. $\cos 9\theta - i \sin 9\theta$.   2. $\cos\dfrac{3\pi}{4} - i \sin\dfrac{3\pi}{4} = -\dfrac{1}{\sqrt{2}}(1+i)$.

3. $\cos\dfrac{5\pi}{6} - i \sin\dfrac{5\pi}{6}$.

4. (i) $1\left(\cos\dfrac{\pi}{4} + i \sin\dfrac{\pi}{4}\right)$, $1\left(\cos\dfrac{5\pi}{4} + i \sin\dfrac{5\pi}{4}\right)$;

    (ii) $\sqrt{5}\,(\cos 26°\ 34' + i \sin 26°\ 34')$, $\sqrt{5}\,(\cos 206°\ 34' + i \sin 206°\ 34')$;

    (iii) $\sqrt{13}\,[\cos(-33°\ 41') + i \sin(-33°\ 41')]$, $\sqrt{13}\,(\cos 146°\ 19' + i \sin 146°\ 19')$;

    (iv) $5\,(\cos 53°\ 8' + i \sin 53°\ 8')$, $5\,(\cos 233°\ 8' + i \sin 233°\ 8')$;

    (v) $1\left(\cos\dfrac{\pi}{4} + i \sin\dfrac{\pi}{4}\right)$, $1\left(\cos\dfrac{5\pi}{4} + i \sin\dfrac{5\pi}{4}\right)$.

5. (i) $1\,(\cos 0 + i \sin 0)$, $1\left(\cos\dfrac{2\pi}{3} + i \sin\dfrac{2\pi}{3}\right)$, $1\left(\cos\dfrac{4\pi}{3} + i \sin\dfrac{4\pi}{3}\right)$;

    (ii) $1\left(\cos\dfrac{\pi}{6} + i \sin\dfrac{\pi}{6}\right)$, $1\left(\cos\dfrac{5\pi}{6} + i \sin\dfrac{5\pi}{6}\right)$, $1\left(\cos\dfrac{3\pi}{2} + i \sin\dfrac{3\pi}{2}\right)$;

    (iii) $1{\cdot}12\left(\cos\dfrac{\pi}{12} + i \sin\dfrac{\pi}{12}\right)$, $1{\cdot}12\left(\cos\dfrac{3\pi}{4} + i \sin\dfrac{3\pi}{4}\right)$, $1{\cdot}12\,(\cos\dfrac{17\pi}{12} + i \sin\dfrac{17\pi}{12})$;

    (iv) $1{\cdot}71\,(\cos 17°\ 43' + i \sin 17°\ 43')$, $1{\cdot}71\,(\cos 137°\ 43' + i \sin 137°\ 43')$,

        $1{\cdot}71\,(\cos 257°\ 43' + i \sin 257°\ 43')$.

7. $1{\cdot}27 \pm i\,(0{\cdot}79)$, $-1{\cdot}27 \pm i\,(0{\cdot}79)$.

8. $\cos\dfrac{\pi}{6} + i \sin\dfrac{\pi}{6}$, $\cos\dfrac{\pi}{2} + i \sin\dfrac{\pi}{2} = i$, $\cos\dfrac{5\pi}{6} + i \sin\dfrac{5\pi}{6}$, $\cos\dfrac{7\pi}{6} + i \sin\dfrac{7\pi}{6}$, $\cos\dfrac{3\pi}{2} + i \sin\dfrac{3\pi}{2} = -i$,

    $\cos\dfrac{11\pi}{6} + i \sin\dfrac{11\pi}{6}$.

9. $\cos 5\theta \equiv 16\cos^5\theta - 20\cos^3\theta + 5\cos\theta$; $\tan 5\theta \equiv \dfrac{5t - 10t^3 + t^5}{1 - 10t^2 + 5t^4}$ where $t = \tan\theta$.

10. $\cos n\theta \equiv c^n - {}_nC_2 c^{n-2} s^2 + {}_nC_4 c^{n-4} s^4 - \ldots$, the expression ending in $(-1)^{n/2} s^n$ or
$(-1)^{(n-1)/2}\,{}_nC_{n-1}\,cs^{n-1}$ according as n is even or odd; $\sin n\theta \equiv {}_nC_1 c^{n-1} s - {}_nC_3 c^{n-3} s^3 + {}_nC_5 c^{n-5} s^5$
$- \ldots$, the expression ending in $(-1)^{(n-2)/2}\,{}_nC_{n-1}\,cs^{n-1}$ or $(-1)^{(n-1)/2} s^n$ according as n is even or odd.
$[c = \cos\theta,\ s = \sin\theta.]$

11. $\sin^6\theta \equiv (10 - 15\cos 2\theta + 6\cos 4\theta - \cos 6\theta)/32$; $\cos^6\theta \equiv (10 + 15\cos 2\theta + 6\cos 4\theta + \cos 6\theta)/32$.

12. (i) $\left(10\theta - \dfrac{15\sin 2\theta}{2} + \dfrac{3\sin 4\theta}{2} - \dfrac{\sin 6\theta}{6}\right)\Big/32$;

    (ii) $\left(35\theta + 28\sin 2\theta + 7\sin 4\theta + \dfrac{4\sin 6\theta}{3} + \dfrac{\sin 8\theta}{8}\right)\Big/128$;

780

(iii) $\left(2\theta - \dfrac{\sin 2\theta}{2} - \dfrac{\sin 4\theta}{2} + \dfrac{\sin 6\theta}{6}\right)\Big/32.$

13. $\cos(\alpha + \beta - \gamma - \delta) + i \sin(\alpha + \beta - \gamma - \delta).$

14. $0, \pm i\sqrt{3}, \pm i/\sqrt{3}.$ **15.** $\pm(3+i), \pm(1-3i).$

18. The given expression equals $2^{n+1}\cos\dfrac{2n\pi}{3}.$

19. $\cos 6\theta \equiv 32c^6 - 48c^4 + 18c^2 - 1,\ \sin 6\theta/\sin\theta \equiv 32c^5 - 32c^3 + 6c$ where $c = \cos\theta.$

20. $\left(z^2 - 2z\cos\dfrac{4\pi}{7} + 1\right), \left(z^2 - 2z\cos\dfrac{6\pi}{7} + 1\right).$

21. (i) $2^n\cos^n\dfrac{\theta}{2}\left(\cos\dfrac{n\theta}{2} + i\sin\dfrac{n\theta}{2}\right);$ (ii) $\cos(n\pi - 2n\theta) + i\sin(n\pi - 2n\theta).$

Exercise 13.2 (p. 411)

1. (i) $\sqrt{2}e^{i\pi/4};$ (ii) $\sqrt{2}e^{-i\pi/4};$ (iii) $e^{i\pi/2};$ (iv) $e^{-i\pi/2};$ (v) $2e^{-i\pi/3}.$

2. (i) $\dfrac{1}{\sqrt{2}} - \dfrac{i}{\sqrt{2}};$ (ii) $-1+i0;$ (iii) $-1+i0;$ (iv) $\dfrac{e}{2} + i\dfrac{e\sqrt{3}}{2};$ (v) $e^x\cos y + ie^x\sin y.$

3. $2\cos\tfrac{1}{2}\theta e^{i(\theta + 4n\pi)/2}.$ **4.** $2\cos\theta e^{i2n\pi}.$

5. (i) $r = 1,\ \theta = \tfrac{1}{4}\pi, \tfrac{1}{2}\pi, \tfrac{3}{4}\pi, \tfrac{5}{4}\pi, \tfrac{3}{2}\pi, \tfrac{7}{4}\pi;$ (ii) $34 + 6\cos\theta - 24\cos^2\theta.$

6. (i) $z = e^{i\theta},\ \theta = 0, \tfrac{2}{5}\pi, \tfrac{4}{5}\pi, \tfrac{6}{5}\pi, \tfrac{8}{5}\pi.$

7. (i) $i, \tfrac{1}{2}(\pm\sqrt{3} - i).$

Exercise 13.3 (p. 416)

1. $(2, 1).$ **2.** Centre $[(bk^2 - a)/(k^2 - 1), 0]$ radius $k|a - b|/(k^2 - 1).$

3. (i) $v^2 = 4(1 - u);$ (ii) $v^2 = 4(1 + u);$ (iii) $u \geqslant 0.$

4. $\tfrac{1}{2}(-i \pm \sqrt{3}).$

5. $x = u/(u^2 + v^2),\ y = -v/(u^2 + v^2);\ u = (\cos\theta)/r,\ v = -(\sin\theta)/r;$
  (i) $u^2 + v^2 + 2u + v = 0;$ (ii) circle centre the origin, radius $\tfrac{1}{2};$
  (iii) straight line $u = \tfrac{1}{2}$ from $u = 0$ to $-\infty.$

6. $(1/\sqrt{2}, \tfrac{1}{4}\pi), (1/\sqrt{2}, 5\pi/4).$

7. $r = \tfrac{2}{3}\cos\theta;\ 0, (\tfrac{1}{3}, \pi/3), (\tfrac{1}{3}, 5\pi/3).$

9. Straight line $v = 5;$ circle $u^2 + (v - 3)^2 = 4.$

Miscellaneous Exercise 13 (p. 417)

1. $-1.$

2. $z = 1 - i\sqrt{3},\ z^2 = -2(1 + i\sqrt{3}),\ z^{-1} = \tfrac{1}{4}(1 + i\sqrt{3});\ (2, -\tfrac{1}{3}\pi), (4, -\tfrac{2}{3}\pi), (\tfrac{1}{2}\sqrt{2}, \tfrac{1}{3}\pi).$

3. $2(\cos 11\pi/6 + i\sin 11\pi/6);\ 512.$

5. $2^{1/3}(\cos\tfrac{2}{9}k\pi + i\sin\tfrac{2}{9}k\pi),\ k = 1, 2, 4, 5, 7, 8.$

6. (i) $1;$ (ii) within the unit circle $|z| = 1.$

7. (i) $(\sqrt{2}, 45°), (\sqrt{2}, 8°\,8'), (1, 36°\,52').$ **9.** (ii) $x = \tfrac{1}{2}.$

10. (i) $(1, \tfrac{1}{6}\pi), (1, \tfrac{1}{3}\pi), (1, \tfrac{7}{6}\pi), (1, \tfrac{4}{3}\pi);$ (ii) $-1 + 2i, 3 - 2i.$

11. (i) $z = i\sqrt{3};$ (ii) $z = 2 + i;$ (iii) $z = 1$ or $-1 + 2i.$

12. (i) $1 - i, -1 \pm 2i;$ (ii) $\tfrac{1}{2}(3 + i\sqrt{3}); \tfrac{1}{6}\pi, \tfrac{2}{3}\pi.$

13. The line $x = 1.$ (a) Circle centre O radius 1; (b) ellipse foci $(0, 1), (0, -1)$; (c) $y = 0$ for $|x| > 1.$

14. (a) and (b). $(128 - 104\sqrt{2})/105.$

16. $(1 - z^n)/(1 - z).$

17. (i) $\omega = \sqrt{\left[\dfrac{(CR_1^2 - L)}{LC(CR_2^2 - L)}\right]};$ (ii) $\tfrac{1}{5}(3 \pm 4i), \tfrac{1}{2}(1 \pm 3i).$

18. (i) $-5\pi/6.$

20. Rotation through $\theta;$ $\mathbf{M} = \begin{pmatrix} \cos\theta & -\sin\theta \\ \sin\theta & \cos\theta \end{pmatrix}.$ Half-turn about $(\sqrt{3}, 0).$

Exercise 14.1   (p. 426)

1. $(-4\mathbf{i} + 12\mathbf{j} - 3\mathbf{k})/13$; $\cos^{-1}(-1/\sqrt{3})$.   **3.** $a = b = c = -1$; $-\tfrac{2}{3}\sqrt{2}$.
7. 8; $-(2\mathbf{i} + 2\mathbf{j} + 3\mathbf{k})$; $\cos^{-1}(8/9)$; $\tfrac{1}{2}\sqrt{17}$.

Exercise 14.2   (p. 437)

1. $\mathbf{r} = 4\mathbf{i} + 2\mathbf{j} + 5\mathbf{k} + t(5\mathbf{i} + \mathbf{j} + 7\mathbf{k})$; $\dfrac{x-4}{5} = \dfrac{y-2}{1} = \dfrac{z-5}{7}$.

2. $\mathbf{r} = 3\mathbf{i} - 4\mathbf{j} + \mathbf{k} + s(-\mathbf{i} + 6\mathbf{j} - 2\mathbf{k}) + t(-6\mathbf{i} + 5\mathbf{j} + \mathbf{k})$; $16x + 13y + 31z = 77$.

3. $\mathbf{r} = 2\mathbf{i} - \mathbf{j} + 2\mathbf{k} + t(2\mathbf{i} - 2\mathbf{j} + \mathbf{k})$; $\dfrac{x-2}{2} = \dfrac{y+1}{-2} = \dfrac{z-2}{1}$; $2\mathbf{i} - \mathbf{j} + 2\mathbf{k}$.

4. $2x + 3y - 3z - 7 = 0$.   **6.** $\dfrac{x-1}{1} = \dfrac{y-2}{-1} = \dfrac{z-3}{2}$.

7. $7x - y - 5z = 0$.   **8.** $x + 2z - 4 = 0$.   **9.** $\sqrt{(101/17)}$.
10. $(\mathbf{r} - 2\mathbf{i} - 3\mathbf{j} + \mathbf{k}).(-3\mathbf{i} + 4\mathbf{j} + 2\mathbf{k}) = 0$; $3x - 4y - 2z + 4 = 0$.
11. $4/(7\sqrt{29})$.   **12.** $(1/\sqrt{2}, -1/\sqrt{2}, 0)$; $x + y - 5z = 0$.
13. A_1 and A_2 are $(0, 2, 1)$ and $(0, -\tfrac{2}{5}, -\tfrac{1}{5})$.
14. $7x - 8y + 3z = 0$; $x = (122/13)$, $y/23 = -z/20$.
15. $(1, 0, 0)$, $(0, 1, 0)$, $(0, 0, 1)$; centre $(\tfrac{1}{3}, \tfrac{1}{3}, \tfrac{1}{3})$, radius $\sqrt{(2/3)}$.
16. Radius $\tfrac{5}{2}$, centre $(\tfrac{3}{2}, 0, 2)$.
17. Centre $(1, 2, 0)$, radius 2; $x^2 + y^2 + z^2 + 3x + y - 5\sqrt{2}z - 4 = 0$; $x + y - \sqrt{2}z - 1 = 0$.
18. $(0, 1, 1)$.   **20.** $\pm Ra/|\mathbf{a}|$; $(\pm Rl/k, \pm Rm/k, \pm Rn/k)$ where $k = \sqrt{(l^2 + m^2 + n^2)}$.
21. $\mathbf{r}_0 \pm Ra/|\mathbf{a}|$; $(x_0 \pm Rl/k, y_0 \pm Rm/k, z_0 \pm Rn/k)$ where $k = \sqrt{(l^2 + m^2 + n^2)}$.
22. $(\mathbf{r}_1 - \mathbf{r}_0).(\mathbf{r} - \mathbf{r}_0) = R^2$.
23. $(\mathbf{r} - \mathbf{r}_1)^2 = (\mathbf{r}_1.\mathbf{n} + D)^2/n^2$; $(x - x_1)^2 + (y - y_1)^2 + (z - z_1)^2 = (\alpha x_1 + \beta y_1 + \gamma z_1 + D)^2/(\alpha^2 + \beta^2 + \gamma^2)$.
24. $(2, 3, 1)$; (i) $7x + 4y - 5z - 21 = 0$; (ii) $x - 8y - 5z + 27 = 0$; (iii) $7x - 2y - 8z = 0$.
25. $\mathbf{i} + 2\mathbf{j} + \mathbf{k}$.
26. (i) $\sqrt{(101/17)}$; (ii) $\mathbf{r}.(-3\mathbf{i} + 4\mathbf{j} + 2\mathbf{k}) = 4$; $3x - 4y - 2z + 4 = 0$.
28. $(\mathbf{r} - \mathbf{p} - \mathbf{s}).\mathbf{p} = 0$; $x - 4y = 2$; $4/17$.

29. $(1/\sqrt{2}, -1/\sqrt{2}, 0)$; $4x + 4y + 7z = 0$; $\dfrac{x}{7} = \dfrac{y}{7} = \dfrac{z}{-8}$.

30. $3x - y + z - 2 = 0$.   **31.** $x + 2y + 2z - 6 = 0$; $y - z = 0$.
32. (i) $2x - 2y - 3z - 1 = 0$; (ii) 3.   **33.** $\tfrac{1}{2}\sqrt{3}$; $x + 3y - 3z = 0$.
34. $x^2 + y^2 + z^2 - 2ax - 2by - 2cz = 0$; centre (a, b, c), radius $\sqrt{(a^2 + b^2 + c^2)}$;
$x^2 + y^2 + z^2 - 2x - 2y + 2(4 \pm 3\sqrt{2})z = 0$.

Exercise 14.3   (p. 442)

1. Three parallel lines with direction ratios $1: -2: 1$.
2. A common point $(4, -1, -7)$.
3. A common line of intersection $\dfrac{x-1}{1} = \dfrac{y-2}{-1} = \dfrac{z}{2}$.

4. Two of the planes (first and third) are parallel.
5. Three parallel lines with direction ratios $1:2:1$.
6. A common point $(2, -1, -1)$.   **7.** The planes are all parallel.

8. A common line of intersection $\dfrac{x+1}{4} = \dfrac{y-2}{3} = \dfrac{z}{-2}$.

9. A common point $(-3, 1, 0)$.
10. Three parallel lines with direction ratios $1:2: -1$.
11. (a) One solution, planes meet at $(0, -26, -17)$; (b) infinite number of solutions, planes meet in line
$3x = z + 17$, $3y = 5z + 7$; (c) no solutions.
12. $x/(3ab - 16a + 2b + 12) = y/(-8a - 8b + 36) = z/(-ab - 6b + 20) = 1/(2a + 7)(2 - a)$.
  (i) Infinite number of solutions, planes meet in line $x = 1 - z$, $y = \tfrac{1}{2}(1 + 2z)$;
  (ii) no solutions, planes meet by pairs in three parallel lines.

13. (a) Planes meet at $(3, 2, 1)$; (b) infinite solution set, planes meet in line $x = 17z - 14$, $y = -10z + 12$; (c) no solutions, planes meet by pairs in three parallel lines.

14. $\lambda = -1, 2$; when $\lambda = -1$, there is an infinite solution set if $k = 3$, planes meet in line $x = -11z - 3$, $y = 2z + 2$.

15. (i) Planes meet at $(2, 1, -3)$; (ii) three planes meet in a line.

16. $x = -\frac{2}{3}z + 2$, $y = \frac{2}{3}z - 1$; $x = 6$, $y = -5$, $z = -6$ and $x = -6$, $y = 7$, $z = 12$.

Exercise 14.4(a) (p. 447)

1. (i) $\frac{5}{2}\sqrt{3}$; (ii) $(-i + j + k)/\sqrt{3}$; $\frac{5}{3}$.   2. 12.

3. $5x - 7y + 11z + 13 = 0$.   4. $bc/\sqrt{(b^2 + c^2)}$, $ca/\sqrt{(c^2 + a^2)}$, $ab/\sqrt{(a^2 + b^2)}$.

5. $\sqrt{(5/7)}$.   6. $7/3$.   7. $(2, -1, 4)$, $(\frac{7}{2}, 0, \frac{7}{2})$.   8. $3\sqrt{2}$.

Exercise 14.4(b) (p. 449)

1. (i) $15/4$; (ii) -33; (iii) $\frac{1}{7}(3i - 2j + 6k)$; (iv) $\frac{3}{2}\sqrt{210}$; (v) $63i + 21j$.

2. (iii) $\lambda = -\frac{4}{5}$; $c = \frac{1}{5}(-3i + 4j + 3k)$.

3. (i) $a \times b/a^2 + pa$, where p is a variable scalar;
   (ii) $[\lambda^2 a + (a.c)c + \lambda(c \times a)]/[\lambda(\lambda^2 + c^2)]$;
   (iii) $[aw + (v \times w) \times u]/[a(a + u.v)]$ when $a + u.v \neq 0$.
   $pv + w/a$, where p is a variable scalar, when $a + u.v = 0$.

5. $m = -(a.b)/a^2 = 0$, $x = (a - a \times b)/a^2 = -(3i + 4j + 5k)/5$.

6. (i) 0; (ii) a; (iii) $k(a - b)$, $k \in \mathbb{R}$; (iv) $pa + qb$; $p, q \in \mathbb{R}$.

Miscellaneous Exercise 14 (p. 450)

1. $-2i - j + 3k$; $i - 3j + 2k$; $7(i + j + k)$; $7\sqrt{3/2}$; $2\sqrt{3/3}$; $7/3$.

2. $81/2$; C lies on a line parallel to AB; 27.

3. (a) $r = -i + 4j + 2k + s(i + 2j - 3k) + t(4i + 3j - 8k)$;
   (b) $\dfrac{x+1}{1} = \dfrac{y-4}{2} = \dfrac{z-2}{-3}$; (c) $\dfrac{x-3}{2} = \dfrac{y-7}{-1} = \dfrac{z+6}{-2}$; $i + 8j - 4k$.

4. (a) $\frac{1}{4}\pi$; (b) $\cos^{-1}(-1/\sqrt{6})$.

5. (a) $x + 2y + 2z - 7 = 0$; (b) $4\frac{1}{2}$; (c) 6; (d) $r = 2i + 2j + \frac{1}{2}k + t(i + 2j + 2k)$; (e) $(1, 0, -\frac{3}{2})$.

6. $\dfrac{x-14}{2} = \dfrac{y-5}{2} = \dfrac{z-2}{-1}$; $\frac{1}{6}\sqrt{2}$; $x - 2y - 2z = 0$.

7. $\frac{1}{2}\sqrt{2}$, $-\frac{1}{2}\sqrt{2}$, 0; $4x + 4y + 7z = 0$; $\dfrac{x}{7} = \dfrac{y}{7} = \dfrac{z}{-8}$.

8. (c) $\dfrac{x}{1} = \dfrac{y+1}{1} = \dfrac{z-1}{-5}$.   9. $a(i + j) + (2a - 1)k$.   10. $\pm\frac{1}{4}$.

11. $x - 2y + z = 0$; $(5, 7, 9)$.

13. $r.(i - j) = 0$; $(0, 0, \pm a)$, $(\frac{1}{2}a, \frac{1}{2}a, \pm\frac{1}{2}\sqrt{2}a)$.   14. $\sqrt{6}d$.

16. (a) $\dfrac{x}{2} = \dfrac{y-1}{-3} = \dfrac{z+1}{2}$,   (b) $2x - 3y + 2z + 5 = 0$; $(1, 3, 1)$.

Exercise 15.1 (p. 456)

1. 4.   2. 789.   3. 0.   4. -3.   5. 0.   6. -8.   7. -182.

8. -345.   9. $xy(y - x) - x - 2y$.   10. $-14abc$.   11. -342.   12. 342.

13. 684.   14. -1026.   15. -342.   16. $-(y - z)(z - x)(x - y)$.

17. $(x - 1)(y - 1)(x - y)(x + y + 1)$.   18. $-(y - z)(z - x)(x - y)(yx + zx + xy)$.

19. $-(b - c)(c - a)(a - b)$.   20. $(a + b + c)(a^2 + b^2 + c^2 - bc - ca - ab)$.

21. $xyz(x^2 + y^2 + z^2 - yz - zx - xy)$.   22. $(x + a)(x^2 + ax + cx + 3ac - 2c^2)$.

23. $(x + y + z)(x^2 + y^2 + 3yz + 3zx + xy)$.   24. $-4abc$.   25. 1, $-6 \pm \sqrt{67}$.

26. -10, $\frac{1}{2}(-5 \pm \sqrt{13})$.   27. -4.   28. -5, $\frac{1}{2}(3 \pm \sqrt{13})$.

29. (a) 2, 5; (b) $xyz(y - z)(z - x)(x - y)$.

30. (i) $(a+b+c)(a-b-c)(b-a-c)(c-a-b)$;   (ii) $1, \frac{1}{2}(-1 \pm \sqrt{143}i)$.

32. $-(a^2+b^2+c^2)(a+b+c)(b-c)(c-a)(a-b)$.

33. $(a_1+a_2+a_3+a_4)(a_1-a_2+a_3-a_4)\left[(a_1-a_3)^2+(a_2-a_4)^2\right]$.

Exercise 15.3 (p. 463)

1. $\begin{pmatrix} 1 & -2 & -3 \\ 3 & -5 & -9 \\ -2 & 4 & 7 \end{pmatrix}$.   **2.** $(-1 \quad 2)$.   **3.** $ad \ne bc$.   **4.** $x, t = 1 \pm \sqrt{(-yz)}$.

5. $\mathbf{E}_1 = \begin{pmatrix} 1 & 0 & 0 \\ 0 & 1 & 0 \\ -2 & 0 & 1 \end{pmatrix}$, $\mathbf{E}_2 = \begin{pmatrix} 1 & -1 & 0 \\ 0 & 1 & 0 \\ 0 & 0 & 1 \end{pmatrix}$, $\mathbf{E}_3 = \begin{pmatrix} 1 & 0 & 0 \\ 0 & 1 & -2 \\ 0 & 0 & 1 \end{pmatrix}$,

$\mathbf{A}^{-1} = \begin{pmatrix} 1 & -1 & 0 \\ 4 & 1 & -2 \\ -2 & 0 & 1 \end{pmatrix}$.

7. $\begin{pmatrix} 0 & 0 & 0 \\ 2 & 0 & 0 \\ 0 & 1 & 0 \end{pmatrix}$.   **8.** ± 1.

9. (b) 1; $\begin{pmatrix} (\cosh\theta + k\sinh\theta) & -(1+k)\sinh\theta \\ (k-1)\sinh\theta & \cosh\theta - k\sinh\theta \end{pmatrix}$.

10. (i) $p = 10, q = -1$; (ii) $\frac{1}{6}\begin{pmatrix} -3 & 3 & 3 \\ 7 & -5 & -3 \\ 5 & -1 & -3 \end{pmatrix}$; $x = \frac{1}{2}(a+b), y = -\frac{1}{2}(a+b), z = \frac{1}{2}(a-b)$.

11. $\frac{1}{2}\begin{pmatrix} 3 & -2 \\ 1 & 0 \end{pmatrix}$.   **12.** $\begin{pmatrix} a^2 & ab & ac \\ ab & b^2 & bc \\ ac & bc & c^2 \end{pmatrix}$; $\begin{pmatrix} a^2 & ab & ac \\ ab & b^2+1 & bc+2 \\ ac & bc+2 & c^2+4 \end{pmatrix}$

13. $\mathbf{K}^n = \begin{pmatrix} 1 & kn \\ 0 & 1 \end{pmatrix}$; $\mathbf{M}^n = \begin{pmatrix} 1 & n & n & \frac{3}{2}n(n-1) \\ 0 & 1 & 0 & n \\ 0 & 0 & 1 & 2n \\ 0 & 0 & 0 & 1 \end{pmatrix}$.

Exercise 15.6 (p. 470)

1. $-\frac{1}{82}\begin{pmatrix} -10 & -7 & -9 \\ -4 & -11 & 21 \\ -22 & 1 & 13 \end{pmatrix}$; $x = -5, y = 3, z = -4$.

2. (i) $x = 1, y = z = 0$; (ii) $x = 3z - 3, y = 2z - 1$.   **3.** $x = 4, y = -2, z = 3$.

4. $x = -a - 7b + 12c, y = -a - 5b + 9c, z = a + 4b - 7c$; $\begin{pmatrix} -1 & -7 & 12 \\ -1 & -5 & 9 \\ 1 & 4 & -7 \end{pmatrix}$; $k = 14$.

5. $\mathbf{A} = \begin{pmatrix} 1 & 0 & 0 \\ 3 & 1 & 0 \\ 4 & \frac{3}{4} & 1 \end{pmatrix}$, $\mathbf{B} = \begin{pmatrix} 1 & 2 & -3 \\ 0 & -4 & 14 \\ 0 & 0 & \frac{9}{14} \end{pmatrix}$; $x = 3, y = -2, z = -1$.

6. Unless $a = -2$ or 1, unique solution $x = (p + ap - q - r)/(a + 2)(a - 1)$,
$y = (q + aq - p - r)/(a + 2)(a - 1)$, $z = (r + ar - p - q)/(a + 2)(a - 1)$;
if $a = -2$, soluble only if $p + q + r = 0$, when $x = \frac{1}{3}(3z + r - p)$, $y = \frac{1}{3}(3z + r - q)$;
if $a = 1$, soluble only if $p = q = r$, when $x + y + z = p$.

7. $al + bm + cn = 0$; $x = (b + lz)/n$, $y = (-a + mz)/n$; yes.

8. $x_1 = \frac{1}{2}y_2 - x_3$, $x_2 = -y_1 + \frac{1}{2}y_2 + x_3$.

9. $k = 1$, e.g. $\begin{pmatrix} w \\ x \\ y \\ z \end{pmatrix} = \lambda \begin{pmatrix} -1 \\ 1 \\ 1 \\ 0 \end{pmatrix} + \mu \begin{pmatrix} -1 \\ 2 \\ 0 \\ 1 \end{pmatrix}$.

10. $-\frac{1}{8} \begin{pmatrix} -12 & 0 & 4 \\ -4 & 2 & -2 \\ 0 & 2 & -6 \end{pmatrix}$; $x = -2$, $y = 3\frac{1}{4}$, $z = 8\frac{1}{4}$.

Exercise 15.8 (p. 479)

1. (i) CW rotation of $90°$ about O; (ii) half-turn about O; (iii) identity.

2. (ii) $\begin{pmatrix} 2 & 1 \\ 1 & 2 \end{pmatrix}$.

4. Identity, CW rotation of $\pi/2$ about O, reflection Ox, reflection in $y = x$.

5. $\pm \tan^{-1}(\frac{1}{3})$.

6. $\begin{pmatrix} 1 & 2 & -2 \\ 3 & 0 & 3 \\ 2 & -2 & 5 \end{pmatrix}$; (a) $x - y + z = 0$, (b) $(-1, 3, 4)$, (c) $\dfrac{x}{2} = \dfrac{y}{-3} = \dfrac{z}{-2}$.

7. $x' = y + 1$, $y' = x - 1$; (a) $x' = -x$, $y' = -y - 2$; (b) $x' = -x$, $y' = -y.180°$.

8. $T_1 = \begin{pmatrix} \cos\alpha & -\sin\alpha \\ \sin\alpha & \cos\alpha \end{pmatrix}$; $T_2 = \begin{pmatrix} 1 & 0 \\ 0 & -1 \end{pmatrix}$; $\begin{pmatrix} \frac{7}{25} & \frac{24}{25} \\ \frac{24}{25} & -\frac{7}{25} \end{pmatrix}$.

9. (a) $\begin{pmatrix} \cos 2\theta & \sin 2\theta \\ \sin 2\theta & -\cos 2\theta \end{pmatrix}$; (b) $\begin{pmatrix} \cos\theta & -\sin\theta \\ \sin\theta & \cos\theta \end{pmatrix}$.

$T_1 = \begin{pmatrix} 0 & 1 \\ 1 & 0 \end{pmatrix}$, $T_2 = \begin{pmatrix} -\frac{1}{2} & \frac{1}{2}\sqrt{3} \\ \frac{1}{2}\sqrt{3} & \frac{1}{2} \end{pmatrix}$, $T_2T_1 = \begin{pmatrix} \frac{1}{2}\sqrt{3} & -\frac{1}{2} \\ \frac{1}{2} & \frac{1}{2}\sqrt{3} \end{pmatrix}$;

rotation through $\pi/6$ about O.

$T_3T_2T_1 = \begin{pmatrix} \sqrt{3} & -1 \\ 1 & \sqrt{3} \end{pmatrix}$, $T_3T_3T_2T_1 \begin{pmatrix} 0 \\ 1 \end{pmatrix} = \begin{pmatrix} -1 & -\sqrt{3} \\ \sqrt{3} & -1 \end{pmatrix}$.

10. $M_1 = \begin{pmatrix} \cos\alpha & -\sin\alpha \\ -\sin\alpha & -\cos\alpha \end{pmatrix}$, $M_2 = \begin{pmatrix} \cos\alpha & \sin\alpha \\ \sin\alpha & -\cos\alpha \end{pmatrix}$.

11. $T_2 = \begin{pmatrix} 1 & 0 \\ 0 & -1 \end{pmatrix}$; $T_\theta = \begin{pmatrix} \cos 2\theta & \sin 2\theta \\ \sin 2\theta & -\cos 2\theta \end{pmatrix}$; $2(\phi - \theta)$.

12. $2(\beta - \alpha)$.

13. (a) $\begin{pmatrix} 0 & 1 \\ 1 & 0 \end{pmatrix}$, (b) $\begin{pmatrix} \frac{1}{2} & -\frac{1}{2}\sqrt{3} \\ \frac{1}{2}\sqrt{3} & \frac{1}{2} \end{pmatrix}$, (c) $\begin{pmatrix} 2 & 0 \\ 0 & 2 \end{pmatrix}$; $\pm 2 + \sqrt{3}$.

Exercise 15.9 (p. 483)

1. $\lambda^2 - 7\lambda + 10 = 0$; $\lambda = 2$, $\begin{pmatrix} 1 \\ -2 \end{pmatrix}$; $\lambda = 5$, $\begin{pmatrix} 1 \\ 1 \end{pmatrix}$.

2. $\lambda = 1$, $\begin{pmatrix} 5 \\ -3 \end{pmatrix}$; $\lambda = 5$, $\begin{pmatrix} 2 \\ -1 \end{pmatrix}$; $\begin{pmatrix} 3 \\ -2 \end{pmatrix} = \begin{pmatrix} 5 \\ -3 \end{pmatrix} - \begin{pmatrix} 2 \\ -1 \end{pmatrix}$.

3. -1, $\frac{1}{2}(1 \pm i\sqrt{3})$.   **4.** 2, $\begin{pmatrix} 0 \\ 1 \\ -1 \end{pmatrix}$; -1, $\begin{pmatrix} 1 \\ -1 \\ 0 \end{pmatrix}$; 3, $\begin{pmatrix} 1 \\ -1 \\ 1 \end{pmatrix}$.

5. -2, $\frac{1}{\sqrt{318}}\begin{pmatrix} 11 \\ 1 \\ -14 \end{pmatrix}$; 1, $\frac{1}{\sqrt{3}}\begin{pmatrix} -1 \\ 1 \\ 1 \end{pmatrix}$; 3, $\frac{1}{\sqrt{3}}\begin{pmatrix} 1 \\ 1 \\ 1 \end{pmatrix}$.

6. -2, 1, 3; line is invariant; $\mathbf{A}^{-1} = -\frac{1}{6}\begin{pmatrix} -4 & 7 & -5 \\ 2 & -5 & 1 \\ 2 & -8 & 4 \end{pmatrix}$.

8. $\begin{pmatrix} -1 \\ 2 \end{pmatrix}$, $\begin{pmatrix} -2 \\ 1 \end{pmatrix}$; $\begin{pmatrix} -\frac{3}{5} & \frac{4}{5} \\ \frac{4}{5} & \frac{3}{5} \end{pmatrix}$.

9. 4, $\begin{pmatrix} 2 \\ 5 \end{pmatrix}$; 7, $\begin{pmatrix} 1 \\ 1 \end{pmatrix}$. $(C_1 e^{2t} + C_2 e^{-2t})\begin{pmatrix} 2 \\ 5 \end{pmatrix} + (C_3 e^{7t} + C_4 e^{-7t})\begin{pmatrix} 1 \\ 1 \end{pmatrix}$.

Exercise 15.10 (p. 487)

1. 7, $\begin{pmatrix} 1 \\ 1 \end{pmatrix}$; -1, $\begin{pmatrix} 1 \\ -1 \end{pmatrix}$; $\frac{1}{\sqrt{2}}\begin{pmatrix} 1 & 1 \\ -1 & 1 \end{pmatrix}$.   **2.** $\begin{pmatrix} \sqrt{(5/6)} & -\sqrt{(1/6)} \\ \sqrt{(1/6)} & \sqrt{(5/6)} \end{pmatrix}$.

3. $\mathbf{X} = \begin{pmatrix} 4 & 1 \\ -1 & -1 \end{pmatrix}$, $\mathbf{X}^{-1} = -\frac{1}{3}\begin{pmatrix} -1 & -1 \\ 1 & 4 \end{pmatrix}$.

Exercise 15.11 (p. 489)

1. $\mathbf{A}^{-1} = \begin{pmatrix} 0 & -3 & 0 \\ 0 & -3 & 6 \\ -2 & 2 & -4 \end{pmatrix}$, $\mathbf{A}^2 = \begin{pmatrix} 7 & -3 & 0 \\ 0 & 4 & 6 \\ -2 & 2 & 3 \end{pmatrix}$, $\mathbf{A}^3 = \begin{pmatrix} -6 & 14 & 21 \\ 14 & -6 & 0 \\ 7 & -7 & -6 \end{pmatrix}$

2. 2, 3, 5; $\mathbf{A}^3 = \begin{pmatrix} -53 & -61 & 178 \\ 56 & 64 & -56 \\ -61 & -61 & 126 \end{pmatrix}$, $\mathbf{A}^{-1} = \frac{1}{40}\begin{pmatrix} 22 & 2 & -14 \\ 10 & 10 & 10 \\ 2 & 10 & 6 \end{pmatrix}$.

3. (a) $\mathbf{M} = \begin{pmatrix} 4 & -2 & -2 \\ 1 & 1 & -1 \\ 2 & -2 & 0 \end{pmatrix}$, $\mathbf{M}^{-1} = \frac{1}{4}\begin{pmatrix} -2 & 4 & 4 \\ -2 & 4 & 2 \\ -4 & 4 & 6 \end{pmatrix}$; (b) $\frac{x}{2} = \frac{y}{1} = \frac{z}{1}$.

Miscellaneous Exercise 15 (p. 489)

1. $3\mathbf{A} + 4\mathbf{I}$.   **3.** 1, $\begin{pmatrix} 1 \\ -2 \\ 2 \end{pmatrix}$; 2, $\begin{pmatrix} 0 \\ 1 \\ -2 \end{pmatrix}$; 3, $\begin{pmatrix} 0 \\ 0 \\ 1 \end{pmatrix}$; $\begin{pmatrix} 1 \\ 1 \\ 1 \end{pmatrix} = \mathbf{a} + 3\mathbf{b} + 5\mathbf{c}$.

4. (i) (a) $x = (p-1)/(p^2+3)$, $y = 2/(p^2+3)$, $z = -(p+1)/(p^2+3)$;
(b) $x = z$, $y = 1 + z$.

5. $55(k-1)$. (a) $x = -3$, $y = 2$, $z = 3$; (b) $\dfrac{x}{10} = -y = -\dfrac{z}{11}$, three planes meet in this line.

6. $\mathbf{L} = \begin{pmatrix} 1 & 0 & 0 \\ 2 & 1 & 0 \\ 3 & 2 & 1 \end{pmatrix}$, $\mathbf{U} = \begin{pmatrix} 3 & 1 & 2 \\ 0 & -1 & 1 \\ 0 & 0 & -3 \end{pmatrix}$; $\mathbf{L}^{-1} = \begin{pmatrix} 1 & 0 & 0 \\ -2 & 1 & 0 \\ 1 & -2 & 1 \end{pmatrix}$.

7. $55(k-1)$; $x=y=z=0$; $x=-2$, $y=1$, $z=3$.

8. $x=4+a$, $y=-2$, $z=-1$; $x=\frac{1}{2}(17-23z)$, $y=\frac{1}{2}(3z-1)$.

9. (i) $\begin{pmatrix} \frac{3}{2} & 0 & 0 \\ 0 & \frac{5}{4} & 0 \\ -\frac{1}{2} & 0 & 2 \end{pmatrix}$; (ii) $4(1-k)(1+k^2)$; $x=-9k$, $y=5k$, $z=7k$.

10. (i) $\begin{pmatrix} 1 & 2 \\ 0 & -1 \end{pmatrix}$, $\begin{pmatrix} -1 & -2 \\ 0 & 1 \end{pmatrix}$, $\begin{pmatrix} 1 & 0 \\ 0 & 1 \end{pmatrix}$, $\begin{pmatrix} -1 & 0 \\ 0 & -1 \end{pmatrix}$;

(ii) $\begin{pmatrix} 8 & 0 & 0 \\ 0 & 8 & 0 \\ 0 & 0 & 8 \end{pmatrix}$; $x=5$, $y=-3$, $z=-1$.

11. $\begin{pmatrix} 1 \\ 1 \\ -1 \end{pmatrix}$, $\begin{pmatrix} 1 \\ 1 \\ -2 \end{pmatrix}$, $\begin{pmatrix} 0 \\ 1 \\ -1 \end{pmatrix}$; $\begin{pmatrix} 1 & 1 & 0 \\ 1 & 1 & 1 \\ -1 & -2 & -1 \end{pmatrix}$, $\begin{pmatrix} 1 & 1 & 1 \\ 0 & -1 & -1 \\ -1 & 1 & 0 \end{pmatrix}$;

$\mathbf{C} = \begin{pmatrix} 1 & 0 & 0 \\ 0 & 2 & 0 \\ 0 & 0 & 3 \end{pmatrix}$, $\mathbf{B} = \begin{pmatrix} 1 & -1 & -1 \\ -2 & 2 & -1 \\ 2 & 0 & 3 \end{pmatrix}$.

12. $\begin{pmatrix} \sqrt{2} \\ 1 \end{pmatrix}$, $\begin{pmatrix} -\sqrt{2} \\ 1 \end{pmatrix}$.

13. $a=2$, $b=-\frac{4}{5}$, $c=\frac{3}{5}$; $x_1=\frac{1}{5}$, $x_2=1$, $x_3=-\frac{2}{5}$.

15. (i) $x=1$, $y=-2$, $z=1$; (ii) no solution; (iii) $x=z-\frac{1}{2}$, $y=\frac{3}{2}-3z$.
(i) 8; (ii) 16.

17. ±1.   **18.** ACW rotation of θ about O; reflection in $y=x\tan\phi$.
19. (i) $38x=-95y=10z$; (ii) $10x-13y-4z=0$; (iii) $y=0$.

Exercise 16.2 (p. 497)

1. $x+\frac{1}{2}x^2+\frac{1}{2}x^3+\frac{13}{24}x^4$.   **2.** $x+\frac{1}{6}x^3$.   **5.** $1+\frac{1}{2}x^2-\frac{1}{6}x^4$.

Exercise 16.3 (p. 501)

2. $32(2x^3+12x^2+9x-15)e^{2x}$.   **6.** (i) $120/x$; (ii) $-512e^{2x}\cos 2x$.

8. 1.   **9.** $1-\dfrac{x^2}{2^2}+\dfrac{x^4}{2^4(2!)^2}-\dfrac{x^6}{2^6(3!)^2}+\dots+\dfrac{(-1)^n x^{2n}}{2^{2n}(n!)^2}\dots$.

Exercise 16.6 (p. 508)

2. (i) $\pi/6$; (ii) $5\pi/6$; (iii) $-\frac{1}{4}\pi$; (iv) $\frac{1}{2}\pi$; (v) $-\frac{1}{4}\pi$; (vi) $\frac{3}{4}\pi$.
3. (i) $\mathrm{Tan}^{-1}(-3)$, second; (ii) $\sin^{-1}(63/65)$, first; (iii) $\sin^{-1}(7/25)$, first; (iv) $\frac{1}{2}\pi$;
(v) $\tan^{-1}1=\frac{1}{4}\pi$, first.

4. $\mathrm{Cot}^{-1}x=(n+\frac{1}{2})\pi-\alpha$; $\sin\alpha=\dfrac{(-1)^n x}{\sqrt{(1+x^2)}}$, $\cos\alpha=\dfrac{(-1)^n}{\sqrt{(1+x^2)}}$.

5. $x=\frac{3}{4}$.   **7.** $\frac{1}{2}\pi+\alpha$.
10. (i) $2/\sqrt{(1-4x^2)}$; (ii) $-1/\sqrt{(4-x^2)}$; (iii) $1/(1+x^2)$; (iv) $a/(a^2+x^2)$; (v) $-1/[x\sqrt{(x^2-1)}]$;
(vi) $\cos x/(1+\sin^2 x)$; (vii) $2\sec^2 x/(1+4\tan^2 x)$; (viii) -1; (ix) -2;
(x) $2e^{2x}/(e^{4x}+2e^{2x}+2)$.
11. (i) $-2/\sqrt{3}$; (ii) $1/8$; (iii) ∞; (iv) $-2/\sqrt{3}$.

12. (i) $\dfrac{1}{x\sqrt{(x^2-1)}}$; (ii) $\dfrac{-1}{x\sqrt{(x^2-1)}}$; (iii) $-\dfrac{1}{1+x^2}$.

14. $2x - y\sqrt{3} - 1 + \pi\sqrt{3}/6 = 0$, $2x + y\sqrt{3} - 1 - 5\pi\sqrt{3}/6 = 0$; $\pi^2\sqrt{3}/8$.

15. $+0.064$.   **17.** 6 radians per min; zero.

18. (i) 1 rad s^{-1}; (ii) $4/17$ rad s^{-1}.   **19.** $\ln[x + \sqrt{(x^2-1)}]$.

20. (i) $x\cos^{-1}x - \sqrt{(1-x^2)}$; (ii) $x\tan^{-1}x - \tfrac12\ln(1+x^2)$;

(iii) $x\sec^{-1}x - \ln[x + \sqrt{(x^2-1)}]$; (iv) $x\cot^{-1}x + \tfrac12\ln(1+x^2)$.

21. (i) $\tfrac12\pi a^2$; (ii) $b = \tfrac13 a$ or $\tfrac12 a$.   **22.** $\tfrac14\pi(\pi+2)a^3$.

23. (i) $\pi/6$; (ii) $\pi/6\sqrt3$; (iii) $\pi/9$; (iv) $\tan^{-1}(\tfrac12) \approx 0.46$; (v) $\pi/6$.

24. (i) $\tan^{-1}(x+2)$; (ii) $\dfrac{2}{\sqrt{15}}\tan^{-1}\left(\dfrac{4x-3}{\sqrt{15}}\right)$; (iii) $\sin^{-1}\left(\dfrac{x+1}{\sqrt2}\right)$;

(iv) $\dfrac{1}{\sqrt3}\sin^{-1}[\tfrac12(x+1)\sqrt3]$; (v) $\dfrac{1}{\sqrt{15}}\tan^{-1}\left[\dfrac{(x+1)\sqrt5}{\sqrt3}\right]$.

26. $\tfrac14\pi - \tfrac12\ln2$.   **27.** $\tfrac12(x^2+1)\tan^{-1}x - \tfrac12 x$.

28. $\ln(x-2) - \ln(x+2) + \tfrac12\tan^{-1}(\tfrac12 x)$.

29. $\tfrac12\tan^{-1}(15/17)$.   **30.** $\pi/12$.   **31.** $\sqrt2 - 1$.

Exercise 16.7 (p. 515)

11. $13/12$, $5/13$.   **12.** $\tfrac14\sqrt7$, $3/\sqrt7$.

13. $\ln2$, $-\ln4$.   **14.** $\pm\tfrac12\ln3$.   **15.** $\ln4$.

17. (i) $2\sinh(2x+5)$; (ii) $x\cosh x + \sinh x$; (iii) $2\cosh2x\cosh3x + 3\sinh2x\sinh3x$;

(iv) $e^x(\tanh x + \operatorname{sech}^2 x)$; (v) $\tfrac12(1 - 1/x^2)$, [note $\cosh\ln x \equiv \tfrac12(x + 1/x)$];

(vi) $-[(1 + x\tanh x)\operatorname{sech}x]/x^2$; (vii) $\cos x.\cosh(\sin x)$; (viii) $\operatorname{sech}x$;

(ix) $4\operatorname{cosech}4x$; (x) $(\cosh x + \sinh x)e^{\sinh x + \cosh x}$; (xi) $-\operatorname{sech}x$; (xii) 0.

18. $bx\cosh t_1 - ay\sinh t_1 = ab$; $ax\sinh t_1 + by\cosh t_1 = (a^2 + b^2)\sinh t_1\cosh t_1$.

19. $PT = [c\cosh^2(x/c)]/\sinh(x/c)$; $PG = c\cosh^2(x/c)$; $TN = c\coth(x/c)$;

$NG = c\sinh(x/c)\cosh(x/c)$; (i) $NZ = c$.

20. $A = 2$, $B = 4$.   **21.** 328; 1310.   **22.** 5 (when $x = -\ln5$).

23. When $|n| > 1$, a minimum $(n^2 - 1)$ at $x = \ln\{(n-1)/(n+1)\}$;

when $|n| < 1$, a minimum $(1 - n^2)$ at $x = \ln\{(1-n)/(1+n)\}$;

when $n = \pm1$ the function is $2e^{\pm x}$ and has no stationary values.

24. (i) $\tfrac12\sinh2x$; (ii) $\ln\cosh x$; (iii) $\ln\sinh x$; (iv) $\tanh x$; (v) $-\operatorname{sech}x$;

(vi) $-\operatorname{cosech}x$; (vii) $\tfrac12(x + \tfrac12\sinh2x)$; (viii) $(\cosh6x + 3\cosh2x)/12$;

(ix) $x\cosh x - \sinh x$; (x) $\tfrac12(\tfrac12 e^{2x} + x)$.

25. (i) $1 - e^{-1}$; (ii) $(\sinh3 + 9\sinh1)/12$; (iii) $(\sinh3 + 3\sinh1)/12$;

(iv) $2\tan^{-1}(e^2) - \tfrac12\pi$; (v) $\tanh1 - \ln\cosh1$.

26. $e - 1 - 2\tan^{-1}e + \tfrac12\pi$.   **27.** $1 - e^{-1}$.

28. (i) 3.31; (ii) 0.707.   **29.** $2\pi(1 - \tanh1) = 4\pi/(1 + e^2)$.

Exercise 16.11 (p. 521)

2. $\pm\ln(2 + \sqrt3)$.   **3.** $\pm\ln[\tfrac13(4 + \sqrt7)]$.

4. $x = -\ln2$, $y = \ln(5/2)$ or $x = \ln(5/2)$, $y = -\ln2$.

5. (i) $\tfrac12\ln\left(\dfrac{1+x}{1-x}\right)$; (ii) $\pm\ln\left[\dfrac{1 + \sqrt{(1-x^2)}}{x}\right]$; (iii) $\ln\left[\dfrac{1 + \sqrt{(x^2+1)}}{x}\right]$;

(iv) $\ln\left[\dfrac{1 + \sqrt{(1+x^2)}}{x}\right]$; (v) $\pm\ln[1 + x^2 + \sqrt{(x^4 + 2x^2)}]$.

6. (i) $-1/(x^2 - 1)$; (ii) $-1/[x\sqrt{(1-x^2)}]$; (iii) $-1/[x\sqrt{(1+x^2)}]$.

7. (i) $\cosh^{-1}x + x/\sqrt{(x^2-1)}$; (ii) $1/(2x)$; (iii) $\sec x$;

(iv) $-1/[x\sqrt{(1-x^2)}]$; (v) $2\cosh2x/\sqrt{(\sinh^2 2x - 1)}$.

8. (i) $9[\ln(1 + \sqrt2) + \sqrt2]/2 \approx 10.3$; (ii) $8[2\sqrt3 - \ln(2 + \sqrt3)] \approx 17.2$.

9. (i) $\tfrac12[x\sqrt{(1+x^2)} - \sinh^{-1}x]$; (ii) $\sqrt{(x^2-1)}/ - \sec^{-1}x$.

10. (i) $x \sinh^{-1} x - \sqrt{(x^2 + 1)}$; (ii) $x \cosh^{-1} x - \sqrt{(x^2 - 1)}$; (iii) $x \tanh^{-1} x + \frac{1}{2}\ln(1 - x^2)$;
(iv) $\frac{1}{2}x^2 \cosh^{-1} x - \frac{1}{4}x\sqrt{(x^2 - 1)} - \frac{1}{4}\cosh^{-1} x$.

11. $ab[2\sqrt{3} - \ln(2 + \sqrt{3})]$.

12. (i) $\ln[3(1 + \sqrt{2})/(2 + \sqrt{13})]$; (ii) $\frac{1}{2}\ln[(8 + \sqrt{63})/(4 + \sqrt{15})]$;
(iii) $\ln[(3 + \sqrt{10})/(2 + \sqrt{5})]$; (iv) $(1/\sqrt{2})\ln[(\sqrt{18} + \sqrt{15})/(\sqrt{8} + \sqrt{5})]$;
(v) $\ln[\{a + b + \sqrt{(a^2 + 2ab + 2b^2)}\}/\{2a + \sqrt{(4a^2 + b^2)}\}]$.

13. $2a^2 \ln(1 + \sqrt{2})$.

15. $\sinh^{-1}\{(x - 1)/\sqrt{2}\}$.

16. $\sin^{-1}\{(x + 2)/\sqrt{5}\}$.

17. $\cosh^{-1}\{(x - 1)/\sqrt{6}\}$.

18. $\dfrac{1}{\sqrt{17}} \ln\left(\dfrac{4x + 5 - \sqrt{17}}{4x + 5 + \sqrt{17}}\right)$.

19. $\dfrac{1}{\sqrt{13}} \ln\left(\dfrac{\sqrt{13} + 1 + 2x}{\sqrt{13} - 1 - 2x}\right)$.

20. $\frac{1}{2}\tan^{-1}\left[\frac{1}{2}(x + 1)\right]$.

21. $\dfrac{1}{\sqrt{2}} \sin^{-1}\left\{\dfrac{(x + 1)\sqrt{2}}{\sqrt{3}}\right\}$.

22. $\dfrac{1}{\sqrt{3}} \cosh^{-1}\left[\dfrac{(x - 1)\sqrt{3}}{\sqrt{8}}\right]$.

23. $\dfrac{1}{2\sqrt{14}} \ln\left(\dfrac{\sqrt{7} + (x + 1)\sqrt{2}}{\sqrt{7} - (x + 1)\sqrt{2}}\right)$.

24. $\sinh^{-1}\left[\dfrac{(2x + 1)}{\sqrt{3}}\right]$.

Exercise 16.12 (p. 523)

1. $\frac{1}{2}\ln(x^2 + 5x + 1) - \dfrac{1}{2\sqrt{21}} \ln\left(\dfrac{2x + 5 - \sqrt{21}}{2x + 5 + \sqrt{21}}\right)$.

2. $\frac{1}{2}\ln(2x^2 + x + 3) + \dfrac{5}{\sqrt{23}} \tan^{-1}\left(\dfrac{4x + 1}{\sqrt{23}}\right)$.

3. $3\sqrt{(x^2 + x + 1)} + \frac{1}{2}\sinh^{-1}\left(\dfrac{2x + 1}{\sqrt{3}}\right)$.

4. $5\sqrt{(x^2 - 3x - 1)} + \dfrac{15}{2}\cosh^{-1}\left(\dfrac{2x - 3}{\sqrt{13}}\right)$.

5. $\frac{3}{4}\ln(2x^2 + x + 5) + \dfrac{13}{2\sqrt{39}} \tan^{-1}\left(\dfrac{4x + 1}{\sqrt{39}}\right)$.

6. $\frac{1}{8}\ln(4x^2 + 2x - 3) - \dfrac{1}{8\sqrt{13}} \ln\left(\dfrac{4x + 1 - \sqrt{13}}{4x + 1 + \sqrt{13}}\right)$.

7. $-\sqrt{(1 - 3x - x^2)} - \frac{1}{2}\sin^{-1}\left(\dfrac{2x + 3}{\sqrt{13}}\right)$.

8. $\dfrac{6}{\sqrt{5}} \ln\left(\dfrac{\sqrt{5} - 2 + x}{\sqrt{5} + 2 - x}\right) - \dfrac{5}{2}\ln(1 + 4x - x^2)$.

9. $\sqrt{(x^2 + 6x + 4)} - \cosh^{-1}\{(x + 3)/\sqrt{5}\}$.

10. $3x + \ln(x^2 + x + 1) + \dfrac{4}{\sqrt{3}} \tan^{-1}\left(\dfrac{2x + 1}{\sqrt{3}}\right)$.

Exercise 16.13 (p. 527)

1. $x \tan x + \ln \cos x$.

2. $\frac{1}{2}\sec x \tan x - \frac{1}{2}\ln(\sec x + \tan x)$.

3. $x \ln x - x$.

4. $\frac{1}{2}x^2 \ln x - \frac{1}{4}x^2$.

5. $x \tanh^{-1} x + \frac{1}{2}\ln(1 - x^2)$.

6. $x \sinh^{-1}\left(\dfrac{x}{a}\right) - \sqrt{(a^2 + x^2)}$.

7. $x \sinh^{-1}\left(\dfrac{a}{x}\right) + a \sinh^{-1}\left(\dfrac{x}{a}\right).$ **8.** $\frac{1}{3}(x^3 - 3x^2 + 3x) \ln x - \dfrac{x^3}{9} + \dfrac{x^2}{2} - x.$

9. $(x + \frac{1}{3}x^3) \tan^{-1} x - \frac{1}{6}x^2 - \frac{1}{3} \ln (1 + x^2).$

10. $-\sqrt{(1 - x^2)} \sin^{-1} x + x.$ **11.** $\frac{1}{4}x^2 \{2 (\ln x)^2 - 2 \ln x + 1\}.$

12. $\dfrac{(3x \sin^3 x + 3 \cos x - \cos^3 x)}{9}.$ **13.** $x \sec^{-1}\left(\dfrac{x}{a}\right) - a \cosh^{-1}\left(\dfrac{x}{a}\right).$

14. $\frac{1}{2}a^2 \sin^{-1}\left(\dfrac{x}{a}\right) - \frac{1}{2}x \sqrt{(a^2 - x^2)}.$

15. $\dfrac{1}{a}(1 - a \cos x) \ln (1 - a \cos x) + \cos x.$

16. $\frac{1}{4}(2x^2 - 1) \sin^{-1} x + \frac{1}{4}x \sqrt{(1 - x^2)}.$

17. $-\frac{1}{2}e^{-x^2}.$ **18.** $\frac{1}{2} \tan^2 x.$ **19.** $\tan x + \frac{1}{3} \tan^3 x.$ **20.** $\frac{1}{2} (\ln x)^2.$

21. $\frac{1}{5} \sin^5 x - \frac{1}{7} \sin^7 x.$ **22.** $-\frac{1}{4} \ln (1 - x^4).$ **23.** $\frac{1}{6} \tan^{-1} (\frac{1}{3}x^2).$

24. $\frac{1}{3}(a^2 + x^2)^{3/2}.$ **25.** $\frac{1}{3} \ln (5 + 3e^x).$

26. $-\frac{1}{3}a^2(a^2 - x^2)^{3/2} + \frac{1}{5}(a^2 - x^2)^{5/2}.$

27. $-\frac{1}{12} \tan^{-1} (\frac{3}{4} \cos x).$ **28.** $\dfrac{1}{a} \sinh^{-1} (e^{ax}/b).$

29. $c \sin^{-1}\left(\sqrt{\dfrac{x}{c}}\right) - \sqrt{[x(c - x)]}.$

30. $\frac{1}{3} \tan^{-1} (e^{3x}).$ **31** $(n + 1)^{-1} \sinh^{n+1} x + (n + 3)^{-1} \sinh^{n+3} x.$

32. $-\dfrac{1}{b} \ln (a + b \cos x).$ **33.** $\dfrac{1}{\sqrt{6}} \tan^{-1}\left(\dfrac{\sqrt{2} \tan x}{\sqrt{3}}\right).$

34. $\frac{2}{3} \tan^{-1} (3e^x).$ **35.** $\frac{2}{7} \sec^{7/2} x - \frac{2}{3} \sec^{3/2} x.$

36. $\frac{1}{2}a^2 \sin^{-1} (x/a) - \frac{1}{2}(2a + x) \sqrt{(a^2 - x^2)}.$

37. $-(x^2 + 1)/x + \ln [x/(1 - x)^2].$ **38.** $\frac{1}{5} \ln [x^5/(1 + x^5)].$

39. $\dfrac{-2}{(1 + \tan \frac{1}{2}x)}.$ **40.** $x - \dfrac{3}{\sqrt{2}} \tan^{-1}\left(\dfrac{\tan \frac{1}{2}x}{\sqrt{2}}\right).$

41. $\frac{1}{3} \tan^{-1} (3 \tan x).$ **42.** $2 \sqrt{(x - 1)} - 2 \tan^{-1} \sqrt{(x - 1)}.$

43. $-2 \ln (1 - \sqrt{x}).$ **44.** $2x - 3 \ln (2 + \tan \frac{1}{2}x) - 2 \ln \sec \frac{1}{2}x.$

45. $\frac{1}{6} \ln (x + 2 \sqrt{x} + 4) - \frac{1}{3} \ln (2 - \sqrt{x}) - \dfrac{1}{\sqrt{3}} \tan^{-1}\left(\dfrac{1 + \sqrt{x}}{\sqrt{3}}\right).$

46. $2 \sqrt{(x - 1)} + 2 \tan^{-1} \sqrt{(x - 1)}.$

47. $2 \sqrt{(2x^2 + 7x + 3)} - \sqrt{2} \cosh^{-1}\left(\dfrac{4x + 7}{5}\right).$

48. $-\dfrac{1}{\sqrt{6}} \tan^{-1}\left(\dfrac{\sqrt{3}.\cos x}{\sqrt{2}}\right).$ **49.** $3 \tan \frac{1}{2}x - \ln (1 + \cos x).$

50. $\dfrac{1}{\sqrt{2}} \ln \left(\dfrac{\sqrt{2} + \sqrt{(x + 1)}}{\sqrt{2} - \sqrt{(x + 1)}}\right).$ **51.** $\ln (1 + \tan \frac{1}{2}x).$

52. $\ln (\sin x + \cos x).$

53. $\frac{1}{3}(x^2 + x + 1)^{3/2} + \frac{3}{16} \sinh^{-1}\left(\dfrac{2x + 1}{\sqrt{3}}\right) + \frac{1}{8}(2x + 1) \sqrt{(x^2 + x + 1)}.$

54. $\tanh x - \frac{1}{3} \tanh^3 x.$ **55.** $\ln (1 + e^{-1}).$

56. $\frac{1}{2}(b - a)^{-1} \ln (b/a).$ **57.** $7/(3 \ln 2).$

58. $\dfrac{1 + \frac{1}{2} \ln 2}{\sqrt{2}} - 1.$ **59.** $\frac{1}{4}\pi - \frac{1}{2} \ln 2.$

60. $\frac{1}{4}\pi - \frac{1}{2}\ln 2 - \pi^2/32.$

61. $\frac{1}{2}\pi - 1.$

62. $\frac{1}{4}(\pi - 1).$

63. $\dfrac{(\pi^2 + 4)}{16}.$

64. $\dfrac{1}{\sqrt{5}}\ln\left(\dfrac{5 + \sqrt{5}}{5 - \sqrt{5}}\right).$

65. $\dfrac{\pi}{4\sqrt{3}}.$

66. $3 - 2\sqrt{2}.$

67. $\sin^{-1}(\frac{1}{5}).$

68. $\frac{1}{3}\ln 4.$

69. $\frac{1}{2}\ln 3 - \dfrac{1}{2\sqrt{2}}\ln\left(\dfrac{2\sqrt{2}+1}{2\sqrt{2}-1}\right).$

70. $1.$

71. $4(\ln 3 - 1).$

72. $\frac{1}{5}\ln 6.$

73. $\ln 2.$

74. $\dfrac{2}{1-a^2}\tan^{-1}\left|\dfrac{1+a}{1-a}\right|.$

75. $\dfrac{\pi - \alpha}{\sin \alpha}.$

76. $\dfrac{(16 - 5\pi)}{4}.$

77. $\ln(\frac{3}{2}).$

78. $\frac{1}{8}\ln 3.$

79. $\dfrac{\pi}{\sqrt{3}}.$

80. $\frac{1}{2}\pi.$ **81.** $\pi/6 - 2/9.$ **82.** $(\pi - 2\ln 2)/8.$
83. $\frac{1}{6}\ln(4/3) + \pi/(6\sqrt{3}).$ **84.** $\frac{1}{2} - \ln(4/3) + \pi/(3\sqrt{3}).$
85. $\frac{3}{8}\ln(4/3) - \frac{1}{8}.$ **86.** $\frac{1}{12}(\pi - 2 + 2\ln 2).$
87. $\frac{1}{4}x^4 + \frac{1}{3}x^3 + \frac{3}{2}x^2 + 5x + \frac{1}{3}\ln(x+1) + \frac{32}{3}\ln(x-2).$

Exercise 16.14   (p. 531)

1. $(\cos x + \sin x)(\cosh y - i\sinh y).$
2. (i) $\sinh 1 \cos 1 - i\cosh 1 \sin 1$;   (ii) $\cosh 1 \cos 1 - i\sinh 1 \sin 1.$
4. $\sinh 2x/(\cosh 2x + \cos 2y),\ \sin 2y/(\cosh 2x + \cos 2y).$
6. (i) $(2n + \frac{1}{2})\pi \pm i\ln(2 + \sqrt{3})$;   (ii) $2n\pi \pm i\ln(3 + \sqrt{8})$;
   (iii) $(2n - \frac{1}{2})\pi \pm \ln(2 + \sqrt{3}).$
7. $\theta + r\sin\theta = 0,\ \pm\pi,\ \pm 2\pi,\ \ldots.$
9. $z = \frac{1}{2}\ln 3 + i(\frac{1}{2} \pm n)\pi.$

Exercise 16.15   (p. 536)

3. (ii) $a = -5/2,\ b = -3/2.$ **8.** (i) $2, -3$;   (ii) $-3/2.$ **11.** $a = \pm 1/5,\ b = 2.$

Miscellaneous Exercise 16   (p. 539)

1. $x = \ln 3,\ y = \ln 2.$ **3.** $A = c^2(1 - e^{-h/c})$; $V_1 = \pi c^2 h.$
4. $1 - (x\sin^{-1} x)/\sqrt{(1-x^2)}$; $y.$
5. (i) $2\tan^{-1} x$;   (ii) $\frac{1}{2}[(2-x)(x-1)]^{-1/2}$; $1 \leqslant x \leqslant 2.$
8. $x = y = \pm\frac{1}{2}\ln(3 + 2\sqrt{2}).$ **10.** $\ln(7/5).$ **11.** $-\operatorname{cosech} t.$
13. $\ln[\frac{1}{2}(3 + \sqrt{13})],\quad \ln[\frac{1}{2}(1 + \sqrt{5})]$; $\ln(1 + \sqrt{2}) + 1 - \sqrt{2}.$
14. (ii) $\ln 3$;   (iii) $\ln(1 + \sqrt{2}) + 1 - \sqrt{2}.$
16. (i) $x,\ y = \ln(3 \pm \sqrt{6})$;   (ii) $\ln[(11 + 5\sqrt{5})/(6 + 4\sqrt{2})].$
17. $x,\ y = \pm\frac{1}{2}\ln[(5 + \sqrt{24})(3 + \sqrt{8})],\ \pm\frac{1}{2}\ln[(5 + \sqrt{24})/(3 + \sqrt{8})].$
19. $2n\pi \pm \ln(2 + \sqrt{3}),\ (2n \pm \frac{1}{3})\pi i.$
21. $u = \cosh x \cos y,\ v = \sinh x \sin y.$ **25.** $A = 0,\ \omega = 1,\ m = 1,\ B = 1.$
28. (i) $6(-1)^n (n-4)!\, x^{3-n}$;   (ii) $\frac{1}{4}[6^n\cos(6x + \frac{1}{2}n\pi) + 3 \times 2^n\cos(2x + \frac{1}{2}n\pi)]$;
   (iii) $-(n-1)![(1-x)^{-n} + (-1)^{n-1}(1+x)^{-n}]$;
   (iv) $2^n x^2 \cos(2x + \frac{1}{2}n\pi) + n2^n x \cos[2x + \frac{1}{2}(n-1)\pi] + n(n-1)2^{n-2}\cos[2x + \frac{1}{2}(n-2)\pi].$
30. (a) $\frac{1}{3}\pi$;   (b) $0.$ **32.** $-\sinh^{-1}(1/[(x-3)\sqrt{2}]).$
33. Each derivative is $2(x^2 + 2)/(x^4 + 4).$ The functions are equal to each other.

36. (i) $\cosh^{-1}(x-2)$;   (ii) $\frac{1}{3}\ln x - \frac{1}{2}\ln(x-1) + \frac{1}{6}\ln(x-3)$;   (iii) $\frac{1}{32}(\sin 4x - 8\sin 2x + 12x)$.

37. (i) $\frac{1}{25}[3\ln(x+1) - \frac{3}{2}\ln(x^2+4) + 5(x+1)^{-1} - 4\tan^{-1}(\frac{1}{2}x)]$;   (ii) $\pi/\sqrt{7}$;

   (iii) $\frac{1}{2}(\sinh^{-1}2 - \sinh^{-1}1) = \frac{1}{2}\ln[(2+\sqrt{5})/(1+\sqrt{2})]$.

38. (i) $\sin^{-1}x - \sqrt{(1-x^2)}$;   (ii) $\frac{1}{2}(x^2+1)\tan^{-1}x - \frac{1}{2}x$.

39. (i) $\frac{1}{2}\ln(8/5)$;   (ii) $(\pi-2)/8$;   (iii) $\theta + 2/(1+\tan\frac{1}{2}\theta)$.

40. (i) $x + \frac{3}{2}\ln(2x+1) - \ln(x^2+2) - \dfrac{1}{\sqrt{2}}\tan^{-1}(x/\sqrt{2})$;   (iii) $-\cosh^{-1}[(x+1)/x\sqrt{2}]$.

41. (i) $\frac{1}{2}e^x(\cos x + \sin x)$;   (ii) $\sqrt{(x^2+2x+3)} + \cosh^{-1}[\frac{1}{2}(x+1)]$.

42. $(2/\sqrt{3})\tan^{-1}(e^x\sqrt{3})$.

43. (i) $\frac{1}{4}[\ln x - \frac{1}{2}\ln(x^2+4) + 2\tan^{-1}(\frac{1}{2}x)]$;   (ii) $-\sqrt{(4x-x^2)} + 2\sin^{-1}[\frac{1}{2}(x-2)]$;

   (iii) $\frac{1}{2}x^2(\ln x)^2 - \frac{1}{2}x^2\ln x + \frac{1}{4}x^2$.

45. (a) $9(\sqrt{2}-1) + 2\sinh^{-1}1 = 9(\sqrt{2}-1) + 2\ln(1+\sqrt{2})$,   (b) $\frac{1}{3}\tan^{-1}(e^{3x})$.

46. (i) (a) $x^{n-1}[(n+1)\ln x - 1](n+1)^{-2}$,   (b) $\frac{1}{2}(\ln x)^2$; $\frac{1}{4}e^2 - 2e^{-1} + 23/4$.   (ii) $15/2 + 8\ln 2$.

Exercise 17.1   (p. 548)

1. (i) $\to 0$;   (ii) $\to \infty$;   (iii) oscillates boundedly taking the values 1, $\frac{1}{2}$, $-\frac{1}{2}$, -1, $-\frac{1}{2}$, $\frac{1}{2}$, ... in succession;   (iv) oscillates unboundedly taking the values in (iii) multiplied by n;   (v) $\to 0$ if $|x| < 1$, $\to +\infty$ if $x \geqslant 1$, oscillates unboundedly if $x \leqslant -1$.

2. (i) Any oscillating sequence u_n for which $|u_n| \to 0$.

4. If $|a| < 1$, $u_n \to -1$; if $|a| > 1$, $u_n \to 1$; if $a = 1$, $u_n = 0$; if $a = -1$, the odd members of the sequence are undefined.

6. $a_n \to \sqrt{k}$; $a_n \to -\sqrt{k}$.   **7.** For $a = 1$, $x_n \to 2$; for $a = 3$, $x_n \to 2$.

Exercise 17.2   (p. 555)

1. Convergent for $|x| \leqslant 1$, divergent for $|x| > 1$.   **2.** Convergent for all x.

3. Divergent.   **4.** Convergent for $|x| < a$, divergent for $|x| \geqslant a$.

5. Divergent unless $x = 0$.   **6.** Divergent.

8. (i) Absolutely convergent for $|x| < 1$, convergent when $x = 1$, divergent when $x \leqslant -1$, oscillates unboundedly when $x > 1$;   (ii) absolutely convergent for all x;   (iii) absolutely convergent for $|x| < 2$, convergent when $x = -2$, divergent when $x \geqslant 2$, oscillates (unboundedly) when $x < -2$;   (iv) absolutely convergent when $|x| < 1$, convergent when $x = -1$, divergent when $x \geqslant 1$, oscillates unboundedly for $x < -1$;   (v) absolutely convergent for all x.

9. (i) Divergent;   (ii) (absolutely) convergent;   (iii) divergent;   (iv) divergent;   (v) (absolutely) convergent;   (vi) (absolutely) convergent;   (vii) (conditionally) convergent.

10. $e(e-1)$.   **11.** (i) 0;   (ii) e^2.

12. (i) The greater of a and b;   (ii) $-\frac{1}{3}$;   (iii) $1/b$;   (iv) 0;   (v) $[\ln(\frac{5}{4})]/[\ln(\frac{3}{2})]$;   (vi) 0 if $n \neq 0$, ∞ if $n = 0$;   (vii) π;   (viii) $\frac{1}{12}$.

13. $a = -\frac{4}{3}$, $b = \frac{1}{3}$; $\lim = \frac{8}{3}$.

14. (i) e^2;   (ii) -1;   (iii) $-\frac{1}{3}$;   (iv) $\frac{1}{2}$;   (v) $-\ln 2$;   (vi) 0;   (vii) 2;   (viii) e.

15. (a) $\cosh\sqrt{x} - 1$;   (b) $-\ln(1-x)$, $-1 \leqslant x < 1$;   (c) $e^x/(2-e^x)$, $x < \ln 2$.

16. (i) $A = 2$, $B = 2$, $C = 3$.

Exercise 17.4   (p. 561)

7. $x > 0$.   **10.** $(x+1)(2x-1)(x-2)$.   (i) $\{x: -1 \leqslant x \leqslant \frac{1}{2}\} \cup \{x: x \geqslant 2\}$;

   (ii) $\{x: x \leqslant -\ln 2\} \cup \{x: x \geqslant \ln 2\}$.

13. (i) $\{x: \frac{1}{4} < x < \frac{3}{2}\}$;   (ii) $\{x: x < -1\} \cup \{x: 1 < x < 2\}$.

15. Each tends to zero.   **16.** $y = 0$, $e^2y = 4x$.

Exercise 17.5   (p. 567)

1. $C\lambda^n$.   **2.** $2^n + 1$.   **3.** $C_1(\frac{4}{3})^n + C_2(\frac{7}{3})^n$.

4. $\dfrac{1}{2^n}\{C_1(3 + 2\sqrt{2})^n + C_2(3 - 2\sqrt{2})^n\}$.   **5.** 2^{n-2}.

6. $2^{n/2} \sin(n\pi/4)$. **7.** $(n-1)2^{n-2}$. **8.** $3^n + (-9)^n$.

9. $4^n - 3^n + n + 1$. **10.** $2^{-(n-2)} - 1$.

11. If $a \ne b$, $u_n = p(a^n b^N - a^N b^n)/ab(b^{N-1} - a^{N-1})$;

if $a = b$, $u_n = (N-n)pa^{n-1}/(N-1)$.

12. $n^2 2^{2n-5}$.

13. $u_n = (n!) \sum_{r=2}^{n} [(-1)^r/(r!)]$. **14.** 2. **15.** $2^{(4r-1)/3}$.

Exercise 17.6 (p. 572)

1. (a) 0·669, (b) 0·667. **2.** 0·596. **6.** $a + \varepsilon f(a) + \varepsilon^2 f(a) f'(a)$.

Exercise 17.7 (p. 575)

1. (i) No; (ii) no; (iii) no. **2.** (i) Yes; (ii) no.

3. (i) False; (ii) false. **4.** 1.

Miscellaneous Exercise 17 (p. 576)

1. (i) $l = 2$; (ii) (a) $|x| < \sqrt{3}$, (b) $x \in \mathbb{R}$.

2. (ii) $a_n \to 2$; $n = 2\cdot5 \times 10^6$; (iii) (a) 0, (b) 1.

3. (a) $x < 1$, (b) $0 < x < 1$, (c) $x < 1$, (d) $|x| < \frac{1}{2}$, (e) $|x| < 1$, (f) $|x| < \ln 2$.

4. (a) $\frac{2}{3}$, (b) $\sqrt{e} - 1$, (c) $\ln 2$, (d) 32/49.

5. $\Sigma \sin(\theta^r)$ convergent, $\Sigma \cos(\theta^r)$ divergent.

7. $x > 0$, $x < -1$; (a) convergent for $x < 1$, (b) convergent, (c) divergent.

9. $\cosh x > \tanh x > x > \sinh x$. **10.** (ii) $3 < a^2 + b^2 + c^2 < 9$.

12. (ii) $\{x : \frac{1}{4}(6 - \sqrt{38}) < x < \frac{1}{4}(6 - \sqrt{2})\} \cup \{x : \frac{1}{4}(6 + \sqrt{2}) < x < \frac{1}{4}(6 + \sqrt{38})\}$.

15. (i) $\{x : x < -5\} \cup \{x : x > \frac{1}{3}\}$. **16.** $A \approx 0\cdot2838$, $C \approx 0\cdot2$; $B \approx 0\cdot238$.

17. (a) $\{x : x < -2\} \cup \{x : x > 0\}$; (b) $1 < x < 3$; (c) $x \ne 1$; (d) $x \ne 1$;

(e) $-1 \le x < 0$, also when $x = 0$ if $b < 0$; (f) $x \ge 0$.

18. If $k \ne 2$, $u_n = C_1 2^n + C_2 k^n$; if $k = 2$, $u_n = (C_1 + nC_2)2^n$; $u_n = (n-1)2^n$.

19. (a) (i) $u_n = 1 + (-1)^n$; (ii) $u_n = 3n + 2 - 2^{n+1}$;

(iii) $u_n = \cos(n\pi/3) + \sin(n\pi/3)$. **20.** $u_n = (-1)^n (2^{1-n} - 2^{1+n})$.

21. $\alpha = 2$, $\beta = -1$, $\gamma = -\frac{1}{2}$; $A = \frac{2}{3}$, $B = -\frac{1}{6}$.

22. (i) $n = 2m$, $\dfrac{u_{n+1}}{u_n} \to 0$, $u_n^{1/n} \to \sqrt{(ak)}$; $n = 2m+1$, $\dfrac{u_{n+1}}{u_n} \to \infty$, $u_n^{1/n} \to \sqrt{a}$; divergent;

(ii) (a) convergent, (b) convergent for $x < 1$.

23. (i) $p \ge 0$, $p \ge q$; (ii) b_{2n+1} oscillatory, $b_{2n} = 0$; (iii) (a) e^{-6}, (b) 0.

25. (i) False; (ii) true; (iii) true. $2|x|$. No.

Exercise 18.2 (p. 585)

1. (i) $y^2 = 1/(1 - 2x^2)$; (ii) $dz/dx = 2/(1 - 2z)$; $(x+y)^2 + x - y = C$.

2. $y = e^{x^2/2}/(2 - e^{x^2/2})$.

3. $(x+2) \tan\frac{1}{2}(x-y) - x = 0$.

4. $2x^3 + 3x^2 y + y^3 = C$.

5. $y^2 = x^2 \ln(Cx^2)$.

6. $x - y = C \exp[3x/2(x-y)]$.

7. $e^{y/x} = \ln(Cx)$.

8. $y + \sqrt{(x^2 + y^2)} = Cx^2$.

9. $(x-y)^2 = x + y$.

10. $(x - 2y)(3x + y)^2 = 5x$.

11. $(4x - y + 5)^3 (x+y)^2 = C$.

12. $x^2 - 2xy + 8y^2 - 14y = C$.

13. $4x - 8y + C = 5 \ln(4x + 8y + 1)$.

14. $x^2 - xy - y^2 + 5y = C$.

15. $(x+y)e^{-y/x} + x \ln(Cx) = 0$.

Exercise 18.3 (p. 589)

1. $xy = (x-1)e^x + C$.

2. $y = \frac{1}{2}(1 - \cot x)e^x + C \operatorname{cosec} x$.

3. $y = 1 + C/\sqrt{(1+x^2)}$.

4. $y(x-1) = (x-2)e^x + C$.

5. $y = b/a + Ce^{-ax}$.

6. $y = \frac{1}{2}(\sin x + \cos x) + Ce^{-x}$.

7. $r = \theta \cos \theta + C \cos \theta$.

8. $y (\operatorname{cosec} x + \cot x) = Cx + \sin x$.

9. $y = (\sec x + \tan x) \left[C + \ln (1 + \sin x) \right]$.

10. $y = x \ln x + 1 + Cx$.

11. $y (C - x) = \sin x$.

12. $xy \sqrt{(C - 2x)} = \pm 1$.

13. $\frac{1}{4}(5e^{-2} + 1) \approx 0.42 \text{ units s}^{-1}$; $\frac{5}{8}(1 - e^{-2}) \approx 0.54$ units.

14. 0.67 units.

15. $x^3 y = -a^4 \left[1 - \sin (x/a) \right]$.

16. (a) $y = (C + x)e^{-2x}$;   (b) $\dfrac{dz}{dx} = \frac{1}{4} - z^2$; $y (2e^x + 4) = x (e^x - 2)$.

17. $5x^2 y = x^5 - 1$.

18. $y = x (x^3 - 2)$.

19. $y = \tan x - \sqrt{2} \sin x$.

20. $y = \operatorname{cosec} x - \cot x$.

Exercise 18.4   (p. 593)

1. $p = \left[\gamma p_0 v_0 - (\gamma - 1)h \right]^{\gamma/(\gamma - 1)} / (\gamma v_0)$.

2. $\left[k(\alpha b_0 - \beta a_0) \right]^{-1} \ln \left(2 - \dfrac{a_0 \, \beta}{b_0 \, \alpha} \right)$.

5. $x = 18 (e^{14t} - 1)/(9e^{14t} - 1)$.

6. 96.3 g.

7. $(1/k) \ln \left[15/(x - 15) \right]$;   19.9.

8. (i) $35°$;   (ii) 6.8 minutes.

9. $dx/dt = k_1 x - k_2$; $x = k_2/k_1 + (N - k_2/k_1)e^{k_1 t}$.

10. $x = x_0 \exp \left(\left[(a - a')t + \frac{1}{2}(b - b')t^2 \right]/100 \right)$; $x_0 \exp \left((a - a')^2/[200(b' - b)] \right)$.

11. $dx/dt = x (312 - 4x)/3$; $x = 78e^{104t}/(25 + e^{104t})$.

12. 14 000 insects;   2110 insects per day.

13. $dN/dt = -kNt$; $N = N_0 e^{-kt^2/2}$;   184 days.   **14.** $k_1 = \ln (5/2), \; k_2 = \ln (5/4)$.

Exercise 18.5   (p. 596)

1. $y = \sin 2x - 2 \cos 2x$.

2. $y = 2 \cos 3x$.

3. $y = \frac{1}{8} \sin 8x$.

4. $y = \frac{1}{3}(7 \sin 3x - \cos 3x)$.

5. $y = Ae^{nx} + Be^{-nx}$ or $C \cosh nx + D \sinh nx$.

6. $y = 3e^x + e^{-x}$.

7. $y = A + Bx + \sin x$.

8. $y^2 = 2x^2 + 2x + 1$.

Exercise 18.7   (p. 601)

1. $y = Ae^{-x/2} + Be^{3x}$.

2. $y = Ae^{5x/4} + Be^{-x}$.

3. $y = Ae^{3x} + Be^{-3x}$.

4. $y = A + Be^{-5x}$.

5. $y = (Ax + B)e^{2x}$.

6. $y = e^{-x/2} (A \cos \frac{1}{2}x + B \sin \frac{1}{2}x)$.

7. $y = e^{-3x/2} \left[A \cos (\frac{1}{2} \sqrt{7}x) + B \sin (\frac{1}{2} \sqrt{7}x) \right]$.

8. $y = A \exp \left[(-5 + \sqrt{21})x/2 \right] + B \exp \left[(-5 - \sqrt{21})x/2 \right]$.

9. $y = (Ax + B)e^{-3x/2}$.

10. $y = Ax + B + Ce^{-x}$.

Exercise 18.8   (p. 604)

1. $y = Ae^{-2x} + Be^{-3x} + \frac{2}{3}$.

2. $y = A \cos 3x + B \sin 3x + \frac{1}{9}$.

3. $y = Ae^{2x} + Be^{-x} + \frac{1}{10} (\cos x - 3 \sin x)$.

4. $y = (A + Bx + x^2)e^x$.

5. $y = Ae^{-2x} + Be^{-x} - xe^{-2x}$.

6. $y = e^{-x/3} \left[A \cos (x \sqrt{2}/3) + B \sin (x \sqrt{2}/3) \right] + (4 \sin 2x - 11 \cos 2x)/137$.

7. $y = e^{3x/2} \left[A \cos (x \sqrt{3}/2) + B \sin (x \sqrt{3}/2) \right] + \frac{2}{3}$.

8. $y = e^{2x} (A \cos x + B \sin x) + (25x^2 + 65x + 67)/125$.

9. $y = Ae^{2x} + Be^{-x} - \frac{1}{4}(2x + 1)$.

10. $y = (Ax + B)e^{2x} + e^x + \frac{1}{4}(x + 1)$.

11. $y = e^{-2x} (A \cos x + B \sin x) + (\sin 2x - 8 \cos 2x)/65$.

12. $y = A + Be^{5x} - \frac{4}{5}x$.

13. $y = e^{-x} (A \cos x + B \sin x) - (\sin 2x + 2 \cos 2x - 4 \sin x - 2 \cos x)/10$.

14. $y = A + Bx + Ce^{-x} + \frac{1}{2}x^2 + xe^{-x}$.   **15.** $y = (A + Bx + \frac{1}{2}x^2)e^{3x}$.

16. $y = Ae^{2x} + Be^x + (3 \cos 2x - \sin 2x)/20$.

17. $y = A + Be^{x/p} + a(e^x - px + x)/(p^2 - p)$.

18. $y = Ae^{2x} + Be^{-2x} + C \cos 2x + D \sin 2x - x/16$.

19. $x = Ae^{3t} + Be^{-t} - 4[\sin(2t + \tfrac{1}{4}\pi) + 7\cos(2t + \tfrac{1}{4}\pi)]/65$.

20. $x = e^{t}(A \cos t + B \sin t) - (\cos 2t + 2 \sin 2t)/10$.

22. $y = 3e^{-2t} - 3e^{-t} + 3 \sin t + \cos t$.

23. $x = e^{3t} - 3e^{2t}$; $t = \ln 2$.   24. $s = 7{\cdot}35$ units; $v = 1{\cdot}20$ units s^{-1}.

26. $\dfrac{1}{16}$ when $t = \ln 4$.   27. $-ka[\sec(na/2) - 1]/2n^2$.

28. $y = e^{-3t}(4 \cos 4t - \sin 4t)/68 + (4 \sin t - \cos t)/17$.

29. $y = \sin \theta + \sin 3\theta$.   30. $x = e^{-kt} \cos[t\sqrt{(n^2 - k^2)}]/2k^2 + (\sin nt)/(2kn)$.

Exercise 18.10 (p. 610)

1. $20{\cdot}3$.   2. $0{\cdot}14$.   3. $0{\cdot}43$.   6. $0{\cdot}538\,82$.

7. $y = 1/(5 - x)$.   9. $y_1 \approx 0{\cdot}105$, $y_2 \approx 0{\cdot}217$.

Exercise 18.11 (p. 613)

1. $\mathbf{r} = (\mathbf{i} + 2\mathbf{j})e^{4t} - 2(2\mathbf{i} + \mathbf{j})e^{t}$.   2. $\mathbf{r} = \mathbf{b} + \mathbf{c}e^{-\lambda t} + \mathbf{a}t/\lambda$.

3. (i) $\mathbf{v} = i\omega \cos \omega t - j\omega \sin \omega t + \mathbf{k}$, $\mathbf{f} = -i\omega^2 \sin \omega t - j\omega^2 \cos \omega t$; $n\pi/\omega (n = 0, 1, 2 \ldots)$.

4. (i) $\mathbf{r} = (3\mathbf{i} + \mathbf{j})e^{-2t} - (2\mathbf{i} + \mathbf{j})e^{-3t}$;

(ii) $\mathbf{r} = \mathbf{a}t/k - (\mathbf{a} - k\mathbf{V})(1 - e^{-kt})/k^2$;   (iii) $\mathbf{r} = \mathbf{V}te^{-nt}$;

(iv) $\mathbf{r} = \mathbf{A} \cos \omega t + \mathbf{B} \sin \omega t + (\mathbf{a} \cos pt)/(\omega^2 - p^2)$ if $p \neq \omega$;

$\mathbf{r} = \mathbf{A} \cos \omega t + \mathbf{B} \sin \omega t - (\mathbf{a}t \sin \omega t)/(2\omega)$ if $p = \omega$.

Miscellaneous Exercise 18 (p. 614)

1. (i) $y = x + 2e^{-x} - 1$;   (ii) $(d\theta/dt) = -k\theta$.

2. (i) $y = x + 1 + C\sqrt{x}$;   (ii) $y = Ae^{-x} + Be^{2x} + \tfrac{1}{10}(\cos x - 3 \sin x)$.

3. (a) $y - \sin y \cos y = x^2 + 6x + C$;   (b) $3y = xe^{2x} - (e^9 - e^3)xe^{-x}$.

4. $94{\cdot}2\%$; $1{\cdot}69 \times 10^7$ units.

6. $y = e^{-x}(A \cos 2x + B \sin 2x) + (25x^2 - 20x - 2)/125$.

7. (a) 8;   (b) $x = Ae^{-2t} + Be^{-3t} + e^{-t}$.

8. $y = Ae^{x} + Be^{2x} + (6 - 6xe^{x} + e^{-x})/12$.

9. $(dy/dt) = -y/(1 + t)^2$; $y = 10\,e^{-t/(1+t)}$.

10. (a) $u = 2(1 + 3x^3)^{3/2}/(27x^3) + 11/(27x^3)$;   (b) $y(2 - e^{x^2}) = 2(e^{x^2} - 1)$.

11. $15kt = 2\pi[(9V/\pi)^{5/6} - y^{5/2}]$; $t_1/t_2 = 1 - 2^{-1/3} \approx 0{\cdot}794$.

12. $y = e^{-kx}(A \cos kx + B \sin kx) + (2 \sin x + k \cos x)/(k^3 + 4k)$.

13. (i) $y = Ae^{x} + e^{-x/2}[B \cos(x\sqrt{3}/2) + C \sin(x\sqrt{3}/2)] - x^4 - 24x + \tfrac{1}{7}e^{2x}$;

(ii) $y = 2 + C/\sqrt{(1 + x^2)}$.

14. (i) $(x - 1)y = (x + 1)[C + \ln(x + 1)]$;   (ii) $4a/9$.

15. $(H^2/2A)t = -H\sqrt{y} - K \ln(K - H\sqrt{y}) + C$.

16. (i) 8 units s^{-1};   (ii) 4 units;   (iii) 1 unit s^{-1}.

17. $y = Ce^{-Rt/L}$; $x = Ce^{-Rt/L} + (E/R)$.

18. (i) $y = x + \cot x + C \operatorname{cosec} x$.

19. (a) $y = Cx^2 - x^2/(1 + x)$;   14/3.

20. (a) $y = (x + 1)/(x - 1)$;   (b) $y = (C + \sec^2 x)/x^2$.

21. $(dm/dt) = -am$; $m(dp/dt) = -bp/(dm/dt)$;   (i) $m = m_0 e^{-at}$;

(ii) $p = p_0(m_0/m)^b$; $p = p_0 e^{abt}$.

22. (i) $y^{-1} = C + \ln[xy/(1 + x)(1 - y)]$;

(ii) $y = e^{-x}(A \cos 2x + B \sin 2x) + \tfrac{3}{5} - \tfrac{1}{2}e^{-x} + (4 \sin 2x + \cos 2x)/17$.

23. (a) $x^2(d^2y/dx^2) - 2x(dy/dx) + (x^2 + 2)y = 0$;

(b) (i) $y = Ae^{-x} + Be^{-2x} + \tfrac{1}{2}x - \tfrac{3}{4}$;   (ii) $x(1 - x)y = (1 - x)\{x + 4\ln[(1 - x)/2] - \tfrac{1}{2}\} + 3$.

24. (a) $x^2y^2 = C(x + y)$.

25. (i) $y = 1 - Ce^{-x^2}$;   (ii) $y = Ae^{-3x} + Be^{-2x} + \tfrac{1}{5}xe^{2x}$.

26. (i) $x = (\tfrac{3}{4} + \tfrac{1}{2}t)e^{-3t} + \tfrac{1}{4}e^{t}$;   (ii) $y = x^4(\ln x - 1) + Cx^3$.

27. (a) $z\,(dz/dx) = (1+x)\,(1+z);\quad y = \ln\,(1+x+y)+\tfrac12 x^2 + C;$
(b) $5y^2 = x^2 - 4.$

28. $y = (x+1)e^x/[3-(x+1)e^x];$ (ii) $y = \cos \ln x.$ **29.** $x^{3/2} + y^{3/2} = 2.$

30. (i) $y = x \ln\,[(x+y)^2/x] + Cx;$ (ii) $y = \sin x + C \cos x;$
(iii) $y = e^{-x}(A \cos 2x + B \sin 2x) + (5x+3)/25.$

31. (i) $y = 3 \cosh 2x + \sinh 2x;$ (ii) $v = \tfrac14 (e^{-2t} + 2t - 1);\quad \tfrac18 (5 - e^{-4}).$

32. (i) $x = e^{-2t}(\tfrac14 t^4 + \ln t + C);$ (ii) $y = A \cos 3x + B \sin 3x;\quad -12.$

33. (i) $(2x - y)^4\,(x + 3y)^3 = C;$ (ii) $y = \pm 1/\sqrt{(Ce^{2x} + x + \tfrac12)};$
(iii) $y = 3(e^{-x} - e^{-2x}).$

34. (i) $y = (x \sin x + \cos x + C)/x^2;$ (ii) $y = 2e^{-5t} \sin 2t.$

35. (i) $y = (x^3 + 2)/x^3;$ (ii) $y = Ae^{-3x} + Be^{-x} + \tfrac{1}{10} (\sin x - 2 \cos x) + \tfrac18 e^x + 1.$

36. (i) $x\,(1 + \cos y) = 2;\quad y = e^{-2x}(A \cos 2x + B \sin 2x) + e^x/26 + e^{-x}/10.$

37. (i) $\tan y = \ln\,[(1 + x)/(1 - x)] + 1;$ (ii) $2 \ln\,(3x - y - 3) + x - y - 3 = C.$

38. (i) $y = C \sin^2 x - 3 \sin x \cos x;$ (ii) $(x - y)^2 = x + y.$

39. (i) $2xy = 5x^2 - 1;$ (ii) $x = e^{-t}(A \cos 3t + B \sin 3t) + (13 + 2e^t)/26.$

40. (i) $y^2 x^2\,(C - 2x) = 1;$ (ii) $y = e^{-x}(A \cos kx + B \sin kx).$

41. $2/3\,\sqrt 3,\ [y = e^{-x} - e^{-3x}].$ **42.** $y = e^{6x} - 8e^x + 6x + 7.$

43. $y = 2 \cos 2x - \cos 4x.$

44. (i) $y(x - 1) = \tfrac12 x^3 - 2x^2 + x \ln x + Cx;$ (ii) $(x - 1) \sin y = \tfrac12 x^3 - 2x^2 + x \ln x + Cx.$

45. $y = x \sin\,(\ln x + C).$

47. (i) $(dv/dt) = k\,(V - v);\quad v(dv/ds) = k\,(V - v);\quad v = V(1 - e^{-kt});$
$s = (V/k) \ln\,[V/(V - v)] - (v/k).$

48. (i) $k = \tfrac13 \ln 2;$ (ii) $18/7.\quad [x = 2a\,(e^{kat} - 1)/(2e^{kat} - 1)].$

49. $x = x_0 \exp\,([(a - a')t + \tfrac12 (b - b')t^2]/100);\quad x_0 \exp\,((a - a')/[200\,(b' - b)]).$

50. (a) $(y - 1)\,(dy/dx) + x - 1 = 0;\quad (x - 1)^2 + (y - 1)^2 = C;$
(b) $(dy/dx) = -\sin x\,(1 + 4 \cos x).$

51. (a) $\tfrac13 x;$ (b) $y = \pm e^x/\sqrt{(C + x^2)}.$

Exercise 19.1 (p. 627)

5. $16\pi/15.$ **8.** $\tfrac12 \pi\,(\pi - 2).$ **9.** $7\pi.$ **10.** $\pi\alpha/(\sin \alpha).$

11. $0.$ **12.** (i) $144\,\sqrt 3/35;$ (ii) $0;$ (iii) $360\,448/63;$ (iv) $0.$

14. $0.$ **16.** $0.$

Exercise 19.2 (p. 633)

1, 2, 4. do not converge.   **3.** $1.$ **5.** $\tfrac{1}{12}\pi.$ **6.** $2\pi.$ **7.** $\pi.$

8. $-\tfrac19.$ **9.** $1.$ **10.** $\tfrac16 \ln 4.$ **11.** $\tfrac32 \pi.$ **12.** $\tfrac12 \pi.$

13. $(\sqrt 2 - 1)/a^2.$ **14.** $2\pi.$ **15.** $\tfrac12 \pi.$ **17.** $1/(a^2 + 1).$

20. $\tfrac12 \pi.$ **21.** $4\pi.$ **22.** $1.$ **23.** $\tfrac12 \pi^2.$ **24.** $2 \ln\,(1 + \sqrt 2).$ **25.** $2\pi.$

Exercise 19.3 (p. 639)

1. $au_n = x^n e^{ax} - nu_{n-1};\ \dfrac13\left(x^4 - \dfrac{4x^3}{3} + \dfrac{4x^2}{3} - \dfrac{8x}{9} + \dfrac{8}{27}\right)e^{3x}.$

2. $u_n = -x^n \cos x + nx^{n-1} \sin x - n(n - 1)u_{n-2};\ -x^3 \cos x + 3x^2 \sin x + 6x \cos x - 6 \sin x.$

3. $u_n = (\tan^{n-1} x)/(n - 1) - u_{n-2};\ \tfrac15 \tan^5 x - \tfrac13 \tan^3 x + \tan x - x.$

4. $2u_n = x^2\,(\ln x)^n - nu_{n-1};\ x^2\,\{4\,(\ln x)^3 - 6\,(\ln x)^2 + 6 \ln x - 3\}/8.$

5. $nu_n = \cosh^{n-1} x \sinh x + (n - 1)u_{n-2};\ \tfrac16 \cosh^5 x \sinh x + \tfrac{5}{24} \cosh^3 x \sinh x + \tfrac{5}{16} \cosh x \sinh x + 5x/16.$

6. $(n - 1)u_n = \sec^{n-2} x \tan x + (n - 2)u_{n-2};\ \tfrac14 \sec^3 x \tan x + \tfrac38 \sec x \tan x + \tfrac38 \ln\,(\sec x + \tan x) + C.$

7. $35\pi/256.$ **8.** $16/35.$ **9.** $2/35.$

10. $\tfrac34.$ **11.** $\pi a^6/32.$ **12.** $5\pi/16a.$

17. $[\sin 2(n - 1)\theta]/(n - 1).$ **18.** (iii) $(23\pi/256) + (88/315).$ **19.** $\ln 3;\ \tfrac23 \sqrt 3 - \ln 3.$

Exercise 19.4 (p. 644)

1. $(2 \ln 2)/\pi \approx 0.44$.
2. 0.091.
3. $e^{-1} \approx 0.37$.
4. $\frac{1}{4}\pi a \approx 0.79a$.
5. (i) $14/\pi$;   (ii) $(25\pi + 48)/28$.
6. $\frac{1}{3}\rho a^2$.
7. $\frac{1}{6}l^2$.
8. $\frac{5}{2}$.
9. 0.619.

Exercise 19.5 (p. 647)

1. On the axis, $\frac{1}{4}h$ from the base.   2. $\frac{3}{2}l$ from one end.
3. On the radius of symmetry, $2r/\pi$ from the bounding diameter.
4. On the axis, $\frac{1}{4}h$ from the base.   5. $(\frac{3}{5}a, 0)$.
6. $[0, (4\pi - 3\sqrt{3})\{4\pi - 6 \ln (2 + \sqrt{3})\}^{-1}]$.
7. $(256a/315\pi, 256a/315\pi)$.   8. $(9/5, 9/5)$.   9. $4/5$.
10. $(3e^4 + 1)/[2(e^4 - 1)]$.   11. $(e^2 - 1)/[4(e - 2)]$.   12. $\frac{5}{8}a$.

Exercise 19.6 (p. 650)

1. $4r/3\pi$ from each bounding radius.
2. $128\pi a^3/15$.
3. $8\sqrt{2\pi} \sin (\frac{1}{4}\pi + \theta)a^3$; $8\sqrt{2\pi}a^3$, $8\pi a^3$.
4. $36\pi^2$.   5. 6π, 9π.
6. $(e - 1)^2/(2e)$; $[2/(e - 1)^2, (e^4 - 4e^2 - 1)/(8e(e - 1)^2)]$; $2\pi/e$.
7. $9c/[2(15 - 16 \ln 2)]$.   8. 7.4.

Exercise 19.7 (p. 656)

1. $\frac{4}{3}Ma^2$, $(M = 2\pi\rho a^3)$.
2. $\frac{2}{7}Ma^2$, $(M = \frac{1}{4}\pi\rho a^3)$.
3. $2M/9$, $(M = 2)$.
4. $I_x = M/(8 \ln 2)$, $I_y = 3M/(2 \ln 2)$, $(M = \ln 2)$.
5. (i) $\frac{3}{10}Mr^2$; (ii) $M(3r^2 + 2h^2)/20$.
6. $[2(a^5 - b^5)]^{1/2}[5(a^3 - b^3)]^{-1/2}$.
7. (i) $10\,ml^2$; (ii) $20\,ml^2$.
8. $\frac{1}{2}Ml^2$, $(M = \frac{1}{2}\rho l^2)$.
9. $\frac{1}{2}R\sqrt{7}$.   10. $\frac{1}{4}\sqrt{5}$.
11. (i) $\frac{1}{2}Ma^2$; (ii) $\frac{2}{3}Ma^2$; (iii) $\frac{1}{3}Ma^2$.

Miscellaneous Exercise 19 (p. 661)

1. (a) 0; (b) $A = 1$, $B = 1$, $\frac{1}{4}(x^4 \tan^{-1} x - \frac{1}{3}x^3 + x - \tan^{-1} x)$; $\frac{1}{6}$.
3. (a) (i) $\frac{1}{2}\pi$, (ii) $\frac{1}{4}(\pi - 1)$; (b) $\frac{1}{4}\pi$.
5. (i) $2a/\pi$; (ii) $\frac{1}{4}\pi a$.
6. $2(\ln 2)/\pi$.   7. $\frac{1}{2}\pi$.   8. $3\pi/16$.
10. 2; 1.   11. (i) 0; (ii) $\pi(\pi - 2)/8$.
14. $1 - \frac{1}{4}\pi$.
15. $\frac{1}{2}\pi/|\sin \theta|$; (i) $\pi/\sqrt{2}$, (ii) $\pi/\sqrt{2}$.
16. $b^2C + a^2S = \frac{1}{2}\pi$; $C = \pi/[2b(a + b)]$, $S = \pi/[2a(a + b)]$.
17. 16.   18. $\frac{1}{2}(\pi^3 - 24\pi + 48)$.
19. (i) $[7\sqrt{2} + 3 \ln (1 + \sqrt{2})]/8$; (ii) 1, $\frac{2}{3}(4\sqrt{2} - 5)$.
20. $u_n - u_{n-1} = 0$; $n(n + 1)\pi$.
21. $[7\sqrt{2} + 3 \ln (1 + \sqrt{2})]/16$.
22. $(2n - 3)(2n - 5) \ldots 3.1\pi/[2^n(n - 1)!]$.
23. $\frac{1}{2}\pi$.
24. (i) $\frac{1}{8}\pi(\pi + 2)$; (ii) $\pi \ln 2$; $[(2 \ln 2)/\pi, (\pi + 2)/(4\pi)]$.
26. 7.41.   27. $17a/6$.   28. $2555a^2/558$.   29. $3\sqrt{3}a^2$.
30. (i) $\frac{1}{4}\pi ab$; $(4a/3\pi, 4b/3\pi)$; (ii) $\frac{2}{3}\pi a^3$, $\frac{3}{8}a$ from base.
31. $8/7$.   32. $\frac{4}{3}\pi ab^2$.   33. $\frac{3}{8}a$ from base.   36. 3.988; $0.997(1)$.
37. $v = 32t$; $v = 8\sqrt{5}$; $\frac{1}{2}V$; $\frac{2}{3}V$.
38. $\sqrt{[b(4a + b)/6]}$.
39. $\frac{1}{4}Ma^2$; $Ma^2(15\pi - 32)/12\pi$.
41. $\pi(1 - \ln 2)$; $\frac{1}{2}(3 - 4 \ln 2)/(1 - \ln 2)$.
42. π; on Ox, distant $\frac{1}{4}(\pi - 2 \ln 2)$ from O; $\sqrt{(2/3)}$.   43. (i) 0; (ii) $\frac{1}{2}\pi$.
44. (i) $\frac{1}{8}\pi$; (ii) $[\sqrt{(1 + a^2)} - 1]/a$; (iii) $\frac{1}{4}\pi$; (iv) $\frac{1}{2}\ln 2$; (v) $(1 + a^2)^{-1}$.
48. $\frac{3}{10}Ma^2$; $M(3a^2 + 2h^2)/20$.
49. (ii) (a) $+$, (b) 0, (c) $+$.

Exercise 20.2 (p. 675)

1. $\frac{1}{2}[\sqrt{2} + \ln (1 + \sqrt{2})]$.
2. $(13\sqrt{13} - 8)a/27$.
3. $[\sqrt{2} + \ln (1 + \sqrt{2})]a$.
4. $\ln (1 + \sqrt{2})$.
5. $2(\tan^{-1} e - \frac{1}{4}\pi)$.
6. $8/\sqrt{3}$.

Exercise 20.3 (p. 678)

1. $\pi(1 + \frac{1}{2}\sinh 2)$. 2. $64\pi a^2/3$. 3. $6\pi a^2/5$. 4. $8\pi(2\sqrt{2} - 1)/3$.
5. $384(1 + \sqrt{2})\pi/5$. 6. $\pi[\ln(1 + \sqrt{2}) - \ln(1 + \sqrt{(1 + e^2)}) + 1 + \sqrt{2} - \sqrt{(1 + e^2)/e^2}]$.
7. $2\pi(e - 1)^2/(e^2 + 1)$. 8. $\frac{1}{2}\pi\sqrt{2}[3\sqrt{10} - \sqrt{2} - \ln(3 + \sqrt{10}) + \ln(1 + \sqrt{2})]$.

Exercise 20.4 (p. 680)

1. $4\pi^2\,cr$. 2. $48\pi^2$. 3. $24\pi a^2$. 4. $12\pi a^2$. 5. $\pi a^2(4\pi - 1)/16$.
7. $2\pi a^2[2(\cos^3\theta + \sin^3\theta) + \pi\cos\theta\sin\theta(\cos\theta + \sin\theta)]$; $\pi\sqrt{2}(2 + \pi)a^2$.

Exercise 20.5 (p. 683)

1. $\frac{3}{2}\pi a^2$. 2. $\frac{1}{4}a^2$. 3. $\pi a^2/16$.
4. $\frac{1}{2}\pi a^2$ when n is even; $\frac{1}{4}\pi a^2$ when n is odd. 5. $1:7:19:37$.
7. $r = 2a(1 - \cos\theta)$; $64\sqrt{3}a^2/27$.
8. $\frac{1}{4}(3\pi - 8)a^2$. 9. $\frac{1}{2}(4 - \pi)a^2$.

Exercise 20.7 (p. 687)

1. $[a(e^{2k\pi} - 1)\sqrt{(1 + k^2)}]/k$. 2. $8a$. 3. $\frac{3}{2}\pi a$. 4. $(27 + 10\ln 4)a/20$.
5. $(\pi\sqrt{2}a)/8$. 6. $4\pi a^3/15$. 7. $2\sqrt{2\pi}(e^{2\pi} + 1)/5$.

Exercise 20.9 (p. 690)

1. (i) $y = 7x$; (ii) $y = x$; (iii) $y = x$; (iv) $y = x$; (v) $y = x$.
2. (i) Maximum at $(0, 1)$, minimum at $(8, -255)$, inflexion at $(4, -127)$;
   (ii) minimum at $(3/2, -27/16)$, inflexions at $(0, 0)$, $(1, -1)$;
   (iii) no stationary points, inflexion at $(0, 0)$;
   (iv) minimum at $(0, 0)$, inflexions at $(\pm a/\sqrt{3}, \frac{1}{4}a)$;
   (v) maxima at $[(2n + \frac{1}{4})\pi, \sqrt{2}]$, minima at $[(2n + \frac{5}{4})\pi, -\sqrt{2}]$, inflexions at $[(n + \frac{3}{4})\pi, 0]$.
3. $\frac{1}{2}\pi$. 4. $(\pm 1/\sqrt{2}, e^{-1/2})$.
5. Maximum at $(1, e^{-1})$, inflexion at $(2, 2e^{-2})$; $x + e^2 y - 4 = 0$.
6. $\cot(\frac{1}{2}\theta)$, $-\dfrac{1}{2a}\operatorname{cosec}^2(\frac{1}{2}\theta)\operatorname{cosec}\theta$; stationary points at $[2a, a(2n + 1)\pi]$.
7. $\{(n + \frac{1}{2})\pi, (n + \frac{1}{2})\pi\}$.
8. (i) Inflexion at $x = 0$; (ii) inflexion at $x = l$; (iii) inflexions at $x = \pm\frac{1}{2}l$;
   (iv) inflexions at $x = \pm l/\sqrt{12}$.
9. Minima at $[(2n + \frac{1}{2})\pi, -5]$, maxima at $[(2n - \frac{1}{2})\pi, 3]$, inflexions at $[(n\pi + (-1)^n\pi/6), -3/2]$.
11. One stationary point if $c \geqslant 1$, three if $c < 1$; $(\pm 1, 3/e)$.
12. $(-\sqrt{3}, -\frac{3}{4}\sqrt{3})$, $(0, 0)$, $(\sqrt{3}, \frac{3}{4}\sqrt{3})$.

Exercise 20.11 (p. 697)

1. $-1/(4a\sqrt{2})$. 2. $-ab\sqrt{[8/(a^2 + b^2)^3]}$.
3. $-ab/(a^2\sinh^2 1 + b^2\cosh^2 1)^{2/3}$. 4. $-1/\sqrt{8}$.
5. $1/(6a\sqrt{2})$. 6. $3/(80a\sqrt{10})$. 7. $1/c$.
8. -1. 9. $22/(13\sqrt{13})$. 10. $-\sqrt{2}/18$.
11. $\frac{1}{2}a$. 12. $\frac{1}{2}a$. 13. $\frac{1}{2}$.
14. $\frac{1}{2}c$. 15. b^2/a. 22. $x^2 + y^2 - y = 0$.

Exercise 20.12 (p. 699)

1. (i) $\frac{1}{3}\pi$; (ii) $\tan^{-1}(\frac{1}{3})$; (iii) $\tan^{-1}(\sqrt{3}/5)$; (iv) $\tan^{-1}\alpha$;
   (v) at $(1, 7\pi/12)$, $\phi = \tan^{-1}(2/3\sqrt{3})$; at $(5, 7\pi/12)$, $\phi = \pi - \tan^{-1}(2/3\sqrt{3})$.
3. $\frac{1}{2}\pi$. 5. $90°$; $51°\,50'$ and $308°\,10'$. 6. $\frac{1}{2}\pi$.

8. $\frac{3}{4}\pi$; the pole and $[2\sqrt{2/3}, \cos^{-1}(\sqrt{\frac{1}{3}})]$, $[2\sqrt{2/3}, \pi\pm\cos^{-1}(\sqrt{\frac{1}{3}})]$, $[2\sqrt{2/3}, 2\pi-\cos^{-1}(\sqrt{\frac{1}{3}})]$.
10. $\phi = \frac{1}{2}\pi - 2\theta$; $\theta = \frac{1}{4}\pi$, $3\pi/4$ are asymptotes to the curve.
11. $\frac{2}{3}\pi$.   **12.** $s = 4a\sin\frac{1}{3}(2\psi - \pi)$.   **13.** $p^2 = r^4/(a^2 + r^2)$; $3/(a2\sqrt{2})$.
14. $p^2 = a^2r^2/(a^2 + r^2)$; $1/(a2\sqrt{2})$.   **16.** $p^2 = a^2r^2/(2a^2 - r^2)$; $2/a$.
18. $\pi/4$, $5\pi/4$, $9\pi/4$, $13\pi/4$.

Exercise 20.13 (p. 701)

1. $b^2x^2 + a^2y^2 = a^2b^2$.   **2.** $xy = c^2$.   **3.** $x^2 - y^2 = a^2$.
4. $(x - 2a)^3 = 27a^2y$.   **5.** $x = -\frac{1}{2}at^2(9t^2 + 2)$, $y = \frac{4}{3}at(3t^2 + 1)$.
6. $(ax)^{2/3} + (by)^{2/3} = (a^2 - b^2)^{2/3}$.

Miscellaneous Exercise 20 (p. 702)

1. (i) $\frac{3}{4}$; (ii) $\ln 2 - \frac{1}{3}$; (iii) $\frac{3}{4}$.   **2.** $(0, 0)$;
3. $\frac{1}{3}(2\sqrt{2} - 1)$; $(2\pi/15)(1 + \sqrt{2})$.   **5.** $(0, \frac{5}{8})$.
6. (b) $\frac{1}{6}\pi^3a^2$; (c) $\frac{1}{2}a[T\sqrt{(1 + T^2)} + \ln\{T + \sqrt{(1 + T^2)}\}]$.
7. $8(2\sqrt{2} - 2)\pi a^2/3$; $\frac{4}{3}(2\sqrt{2} - 1)a/[\sqrt{2} + \ln(1 + \sqrt{2})]$.
9. (i) $8 + 2\ln 3$; (ii) $8(19 - 9\ln 3)/3$.   **12.** (ii) $3 + \frac{1}{8}\ln 2$.
13. $5\cdot87a$.   **14.** π; $\frac{1}{10}\pi(173 - 36\sqrt{3})$.
15. $2\sqrt{2} r\sin\theta = a$; (a) $\frac{1}{4}a^2(\sqrt{3} - 1)$, (b) $\frac{1}{4}a^2(\sqrt{3} + 3)$.
16. (ii) $\frac{1}{3}a^2\tan\frac{1}{2}\alpha(3 + \tan^2\frac{1}{2}\alpha)$.   **17.** $12a$; $18\pi^2a^2$.
18. $s = c\sinh 1$; $S = \pi c^2(1 + \cosh 1\sinh 1)$; $V = \frac{1}{2}\pi c^3(1 + \cosh 1\sinh 1)$.
19. $t = \ln[(s + \sqrt{(s^2 + 2)})/\sqrt{2}]$.   **20.** Inflexions at $(\pm\frac{1}{2}\pi a, a)$, $(\pm\frac{3}{2}\pi a, a)$.
22. $64\pi a^2/3$; $5\pi^2a^3/8$.   **24.** $17\sqrt{17}/4$.
26. (i) $k = -\frac{1}{3}$; (ii) inflexion at $x = 0$, minimum at $x = \frac{1}{2}\pi$.
27. $22x - y = 7$ at $(1, 15)$ meets curve again at $(7, 147)$; $32x + y = 128$ at $(4, 0)$
 meets curve again at $(-2, 192)$.
28. $3a\sin t\cos t$; $\frac{1}{3}\pi$.
29. $y - \sqrt{2x} + 2\sqrt{2} = 0$; $-\frac{1}{2}\sqrt{2}$; $\rho = 9\sqrt{3}/2$.
30. $-\operatorname{cosech} t$; $a\sinh t$.
31. $-xe^{-x}$, $(x - 1)e^{-x}$; maximum e^{-1} at $x = 0$, inflexion at $(1, 2e^{-1})$.
32. $\tan t$, $(\sec^3 t)/(at)$; $x\cos t + y\sin t - a(1 - \cos t) = 0$.
33. $8x - y - 6 = 0$, $8x + y + 6 = 0$; $(-\frac{1}{3}, -8\frac{2}{3})$, $(\frac{1}{3}, -8\frac{2}{3})$.
35. (i) $y = \pm x$, $x + a = 0$; (ii) $y = \pm x$, $y = 0$; $\frac{1}{4}a$.
36. (i) $\pm\frac{1}{2}$; (ii) $64a^2/3$; (iii) $2[2\sqrt{5} + \ln(2 + \sqrt{5})]a$.
37. $(t^2 + 2t)/(t + 1)$.   **38.** Minimum.   **39.** $s = 4a\sin\psi$.
40. $\frac{1}{8}$.   **41.** (ii) $\frac{1}{4}\pi$, $\frac{5}{4}\pi$; (iii) $r = e^\theta$.   **42.** $\pi^2a^3/4\sqrt{2}$.
43. $20a^2/3$; $8\sqrt{2a}/5$.   **44.** $\frac{4}{9}a\sqrt{3}$; greatest values.

Exercise 21.1 (p. 711)

1. (i) $42\cdot0°$; (ii) $48\cdot9°$; (iii) $70\cdot8°$; (iv) $90°$.
2. Only (i), (ii) and (iv) represent line-pairs.
  (i) $x - y + 1 = 0$, $2x + y - 5 = 0$; (ii) $3x - 2y - 7 = 0$, $x + y + 4 = 0$;
  (iv) $2x - y - 1 = 0$, $x + y + 2 = 0$.
3. $a + b = 0$.   **4.** $3x^2 - 8xy - 3y^2 = 0$.
5. $(1 + 2gl + cl^2)x^2 + 2(fl + gm + clm)xy + (1 + 2fm + cm^2)y^2 = 0$;
  $[(la - lb + 2mh)/2(2lmh - am^2 - bl^2)$, $(mb - ma + 2hl/2(2lmh - am^2 - bl^2)]$.

Exercise 21.2(a) (p. 718)

9. $y = x$, $y = 2x$.   **10.** $y + 1 = 0$, $y + x = 3$.

11. $y = 0$.   **12.** $x = \pm\sqrt{2}$, $y = x$.
13. $4y + 8x + 1 = 0$, $4y - 8x - 1 = 0$.   **14.** $x + 1 = 0$, $y = 1$.

15. $y = 0$, $y = x$, $y + x = 0$.
18. $x = \pm 1$, $y = \pm 1$.

16. $y + x - 1 = 0$, $y - x + 1 = 0$.
21. $\theta = \pm \frac{1}{3}\pi$.

Exercise 21.2(b)   (p. 721)

4. $\pi - \frac{1}{2}\theta$.

Exercise 21.6   (p. 729)

1. (i) $3x - y + 4 = 0$;   (ii) $10x + 2y - 23 = 0$.
2. (i) $(9/5, 3)$;   (ii) $(-8a, -2)$;   (iii) $(5, -1)$;   (iv) $(14, 17)$;   (v) $(-3, 1)$.
3.   (i) $9x^2 - 24xy + 8y^2 + 6x + 8y - 7 = 0$;   (ii) $xy = 0$;
   (iii) $31x^2 + 40xy + 10y^2 + 40x + 20y + 10 = 0$;
   (iv) $22x^2 + 4xy - 3y^2 - 40x - 10y + 15 = 0$;
   (v) $x^2 - 5xy - y^2 + 20x + 8y - 16 = 0$.
5. $y^2(x + 2a) + 4a^3 = 0$.   **6.** $b^2x^2 + a^2y^2 = 4a^2b^2$.
8. $y_1 y = 2a(x + x_1)$; $k = -2ab$.

Exercise 21.7   (p. 732)

1.   (i) Hyperbola, $(13/16, -1/16)$;   (ii) ellipse, $(-6/43, -35/387)$;
   (iii) parabola;   (iv) ellipse, $(3, 2)$.
2. $(1, 1)$; $11x + 16y - 70 = 0$; $16x - 11y + 1 = 0$.
3. $(13/16, -1/16)$; $8x + 24y - 5 = 0, 6x - 2y - 5 = 0$;
   $4x - 28y - 5 = 0, 4x + 4y - 3 = 0$.
4. $x'^2 + 4x'y' + y'^2 + 3 = 0$; $x' + y' = 0$, i.e. $x + y = 2$.
5. $(0, -36/23)$.

Exercise 21.8   (p. 734)

2. $x^2 + y^2 + 3x + 3y = 0$.

Exercise 21.9   (p. 735)

2. $(-9g - 16a, 27f)$.
3. $t^3(x^2 + y^2) - c(1 + 3t^4)x - ct^2(3 + t^4)y + 3c^2t(1 + t^4) = 0$.
4. $[\frac{1}{2}b, (b^2 - 2a^2)/4a]$; $\frac{1}{2}a$.

Exercise 21.10   (p. 738)

1.   (i) $9x - 7y + 5 = 0$, non-intersecting, $x^2 + y^2 + 4x - 6y + 1 + \lambda(9x - 7y + 5) = 0$;
   (ii) $9x - 2y = 0$, intersecting, $2x^2 + 2y^2 - 5x + 4y + \lambda(9x - 2y) = 0$;
   (iii) $16x - 7y - 2 = 0$, non-intersecting, $x^2 + y^2 + 5x - 2y - 1 + \lambda(16x - 7y - 2) = 0$;
   (iv) $ax + by = 0$, intersecting if $c < 0$, touching if $c = 0$,
      non-intersecting if $c > 0$, $x^2 + y^2 + c + \lambda(ax + by) = 0$.
2. $(2, -8)$, $(-89/29, 111/29)$.   **3.** $(-1, 4)$ $(11/2, -5/2)$.
4. $(0, 6)$; $x^2 + y^2 - 12y - 1 = 0$.
5. $(2/5, -6/5)$; $x^2 + y^2 - 4x + \lambda(3x + y) = 0$.
6. $(2, 1)$, $(-3, -2)$; $x^2 + y^2 - 23x + 41y = 0$.
7. Both $(12/5, -16/5)$; the circles touch at $(12/5, -16/5)$.
8. $x^2 + y^2 - 10x + 10y = 0$, $49x^2 + 49y^2 + 550x - 290y + 650 = 0$.
9. $(\pm 2k, 0)$.
10. $4(x^2 + y^2) + 2x + 20y + 19 = 0$, $4(x^2 + y^2) - 10x + 16y + 1 = 0$;
   $15(x^2 + y^2) - 32x - 4y + 12 = 0$.
11. $(2, -4)$; $x^2 + y^2 + 6x + 8y = 0$.
12. $x^2 + y^2 - 2x + 6y + 9 = 0$, $x^2 + y^2 - 8x + 12 = 0$.

Miscellaneous Exercise 21 (p. 739)

1. $y - 3x = 0$, $y + 3x = 0$.
2. $16\pi(3 \ln 2 - 2)/3$.   3. a^2.   5. $x = 8$, $4y = \pm 3x$.
6. 8; $(0, \frac{3}{8}\pi)$.   7. $\sqrt{(k\pi)}$; 0.
9. $2/a$; $(\frac{1}{3}a \sqrt[3]{2}, \frac{1}{3}a \sqrt[3]{4})$; $-6/(a^3 \sqrt{2})$.   10. $\frac{1}{12}\pi a^3 \sqrt{2(8 \ln 2 - 3)}$.
11. $y = 0$, $y = \pm \sqrt{3}x$. $(\frac{2}{3}, 0)$, $(\frac{2}{3}, \pm\frac{2}{3}\sqrt{3})$; $\frac{9}{2}$, $-\frac{1}{4}(9 \pm 4\sqrt{3})$.

12. $\frac{1}{2}n(e - 2)$.   13. $\begin{pmatrix} \frac{1}{2}\sqrt{2} & -\frac{1}{2}\sqrt{2} \\ \frac{1}{2}\sqrt{2} & \frac{1}{2}\sqrt{2} \end{pmatrix}$; $Y^2 = \sqrt{2}X$.

15. $k[(x - h)^2 - (y - k)^2] + 2(a - h)(x - h)(y - k) = 0$.
16. $\alpha x'x + \beta y'y = 1$.   17. $x = a/e$.   18. $x^2 + y^2 + 4x + 10y + 4 = 0$.
19. $ax^2 + 2hxy + by^2 + 2(gx + fy)(px + qy) + c(px + qy)^2 = 0$.
20. $(a\beta - b\alpha)xy - f\alpha x + g\beta y = 0$.
21. $x^2 + y^2 - 3x - 3y + 4 = 0$, $x^2 + y^2 - 2x - 3y - 4 = 0$,
    $x^2 + y^2 - x - 3y - 12 = 0$; $x = 8$.
23. $(-1, 1)$, $(2, -3)$; $49(x^2 + y^2) - 202x + 302y + 648 = 0$, $x^2 + y^2 + 2x - 2y + 2 = 0$.
24. $y^2 = 2a(x - a)$; focus $(\frac{3}{2}a, 0)$, directrix $x = \frac{1}{2}a$; limit directrix $x = 2a$.
25. $t^3x - ty + c(1 - t^4) = 0$.

Exercise 22.2 (p. 744)

1. (a) S; (b) R, S, T; (c) S; (d) R, S, T; (e) R (assuming coincident lines to be parallel), S, T; (f) none. (b), (d) and (e) are equivalence relations.
2. {Birmingham, Cardiff, Glasgow}, {Chicago}, {Tokyo}, {Paris, Berlin, Madrid, Rome}.
5. (b) {all triangles with the same angles, e.g. 84°, 61°, 35°};
   (f) {all vectors of equal magnitude, e.g. 5}.
6. $x \in \{1, 2, 3, 4, 5, 6\}$, $(2, 4)$, $(3, 5)$, $(4, 2)$, $(5, 3)$.
7. (a) S; (b) R, T; (c) T; (d) T; (e) T.

Exercise 22.5 (p. 748)

1. (i) {lines in the same direction}; (iii) {complex numbers with the same modulus}; (iv) {circles of equal radius}.
3. Q.   4. (i) $x \equiv 55$ (mod 71); (ii) $x \equiv 61$ (mod 139).
5. $A = 5$, $B = -23$; (F) $12h - 58m - 39s$, (S) $11h - 33m - 39s$.   6. 38 (mod 77).
7. $a + 10b + 100c + \ldots$ divisible by 11 if $a - b + c - \ldots = 0$ (mod 11).
8. (a) $(x, y) = (0, 1)$, $(1, 3)$, $(2, 0)$, $(3, 2)$, $(4, 4)$ (mod 5); $(3, 2)$ (mod 5); (b) 0, 1, 2, 4.
11.   (i) Symmetric; (ii) equivalence relation, only class is S;
      (iii) reflexive, transitive, order relation, $2 < 4 < 8 < 16$, $3 < 9 < 18$, $2, 3 < 6 < 12 < 18$, etc;
      (iv) equivalence relation, {2, 4, 8}, {3, 9}, {5}, {7}.

Exercise 22.6 (p. 751)

1. (i) Yes; (ii) Yes; (iii) Yes; (iv) Yes.
3. (i)

	1	2
1	1	2
2	2	1

(ii)

	0	1
0	0	1
1	1	0

(a) $(1, 0)$, (b) $(1, 1)$, (c) $(2, 0)$, (d) $(2, 1)$;

	a	b	c	d
a	a	b	c	d
b	b	a	d	c
c	c	d	a	b
d	d	c	b	a

;(1, 0).

4. (a) (i) Yes, (ii) no, (iii) no; (b) (i) no, (ii) no, (c) no.

5.

	(1, 0)	(i, 0)	(0, 1)	(0, i)
(1, 0)	(1, 0)	(i, 0)	(0, 1)	(0, i)
(i, 0)	(i, 0)	(−1, 0)	(0, i)	(0, −1)
(0, 1)	(0, 1)	(0, −i)	(−1, 0)	(−i, 0)
(0, i)	(0, i)	(0, 1)	(i, 0)	(−1, 0)

; (ii) (1, 0).

6. (i) Yes; (ii) yes; (iii) no.

7.

*	a	b,
a	a	a
b	b	b

*	a	b
a	a	b
b	a	a

8. (a)

+	E	0,
E	E	0
0	0	E

(b)

−	E	0,
E	E	0
0	0	E

(c)

×	E	0;
E	E	E
0	E	0

no; yes.

9. (i) Yes, (ii) yes. Neutral element $= -2$, $a^{-1} = -(a+4)$. (a) No, (b) yes.
10. (a) Yes, (b) no, (c) no, (d) none.

11.

G	−2	−1	0	1	2
−2	−2	−1	0	1	2
−1	−1	−1	0	1	2
0	0	0	0	1	2
1	1	1	1	1	2
2	2	2	2	2	2

; yes, yes, yes, −2, none.

12. (i)

	0	2	4	6;
0	0	3	6	9
2	1	4	7	10
4	2	5	8	11
6	3	6	9	12

no, no, no, no, none;

(ii)

	0	4	9
0	0	0	0
4	0	4	6
9	0	6	9

; no, no, yes, no, none;

(iii)

	0	5	−5
0	10	5	15
5	5	0	10
−5	15	10	20

; no, no, yes, no, none.

Exercise 22.9 (p. 756)

1. (a) (iii); (b) (i), (ii), (iii); (c) (i), (ii), (iii);
(d) (iii); (e) (i), (iii); (f) (i); (g) (iii); (h) none; (j) (i), (iii).

2. (a) Yes, (b) no, (c) no.

3. (a), (b), e(i), (f), (h), (j) are groups; (c) not closed, not associative, no identity element; (d) same; (e) (ii) not closed; (g) not all matrices have inverses; (k) 5 has no inverse.

5.

	1	4	7	13	.
1	1	4	7	13
4	4	1	13	7
7	7	13	4	1
13	13	7	1	4

6. A (a) no;   (b) yes;   (c) yes;   B (a) yes;   (b) yes;   (c) no.

9.

	0	1	2	3	4	.
0	0	1	2	3	4
1	1	2	3	4	0
2	2	3	4	0	1
3	3	4	0	1	2
4	4	0	1	2	3

10. $b - x \,\forall\, b$.

Exercise 22.11   (p. 760)

1. $\{I, R, R^2, R^3\}$; $\{I, R^2\}$.   **2.** No.   **3.** Yes.   **4.** No.
5. No; yes.   **6.** {Each element, ϕ}.
7. $\{1\}, \{1, 9\}, \{1, 3, 7, 9\}$.   **11.** Yes.
12. Yes.   **14.**

	e	a	b	c	; $\{1, i, -1, -i\}$,
e	e	a	b	c
a	a	b	c	e
b	b	c	e	a
c	c	e	a	b

15.

	I	T_1	T_2	T_3	.
I	I	T_1	T_2	T_3
T_1	T_1	I	T_3	T_2
T_2	T_2	T_3	I	T_1
T_3	T_3	T_2	T_1	I

16. (ii) a.   **17.** (i) (1, 0);   (ii) $(1/a, -b/a)$.

19.

Group	Subgroup of T	Subgroup of U
V	✓	×	✓
W	×	×	×
X	✓	✓	✓
Y	✓	✓	✓

21.

	1	5	7	11	13	17	19	23
1	1	5	7	11	13	17	19	23
5	5	1	11	7	17	13	23	19
7	7	11	1	5	19	23	13	17
11	11	7	5	1	23	19	17	13
13	13	17	19	23	1	5	7	11
17	17	13	23	19	5	1	11	7
19	19	23	13	17	7	11	1	5
23	23	19	17	13	11	7	5	1

22. $p \to q$; $p \to q$.

Exercise 22.12 (p. 765)

1. $\{1, 4, 11, 14\}$.   **3.** $n \in \{1, 2, 3, 4, 6, 12\}$; g, g^5 are generators of G, each have order 6, g^2 and g^4 have order 3; g^3 has order 2; $g^0 = e$ has order 1.

5. $f^5\ \ f^4\ \ f^3\ \ f^2\ \ f$.   **7.** $\{2, 4, 8, 16, 14, 10\}$.

	f^5	f^4	f^3	f^2	f
f^5	f^5	f^4	f^3	f^2	f
f^4	f^4	f^3	f^2	f	f^5
f^3	f^3	f^2	f	f^5	f^4
f^2	f^2	f	f^5	f^4	f^3
f	f	f^5	f^4	f^3	f^2

10. $\{T^2, T^4, T^6\}, \{T^3, T^6\}$.   **11.** $a = 1$, b and c arbitrary, $d = -1$.

Exercise 22.13 (p. 769)

1. $\{(abcd), (badc)\}, \{(abcd), (cdab)\}, \{(abcd), (dcba)\}$.
2. $p_1 = (abc), p_2 = (cab), p_3 = (bca), p_4 = (bac), p_5 = (cba), p_6 = (acb)$.

	p_1	p_2	p_3	p_4	p_5	p_6
p_1	p_1	p_2	p_3	p_4	p_5	p_6
p_2	p_2	p_3	p_1	p_5	p_6	p_4
p_3	p_3	p_1	p_2	p_6	p_4	p_5
p_4	p_4	p_6	p_5	p_1	p_3	p_2
p_5	p_5	p_4	p_6	p_2	p_1	p_3
p_6	p_6	p_5	p_4	p_3	p_2	p_1

3. $(ABCD), (ACBD), (ACDB), (ABDC), (ADCB), (ADBC)$.

Exercise 22.16 (p. 771)

5. (a) Ring; (b) field; (c) field.

Miscellaneous Exercise 22 (p. 775)

1. (ii) {Points equidistant from O}.   **2.** (a) M_1, yes; M_2, no.
3. (i) $x = 3 + 24k$, $y = -(3 + 60k)$ $(k \in \mathbb{Z})$; (ii) no solution.
6. $\{(0, 0)\}$, $\{(0, 1), (0, 2)\}$, $\{(1, 0), (2, 0)\}$, $\{(1, 1), (2, 2)\}$, $\{(1, 2), (2, 1)\}$;

*	(1, 0)	(1, 1)	(1, 2)
(1, 0)	(1, 0)	(1, 1)	(1, 2)
(1, 1)	(1, 1)	(1, 2)	(1, 0)
(1, 2)	(1, 2)	(1, 0)	(1, 1)

7. (a) $x \equiv 4 \pmod 7$; (b) $x \equiv 3 \pmod{23}$; (c) $x \equiv 4 \pmod{11}$.
8. (i) Yes, (ii) no; (i) no, (ii) yes, (iii) no.
9. (i) 3, (ii) 4; 3.   **12.** $b - x$ $(b \in \mathbb{R})$.
15. $x = -i$, $y = z = t = 0$.
17. K is of order n^2, identity element (e, f), $(x, y)' = (x', y')$.
18. (i), (ii), (iii); q, r, t.
19. $I = \begin{pmatrix} 1 & 0 \\ 0 & 1 \end{pmatrix}$, $J = \begin{pmatrix} 0 & 1 \\ -1 & 0 \end{pmatrix}$; $\{z : |z| = 1\}$; $\{\theta = 0, \theta = \frac{2}{3}\pi, \theta = \frac{4}{3}\pi\}$.
20. (i) $e = \frac{1}{2}$, $x^{-1} = 1/(4x)$; (ii) $\begin{pmatrix} 1 & 0 \\ 0 & 1 \end{pmatrix}, \begin{pmatrix} -1 & 0 \\ 0 & -1 \end{pmatrix}$.
23. 6 is of order 2, 2 and 4 of order 3, 3 and 5 are generators; proper subgroups $\{1, 6\}, \{1, 2, 4\}$.
26. (i) Ring; (ii) ring; (iii) field; (iv) neither; (v) neither.
28. Basis could be $(1, 2, -1), (3, -1, 2)$; dimension 2; $a = -17/7$.

Index

2

4

5

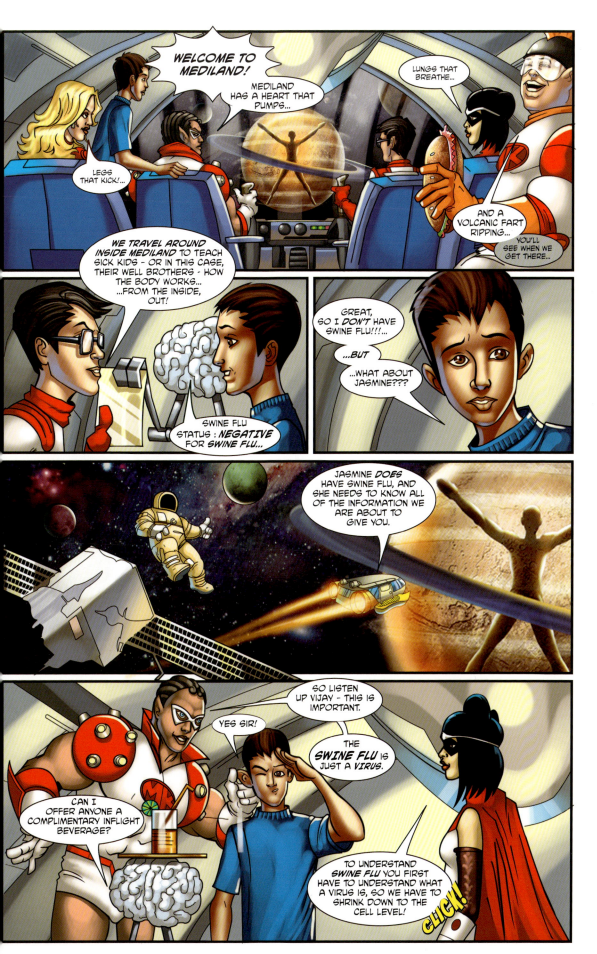

7

8

THE *FLU* CAUSES INFECTION IN BOTH *ANIMALS* AND *HUMANS*.

THERE ARE THREE TYPES OF FLU VIRUSES, *A, B,* AND *C.*

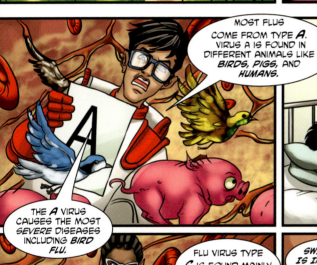

MOST FLUS COME FROM TYPE *A*. VIRUS A IS FOUND IN DIFFERENT ANIMALS LIKE *BIRDS, PIGS,* AND *HUMANS*.

THE *A* VIRUS CAUSES THE MOST *SEVERE* DISEASES INCLUDING *BIRD FLU.*

FLU VIRUS TYPE *B* IS USUALLY FOUND ONLY IN *HUMANS.*

FLU VIRUS TYPE *C* IS FOUND MAINLY IN *DOGS, PIGS,* AND *HUMANS.*

SWINE FLU IS INFLUENZA TYPE A.

THE MEAN TYPE!

THE FLU MAINLY INFECTS YOUR *NOSE, THROAT* AND *LUNGS*... SO LET'S HEAD TO THE LUNGS!

CLICK!

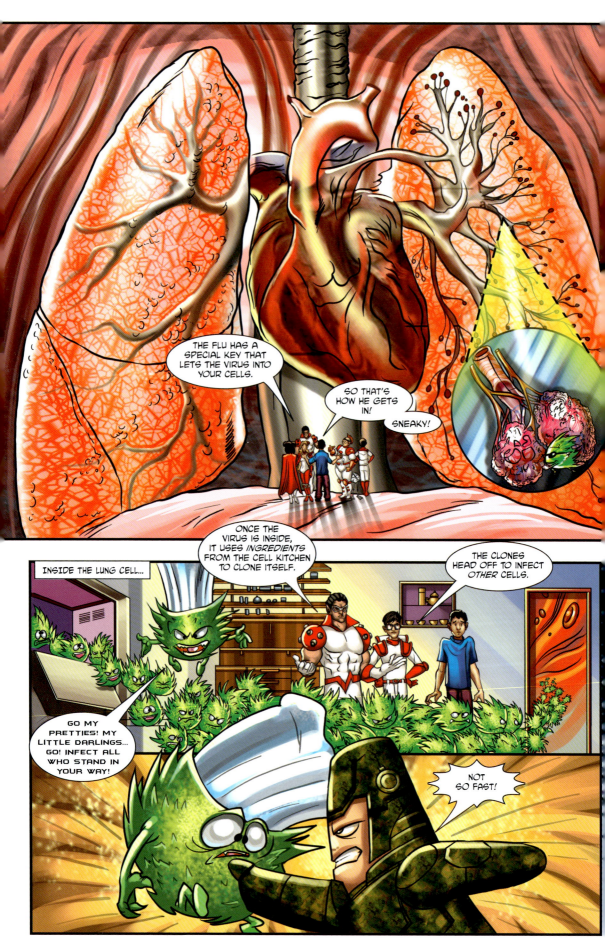

11

12

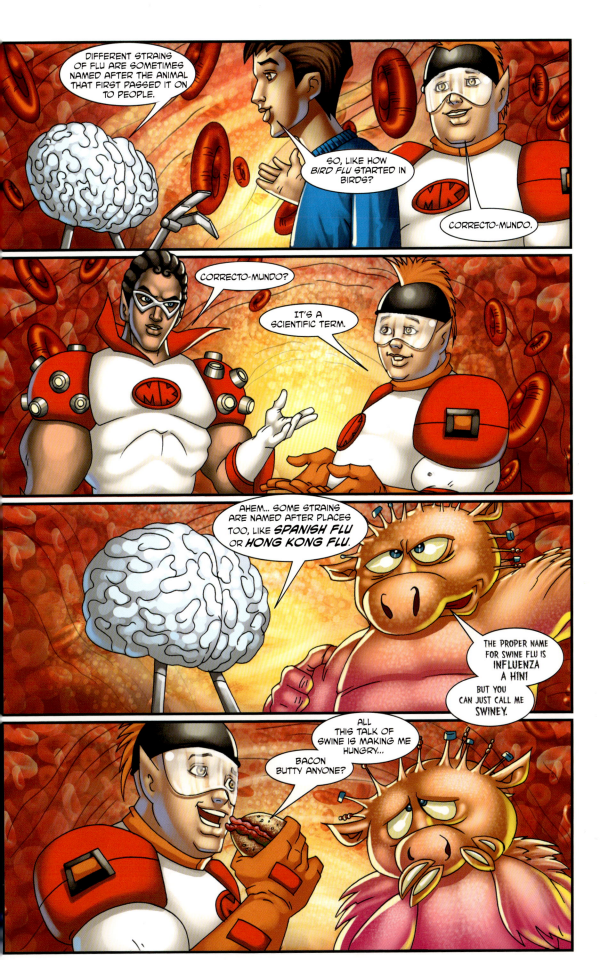

14

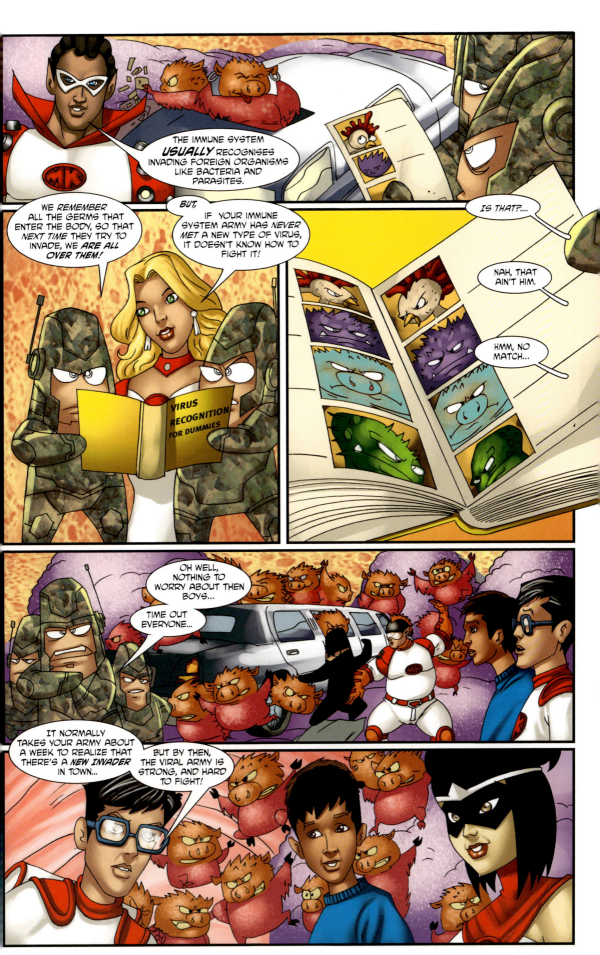

18

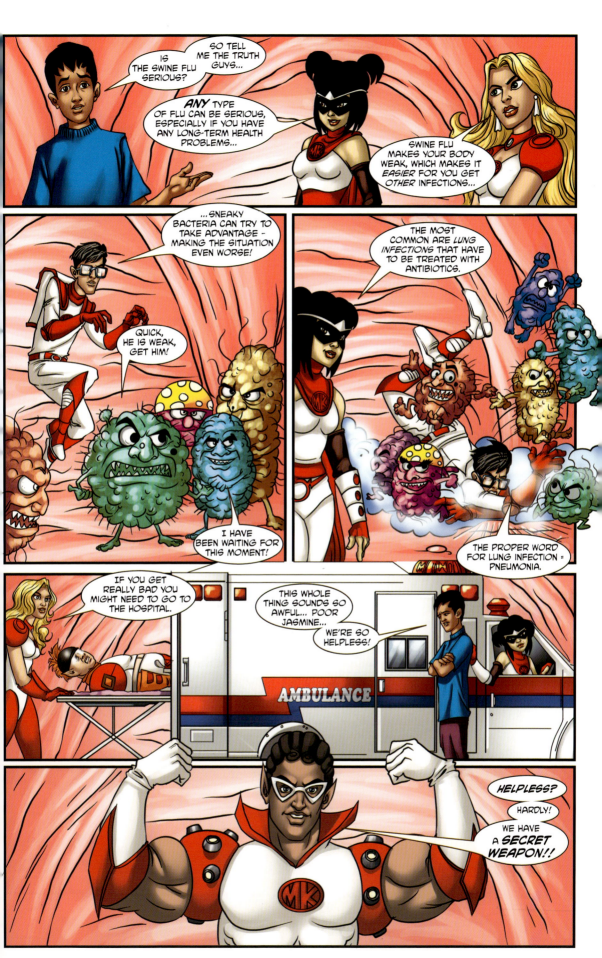

19

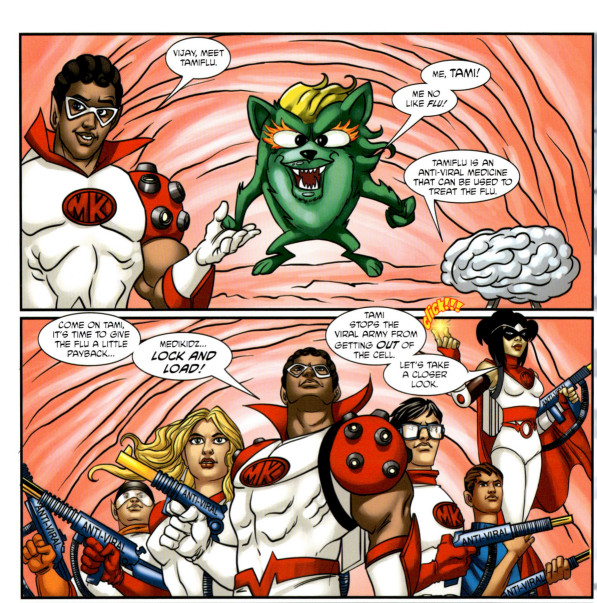

ANOTHER WAY DOCTORS ARE TRYING TO STOP THE SWINE FLU VIRUS IS BY MAKING A *VACCINE.*

I AM CLOSE TO PERFECTING ONE AS WE SPEAK!

SOME VACCINES ARE MADE OF A WEAKENED VIRUS...

..IT'S TOO WEAK TO HARM YOUR BODY,

BUT IT'S STILL ALIVE SO THAT YOUR IMMUNE SYSTEM ARMY ADDS IT INTO THEIR VIRUS RECOGNITION MANUAL!

...TOO...WEAK... TO...MOVE...

NEXT TIME ONE OF YOUR KIND COMES IN, WE'LL BE READY TO FIGHT!

SWINE FLU VACCINES NORMALLY USE *DEAD* SWINEYS INSTEAD OF *WEAK* ONES.

THEY WORK IN THE SAME WAY, TO PRIME THE IMMUNE SYSTEM FOR BATTLE, IN CASE OF LATER INFECTION!

22

24

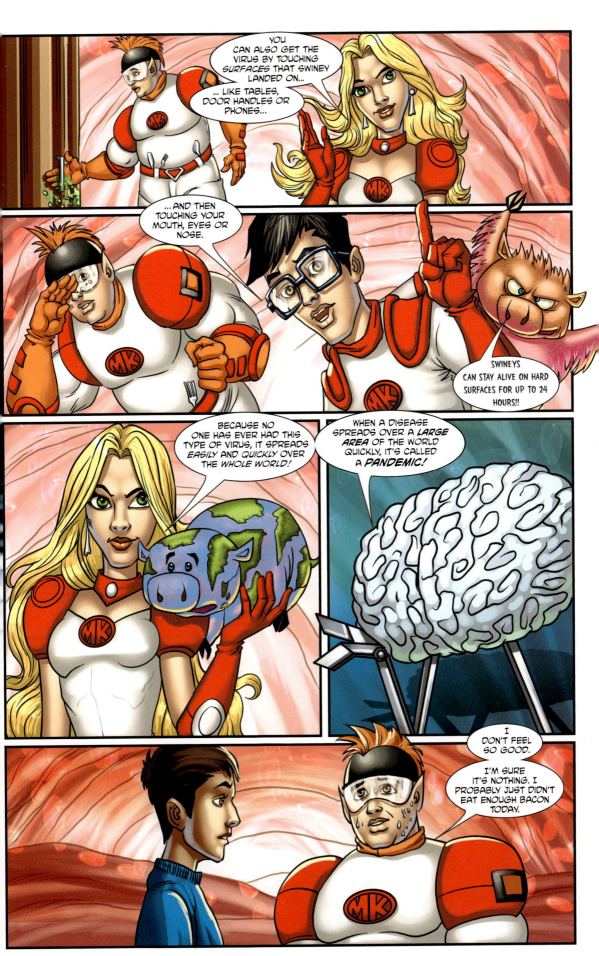

25

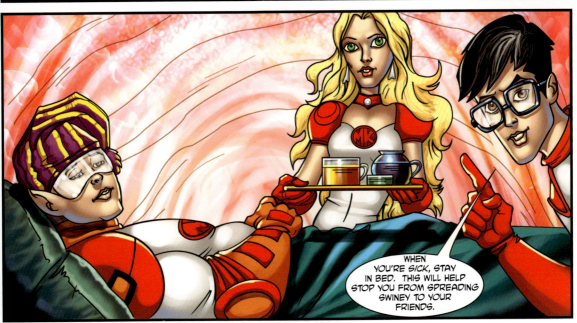

27

28

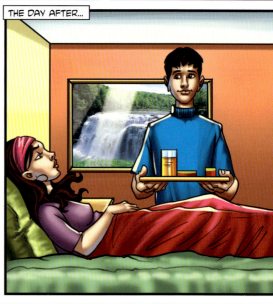

30

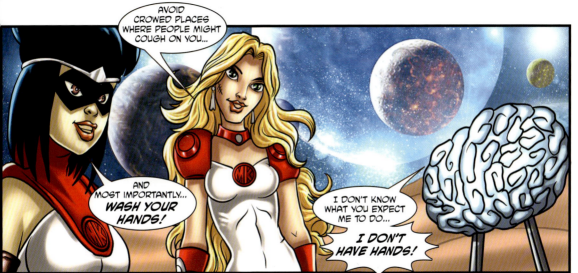